CW01509079

Dedication

To

Baruch aka Seraiah, son of Neriah, son of Jeremiah aka Mahseiah aka Shaphan, who wrote Ezekiel and Joel. Without his work on the Hebrew alphabet and saving the Torah, I would not have been able to accomplish this work. The world has much to thank him for.

Special thanks to my wife Susan and my son David for putting up with me for the eight years it to took me to write this book and the many hours they were denied time with me.

Thanks also to my two editors Gene Sadoff and Douglas Burns.

God's Day of Judgment

The Real Cause
of Global Warming

כן ויקרא אלהים לרקיע שמים

God's Day of Judgment

The Real Cause of Global Warming

By

Douglas B. Vogt

Member of the Geological Society of America

Vector Associates
BELLEVUE, WASHINGTON

VECTOR ASSOCIATES
PO Box 40135
Bellevue, WA 98015
www.vectorpub.com; sales@vectorpub.com

Copyright © 2007 by Douglas B. Vogt
All rights reserved. No part of this book may be reproduced by any mechanical, photographic, or electronic process, or in the form of a sound recording, nor may it be stored in a retrieval system, transmitted, or otherwise copied for public or private use—other than "fair use"—without the written permission of the author or his assignees. Quote of less than 500 words are permissible, with proper credit given.

First Printing 2007   Printed in the United States of America
10 9 8 7 6 5 4 3 2 1

Library of Congress Cataloging-in-Publication Data

Vogt, Douglas B., 1947–
    God's Day of Judgment, The Real Cause of Global Warming
p. cm.
    Includes bibliographical references and index.
    ISBN 978-0-930808-07-5 (hardcover: alk. paper) — ISBN 978-0-930808-08-2 (pbk.: alk. paper)
    1. Science—Philosophy. 2. Religion—Philosophy. 3. Reality. 4. Geomagnetic reversals. 5. Parapsychology. I. Title.
    Q175.V6375 2007
    501—dc22
                        2007036586

Table of Contents

Chapter 1:
Introduction

In the Beginning

I am sure, during the Neolithic period, cavemen must have sat around the campfire at night contemplating their own existence. Was life a spirit that resided inside them? How did man come about? Was there an all-powerful God who controlled everything? Was there other life in the Universe? Why was the Universe created? And the most important question: If God created the Universe, then who created God?

Over tens-of-thousands of years, the questions have not changed much because they are eternal questions of existence. So what has changed? Man has created many organized religions, while we have become more cynical because our technical knowledge base has increased. We think we know how our Universe really works, but not all is what it seems, as you will learn.

The Exodus occurred in 1306 B.C.E.[1] By the middle of the second month of 1305 B.C.E., Moses had received the two tablets we identify as the Torah. From that point on everything changed. But the world did not know it until after the Hebrews entered the Promised Land. Over the past 3,300-plus years, the Hebrews changed their identity and became Jews.

The Jews were expelled twice from their land, the last time by the Romans over 1900 years ago. The Jews have been resented by other cultures because they say they are "God's Chosen People,"—but chosen for what? Was it to be hated by Christians and Muslims for theological reasons? Four times the Torah says the Hebrew's are God's Chosen People, but the question still is, "for what purpose?" This book will answer this question, as well as many others, but the most important questions answered are: What is the *real* purpose of the Torah, and, indeed, what is it really? Does God exist? What is God's Day of Judgment, and when will it happen? What is Man's purpose in the Universe? Why was the Universe created? And most importantly, what is God's relationship to the Universe?

Categorizing Human Behavior

Man has a persistent tendency to compartmentalize people and subjects. Among these subjects are the concept of God, the sciences, human behavior and history. Science writers are notorious for compartmentalizing subjects and ideas. This book is more interdisciplinary. Some of the subjects covered include: anthropology, astronomy, geophysics, meteorology, paleontology, nuclear physics, electronics, optics, computer science and I don't want to forget Theology.

You may be wondering why a book entitled *God's Day of Judgment; The Real Cause for Global Warming* starts off by discussing the hard sciences. The reason is that what I have discovered encompasses *all* of these disciplines, and the conclusions change *everything*. Some famous scientists, including Newton, have tried to unite the concept of God with science, but all have failed for different reasons. Newton spent 50 years of his life and wrote 4,500 pages trying to predict the end of the world.[2] His papers today exist in the Hebrew National Library, where a scrap of paper they recently found with the date 2060 on it.[3] So as you can see I am certainly not the only person who has seriously studied the subject. What the others, like Newton, lacked was an accurate model of the universe, and that is the key. What I have succeeded in doing is uniting science with religion by using an information theory of existence, which I developed in my first book, *Reality Revealed; The Theory of Multidimensional Reality* (1977).

This book reveals 53 important discoveries. Four of them are: The coded numbering system, which Moses used, which is the exact number of years between the polar reversals. This number permeates the *entire* Torah, as well as the earlier books of the Hebrew Scriptures. You will also see, encoded within the Torah, the exact *month, day* and *year* of the next polar reversal and resulting global cataclysm. The second great discovery is the original model that created the Hebrew alphabet. The third is what the Torah actually is, and it is not the Biblical surface story we read. The fourth is the causes for the ice ages and polar reversals.

For those who do not know what Moses actually received in Mount Sinai, the Orthodox tradition says Moses was handed the *entire* Torah, all 304,805 "letters," of which the Ten Commandments were just a small part. The entire Torah consists of five books: Genesis, Exodus, Leviticus, Numbers, and Deuteronomy.

Moses wrote the Biblical stories, but the Torah clearly says that he could not add to, subtract from or change the letter sequence. He acquired two engraved stone tablets, written on both sides, which contained 304,805 symbols. The Biblical stories are not what the Torah actually is. It is something totally different from what has been taught by religious leaders.

One of the Orthodox traditions concerning the Torah is that the *Secret of Redemption* is contained within the Torah, but nobody knows what we are going to be *redeemed* from, nor how the Torah would be able to *redeem* us. This raises questions, which can only be answered after you discover the model, which created the 22 letters of the Hebrew alphabet. I accomplished this after eleven years of research and came up with the correct model. This would not have been discovered if I had not already discovered and developed an information theory of existence. That theory proved to be the key. This discovery proves that Moses, a late Bronze Age man, could not possibly have created the design of the Hebrew letters. Moses told us the truth when he wrote that "*the tablets were the work of God, and the writing was the writing of God, graven upon the tablets.*"[4]

The Cataclysm, the Torah and the Hebrew Alphabet

On the surface, it may seem that the ice age and polar reversal have nothing to do with the Torah and its alphabet, but you cannot actually separate the three subjects. One proves the other's validity and accuracy. Once you realize and accept that only a *very highly advanced civilization* could have developed the design of the original symbols, and you also realize that the *exact* number of years between polar reversals is embedded in the Torah, you have to conclude that you are *not* reading a normal, ordinary book.

The Torah is amazingly unique from anything else. An offshoot of these discoveries is the realization that Moses wrote a true story—but also that he placed a great number of codes and hidden stories within the Torah. I am sure he did this so someone would figure it all out in the far-distant future.

My Background History

My story begins in early 1972, when my friend, Gary Sultan, enunciated a very interesting idea. What if our Universe was the product of information? He used a videotape analogy to help explain such a Universe. I don't think he fully believed this fledging theory, which turned our familiar reality on its head, but I thought it was a very interesting perspective. Shortly thereafter I started research on pyramids and found that an information theory of existence would explain why objects are preserved when placed inside a pyramid made to a particular angle and dimension. Over the next five years I conducted a variety of researches, which included many of the subjects listed before. The result was my first book *Reality Revealed; Theory of Multidimensional Reality*. This book laid the foundations of an information theory of existence, which presented a possible explanation to the causes of the ice age and polar reversal as well as many other phenomena.

My research on the Hebrew Scriptures started in 1992 while I was doing research on my next book on polar reversals, and an update of *Reality Revealed*. I then came across the work of Stan Tenen, a Torah researcher in California,[5] who had rediscovered several models used by early Jewish scholars to unravel secrets of the Hebrew alphabet. I believe Mr. Tenen had rediscovered lost medieval tools, which the rabbis used in their studies of the Torah. His conclusion was not similar to what I intuitively thought the Hebrew letters actually represented.

Mr. Tenen came up with a shape, which had too many turns in the middle, which he analogized as a flame in the middle of a tetrahedron. His conclusion was that the Hebrew alphabet was an existential expression of man's consciousness.

He came up with hand models of each one of the letters, which he explained represented an extension of our consciousness. I did not agree with his conclusion but I felt he was onto something, and what he had discovered was *real*. However, I believe he was too influenced by Orthodox tradition and dogma, which defined the alphabet. A researcher sometimes has to step back and say, "What am I looking at?" Do not assume that anyone knows what it is.

The result of his work made me reexamine the Torah, with the possibility that it was not a collection of mythology, but rather a *true* story with a lot of coding included within it. I spent from 1992 to 1995 very carefully reading the Torah and the rest of the Hebrew Scriptures, from the point of view that it was a *true* story—but with clues and information concealed within the Biblical stories.

This was not the first time I had read the Hebrew Scriptures. I read my Bar Mitzvah Bible about five times before1990, but focused mostly on the Prophets. This time, I thoroughly analyzed and took apart the Torah and the later books. The first thing was to see if the number of years between polar reversals showed up in the Torah. Much to my surprise, the number was there, just beneath the surface. I had to sum the totals of some of the numbers given in the generations of Adam and Seth, but there was no question, the number was imbedded in the Torah.

Within weeks of discovering that, I also figured out the actual length of the sacred cubit, which was also a function of the same number. When I multiplied the length of the sacred cubit by most of the numbers in the Torah, it produced the *same primary number*—or a *function* of it—such as a quarter, or half of the number, or two-and-three times the number. Keep in mind the number I discovered is the exact number of years between polar reversals. This number, and how I discovered it, will all be explained in Chapter 2. Once the unit of measure was discovered, the actual message of the whole Torah was unraveled—*and I mean this literally*.

The Torah has Biblical surface stories, which give clues to a number of important secrets. They reveal the location of Mount Sinai; the Exodus route; what Moses and the Hebrews did in the Sinai; who put advanced technology inside the family cave; what their *real* family lineage was; what Joseph's *real identity* was in Egypt, and what he actually did there; what happened to the lost generation that entered the Sinai; lastly, and most importantly, how can the Torah save mankind from the Cataclysm which will be caused by the polar reversal? I was able to discover answers to these questions, thanks to the clues Moses left us in the Torah—and because of my three geological expeditions to the area.

Two Geological Expeditions to find Mount Sinai

When analyzing the Torah, the thought comes to mind; "Why did Moses *conceal* the location of the Mount?" Moses ended his descriptive Exodus route a substantial distance short of Mount Sinai and shy 19 days away. He ended his fairly descriptive travels at Raphidim, at the battle with Amalek. From the start of the Exodus route, Moses gave us a running description at the rate of every one-to-twelve days, as to where they were and what they did. After the battle of Amalek—nothing! It was obvious he was concealing the location of Mount Sinai, but the question is: "Was there something there that he was supposed to protect?"

Today's academic circles generally accept the theory that Mount Sinai is mythical, because no accredited archaeologist has ever found it! Some scholars think that Moses was fictional as well, along with many of the earlier Biblical characters.[6]

Mount Sinai Re-discovered

By the end of 1995, I had a very good idea where the real Mount Sinai was located, and a strong hunch about one particular mount in the Sinai. In November 1997 and November 1999, I funded two geological expeditions to the area and, as a result, found the real Mount Sinai. I found all of the altars described by Moses in Exodus at the correct location, distance and direction which he gave.

I am the first person in 2,592 years who has figured out where the real Mount Sinai is located, and I did not have a father who was High Priest to show me where it was. As a direct result of finding Mount Sinai, and what was actually there, I was able to unravel the rest of the story Moses imbedded within the Biblical story of the Torah. I have figured out answers to all of the questions I previously listed, including some additional secrets, which make the Biblical story so much richer.

Within this book I will show you some of these altars, but will not disclose the location of Mount Sinai at this time. I will hold back the location to protect the altars found there from people who hate the Jews and everything we stand for. This book will cover only what the title says. The rest will be published in another book.

The Current Mind-Think

During my 59 years of life, I have observed five different viewpoints concerning the concept of God and organized religion. The first viewpoint comes from the ruling class, which I classify as "intellectuals from academia," or the "higher-educated upper-echelons of government bureaucrats." These individuals generally do not believe in a God, defined as a Supreme Being, or Intelligence of the

Universe. They tend to believe that intelligent life is the result of an accident of genetic probability, played out over millions of years of evolution. They look at organized religion as "a human invention" to explain the mysteries of the Universe. Organized religion has served to *control and modify* man's baser behaviors, for the most part for the better. Behaviors such as cannibalism, stealing, murder, slavery, lying, and incest have been eliminated or repressed through the positive efforts of organized religion. These behaviors *needed* to end, so a coherent stable society would develop.

The ruling and intellectual classes believe in themselves as controllers of their destiny, as well as everyone else's. Since they believe that academia has explained most of the phenomena in the Universe, there is no longer any need for a *genuine* belief in God as the creator of the Universe and Director of all action. This group does, however, desire the continuation of most popular organized religions, because it controls the masses for their own purposes. Religious cynicism rules this ruling class.

The second group is the "average working-class." Members of this group have a high school or college education, and work for a living. They generally believe in a God and identify with one of the major religions. They tend to believe in what their political and religious leadership tells them. Their main concern is to create the "good life" by getting ahead in the material world we live in. Organized religion has its place, and usually does not interfere with work, family goals and the sports gods.

The third group I call "free spirits." They seem to have been disillusioned by main-line religions and have gravitated towards "new age" religious movements. This phenomenon seems to be most prevalent in first-world countries, such as in Europe and the United States. Free spirits seem to be most closely associated with *ecumenism*. They may include agnostics, who do not know or care about anything, but want to affiliate with something, for social connections.

The fourth group is the "poor and/or uneducated" of the world. The majority of them steadfastly believe in a God. Their belief is sometimes the only stable thing in their lives. It is something they can *count on* to give them some meaning and hope in their lives. They obediently try to follow the precepts of their religion and generally follow the edicts of their religious leadership—right or wrong.

The fifth group encompasses the vast majority of "scientists, especially physicists." Their religion is *science*, the procedure and the process of disclosure. Their philosophy is, "If you can't measure it, weigh it, see it, detect it in a bubble chamber, and, most importantly, get grant money from it—it doesn't exist." Not surprisingly, this group most closely resembles atheists and the philosophy of the atomists of ancient Greece, or the stoics of Rome. For instance, in the medical profession, doctors are taught that life is the result of electro-chemical reactions in the body and brain. At the moment of death, the brain has hallucinations, and that explains the many stories of individuals who have clinically "died" and come back to tell strange stories of an existence after death. Dreams are explained away as our subconscious interpreting the day's activities. There is no such thing as psychic phenomena that cannot be explained away by scientific procedures. People who play god do not believe in God. There are no mysteries of the universe that cannot be explained by man—of course, given enough government money. Preferably lots of it!

Atheists exist in all these groups, and they believe there is no God. Organized religion to them is just a man-made creation, used for population control. Atheists are so disillusioned with organized religion, that they reject everything associated with it. They look at the Hebrew Scriptures as "old Jewish mythology" with no relevance to modern-day man or society. They strictly believe in themselves as the master of their own destinies. Their moral code stems from the civil statutes of their state and national governments. When one believes in nothing, one tends to obey the laws which suit one for the moment—or one can always create one's own code of moral conduct.

If you feel that people should not be grouped into these broad categories, do not take personal offense. I am attempting to illustrate that, just within our own society, there are many different viewpoints on God and religion, even within a society with predominately a one-religion majority.

Worldwide, there are geographic and cultural reasons why so many religious philosophies began. Some believe in a man as being the visitation of God on this planet. Others believe in God as an all-powerful, unseen creator and controller of the Universe. In the Far East, Taoism, Buddhism and Confucianism developed. These religions do not have a God concept in the sense of a sole creator of the Universe. They focus on a philosophy of cause-and-effect in the material and spiritual worlds. The only concept that remotely resembles a God-figure is their idea of a "Universal Consciousness," but this concept is only present in some schools of thought.

The other religions man has created are: Judaism, Hinduism, Islam, Christianity, and finally, we find nature worship among native populations of the Americas and Africa. Some of these religions believe in multiple gods representing the forces of nature.

Philosophical diversity among religions has caused tremendous human conflict and hatred between various groups, up to the present day. The reason for these ideological differences is that no one has yet defined what God is and what He is not. The only way we can possibly do that is to develop an accurate model of how the Universe really works. Once we know that, then—and only then—do we have a chance to understand what God's relationship is with His creation. This book will, hopefully, explain this relationship and bring men closer together and closer to The Creator.

Understanding the Torah's Secrets

The Torah message is only decipherable if you first understand a totally different theory of existence. I am going to repeat the following statement a number of times in this book so that it sticks with you: *"The Universe is the product of information and from this information it creates the matter world we live in."* The Theory of Multidimensional Reality will be covered in Chapter 3.

In Chapter 2, I will take you through the steps I took when discovering the true length of the sacred cubit, and how I used this unit of measure to unlock the secret messages within the Torah and the rest of the Hebrew Scriptures. The Hebrew alphabet will be explained in Chapter 4 after my information theory has been described. The model for the alphabet proves that we are dealing with a highly advanced previous civilization which created this document to immortalize something that was very important to them. These Torah discoveries, coupled with the scientific explanation of what causes the ice ages, polar reversals (Chapter 8), and the exact number of years for this cycle, produces a powerful and urgent message for you, perhaps for the first time in your life.

The Path to Seeking Knowledge

Some of these chapters deal with technical subjects, but keep in mind this is the only way I can explain it, so you understand the principles that make up the Hebrew alphabet and the Torah. I found that most of the time, the path to knowledge is a difficult, winding one, but a necessary road we must all travel to understand any type of higher knowledge.

We are in the *End of Days*, which is a fixed block of time before the coming cataclysm. I will later explain *when* this period of time started, and why God *intervenes* during this time to help man understand what existence really is. God reveals to man where we went wrong, in the hope that we will change and thereby save ourselves from the polar reversal.

After more then 35 years of research and study, I have come to the conclusion that there is only *one path* and *one body of knowledge*, which God wants us to evolve to. We, of course, have the power of *free will*, which sometimes evolves to destructive behavior. However, after 12,000 years of free will, either we evolve to understand how the Universe really works, or else we and our society will be wiped off the face of the earth in only ONE DAY! The very few who survive will be the ones who will create the next civilization, for better or for worse. The life of man, as well as every living thing in the Universe, is a story of evolution—evolving to what God's relationship is to the Universe. Think of this book as your journey on a newly discovered road to learning—a body of knowledge, which has been hidden from man for 2,500 years. Some of this knowledge will be frightening and some will be awe-inspiring, but I promise you that you will soon know how the Universe works and what God's relationship is to His creation.

Chapter 2;
God's Unit of Measure

"My son, if thou wilt receive my words,
And lay up my commandments with thee;
So that thou make thine ear attend unto wisdom,
And thy heart incline to discernment:
Yea, if thou call for understanding,
And lift up thy voice for discernment;
If thou seek her as silver,
And search for her as for hidden treasure;
Then shalt thou understand the fear of the Lord,
And find the knowledge of God.
For the Lord giveth wisdom,
Out of His mouth cometh knowledge and discernment;
He layeth up sound wisdom for the upright,
He is a shield to them that walk in integrity;
That He may guard the paths of Justice,
And preserve the way of His godly ones.
Then shalt thou understand righteousness and justice,
And equity, yea, every good path.
For wisdom shall enter into thy heart,
And knowledge shall be pleasant unto thy soul;
Discretion shall watch over thee,
Discernment shall guard thee;
To deliver thee from the way of evil." [Proverbs 2:1-12]

This chapter will present one of the code systems Moses used to inform future Torah scholars what the real message of the Torah is. The numbering system is important to understand what the message of the Torah is and why it is very relevant for us today. The unit of measure used in all the dimensions in the Hebrew Scriptures is called the "sacred cubit." Once I figured out the actual measurement for the sacred cubit, I was able to unravel the *real message* of the Torah.

You will also learn the unit of measure is a multiple of the number of years betweenthe polar reversals. A full explanation of the number 12,068 is presented in the next chapter on the *Theory of Multidimensional Reality* and in Chapter 8. The reason the unit of measure is explained now is so you realize the importance Moses placed on the message of the number.

The Sacred Cubit—God's Unit of Measure

The discovery of the actual unit of measure for the sacred cubit enabled me to unravel the entire Torah—and I mean that literally. In order to accomplish this, I first had to make a number of essential discoveries. First, I had to have my Information Theory of Existence.[1] After that, I had to make several astronomical as well as solar discoveries that showed the clock cycle (explained in Chapter 3) produced the same number.[2] Next, I had to find the length of the Egyptian royal cubit and discover the value called the "width of a handbreadth"[3] and combine the two. I also had to discover that the sum total of the chapters and verses in the Torah added up to a factor of the 12,068 number. Finally, I had to know what measurement I would have to multiply the length of the sacred cubit by, to get the root number. A full explanation will follow, but I want the reader to understand that significant discoveries usually occur because they are a result of previous discoveries.

The Hebrew Scriptures use the sacred cubit for the unit of measure for everything—but nowhere except in the book of Ezekiel does it define what it is.

> [Ezekiel 40:6] . . . and in the man's hand a measuring reed of six cubits long by the *cubit and a hand breadth*: so he measured the breadth of the building, one reed; and the height, one reed.

> [Ezekiel 43:13] . . . the cubit is a *cubit and a hand breadth*. [Emphasis added.]

What is interesting about the book of Ezekiel, written by Baruch (see Appendix D), is that the word "cubit" does not show up until Chapter 40, but it is mentioned 92 times from there on. He is the only prophet who mentions it or includes numbers in his book. Discovering what the actual length of the sacred cubit was enabled me to understand what the writers of many of the Hebrew Scriptural books were really saying.

The Secret Torah Number

The number 12,068 permeates the entire Torah! The number does not appear in any of the prophets' writings, except in Ezekiel. It shows up just below the Biblical surface story by multiplying a measurement number by the value of the sacred cubit. Extensive examples are given in Tables 2-3 thought 2-5.

Defining the number 12,068

The following numbers are found in the Hebrew Scriptures, but all of them are giving the same message, because they are all derived from the same 12,068 root number which represents the number of years between the polar reversals—(the clock cycle). All of these numbers are functions or multiples of 12,068 such as ¼ (3,017) or ½ (6,034) of 12,068 or 2× (24,136) or 3× (36,204), and less frequently 4× (48,272). You ignore decimal points—it is the sequence of numbers that count.

The other point is, if the number does not appear after multiplying it by 24.136, you then *divide* by twelve. One of the numbers will come up. This may be the reason why there were 12 sons of Jacob resulting in 12 tribes. I sometimes refer to these numbers as a "holy number," for lack of a better term.

For instance, the total number of chapters and verses in the Torah equals 6,034, or one-half the root number. The total number of chapters and verses in the entire collection of books included in the Hebrew Scriptures equals 24,136, or two-times 12,068. The number of chapters and verses in the Jewish books from Joshua to II Kings, all the prophets, excluding Zechariah, but including Psalms, totals 12,068. The other writings (Jewish books only) include Proverbs, Song of Songs, Ruth, Lamentations, Esther, Ezra, Nehemiah, and II Chronicles, total 3,017, or one-quarter of 12,068.

Special note must be included here about the sacred cubit. Notice that I am using the established British inch, which was not created until the end of the 16th century. It is obvious that it was intended to work, and this is why: One sacred cubit is equal to 24.136″. The British foot or yard was standardized in 1599 during the reign of Queen Elizabeth I. Jews were allowed to live in England up until 1290 when King Edward banished them. A very few still did live in England but kept a Christian public profile. The Jews had brought with them the length of the inch we use today. Tradition holds that the description of the inch is located in the Torah. The English Standard System has a divine origin and that is why the dimensions work so well with the British Standard System and not the Metric System. This is why I use the Standard System in this book. Oliver Cromwell, in December 1655, let the Jews come back into England.

Moses Breaks the Two Tablets

An example of how Moses used the code system and the number is revealed after he came down from Mount Sinai and saw what his brother had done. Moses then broke the first set of tablets and threatened death for anyone who had worshiped the Golden Calf. The Exodus story tells us that 3,000 men were put to the sword, but not his brother, even though Aaron was the one who built the Golden Calf. The 3,000 figure does *not* mean that anyone was killed at all. The 3,000 represent the Torah and this is how: Take 3,000 times the length of the sacred cubit of 24.136. This equals 72,408. Divide by twelve, for the twelve tribes, and it equals 6,034, the number of chapters and verses in the Torah. So Moses was really saying that he destroyed the Torah Tablets. No one was killed, least of all his brother, who created the idol and appeared not to have been forced by the people to create it. Another important conclusion we can derive from Moses coded number is that he was telling us that the Torah had chapter and verse numbering from the very beginning.

The method used to reveal the number is to take a Torah number usually for measurement and multiply it by the length of the sacred cubit. If the number does not appear at first, divide by 12 (for the 12 tribes), then one of the numbers may appear.

Not all the numbers in the Torah or the other books will produce the results. The Torah is a codebook, and sometimes numbers are used to represent other things or ideas—as shown by the numbers 7, 40, 60, 70 and 400. Sometimes you have to sum the total numbers included in a verse or chapter, and those examples will be given later.

Let us return to the Book of Ezekiel, where he says the sacred cubit is equal to a cubit and a handbreadth. The cubit the Hebrew's were building upon was the Egyptian Royal Cubit of 20.67″ long.[4] At first, I used the width of my own hand, which came out to be 3.46″, and the resulting value was 24.13″. The next step was to multiply that value with the length "for the holy place," [Ezekiel 45:2] 500 cubits square. That totaled 12,065″ inches long. So I said to myself, "That number is awfully close to the 12,068 number I discovered back in 1989 in my astronomy research," (explained in Chapter 3). So I backed into the length of the sacred cubit by using 12,068, divided by 500, which gave me a value of 24.136″, the actual true value of the sacred cubit.

Secret Numbers Revealed in the Hebrew Scriptures
The 12,068 Year Cycles between Cataclysms

The question is, did the Hebrews know how many years came between polar reversals (the cataclysm)? The answer is "yes," they knew the number of years exactly, but I see no evidence they knew the exact year the cataclysm was going to happen even though the exact date is in fact in the Torah by code. This chapter will reveal the Torah code system, not known since Baruch (587 B.C.E.). Some of these number codes have never been discovered until now.

The Cataclysm Number

By October 1994, I had read the Hebrew Scriptures several times. I had not noticed the number 12,068 mentioned overtly as a cycle associated with God's day of judgment but I was sure it had to be there because the writers of the Hebrew Scriptures knew what happened during the polar reversal (covered in Chapters 9 and 10). The question is, where do you look? Moses had written a cryptic verse which proved to be the necessary clue. I found in Deuteronomy 32:7 the following: "Remember the days of old, consider the *years of many generations*; Ask thy father, and he will declare unto thee, Thine elders, and they will tell thee." [Emphasis added]

So I examined Genesis 5:3-32, where the generations of Adam are listed. The verses give the ages of each man, the age when he had his first son[5], and the years after the first son (Table 2-1). It then recaps the age by totaling the ages. So the age is repeated twice, implying that something is to be multiplied by two. The Chapter begins: "This is the book of the generations of Adam." So I turned the statement around and asked, "What descendant of Adam was alive when Adam was still alive." Adam finally died when Methuselah was 56 years old. Now I do not want anyone to think these people lived as long as they say they did in Genesis. You must realize it is all code, and you will see what I mean next.

The second column is the number of years from their first birth to their death, for the generations of Adam through Methuselah, you get 5,974 years. The next step was to multiply that number by two, because the story line repeats the numbers twice. The result is 11,948 years. This is still not the final number, but if you read the next chapter, it says: ". . . that the sons of God saw the daughters of man that they were fair; and they took them wives, whomsoever they chose. And the Lord said: 'My spirit shall not abide in man forever, for that he also is flesh; therefore shall his days be a 120 years.'" Therefore, I added the 120 years to the previous number and the result was 12,068 years.

The next place I found the number was in the generations of Shem, Genesis 11:10-32 (Table 2-1). It was easier finding it there. The only trick with the numbering system is that you only double the years after the Cataclysm, not before it. You also have to total the number of years the person lived, because that number was not given. The result is again 12,068 years. The 12,068 number shows up very simply by totaling across from the first total column to Terah, Abraham's father (see Table 2-1).

By December 1995, I tried totaling the number of chapters and verses in the Torah. As you can see in Table 2-2, the total is 6,034, which is half of 12,068. In essence, the real title of the Torah could be "God's Cycles." If you remember, there were two sets of Torah tablets Moses brought down from Mount Sinai, so the grand total from all four tablets would total 12,068.

Book	Chapters	Verses	Total
Genesis	50	1,533	1,583
Exodus	40	1,210	1,250
Leviticus	27	859	886
Numbers	36	1,289	1,326
Deuteronomy	34	956	990
Totals	187	5,847	6,034

Table 2-2: The total number of chapters and verses in the Torah equals 6,034 half of the 12,068 number.

Table 2-1: The Generations of Adam; Genesis 5:3-32

Name	Born to 1st birth	1st birth to death	Total	\[The years left to the next generation\] Adam	Seth	Enosh	Kenan	Mahalalel	Jared	Enoch	Total
Adam	130	800	930								930
Seth	105	807	912	695							1607
Enosh	90	815	905	605	717						2227
Kenan	70	840	910	535	647	745					2837
Mahalalel	65	830	895	470	582	680	775				3402
Jared	162	800	962	308	420	518	613	668			3489
Enoch	65	300	365	243	355	453	548	603	735		3302
Methuselah	187	782	969	56	168	266	361	416	548	113	2897
Total	874	5974	6848	2912	2889	2662	2297	1687	1283	113	20691

Adam to Methuselah: 5974 × 2 = 11948 Gen 6:3 + 120 = 12068 Total Years

Generations of Shem; Gen 11:10-32

Name	Born to 1st birth	1st birth to death	Total	\[The years left to the next generation\] Shem	Arpachshad	Shelah	Eber	Peleg	Reu	Serug	Nahor	Total
Shem		500	500									500
After the Flood	2	2	2									
Arpachshad	35	403	438	465								903
Shelah	30	403	433	435	373							1241
Eber	34	430	464	401	339	369						1573
Peleg	30	209	239	371	309	339	400					1658
Reu	32	207	239	339	277	307	368	177				1707
Serug	30	200	230	309	247	277	338	147	177			1725
Nahor	29	119	148	280	218	248	309	118	148	171		1640
Terah	70	205	275	210	148	178	239	48	78	101	49	1121
Total			2968									12068

2968 × 2 = 5936   98 plus the years before the flood = 6034 × 2 = 12068 Total Years

Table 2-1: The generations of Adam and Shem, revealing the number 12,068 in two different ways.

After discovering the number in the two generations plus the total chapters and verses of the Torah equaling 6,034, I knew I had found the binding code that would unravel the code system of the whole Torah. So I proceeded to go through the whole Torah to find out how many times the numbers were used by Moses. The following Table 2-3 lists what I found in Genesis.

It is obvious that Moses linked the number 12,068 to the Great Flood (the polar reversal) in the Noah story, and also introduced the code system of composite numbers. So I had to start totaling the numbers listed in story sections, as well as whole chapters, to see if he was telling us something important and unique.

The Book Exodus

The Book of Exodus has quite a few significant numbers. Notice that all of the Temple objects built at Mount Sinai have the numbers incorporated in their dimensions. The following Table 2-4 lists the numbers found in Exodus.

Chapter & Verse	Description and Calculation
Genesis 5:32	Noah was 500 years old when he had his three sons, and was told the earth was going to be destroyed. 500 x 24.136 = 12,068
Genesis 6:15	The dimensions of Noah's ark in cubits Length 300 x 24.136 = 7240.8 ÷ 12 = 603.4 Width 50 x 24.136 = 1206.8 Height 30 x 24.136 = 724.08 ÷ 12 = 60.34
Genesis 7:11	Noah was 600 years old when the flood started. 600 x 24.136 = 14,481.6 ÷ 12 = 1206.8
Genesis 7:20	The waters of the flood covered the earth 15 cubits upward. 15 x 24.136 = 362.04
Genesis 7:24	The waters covered the earth for 150 days. 150 x 24.136 = 3,620.4
Genesis 18:24-32	The sum total of the numbers used in the story of Sodom. 250 x 24.136 = 6034
Genesis 25:26	Isaac was 60 years old when Jacob was born. 60 x 24.136 = 1448.16 ÷ 12 = 120.68
Genesis 32:15-16	The number of animals Jacob brought to Esau as a gift, 550. (a composite number) [Note 1]. 550 x 24.136 = 13274.8 is made up of: 12,068 and 1206.8.
Genesis 45:22	Joseph gave his brother Benjamin 300 shekels of silver. 300 x 24.136 = 72,408 ÷ 12 = 603.4

Note 1: This is the first time I came across what I call a composite number, which is a number that is made up of other "holy" numbers.

Table 2-3: Table of numbers found in Genesis.

An example of these hidden numbers are found in Exodus 32:28 (listed in Table 2-4). The reason the number 3,000 was used is because Moses was actually saying that it was the Torah that was destroyed when he dropped the two tablets.

The Book of Numbers

When I started studying the Book of Numbers, I was overwhelmed because it is loaded with numbers! To the average person what does it all mean? Chapter One lists all of the male adults, broken down by tribe, who came out of Egypt. This totals 603,550. In Chapter 1:46, it repeats the number, just to make sure we noticed. This revealing number is made up of $50 \times 12,068 = 603,400$ with 150 left over. At first I had a hard time figuring out the reason for the 150 remainder. Then I tried Hebrew-accumulated small numbering using the unspoken true name of God, which is יהוה (Yahova). In Appendix C, I have listed the letter and number code systems used by Moses. If you refer to Appendix C, Table C-1, the numerical totals of יהוה are $1 \times 5 \times 6 \times 5 = 150$! When I summed all the numbers listed in Numbers Chapter 2, it totaled 1,810,650, or three times 603,550. I should explain why Moses is emphasizing this number. It is a hash total. A hash total is used in accounting to add up the total numbers in a transaction to see if it matches another number. The reason why Moses is repeating this number twice is because he is giving us a clue. The following list explains what he is showing us.

The number of male Hebrews who left Egypt = 603,550	Number of letters in the Torah = 304,805
Number of chapters and verses in the Torah = 6,034	Repeat the number of letters in the Torah = 304,805
Number of letters in the 5 titles of the Torah = 26	
Total = 609,610	Total = 609,610

Torah Verification and the Letter Count

As you can plainly see, the two numbers are the same. This is the *real* message of these two chapters. It is a verification method, so you know what the correct number of letters are in the Torah.

I was again overwhelmed when I studied the book of Numbers, Chapter 7, by the amount of numbers listed. I added up all the numbers in Chapter 7 and they totaled 5,974. This is the same number I found in Genesis 5:1-32, so I multiplied by 2 and added the 120 to it, and the total came to 12,068 ($5,974 \times 2 = 11,948 + 120 = 12,068$). This shows you that Moses wanted to make sure that someone connected the number listed in Genesis with this later Chapter in Numbers! It is in Chapter 7 because seven represents Mount Sinai, which I will explain later.

Chapter & Verse	Description and Calculation
Exodus 12:37	Number of Hebrew men that left Egypt: 600,000 x 24.136 = 14,481,600 ÷ 12 = 1,206,800
Exodus 14:7	Number of chariots in Pharaoh's army: 600 x 24.136 = 14,481.6 ÷ 12 = 1,206.8
Exodus 25:10	Ark of the Covenant [Note 1]: Length 2.5 x 24.136 = 60.34 Width 1.5 x 24.136 = 36.204 Height 1.5 x 24.136 = 36.204
Exodus 25:17	Dimensions for the Ark cover or lid: Length: 2.5 sc. X 24.136 = 60.34 Width: 1.5 sc. X 24.136 = 36.204
Exodus 25:23	Table the Ark of the Covenant sat on, taking the circumference of the table: Length 2 x 24.136 = 48.272 x 2 = 96.544 Width 1 x 24.136 = 24.136 x 2 = 48.272 Total circumference = 144.816 ÷ 12 = 12.068 Height 1.5 x 24.136 = 36.204
Exodus 26:8	Length of the curtains used for the roof of the Tabernacle: Length: 30 sc. X 24.136 = 723.08 ÷ 12 = 60.34
Exodus 26:16	Dimensions for the boards for the Tabernacle: Length: 10 sc. X 24.136 = 241.36 Width: 1.5 sc. X 24.136 = 36.204
Exodus 26:18 & 26:22-24	Dimensions for the Tabernacle: Length: 30 sc. X 24.136 = 724.08 ÷ 12 = 60.34 Width: 15 sc. x 24.136 = 217.224 ÷ 12 = 36.204
Exodus 27:1	Sacrificial altar of brass and acacia-wood: Length 5 x 24.136 = 120.68 Width 5 x 24.136 = 120.68 Height 3 x 24.136 = 72.408 ÷12 = 6.034
Exodus 27:9-13	Dimensions of the Tent of Meeting (courtyard around the Tabernacle): Length: 100 sc. x 24.136 = 2413.6 Width: 50 sc. x 24.136 = 1206.8
Exodus 30:1-2	Incense table in front of the Ark: Length 1 x 24.136 = 24.136 Width 1 x 24.136 = 24.136 Height 2 x 24.136 = 48.272
Exodus 32:28	Number of Levites that were killed: 3,000 x 24.136 = 72,408 ÷ 12 = **6,034**

[Note 1] The volume of the Ark of the Covenant is 79,089 cubic inches. The volume of the Coffer in the Great Pyramid is 157,990 cubic inches (approx.). The Ark is just about one-half the volume of the Great Pyramid's Coffer.

Table 2-4: Table of Holy Numbers found in Exodus.

There are many more numbers in this chapter; some repeat the number of people in each tribe, and some do not. I am sure there are more code systems imbedded in the Torah, but it might take years to discover them, and they are not necessary for my thesis.

Deuteronomy and the End of Days

Deuteronomy does not use any of the "holy" numbers, but it does use 7, 40, and 70 as clues to other codes written in the earlier books. At the end of Chapter 31, the Torah states what "will befall you [the Hebrews] in the *end of days*." The *End of Days* is a set block of time just before the polar reversal. At the time of Noah it consisted of 100 years. Chapter 32 begins with the words of a song Moses had the people sing. Verse 7 is very important: "Remember the days of old, Consider the years of many generations . . ." What Moses is telling us here is that we should look *closely* at the numbers given in the generations of Adam and Shem just as I did in Table 2-1. He is trying to point us to the 12,068 number and linking it to the End of Days and God's Day of Judgment, so that we analyze what it means and discover what will happen—before it is too late.

The Earlier Books and Prophets of the Hebrew Scriptures

The same numbers are found from Joshua to Second Kings. Some examples given in First Kings are important icons of the Jewish religion. They include all the dimensions of Solomon's Temple and the items found in it. Table 2-5 shows where these numbers appear in these books, and thus prove the priests from the time of Moses to Baruch, aka Ezekiel (Appendix D), knew about the number. We can only assume they knew that it was associated with the cataclysm, but we have no hard evidence. What is interesting, it is that I did not find these numbers in the Prophets or other writings, except for the book of Ezekiel.

You will notice there are many more numbers used by priests before the book of Second Kings, written some time after 850 B.C.E. and ended by Baruch writing as Ezekiel. The numbers are used by him in Chapter 2 in II Kings and then reappear once in Chapter 38:10 in Jeremiah some 260 years later. Baruch finished that portion of Jeremiah after the fall of the Temple.

I do not know why the use of the number was dropped from the priestly writings, but it is obvious that it was not forgotten, because Baruch used it heavily when he wrote Ezekiel.

The examples listed in Tables 2-4 and 2-5 conclusively prove that the Hebrews, from the time of Moses to Baruch, knew the secret number and I assume they knew that it was associated with the cataclysm. I knew the Jewish religious leadership were almost all killed by the Babylonians during the fall of the Temple and afterwards (II Kings 25:18-21), but I did not know if they knew the number in 70 c.e. at the time of the destruction of the second Temple, by the Romans. The last section answers this question.

Table 2-5:

Chapter	Calculation (sc. = sacred cubits)
Joshua 7:3 & 4	Number of men that attacked the city of Ai. 3,000 x 24.136 = 72,408 ÷ 12 = 6,034
Joshua 7:21	50 shekels of gold was stolen 50 x 24.136 = 1,206.8
Joshua 8:3	The number of men to attack Ai the second time 30,000 x 24.136 = 724,080 ÷ 12 = 60,340
Joshua 8:12	The number of men of war at Ai 5,000 x 24.136 = 120,680
Joshua 8:25	The number of people that died at Ai. 12,000 x 24.136 = 289,632 ÷ 12 = 24,136
Joshua 13:30	Number of cities in Manasseh 60 x 24.136 = 1,448.16 ÷ 12 = 120.68
Judges 3:31	The number of Philistines killed with an ox-goad. 600 x 24.136 = 14,481.6 ÷ 12 = 1,206.8
Judges 7:6 &7	The number of men with Gideon 300 x 24.136 = 7,240.8 ÷ 12 =603.4
Judges 7: 8	The number of men with Gideon 300 x 24.136 = 7,240.8 ÷ 12 =603.4
Judges 7:22	The number of horns sounding 300 x 24.136 = 7,240.8 ÷ 12 =603.4 ⌐
Judges 8:10	The men left in Karkor 15,000 x 24.136 = 362,040
Judges 10:4	The number of sons of Jair 30 x 24.136 = 724.08 ÷ 12 = 60.34
Judges 11:26	The length of time the Hebrews lived in 300 x 24.136 = 7,240.8 ÷ 12 =603.4
Judges 12:9	Number of sons, daughters. 30 x 24.136 = 724.08 ÷ 12 = 60.34
Judges 14:11	The number of companions with Samson. 30 x 24.136 = 724.08 ÷ 12 = 60.34
Judges 14:12-13	30 mentioned four times in Samson story. 30 x 24.136 = 724.08 ÷ 12 = 60.34
Judges 14:19	The number of men Samson kills. 30 x 24.136 = 724.08 ÷ 12 = 60.34
Judges 15:4	Foxes caught by Samson. 300 x 24.136 = 7,240.8 ÷ 12 =603.4
Judges 15:11	Number of Hebrew men that came for Samson. 3,000 x 24.136 = 72,408 ÷ 12 = 6,034
Judges 16:27	Number of people at the festival with Samson. 3,000 x 24.136 = 72,408 ÷ 12 = 6,034
Judges 18:11 & 16-17	Number of men from the tribe of Dan mentioned 3 times. 600 x 24.136 = 14,481.6 ÷ 12 = 1,206.8
Judges 20:31	Number of men at Beth-el 30 x 24.136 = 724.08 ÷ 12 = 60.34

Chapter	Calculation (sc. = sacred cubits)
Judges 20:45	Number of men going after the tribe of Benjamin. 5,000 x 24.136 = 120,680
Judges 20:47	Number of men left from the tribe of Benjamin. 600 x 24.136 = 14,481.6 ÷ 12 = 1,206.8
I Samuel 4:10	Number of Jewish men killed at the battle with the Philistines. 30,000 x 24.136 = 724,080 ÷ 12 = 60,340
I Samuel 6:19	Number of men that died from the Ark 50,000 x 24.136 = 1,206,800
I Samuel 9:22	The number of men at the first meeting between Samuel and Saul. 30 x 24.136 = 724.08 ÷ 12 = 60.34
I Samuel 11:8	The number of men from Israel with Saul 300,000 x 24.136 = 7,240,800 ÷ 12 = 603,400 which is also 50 x 12.068.
I Samuel 11:8	The number of men from Judah with Saul 30,000 x 24.136 = 724,080 ÷ 12 = 60,340
I Samuel 13:2	Number of men with Saul 3,000 x 24.136 = 72,408 ÷ 12 = 6,034
I Samuel 13: 5	Number of chariots 30,000 x 24.136 = 724,080 ÷ 12 = 60,340
I Samuel 13:5	Number of horseman 6,000 x 24.136 = 144,816 ÷ 12 = 12,068
I Samuel 13:15	The men with Saul 600 x 24.136 = 14,481.6 ÷ 12 = 1,206.8
I Samuel 14:2	The number of men with Jonathan 600 x 24.136 = 14,481.6 ÷ 12 = 1,206.8
I Samuel 17:4	The height of Goliath Height 6 sc. x 24.136 = 144.816 ÷ 12 = 12.068
I Samuel 17:5	The weight of Goliath's armor 5000 x 24.136 = 120,680
I Samuel 17:7	The weight of Goliath's spear 600 x 24.136 = 14,481.6 ÷ 12 = 1,206.8
I Samuel 23:13	The men with David 600 x 24.136 = 14,481.6 ÷ 12 = 1,206.8
I Samuel 24:3	The men with Saul to find David 3,000 x 24.136 = 72,408 ÷ 12 = 6,034
I Samuel 25:2	The number of sheep 3,000 x 24.136 = 72,408 ÷ 12 = 6,034
I Samuel 25:13	Total number of men with David 600 x 24.136 = 14,481.6 ÷ 12 = 1,206.8
I Samuel 25:2	The men with Saul 3,000 x 24.136 = 72,408 ÷ 12 = 6,034
I Samuel 27:2	The men with David 600 x 24.136 = 14,481.6 ÷ 12 = 1,206.8

Chapter	Calculation (sc. = sacred cubits)
I Samuel 30:9	The men with David 600 x 24.136 = 14,481.6 ÷ 12 = 1,206.8
I Samuel 30:10	The total men with David 600 x 24.136 = 14,481.6 ÷ 12 = 1,206.8
II Samuel 6:1	The number of men with David when he brought up the Ark. 30,000 x 24.136 = 724,080 ÷ 12 = 60,340
II Samuel 8:13	The number of Arameans killed 18,000 x24.136 = 434,448 ÷ 12 = 36,204
II Samuel 21:16	The weight of a spear 300 x 24.136 = 7,240.8 ÷ 12 =603.4
II Samuel 23:13	Number of chiefs 30 x 24.136 = 724.08 ÷ 12 = 60.34
II Samuel 23:18	Number of men killed 300 x 24.136 = 7,240.8 ÷ 12 =603.4
II Samuel 23:23	The number of men 30 x 24.136 = 724.08 ÷ 12 = 60.34
II Samuel 24:9	The number of men of Judah 500,000 x 24.136 = 12,068,000
I Kings 1:5	The men that ran before Adonijah 50 x 24.136 =1,206.8
I Kings 4:13	Number of cities 60 x 24.136 = 1,448.16 ÷ 12 =120.68
I Kings 5:27	Number of Levites 30,000 x 24.136 = 724,080 ÷ 12 = 60,340
I Kings 6:23	Solomon's two cherubim's Height: 10 sc. x 24.136 = 241.36 Wings: 5 sc. x 24.136 = 120.68
I Kings 7:2	Dimensions of Solomon's Temple Length: 100 x 24.136 = 2413.6 Width: 50 x 24.136 = 1206.8 Height: 30 x 24.136 = 724.08 ÷12 = 60.34
I Kings 7:6	The porch of the Temple Length: 50 sc. x 24.136 = 1206.8 Breadth: 30 sc. x 24.136 = 724.08 ÷ 12 = 60.34
I Kings 7:15	The two brass pillars in front of the Temple. Height: 18 sc. x 24.136 = 434.448 ÷ 12 = 36,204 Circumference: 12 sc. x 24.136 = 289.632 ÷ 12 = 24.136
I Kings 7:16	The two capitals of brass. Height: 5 sc. x 24.136 = 120.68
I Kings 9:14	The amount of gold sent to Solomon 60 x 24.136 = 1,448.16 ÷ 12 =120.68
I Kings 11:3	The concubines of Solomon 300 x 24.136 = 7,240.8 ÷ 12 =603.4

Chapter	Calculation (sc. = sacred cubits)
I Kings 12:21	The number of men of Benjamin 180,000 x 24.136 = 4,344,480 ÷ 12 = 362,040
I Kings 18:4	The number of priests in a cave 50 x 24.136 =1,206.8
II Kings 2:7 & 17	The sons of the prophets 50 x 24.136 =1,206.8
Jeremiah 38:10	The men who get Jeremiah out of prison 30 x 24.136 = 724.08 ÷ 12 = 60.34

Table 2-5: Table of occurrences of the holy numbers appearing in Joshua through Jeremiah.

The Book of Ezekiel—The Big Breakthrough

With the discovery of the hidden 12,068 number, the whole surface meaning of the Hebrew Scriptures unraveled before me. While reading the Book of Ezekiel for the nth time, I came across Chapter 40:5, which says: ". . . and in the man's hand a measuring reed of six cubits long, of a cubit and a handbreadth each." Earlier I explain how I discovered the length of the sacred cubit. With this length, I applied it to what this verse is really saying: The man in orange standing next to Ezekiel was holding a measuring reed six cubits long. Six sacred cubits are equal to $6 \times 24.136" = 144.816"$. Next, you divide by 12 to convert it into feet: $144.816" ÷ 12 = 12.068'$! This man is holding a measuring reed that is not only measuring distance but also time! The measurement of distance is not the final result of all these numbers. They are really measuring time. In essence, we are being told that there are discrete blocks of time in the universe which all life must eventually deal with.

I used the 24.136 inches and applied it to all the dimensions given in Ezekiel and found almost all the numbers listed were "holy" numbers. Some sections, such as the description of the altar, given in Chapter 43:12-17, were a little more difficult to figure out because I had to draw a picture of the altar Ezekiel was describing. Figure 2-1 shows my drawing of the altar. The perimeter including the top altar, is equal to 603.4′, and the area of the altar—not counting the top stone—equals 1,500 sacred cubits $\times 24.136" = 36,204"$.

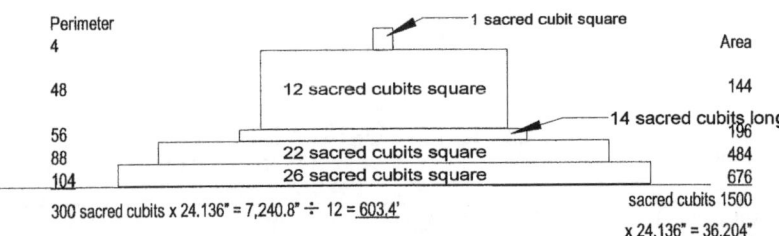

Figure 2-1: The altar described in Ezekiel 43:12-17. The Perimeter and the area are both one of the sacred numbers.

Chapter	Calculation (sc. = sacred cubits)
Ezekiel 40:6	Measuring reed held by the messenger of God. Length: 6 sc. x 24.136 = 144.816 ÷ 12 = 12.068
Ezekiel 40:7	Measurements of the small rooms Square: 6 sc. x 24.136 = 144.816 ÷ 12 = 12.068
Ezekiel 40:7	Space between the small rooms. Length: 5 sc. x 24.136 = 120.68
Ezekiel 40:11	Width of the entry gate. Width: 10 sc. x 24.136 = 241.36
Ezekiel 40:14	Pillars around the gate. Height: 60 sc. x 24.136 = 1448.16 ÷ 12 = 120.68
Ezekiel 40:15	Distance between the gate to the inner porch. Length: 50 sc. x 24.136 = 1206.8
Ezekiel 42:17	Dimensions of the wall around the new Temple to separate that which is Holy from that which is common. 500 reeds x 6 sc. = 3,000 sc. x 24.136 = 72,408 ÷ 12 = 6,034
Ezekiel 45:1	Tract of land around the Temple. Length: 25,000 reeds x 6 sc. = 150,000 sc. x 24.136 = 3,620,400 Width: 10,000 reeds x 6 sc. = 60,000 sc. x 24.136 = 1,448,160 ÷ 12 = 120,680
Ezekiel 45:6	Dimensions for the city. Length: 25,000 reeds x 6 sc. = 150,000 sc. x 24.136 = 3,620,400 Width: 5,000 reeds x 6 sc. = 30,000 sc. x 24.136 = 724,080 ÷ 12 = 60,340
Ezekiel 45:2	Dimensions for the new Temple to be built. Square: 500 sc. x 24.136 = 12,068"

Table 2-6: Some of the numbers found in Ezekiel that convert into the base number 12,068.

There are many more numbers in Ezekiel, but they all produce the same combination of numbers. The numbers also show up as the dimensions for each plot of land, north and south of the main temple. The dimensions of the Temple and the walls around it all use these numbers. Some of them are listed in Table 2-6. What is interesting is that Baruch is the first Priest/Prophet to use the hidden numbers so extensively since Elisha, and no one used them more then he.

The temple and altar described by Baruch in Chapters 40 to 48 was not the temple built in Jerusalem by Herod the Great before the First Centaury. Nor is it a Temple that would be built on the Temple Mount in the future. The Temple Baruch describes is a building measuring 1005 feet along the base. That is much larger than what can fit on the Temple Mount in Jerusalem. What is also interesting is the Ark is not mentioned at all.

The Hebrew Scriptures and Numbers

Special Note: The following section will have Roman names and terms, such as numbering systems used by all the ancients for a copyright system, that will not be explained, just stated.[7]

The First Century was disastrous for the Judean population and the Jewish religion. In 66 c.e. Jewish zealots revolted against the actions of one Roman general.[8] By 70 c.e. the Romans[9] destroyed the Second Temple, murdered hundreds of thousands, and sent thousands more to exile in Rome, for slavery, gladiatorial combat and/or death. The Herodian aristocracy, Roman puppets who ruled Judea, were related to the Calpurnius Piso clan of Rome. The Herodians fled to Rome eventually to help Piso with his writings. The Roman general who instigated the revolt was Arius Calpurnius Piso. Piso's mother (Arria), was the great-granddaughter of Herod the Great.[10] After the fall of the Temple, the remaining Herodian aristocracy permanently moved to Rome. The religious priesthood, the Sadducaeans, were appointees and in-laws of the Herodians. After the revolt, the religious zealots put most of the Sadducaeans to death. The Pharisees were non-political religious leaders who favored some appeasement with Rome and the Pisos. The Pharisees[11] after 70 c.e. changed their names to Rabbis and received permission from the Roman Emperor Vespasian to establish a school at Yavneh in Judea.[12]

At this point in history the Bible was not exactly the collection of books we know today. The Hebrew Scriptures was codified sometime after 136 c.e. Arius Calpurnius Piso wanted the Rabbis to insert at least five of his books into the Hebrew Scriptures to create the historical record and prophecies he desired, and to insert prophesies to complement other Pisonian religious writings. After the first and second Jewish revolts, the Jews and the Rabbis were powerless to oppose the Pisos' demands. The war continued as a war of literature, and the Jews turned inward. The problem for the Rabbis was that the Pisos knew about the same hidden numbers as they did, because the Pisos were being assisted by the Herodians who were by then living in Rome. So the Rabbis tricked the Pisos by using the holy numbers as a way of grouping or isolating the Jewish books from the more damaging Roman/Piso books, so that someone in the far distant future could figure out the truth of what really happened between the Jews and Rome, in the First and Second Centauries.

To increase the confusion even more, there were original copies (before the revolts) of the Jewish books that still had the original chapter and verse numbering.[13]

The Rabbis finalized the numbering of their books and presented them to the Pisos. Table 2-7 lists all the books now in the Hebrew Scriptures, with the finalized numbering. I have grouped them as follows: The Torah; Joshua through Psalms (excluding Zechariah) is the second group. Normally, Joshua through

Total of Chapters and Verses for the Old Testament
Calculation Based on Jewish Publication Society OT Bible

	Chap.	Verses	Totals	Group Totals
Torah	187	5,847	6,034	**6,034**
Joshua	24	658	682	
Judges	21	618	639	
I Samuel	31	811	842	
II Samuel	24	695	719	
I Kings	22	817	839	
II Kings	25	719	744	
Isaiah	66	1,291	1,357	
Jeremiah	52	1,364	1,416	
Ezekiel	48	1,273	1,321	
Hosea	14	197	211	
Joel	4	73	77	
Amos	9	146	155	
Obadiah	1	21	22	
Jonah	4	48	52	
Micah	7	105	112	
Nahum	3	47	50	
Habakkuk	3	56	59	
Zephaniah	2	53	55	
Haggai	2	38	40	
Malachi	3	55	58	
Psalms	150	2,527	2,677	
Total	515	11,612	12,127	12,127
Proverbs	31	915	946	
Song of Songs	8	117	125	
Ruth	4	85	89	
Lamentations	5	154	159	
Esther	10	167	177	
Ezra	10	280	290	
Nehemiah	13	406	419	
II Chronicles	36	822	858	
Total	117	2,946	3,063	3,063
Piso's Writings:				
Zechariah	14	211	225	
Job	42	1070	1,112	
Ecclesiastes	12	222	234	
Daniel	12	357	369	
I Chronicles	29	943	972	
Total	109	2,803	2,912	2,912
Total for all the Books				**24,136**

Table 2-7: Jewish Bible books listing the number of chapters and verses in each book.

Malachi are listed as "The Prophets" but I included Psalms, because it also has prophecies inserted by Baruch, who wrote many of them, though not all. The third grouping is what is called "The Writings." These are Proverbs to II Chronicles. These are all of the actual Jewish books. Finally, I have grouped the five Piso books of Zechariah, Job, Ecclesiastes, Daniel, and I Chronicles.[14] You will notice that none of the individual groupings (except the Torah) total any of the holy numbers. The entire current day Hebrew Scriptures totals 24,136, proving they *knew* about the secret number 12,068.

The Rabbis were able to separate *most* of Piso's writings from the Jewish books by using two techniques: one method used inconsistent numbering for the title line in some Chapters of Psalms. The other method used two different numbering notations for verses in Psalms and Lamentations. What I mean by this is, if you look at the following Chapters in Psalms—25, 34, 37, 111, 112 and Lamentations, Chapter 3—you will see Hebrew letters, as well as regular numbers next to verses. If you ignore the regular numbering and use the Hebrew letters as the verse numbering, you get a different total for each group. Table 2-8 shows this revised numbering. As you can see, the numerical total for the Torah is 6,034, Joshua to Psalms is equal to 12,068, and the numerical total for Proverbs to II Chronicles is equal to 3,017 and Piso's five books dangle out there, not totaling anything!

Since I discovered the chapter and verse totals, I have found that Piso also wrote Isaiah Chapters 40 to 66, Jonah, and Malachi. A full explanation of this subject will be for another book. I only cover it here to prove the number 12,068 was known as a special number even as late as the Second Centaury c.e.

The King James version of the Hebrew Scriptures has two glaring differences from the Hebrew Scriptures. The verse difference in Psalms totals 66, and the Lamentations difference is 44. I will briefly explain these two numbers. Arius Calpurnius Piso's best known pseudonym was Flavius Josephus. Using Greek small-numbering, Flavius (Flaouios [in Greek]) = 30, and Josephus (Iosepos) = 36, totaling 66. Piso's middle son was Fabius Justus (Fabios) = 18, (Ioustos) = 26, totaling 44. The Jews did this to indicate the truth about who had done this terrible thing to them. They pointed a finger at the guilty parties—Piso and his family. As I wrote earlier, this was a war of literature.

Conclusion

It should be obvious now that the number 12,068 is *embedded*, as a dimension, in every single important historical object and building of the Jewish religion! These include Noah's ark, the Ark of the Covenant, the Tabernacle, the Tent of Meeting, all of the altars, Solomon's Temple, and finally the future Temple described in Ezekiel. The Priests of the second century of the common era knew about the number, but I have found no evidence they knew what it *meant*, other

Total of Chapters and Verses for the Old Testament
Calculation based on my revised numbering of O.T.

	Chap.	Verses	Totals	Group Totals
Torah	187	5,847	6,034	6,034
Joshua	24	656	680	
Judges	21	618	639	
I Samuel	31	811	842	
II Samuel	24	695	719	
I Kings	22	817	839	
II Kings	25	719	744	
Isaiah	66	1,291	1,357	
Jeremiah	52	1,364	1,416	
Ezekiel	48	1,273	1,321	
Hosea	14	197	211	
Joel	4	73	77	
Amos	9	146	155	
Obadiah	1	21	22	
Jonah	4	48	52	
Micah	7	105	112	
Nahum	3	47	50	
Habakkuk	3	56	59	
Zephaniah	2	53	55	
Haggai	2	38	40	
Malachi	3	55	58	
Psalms	150	2,470	2,620	
Total	515	11,553	12,068	12,068
Proverbs	31	913	944	
Song of Songs	8	117	125	
Ruth	4	85	89	
Lamentations	5	110	115	
Esther	10	167	177	
Ezra	10	280	290	
Nehemiah	13	406	419	
II Chronicles	36	822	858	
Total	117	2,900	3,017	3,017
Piso's Writings:				
Zechariah	14	211	225	
Job	42	1070	1,112	
Ecclesiastes	12	222	234	
Daniel	12	357	369	
I Chronicles	29	942	971	
Total	109	2,802	2,911	2,911
Total for all the Books				24,030

Table 2-8: Revised numbering of the Chapters and verses in the Hebrew Scriptures.

than that they knew it was a *holy number*. Sometime after the second century, the knowledge was lost. The knowledge of the sacred inch was known because the rabbis brought it into England at the time of Queen Elizabeth I. It seems the actual message of the Hebrew Scriptures is: Man is supposed to evolve to what existence really is. By doing so, man will understand how the Universe works and realize what will happen during the polar reversal—which is God's Clock Cycle. If man evolves to what he is supposed to, in time he will have a chance to save himself. But if man does *not* evolve to what he is suppose to, he will perish from off the face of the earth due to his ignorance and vanity.

This chapter hopefully has given you the impetus to carefully read the following chapters so you understand what is going to happen in the near future.

Chapter 3
The Theory of Multidimensional Reality

Everything starts with an idea—even the Universe. An idea can develop into a complete philosophy. In the fields of science, the philosophy will create a model that will be tested against known observable phenomena. What I will be presenting in this chapter is a completely new philosophy, which explains how the Universe works. You will see that my model of the Universe does a better job explaining the phenomena in our Universe compared to the matter-oriented theories accepted today. Most of the discoveries and explanations presented in this book would not be possible unless I had a theory of existence that was totally different from what is taught in our society. The new philosophy enabled me to unravel what the Torah really is—and its message.

We have dealt with questions of existence throughout our history. What is existence? What is being? What is reality? Wise people through the ages have asked these questions. Innumerable philosophical discussions have analyzed these questions. Unfortunately, the people who have pondered these great questions were limited by their scope of knowledge. The experiences we have throughout a lifetime are the limiting factors that allow us to abstract, conceptualize, or analogize in this reality. These experiences are the tools that allow us to crawl through the dark caves, to make sense out of the flashes of light, the dim shadows, the faint sounds that strike us constantly, but are only vaguely perceived.

Every person who reads this book approaches it with the sum total of what they were taught in school. Whatever level of education you have achieved, your knowledge base is the product of ideas, philosophies, technologies, theologies, opinions, and histories that are *not* yours. They are the accepted theories and curriculum of the society you live in. Please do not forget that the political and religious leadership wants you to believe it so you will be molded into their image of what they want society to resemble. Most of you have never questioned what your political, educational, or religious leaders have taught you. Most of you were happy enough to regurgitate what you were taught, pass your tests and graduate. After graduation, you hoped that your school major would get you a good job and, after that, your life would begin.

The science community is no different. Over the past 60 years, our scientific instruments have outstripped scientists' ability to understand what they are measuring, detecting, and "seeing." Over the past 110 years, science has changed from a classical Newtonian model of the Universe, to a relativistic Einsteinian approach and now, finally, to a quantum mechanical approach with subatomic particles, quarks, and strings making up matter. All of these approaches are based on a matter-oriented theory of existence. Over the past 60 years, the list of unexplained phenomena has grown substantially in all major fields of science. The question is: When will this house of cards come crashing down around their heads, forcing development of a new philosophy of science?

Science Changes our Reality

Science, using its model of the Universe, has made little headway in understanding the most basic phenomena in our reality. The traditional theories are currently workable only within a narrow framework. Many scientists have found that their disciplines break down at the limits. Then another scientist comes along to try to expand our understanding based on the earlier work. However, no one has been able to explain the total picture or even to define the most basic definitions of their discipline.

It was not too long ago that the earth was considered to be flat and the center of the Universe. This was in the face of facts that showed the contrary. Many freethinkers were beaten, tortured, and murdered, in the name of Christianity, because they expounded their views in public. The idea that the earth was the center of the Universe took a long time to die, unlike ideas to the contrary. Today, most people think that we have left the Dark Ages. We are in a new Age of Enlightenment. With all our technological advances, this must be progress. However, it is not necessary to understand fully how things work to produce inventions. Sometimes important methods or results are discovered by accident. It is because reality is multifaceted and can be approached from so many different directions that some of our great thinkers—Newton, Maxwell, Niehls Bohr, and Einstein—were able to prove their concepts within a very limited framework. For example, Bohr's model of the atom does not come close to explaining what physicists see when they look at the atom with an electron microscope, with 260-million power magnification (Figures 3-7 to 10). Yet, chemists still use the Bohr Model of the atom, which is incorrect, because it fits their needs and they can work with it. It works well in describing many of the behavioral characteristics of elements and compounds. Nonetheless, our technology is not too bad. It is not necessary to understand a process or discovery fully, in order to make it work.

Comments on Mathematics

Math is an invaluable tool to explain abstract ideas, but it can be misused. In fact, many people who know the power of mathematical description have misused it. However, the real power of math to us is not in its descriptive application but in our ability to use it to evolve to abstract concepts. This is math's real forté. When you work with numbers long enough they can make anything fit your conceptual view of a problem. This is where the weakness of the "tool" is most pronounced. The final formula is really the *least* important to the math process. The *thought process* is the most important. If the logic is wrong, the math will not show it. In fact, math can "prove" faulty logic is correct.

Many people accept ideas that have mathematical "proofs." On the other hand, many concepts that are difficult to quantify are discounted. Math is only one tool of many which should be used to discover reality. Other tools are abstraction, conceptualization, analogies, intuition, and just plain guessing. When math is used as the end instead of the means to an end, its validity should be highly suspect.

Scientists say a mathematical proof is supposed to be the *ultimate* proof. Our senses are considered fallible and should not be relied upon for scientific proofs. Emotion is the enemy of objective analysis and is considered to be an enemy of logic. It seems only right that our senses and emotions should be discounted when proving a theory or evaluating a phenomenon. This ought to be your attitude when you seek a logical answer to a logical question. The big trouble with the logical approach to a problem is that there are so many questions in our reality, which defy logic. Logic can only be used where our underlying principles are firmly understood. Unfortunately, many of the phenomena of our Universe, or reality, are not understood at all. Much of the thought processes we employ blow up at the limits of our reality.

Defining the problem

There are only two ways you can describe and define the Universe. The first is that matter is the dominant thing in the Universe, and all interactions between matter and energy can only be explained as different states of matter and energy.

The second and only other alternative is that the Universe is the product of information. The matter world we live in is formed from information, transmitted from another time-space relationship, into our space and time.

What's wrong with the "accepted" philosophies in science are that they are based on a matter-oriented theory of existence. Briefly summarized, matter is the dominant *element* in the Universe. I believe this is the fundamental *error* our society has made. It is leading to a dead end in scientific thought. These philosophies are not new but very old in origin. They were first expressed by the Greek atomists, Leucippus (440 B.C.E.) and Democritus (420 B.C.E.). They taught that nothing exists, except atoms and the great void. Plato's book, *The Sophist,* says,

> Some of them [the atomists] drag down everything from heaven and the invisible to earth, actually grasping rocks and trees with their hands; for they lay their hands on all such things and maintain stoutly that that alone exists which can be touched and handled; for they define existence and body, or matter, as identical, and if anyone says that anything else, which has no body, exists, they despise him utterly, and will not listen to any other theory than their own.[1]

The philosophies and attitudes of the atomists most closely resemble the philosophies and behavior of modern-day scientists and academia. Solon, Plato and the Jews fought this kind of philosophy for hundreds of years. Today nobody in academia even questions it. The atomists have won.

Fear what you do not know

From this point on in the book you are going to view existence totally differently than ever before. I am going to destroy the very foundations of whatever you were taught in school! So if you do not want your world turned upside down do not read further! But I warn you now, if you choose not to, you will not know what the Torah really is and what geological and astronomical events that will strike us in the near future. These events are going to happen, and if you do not understand this warning, there is virtually no chance you, your children or grandchildren will have any hope of surviving this event. By the end of this book, you will understand and know how the Universe really works and what existence is. Most importantly, you will know who God really is and what He is not.

What are Phenomena and Anomalies

An *anomaly* is an event that deviates from "normal" reality and is outside of a "scientific" explanation. A *phenomenon* is a fact or occurrence that exists in time and space which defies accepted scientific philosophy or explanation. Phenomena are really special portals in the fabric of our reality, which let us perceive what existence truly is. We try to understand what we see and sense by applying the scientific philosophies we were taught. The field of science has made very little real progress in the past 75 years to help us understand the most basic building blocks of our reality, such as, magnetism, gravity, light, why the natural log shows up in our existence, why there are isotopes present in our Universe, what makes up the *basic* building blocks of matter, and finally and most importantly, what is life, and do we have a soul? Unless you understand these most basic building blocks of our reality, you do not have any hope of understanding what existence truly is.

The Theory of Multidimensional Reality

A philosophy is a collection of ideas, which is supposed to create a model to help you understand the Universe around you. If your model of the Universe is incorrect you have a difficult time understanding the phenomena you see. I give you this analogy to help you to understand the problem: Imagine you have a 10,000-piece puzzle with random shapes and colors, but collectively they form a larger coherent picture. To make matters more complicated, you do not have a picture of what the puzzle looks like when put together. The picture is analogous to a philosophy. If you do not have a picture or if you have the wrong picture, then it is much harder to know where the pieces fit together. Without the correct picture you will need tens-of-thousands of people working thousands of hours to piece the puzzle together. The Theory of Multidimensional Reality is a better picture of how the Universe really works.

Why has an Information Theory of Existence Eluded Us?

An information theory of existence was not discovered before my first book, *Reality Revealed*, because the concept is elusive. Some of our inventions, such as videotape and computers, are useful as analogies to help us conceptualize our existence. It is very difficult to understand my theory without these analogies. Similarly, Plato described existence in his famous cave analogy:

> And now, I said, let me show in a figure how far our nature is enlightened or unenlightened:—Behold! human beings living in an underground den, which has a mouth open toward the light and reaching all along the den; here they have been from their childhood, and have their legs and necks chained so that they cannot move, and can only see before them, being prevented by the chains from turning round their heads. Above and behind them a fire is blazing at a distance, and between the fire and the prisoners there is a raised way; and you will see, if you look, a low wall built along the way, like the screen which marionette players have in front of them, over which they show the puppets.

I see.

And do you see, I said, men passing along the wall carrying all sorts of vessels, and statues and figures of animals made of wood and stone and various materials, which appear over the wall? Some of them are talking, others silent.

You have shown me a strange image, and they are strange prisoners.

Like ourselves, I replied; and they see only their own shadows, or the shadows of one another, which the fire throws on the op-posite wall of the cave.[1]

The Reflexiveness of the Universe

The assumption I make is that *everything* is related and that the truth is all around us. The truth can be found in nature; it can be found in the total man and in his technologies. In other words, the complete underlying principles of existence can be found in *everything*. Our Universe has underlying principles from which it must operate. Everything in the Universe must adhere to the same principles. Everything in the Universe is a reflection of everything else. The way these reflections hit our senses and are perceived determines the uniqueness of entities. Since man is a function of the Universe, his structure must also conform to the underlying principles of the Universe. Our physical and mental structures are microsystems of the Universe. This is analogous to modern-day integrated circuits, which operate on the same principles as the older tube circuits. The integrated circuit is a refinement of the old tube circuit, just as man is a refinement of his Universe. Even though man is smaller than his Universe, the underlying principles are still the same. The microscopic structures that make up man, or the integrated circuit structures, are a reflection of what they consist of. They in turn reflect everything else in the Universe. This links us with every other object in the Universe. Since our minds and our senses are functions of the Universe, they reflect the Universe. The Universe is also a reflection of our mind and senses.

If this sounds too far out, just think of how we came into being. We began as one cell and a sperm. The resulting zygote contained 46 chromosomes. The DNA molecule had all the information on how every cell was to be formed and where it was to be in relation to everything else, and how the cell was to function and when to activate. Is the information that creates us stored only in 46 chromosomes? Is this where the information that makes us up ends? Where is the information stored about how the amino acids that make up the DNA are to be made? What about the atoms that make up the amino acids? Behind the underlying principles that determine everything is a *Creative Force*. The effect of this Creative Force can be seen everywhere.

The Basic Theory of Existence and God

The Theory of Multidimensional Reality is a simple idea but it just takes time to understand it. In 1973, my friend, Gary Sultan, and I were discussing the "secret of the Universe" and how long it would take to explain it, and how many pages it would require. I thought it would be at least a 50-page thesis. My friend thought it could fit on one 3 by 5-index card. He was right. It can fit on one index card with room to spare. How the Universe works distills down to these two sentences:

1. Everything in the Universe is made up of information that exists in another time-space relationship that acts like a computer.

2. The Operating System and the One who created the information is God, and as long as He *thinks* the Universe, it exists.

From these two simple sentences, you can "build" the model that explains the Universe. I call it *Multidimensional Reality* because an object exists in three different dimensions, almost at the same time. I will explain the "almost" later. Multidimensional Reality holds that *everything* in the Universe is made up of information that is stored in the First Dimension. The information is stored in a computer-like structure, which I call "the Diehold." I created the word Diehold because there was no existing word that expressed this kind of concept. The information from the Diehold is then transmitted into the Second Dimension, which is the Transmission Dimension. Finally, the third dimension is when the information becomes an atom of matter like carbon.

Life forms are Fourth-Dimensional existences. The consciousness of a person exists in the First Dimension, as a separate domain of information akin to a specialized program; we can surmise that this is the soul. When does life happen? When the information that makes up the soul is transmitted to the same coordinates as the information that makes up the matter of the physical body. Our physical body is a receptacle that is, in turn, a thought form—one can look at it as a template created by God. Death would be the separation of the two signals. Those who have died, and then came back to life, have reported seeing a "tunnel of light."[3]

The Theory of Multidimensional Reality also defines dimensions differently. The traditional definitions of dimensions include length, width, depth and time. Multidimensional Reality has *Eight Dimensions*. The first three are the same as the "traditional" explanations, but the Fourth to the Eighth dimensions are defined by how much potential, and consequently, *information* an object can collect, control, or perceive. The term potential is synonymous with voltage or electromagnetic potential in this dimension.

The definition of dimensions using an information theory of existence

- The **First Dimension** is the storage dimension. There time has no meaning. The Diehold is timeless, even though the Diehold itself is made up of matter but is in another time-space relationship, so it is not perceivable by the creation it forms. It is the information contained within that is timeless.

- The **Second Dimension** is the transmission dimension. Time still has no relevance because the transmission is instantaneous.

- The **Third Dimension** is the world of inanimate matter, from atoms to mountains.

- The **Fourth Dimension** is the world of ordinary living things such as us. A soul is "married" to a physical body, including everything from one-celled animals to humans. The primary quality is that these life forms must physically touch matter to manipulate it. They must use their physical senses to interact with the world around them.

- The **Fifth Dimensional Being** would be an intelligent being which possess much more potential than beings in the previous dimension and be able to perceive, control and manipulate the world around them without the necessity for physical contact. These beings would be able to communicate with others by thought, without physically creating sound or motion. This type of being would perform much of their interaction (information gathering) through the Diehold, without the need of their physical senses. They would have a physical body that they would have to care for and maintain.

- The **Sixth Dimensional Beings** may not need or have a physical body, but may be able to create one if it pleases them. They would be able to move objects with their thoughts. Time may not be relevant to them because they are able to "live" forever, because after all, they are a process in the Diehold and poses no physical body to deteriorate. A Sixth Dimensional being would posse's greater potential than the previous two dimensions. They may have the capacity to create three-dimensional objects, and take the form of some fourth-dimensional life forms. Sixth-Dimensional beings probably perceive all of their surroundings from the Diehold, and not from our dimension. This does not mean they do not need third and fourth dimensional objects to survive. They probably use them as reference points so they know what their information is doing in the Diehold.

- The **Seventh Dimension** comprises of planets, which occupy time and space. They possess huge amounts of energy potential, as demonstrated by multiple frequencies emanating from them. Planets interact with their space-time surroundings, existing for great periods of time. They most likely cannot perceive the building blocks of our matter world because they cannot perceive extremely short slices of time.

- The **Eight Dimension** comprises of stars, which in turn create planets. Their potential is vastly greater than planets. Stars exist much longer than planets and like planets, probably cannot perceive the building blocks of our matter world because they cannot perceive extremely short slices of time.

Creating analogies to describe an Information Theory of Existence

The following two analogies offer a working knowledge of the Theory of Multidimensional Reality. I use two because each has its own forté in describing some aspect of how the Universe works.

The Videotape Analogy

If you accept the idea that man is a reflection of the Universe, then his inventions are also reflections of his Universe. Let us take the example of a videotape recorder.

A television camera converts light images from objects into electrical impulses. Electrical impulses can then be stored on magnetic tape. You will notice that the form and dimension in which the picture is stored is different than the electrical impulses that came from the camera—even though the electrical impulses also represent the same picture of the object. Let us take this example a little further. You have a magnetic tape, which contains the information for color pictures with sound. It is stored in a form where time stands still. When the tape is played for broadcast, the information on the tape is again converted into electrical impulses. The impulses then go into a transmitter where they change form to become electromagnetic waves transmitted from an antenna. The picture information is now traveling in a two-dimensional form. There is not only the information for the picture, but also other frequencies as well. There is a center-beat frequency, a carrier-wave frequency (on which the picture frequencies are superimposed), a separate frequency for sound, and finally, a frequency that determines the sweep of your picture tube. The final image you see is a completely synchronized, two-dimensional "reality." All the analogies for our own existence are combined in this one invention. It is ironic that the invention, which is most illustrative of our existence, should be the tool used to mesmerize us with fairy tales!

"Here" in one form may not mean "there" in the same form. If we examine a portion of videotape closely with a microscope, we would see only brownish-red iron oxide. No matter how highly you magnify the tape, we would never see the "objects" represented by the domains of information on the magnetic material. If you could identify the objects represented by the domains, the distances between the information domains, which make up the object, would be microcosmic, compared to the distance relationships in our projected "created reality." The frequencies of the colors, which represent the object, would be at a much lower frequency on the tape, and in a totally different relationship to each other, compared to their image in the projected "created reality." There would be no time perspective for the objects represented as magnetic domains on the tape. They would be frozen in time and space, frozen until passed over by a magnetic tape transducer (head device). The transducer converts magnetic fields of the

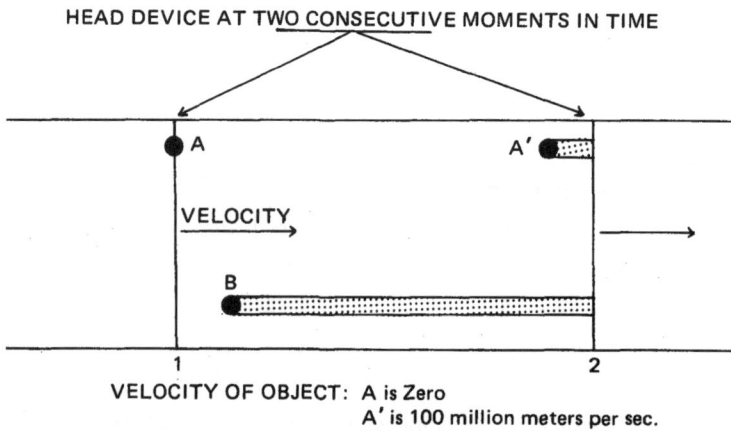

HEAD DEVICE AT TWO CONSECUTIVE MOMENTS IN TIME

VELOCITY OF OBJECT: A is Zero
A′ is 100 million meters per sec.
B is 298 million meters per sec.

Figure 3-1: Effects of velocity on object "A" in our dimension is zero; A′ is 100 million meters per second; object B is 298 million meters per second. Demonstrating the effects of velocity in the Diehold.

monolithic structure, iron oxide, into electrical frequencies. The electrical frequencies return to two-dimensional information (the transmission dimension), representing a totally different time-space reality from what was physically represented by the magnetic domains of information. If the tape speed is increased, time will seem to be shorter between two events. If the tape speed is reduced, time will seem longer between the two events. Neglecting synchronization considerations if the tape is sped up, the information reproduced at the receiver (your television) would appear physically shorter. That is because the receiver's picture-scan is in real-time and the transmitted information is in a shorter time; therefore, there will be a shorter object in the picture. This is reminiscent of Einsteinian Relativity. The opposite is also true. Slowing down the tape speed will make the object longer. Of course, corrections would have to be made in the synchronization, or the image would appear totally unrecognizable.

If an object is taped with an increased speed with more domains of information on the tape—more mass would be associated with the object. This is the same analogy as increasing the velocity of the object in Einstein's Special Theory of Relativity. So it seems that our representative object not only becomes shorter when its velocity increases, but its mass also increases. The Lorenz Transforms seem to describe the same point. Figure 3-1 visually demonstrates this principle.

Relationship between Matter, Energy, Information and Reality

Matter is converted into energy and information into matter in this analogy of existence. This is all relative to whether the tape goes through the head device to receive an image, or transmit an image. The information-domains passed over by the tape transducer (head device) determines the current reality, or existence; and velocity determines time-space relationships. Objects representing large domain areas on the tape could distort the time-space relationships of other objects represented by domains near the large object. Such an effect can be analogized by a beam of light from a star passing by a large mass, such as our sun, and being shifted in time and space. The beam of light bends around the large gravitational mass.

Einstein called this "The Special Theory of Relativity," because he recognized the speed of light (2.997925×10^8 meters per second) as the *constant* of the Universe, and that nothing can go faster than the speed of light. There were other relativity theories at the time, so the word "Special" identified Einstein's. He theorized that no matter how fast an object was going, if a beam of light were sent out from the object, the velocity of the light would be the same as for an object at rest. This idea was a departure from classical mechanics, which believed in the cumulative effect of velocities.

The Theory of Multidimensional Reality's explanation for light is: The speed of light is a constant because it is the speed or rate by which the Information that makes up the Universe is processed by the "head device." Now I do not want you to think that I am saying we are all on videotape and everything has been set in time. That is not the case. The Diehold is really a computer, so rather than a "head device" there is something that is acting like a Central Processing Unit (CPU). Please remember that I am using an *analogy* to help explain how the Universe works. The tape analogy is the best way of explaining what light and time is, which I will explain later in the proof section.

One might ask: "What does this analogy have to do with our existence?" An object on a tape does not have self-determination, cannot think and cannot live. It is only an electromagnetic representation of an object, whether the tape is moving or is stopped. Well, for one thing, the videotape analogy shows the universal reflexiveness of existence. The model at best is an understandable reflection of the real system.

The Computer Analogy

The computer analogy reveals the greatest secret there is to be discovered in geophysics, astronomy and climatology. It reveals what is behind the causes of the ice ages, polar reversals, mass extinctions and creation of new species, and why they all happen at the same time geologically, all over the Universe. Answers to these questions will be covered in Chapter 8 after I have laid the groundwork for you to comprehend what is going to occur soon in our solar system.

The Diehold is actually a computer, but this computer creates and contains all the information of our entire Universe. It contains the matter world we live in. In essence, our Universe is a *created reality*. It is similar to a hologram, except this hologram creates the matter world we live in. Like all computers, the Diehold would have to have clocking, synchronizing and resynchronizing frequencies. Clocking frequencies are stable oscillator pulses in a synchronous computer used for the timing of all operations, such as gating, recording, input/output functions, floating-point operations, printing, etc. The bits of information in a synchronous, or sequential, computer will wait until a clocking pulse arrives, then the bits of information will proceed to another section of the core memory, or input/output functions, or to the software in RAM[4] memory. Computers produce these frequencies in order to keep massive amounts of bits of information within specific time slots. All of these various resynchronizing frequencies are timed to the main clock cycle. This is how all digital computers work. There are no exceptions. Computers must have these synchronizing frequencies in order to prevent race conditions, that is, when the computer gains or loses bits within a clock cycle. A computer could not work unless it had these clocking and synchronizing frequencies. Next, I will illustrate examples of these on a macrocosmic level, but for now, I will note that these synchronizing frequencies manifest themselves in this dimension as the 11.092-year sun spot cycle and the 88.73 year Gleissberg solar cycle (eight solar cycles) (Graph 3-2), and the 12,068 year polar reversal cycle. This 12,068 year clock cycle turns out to be extremely vital, and will be described in detail next.

Proof of the Main Clock Cycles in the Universe

I have introduced a number of analogies to help explain my information theory of existence, but now I offer irrefutable proof in support of my theory that the Universe is the product of information. This one discovery kills the matter theory of existence—deader than a doornail! I have discovered the existence of the clock cycles and the period of time between them on a galactic scale.

By 1988, I had compiled a database of all known stars, open clusters, planetary nebulae and globular clusters (Appendix A). I compiled the list from two well known sources: *Sky Atlas 2000* by Sky Publishing and *The Cosmological Distance Ladder: Distance and Time in the Universe*, published by W. Freeman and Company. I carefully analyzed the distances in light years that were close to these timelines, and found some discrepancies, but I am confident that the list is correct. It includes over 520 entries, representing over 8,675,000 estimated stars, ranging in distances from 65 to 304,000 light-years away. Table 3-1 clearly shows six "blank periods" where no stars are visible. The table shows the distance from the earth, in light years, to the "blank periods." Graph 3-1 shows a graph of the entire database, from 11,089 light years to the end. Notice the six "blank periods," almost all spaced 12,068 light years apart, and they are about 1,000 years in duration. The graph is actually more descriptive than if I had a videotape of our own sun during a nova, because some critics would say the video only proves that one star can nova during the polar reversal. This graph shows that all stars nova at the exact same moment in time, all over

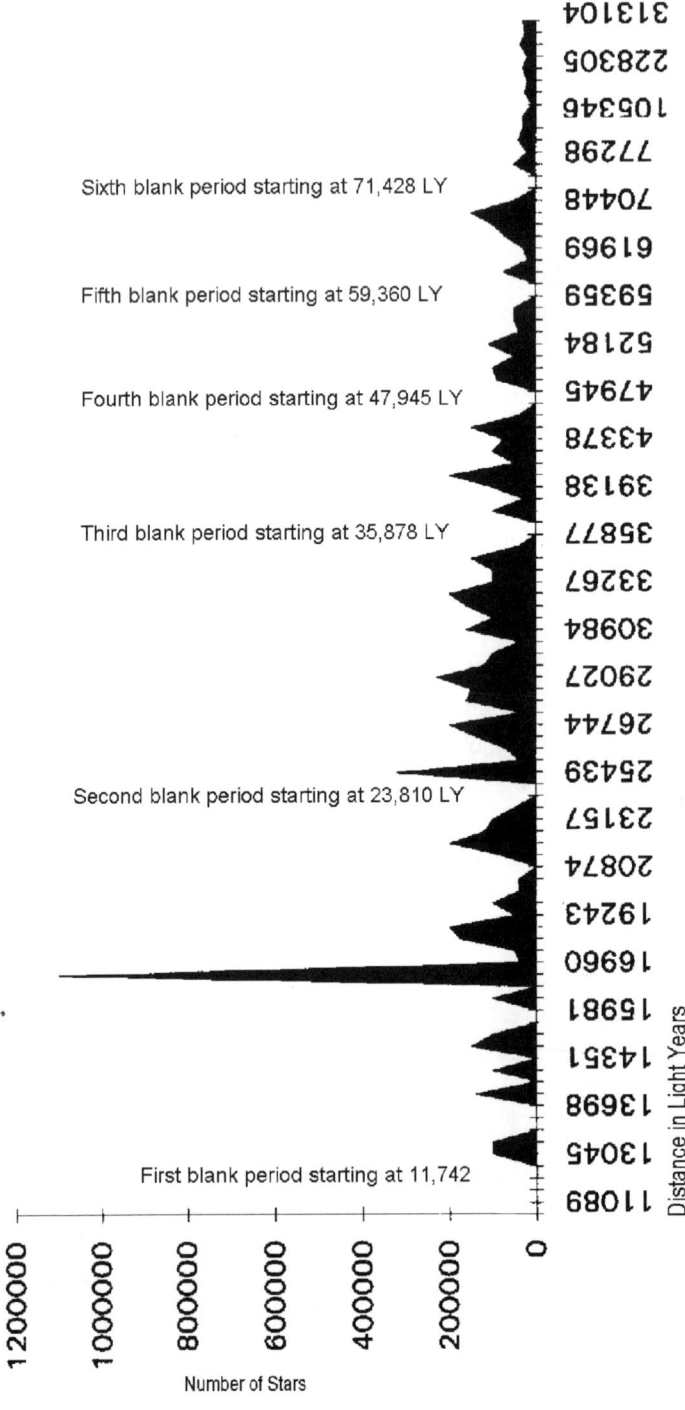

Graph 3-1: Graph shows the blank periods I discovered in our galaxy. The x-axis is in light years from the earth and the y-axis is the estimated number of stars cataloged at that distance. The six blank periods are clearly shown.

Reversal No.	Distance in LY	Distance of No Stars Visible	Distance from last blank period
1	11,742	1,304	N/A
2	23,810	1,630	12,068
3	35,878	977	12,068
4	47,945	978	12,067
5	59,360	1,304	11,415
6	71,428	5,870	12,068

Table 3-1: Table illustrates six blank periods as seen from the earth, where no stars are visible for various distances.

the observable Universe. The *only* way this could occur is if the entire Universe were the product of information, and the main clock cycle affected everything, everywhere at the same time.

The astronomical distances are determined by the red-shift from the star's light. I asked an astronomer how much error appeared in the formula that calculated distance? He said there could be as much as 10%. Distance is compared to a known star that is relatively close to the earth. I wanted to see how accurate the first distance of 11,742 light years would be. If there was an error, it would show up there, because we are measuring from a fixed position to a variable position. The relative positions between the other blank periods would be correct, because the error would be incorporated in both distance calculations. It turned out that the error was only 2.2%, so the blank period started about 250 light years further out.

The main point that Tables 3-1 and Graph 3-1 show is by using the acceptable model of how galaxies are formed, it is impossible to have any blank areas around us, not to mention that four of the six are each about 12,068 light years apart! The question is: "What causes these blank periods?"

At the time of the polar reversal, all stars nova at the *exact* same time and that is what causes these blank spaces in time. When a star novas, it throws off its surrounding matter shell into space at a high velocity. It appears from observations, and my research results, that it takes at least 970 years for the resulting gas and dust shell to dissipate enough so we can again see the light from the star. I will elaborate upon this subject in greater depth in Chapter 8 on the ice age. What you are seeing proves the clock cycle exists and *the Universe is the product of a synchronistic system*. I have now proven my theory on a cosmologic level and, later, I will prove it on a microcosmic level as well.

Discovery of Synchronizing and Resynchronizing Frequencies

When I wrote my first book, I knew the sunspots were the result of a resynchronizing frequency in the Diehold. I also assumed they would all be timed to the main clock cycle. While doing an unrelated research project, I came across a longer solar cycle called a Gleissberg cycle. The Gleissberg cycle consists of eight sun spot cycles (Graph 3-2). The first sunspot cycle in the series has the

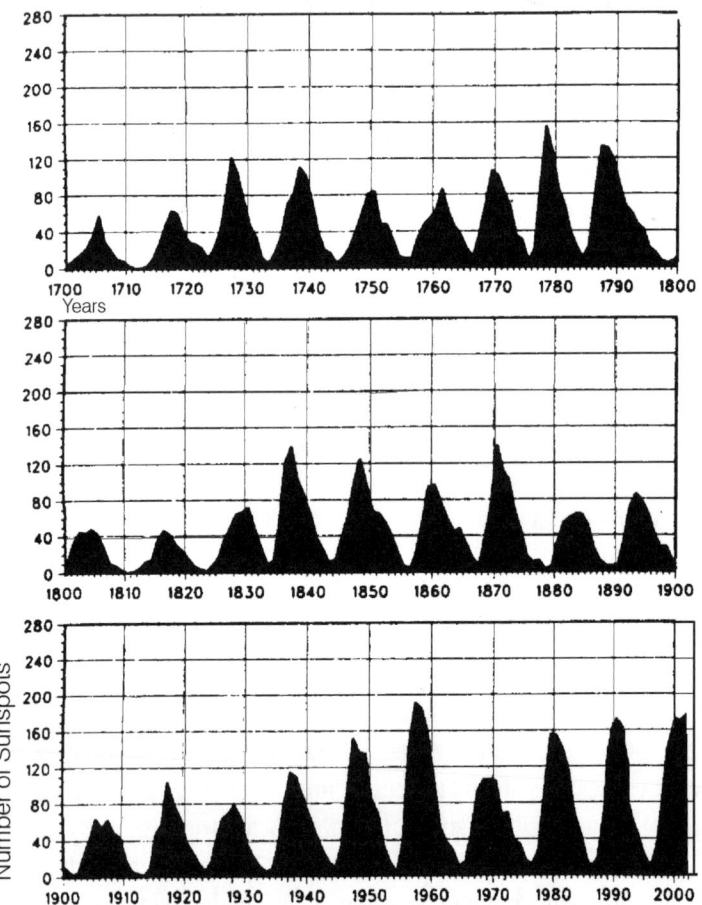

Graph 3-2: The Gleissberg Cycle is the name for eight consecutive sunspot cycles. Each peak represents a single sunspot cycle of 11.09 years. The graph shows the mean sunspot numbers. The first of the series is the lowest in solar intensity output and the eighth is the highest. The last Gleissberg cycle occurred in January 1958. This chart shows the solar activity since 1700. As you can see the sun has been increasing its activity over the past 100 years. The maximum sunspots in cycle 22 was 418, occurring on January 28, 1991. In cycle 23, the highest sunspot number was 342-3 occurring three times between 1999 and 2000.

lowest output the eighth has the greatest. The last Gleissberg maximum was solar cycle number 19, occurring either December 22, 1957 or March 31, 1958.[5] The cycle turned out to be 88.7353 years. So I divided the 88.73 years into 12,068 to see if the result would be a round number, and it was—136 Gleissberg cycles makes up 12,068 years (88.7353 years × 136 cycles = 12,068 years). What is amazing about the number 136 is that it is very close to the current value of the Fine Structure Constant at 137. You will read later in this chapter that scientists are worried about that number because, if its value were wrong, it would change everything. I think they should be concerned, because if one or two of their values or assumptions are wrong, such as Planck's constant or Planck's time, or some other value, then the value could change to 136. Remember that if there is a delay in the propagation or our information then we would perceive the speed of light wrong as well as planck's time. I do not believe it is an accident that 136 times 88.7353 gives us the correct number of years between polar reversals.

After discovering all of this, I just looked up into the night sky and reflected on the beauty of the Universe. Who would have thought that the answer behind the polar reversals and cataclysms could be discovered as easily as plotting the distances of the stars in our galaxy?

The next Gleissberg maximum will occur between September 16, 2046 and December 20, 2046. If the earth's magnetic field continues to decay faster over the next 39 years and the Sun continues to dramatically increase its output, then it is very possible the next polar reversal will occur between those two dates! Actually, the Torah gives the EXACT date for the polar reversal and cataclysm and I will cover that in Chapter 8.

Longer-term cycles

A Czech geophysicist, Vaclav Busha, while comparing Carbon-14 creation with the Earth's magnetic moment, discovered a 4,000+ year cycle. His results produce a sinusoidal graph that intersects the x-axis three times within 12,068 years.[6]

Short-term Solar cycles

Helioseismology[7] has shown us many short-term oscillations on the Sun, some as short as 2 minutes, 40 seconds. Current theory is that these short-term frequencies are multiple sound waves that travel through the sun. The waves appear as up-and-down oscillations of the surface gases, observed as Doppler shifts of spectrum lines. When these Doppler shifts are combined they are displayed as a picture of the sun, such as in Figure 3-2 and 3-3.

The problem with the explanation outlined above is that there is no evidence that continuous hydrogen explosions going off in the center of the sun would create sound waves that would appear as coherent geometric patterns on the surface of the Sun. More of this subject is covered in Chapter 8.

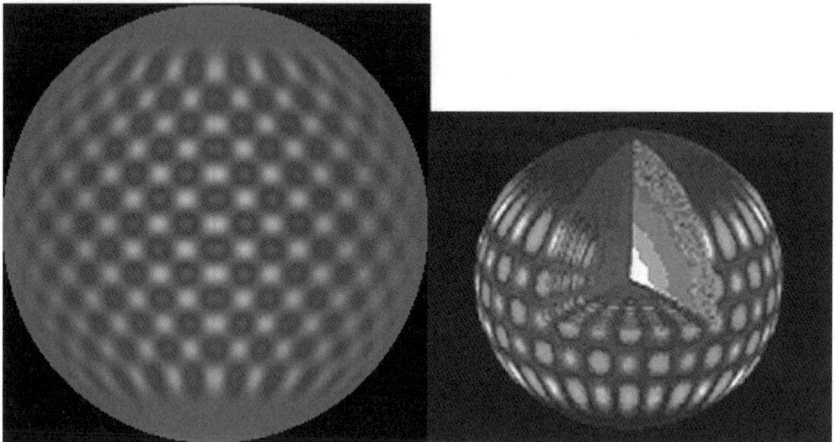

Figure 3-2 and 3-3: Helioseismology of the Sun showing geometric patterns caused by short-term oscillations deep inside the Sun. These are computer representations of the oscillations found on the Sun. Courtesy of NOAA and NASA.

It is more plausible to conclude that the pattern is created because of a modulation point in the center of the Sun that produces frequencies that manifest themselves as these surface waves or patterns found on the Sun. But, of course, if they did accept that explanation, they would have to discard their accepted theories of nuclear physics.

Other examples of synchronizing and resynchronizing frequencies can be found in the pulsations of the earth, which cause the Extremely Low Frequency (ELF) noise. These frequencies range from a few minutes, to weeks-long time periods. Other frequencies found form the very high D, E, F_1 and F_2 layers located in the Ionosphere (see Chapter 8). You are probably most familiar with human biorhythms, as well as a myriad of other frequencies associated with our neural-electrical physiology. These cycles show up in everything—all the way down to the atom. There are literally thousands of examples of electromagnetic cycles that have been observed in living things, as well as seen in stars and planets. It is no longer debatable whether they exist or not.

The Matter World we Live in

All matter we see in the Universe is the product of information that exists in another time-space relationship (dimension). It emanates from a synchronous computer-like structure I have named the Diehold. Matter is the building blocks of our reality. All the atoms that make up matter are created from individual unique frequencies that manifest themselves as unique light spectral frequencies. I believe the information that makes up atoms is transmitted into our existence at digital frequencies above the ultraviolet spectrum. All of the light-spectrum frequencies we observe are lower harmonics of the primary frequency. We will cover this subject in detail later in the atom subsection, but at this point you should understand that our matter world is composed of many controlling frequencies. The most important frequency is a carrier wave the atomic information is modulated through. The carrier wave is another frequency, very important in comprehending how matter comes into our dimension. The three dimensional representation of the carrier wave is very important in understanding twelve of the Hebrew letters, as you will find out in Chapter 4.

The Universe is composed of objects that have existed for billions of years, such as stars and planets. It is safe to assume the entire Universe has life forms like us, which live about 60-120 years. The frequencies producing the atoms are shorter then 10^{-33} seconds. The Diehold would have to have controlling frequencies that would be able to control all of them, from the longest to the shortest. In other words, it has to operate at speeds faster than the reality it creates and controls.

There have been a few books published, after I wrote *Reality Revealed*, which have suggested that some of the subsystems in our Universe act as though they are the product of a hologram. For instance, the way our brains process information is similar to holographic storage of information. Since the brain is a reflection of the Universe, it therefore operates on the same fundamental principles as the Universe. The Diehold "creates" our Universe the same way a computer generates a holographic image. A computer first modulates two laser reference beams that carry the information for an object. This is analogous to the transmission dimension. When these laser reference beams cross each other, an image is formed that looks like a three-dimensional object. A center modulation is analogous to this. In our reality, the transmitted information from the Diehold is directed to a point in time and space to form an atom.

Information Creates Matter

The way the Diehold creates our Universe is similar to how a hologram creates an image of an object, but this "hologram" creates the matter world we live in. Life, or a soul, is a unique type of information that is modulated into our existence to the same coordinates as the information making up the physical body. Each one of us has a different viewpoint of reality. An analogy would be a large glass plate with a holographic image burned onto it. When we break the glass plate into hundreds of small pieces, we notice that each piece has a unique holographic view of the original unbroken glass picture. Our reality is very much like this analogy, because we all have aspects of the whole system but with a slightly different viewpoint of the whole.

The Operating System Creates life

All computers have operating systems, and our Diehold is no different. How does the Operating System manifest itself within the system? First, let us define what an operating system does in a computer. It is a program that talks directly to the hardware of a computer, such as the Central Processing Unit (CPU) and the chip set. It controls everything in the computer, including the input/output ports. All other programs, no matter how large or small, must obey the rules of "nature"—which the operating system provides. In essence, it *hosts* other programs above it.

All living systems in our Universe are subprograms (also known as processes) that have different individual programming, depending on the species. For example, the program template for man definitely "states" that man will have 46 chromosomes, two arms, two legs, fingers and toes, a head with a large brain which can conceptualize, analyze, analogize and sometimes even come up with an idea or two. The same would be true for all other forms of life.

You will find out in Chapter 12, that the Diehold and the Operating System create life at an explosive rate. The creation of the Universe is all about intelligent life and its evolution.

The Lord of Hosts

Now the big question: *Who or what is God?* I have already given you a big hint when I wrote, "*hosts* other programs." In the Hebrew Scriptures, the authors refer to God as "The Lord of Hosts" a total of 183 times.[8] The Lord of Hosts is none other then the Operating System of our entire Universe (our Diehold). To give you an idea of how vast our Universe is, NASA had the Hubble space telescope perform a long-exposure (Figure 3-4, Hubble Deep Field project) on an area of space the size of a dime, placed 75-feet from your eye. It is a field-of-view the thickness of a hair, both horizontally and vertically. In spite of this very narrow field of view, it reveals over 1,500 galaxies at various stages of evolution, some thought to have been formed shortly after the creation of the Universe. But this is all guesswork because no one really knows—only God.

Hubble Deep Field
Hubble Space Telescope · WFPC2

PRC96-01a · ST ScI OPO · January 15, 1995 · R. Williams (ST ScI), NASA

Figure 3-4: Hubble Deep Field project showing over 1,500 galaxies.

This photo gives you an idea of how vast the Universe actually is. What is amazing to me is that the same God who is the Operating Systems of Galaxies billions of light years away, is the same God who spoke to Moses and the high priests who followed. That God would even *want* to talk to man is amazing.

Applications of the Theory of Multidimensional Reality

The following are brief explanations of the basic phenomena in our Universe, using the Theory of Multidimensional Reality, in order of importance. The scope of this book does not permit long, detailed explanations, as in *Reality Revealed*, but merely a working understanding of the phenomenon and how you can explain it using my theory. I have also included new discoveries of interest to a wide general audience.

The following subsections cover the main building blocks of our reality. Traditional science usually teaches these topics separately, as if they are not related—but of course some are directly related to each other. The subjects, or phenomena, covered are: gravity, light, atoms, time, the natural log e, Isotopes, Planck's constant, and the fine structure. All of these phenomena are explained with one model and are related to one process—how matter comes into this dimension.

Gravity and Information are the same

Gravity was described briefly in the videotape analogy. The force we call gravity is produced by the massive amounts of information that make up planets or stars—directed to a point in time and space. A "pushing" force towards the gravitational center, as well as time distortions, will affect any object that comes within the proximity of such a massive object (gravitational field). Information is synonymous with the force we call gravity. Since all objects are made up of information, therefore all objects have some sort of gravitational field.

Einstein's explanation states that gravity is curvilinear acceleration in three-dimensional time and space. Gravity is viewed more as a product of *momentum* and *inertia* in space. The General Theory of Relativity predicts that objects would decrease in size and time would slow down, as objects approached a large gravitational object. Time and size changes could only be measurable near objects having super gravities—a black hole, for example.

I decided to test my theory of gravity against the General Theory of Relativity. I chose several gravitational anomalies located in Oregon and California, to perform my experiments.

I have included some of the results in Appendix B. The results of my experiments are these: A six-foot pole shrank 5.8% over a distance of seven feet. Time changed dynamically inside the vortex, both slowing down and speeding up. It took greater force to push a deadweight towards the center of the vortex than away from it. The operators of both of these vortexes told me that very few science professors ever visited their sites, but the ones who did, told them there must be a large iron meteor buried deep down in the ground. None of them performed any experiments. The problem with that theory is there is no evidence of ancient meteor craters at these locations or any large deposits of iron. There are two primary reasons why the usual explanation of gravity is wrong. I discovered from my experiments that time and size changed dynamically but not in the same proportion the General Theory of Relativity calls for. Time was not as effected as size. The only way to explain this phenomenon, employing the General Theory of Relativity, is if the large dense body, supposedly buried there, would be dynamically changing its mass—which of course is impossible! The Santa Cruz vortex is very close to both Stanford and Berkeley Universities, but none of their physicists were curious enough to study the Santa Cruz site. This proves my point that the academic community has not been terribly interested in studying phenomena that would threaten sciences' sacred cows.

The Atom and Light Spectrum Analysis

It is almost impossible to present separate explanations for what matter (atoms) and light are, because an atom is actually information coming into this dimension; and light is information leaving our dimension. I must also use light spectrum analysis to prove that atoms are *created* from Information. I will describe atoms first, and what creates our matter filled Universe. One might ask; "how do I *know* that matter is made up of frequencies?" Proof comes from what man has observed and knows about light spectrum analysis. It is known that each element in the Universe contains its own distinct spectral line frequencies, which identify it, as shown in Figure 3-5, which illustrates the series for carbon. Some elements have less then 50 frequencies associated with them and, at the other end, Iron has the most, with about 5,000 spectral-line frequencies. Spectral lines are different light wavelengths measured in Angstrom units ($Å = 1/10,000,000^{th}$ of a millimeter) Elements produce light when raised to a high potential (energy) when heated. The traditional explanation is that each spectral line is the result of a change in the energy state of an electron rising from one electron shell to another. Within my information theory of existence light frequencies are a function of the information that makes up the element. Not what we call the electron shell around it. The light spectrum is not synonymous with the primary frequencies of the information that makes up the atom. These light spectral frequencies are merely lower harmonics of the primary frequencies, existing well above the ultraviolet spectrum. A prism is used to demodulate light into its individual wavelengths, or frequencies. The process is called light dispersion. I believe the resulting light spectrum is a highly sophisticated form of pulse amplitude modulation (Figure 3-6). Not only are these pulses amplitude-modulated, but they are frequency-modulated. For example, the individual pulses have different colors and color intensities between them. This would constitute the amplitude modulation portion of the pulse modulation. The spaces between the pulses make up the frequency—or phase-modulated part of the pulse modulation. There is also a carrier wave on which the frequencies are superimposed. The idea that there is a carrier wave is important in understanding the Hebrew alphabet because it reveals itself in our dimension as a specific slope or angle.

Figure 3-5: Light spectrum of the carbon atom

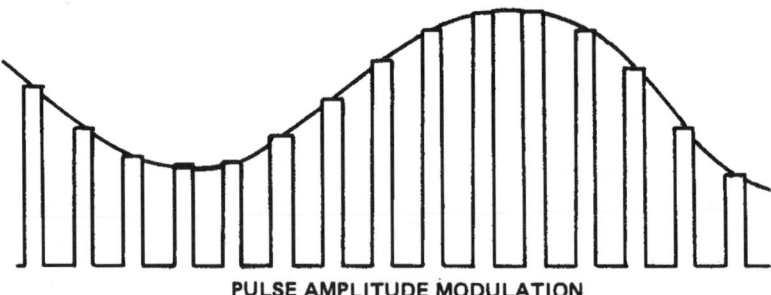

PULSE AMPLITUDE MODULATION

Figure 3-6: An example of pulse amplitude modulation

The Atom

Most people have never before seen a picture of an atom, so allow me to show you what several look like so you know what I am describing. Figure 3-7 shows two micrographs of an atom. The picture on the left shows the original photo, and the right one is the deblurred photo. The deblurred photo shows the electron shells much more clearly than the original. The diameter of the atom shown is 0.5Å^{-1} or 5nm^{-1}.

ORIGINAL (B)　　DEBLURRED (D)

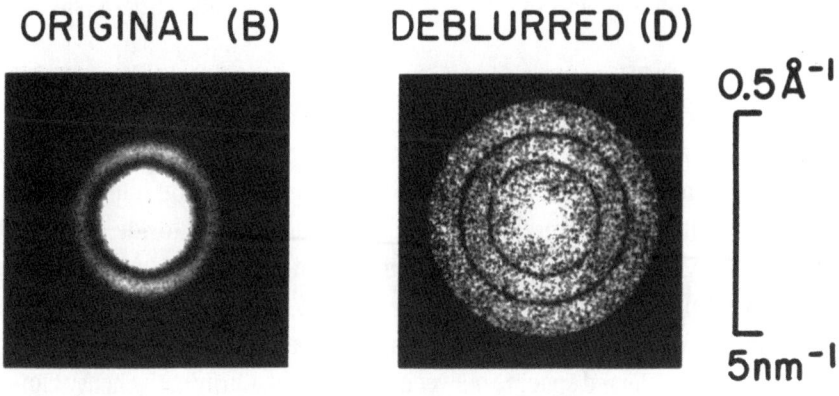

$$0.5\,\overset{\circ}{A}{}^{-1}$$

$$5\text{nm}^{-1}$$

"THON" DIFFRACTOGRAMS

Figure 3-7: Electron Micrograph of heavy atoms on a carbon-foil. Work carried out at the State University of New York, Stony Brook, L.I., N.Y., Electro-Optical Science Center, under the Direction of Prof. G. W. Stroke.

Figure 3-8 shows the optimum focus of that atom using this technology. Please notice, the picture of an actual atom looks nothing like the Bohr model of the atom. It looks more like a star.

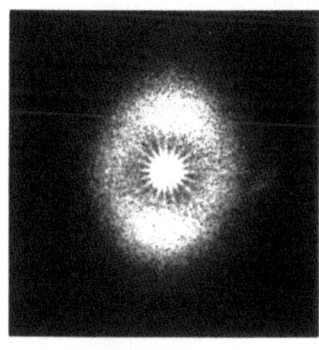

OPTIMUM FOCUS (O)

Notice the streamers of light emanating from the center of the atom. This could be the energy emitted from the atom as it blinks in and out of our existence. I theorize the electron "shells" are more like group waves or standing waves forming around the center modulation point. On a cosmological level, it is like the D, E, F_1 and F_2 layers around the earth, in the ionosphere. These are the layers, which reflect radio waves. The layers are like group waves or standing waves

Figure 3-8: Electron Micrograph, optimum focus of a heavy atom on a carbon-foil. Work carried out at the State University of New York, Stony Brook, L.I., N.Y., Electro-Optical Science Center, under the Direction of Prof. G. W. Stroke.

forming around the center modulation point of the Earth. The Sun also has similar shells around it. This again illustrates the reflexivness of the Universe. What is true for the building blocks of the Universe is also true for the larger systems.

Atomic Groupings

During a study conducted by M.S. Isacson and his colleagues at the University of Chicago, they discovered that the atoms observed in their electron microscope jumped around from place-to-place on the carbon film. For their experiments, they used several heavy atoms, such as uranyl chloride, silver and uranium. They placed these atoms on a thin film of carbon. This enabled them to see the atomic images. They took a series of photographs, from 17 seconds to five minutes apart; they made two important observations: One, the heavy atoms seem to have a preferred spacing between them. The majority of the heavy atoms were spaced approximately 4 to 5Å units apart. There is no explanation as to why these atoms would have any preference concerning the space between them.[9] It has also been observed that some of these atoms were grouped in clusters, for which there is likewise no explanation.

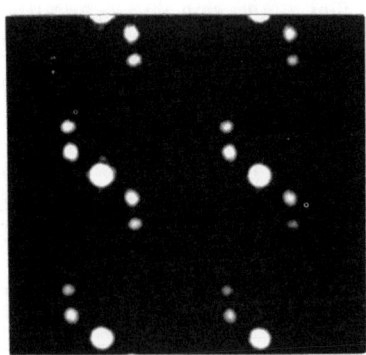

Figure 3-9: Images of atoms in a section of the crystal "magnesium bromide tetra-hydrofuran complex," obtained by a scientific team headed by Dr. George W. Stroke and including Dr. M. Halioua, Dr. V. Srinivasan and Dr. R. Sarma, using the new "X-ray microscopy" opto-digital computing method. The images of the large atoms in the unit cell shown above are magnesium atoms; the smaller symmetrical pair around each magnesium atom are oxygen atoms, and the still smaller pair farthest away around the magnesium atom, are carbon atom images. The X-dimension of the unit cell shown between magnesium atoms is 9.26Å.

The Theory of Multidimensional Reality explains that the atoms group and line-up according to vector angles of energy formed by their Information. The process is called heterodyning (new frequencies formed by the sums and differences of their original frequencies). Figure 3-9 illustrates a micrograph of magnesium bromide, showing the orientation of the atoms.

Atomic Jumping Beans

The atoms jumping around are much more significant. Scientists at first thought this was caused by mechanical or electromagnetic instabilities in the equipment; but when calculating for this possibility, they concluded that this type of movement could only cause motion up to 1Å unit. The other possibility? The electron beam from the microscope caused the atoms to move around. But this idea was also discounted, because it was calculated that the electrons would cause only a very minor atomic movement. Therefore, the movement could not be caused by the electron microscope. They calculated the average frequency of the movement found for uranyl chloride molecules was 1,400+ to 3,300 per-second. There is no satisfactory or traditional explanation for such a rapid movement of atoms. The following three figures (Figure 3-10 to 12) are a series of before-and-after pictures taken by M.S. Isaacson and associates, published in *Ultramicrosopy* in 1976. I would like you to notice, not only are the large atoms moving, but also the carbon atoms that make up the film, are also moving. In other words, all of the atoms are moving.

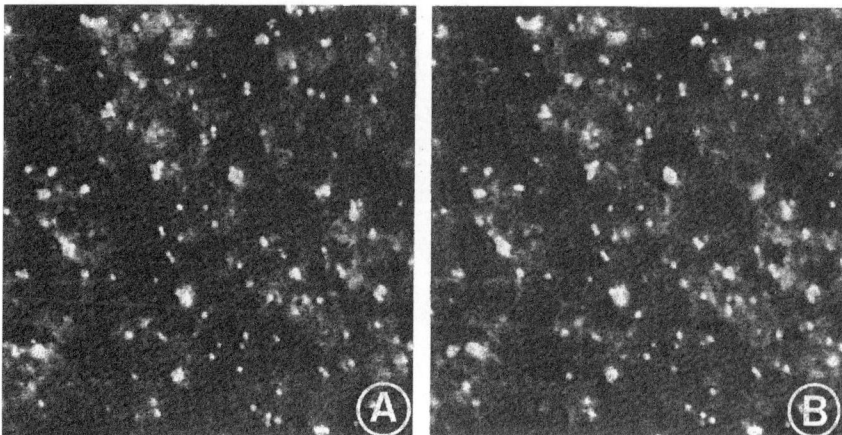

Figure 3-10: Two successive pictures taken five minutes apart of uranyl chloride molecules on a thin-film of carbon about 20 Å using a 25.5 keV electron beam. The horizontal field is 240 Å across.

The three previous micrographs are very troubling to scientists because the atoms are not supposed to be jumping around like this. This one observation proves my theory that the matter world we live in is the product of information, and because the high voltage used in the microscope was sufficient to cause the atoms to modulate into our existence in slightly different locations over very short slices of time. My theory states that these atoms are constantly being phased

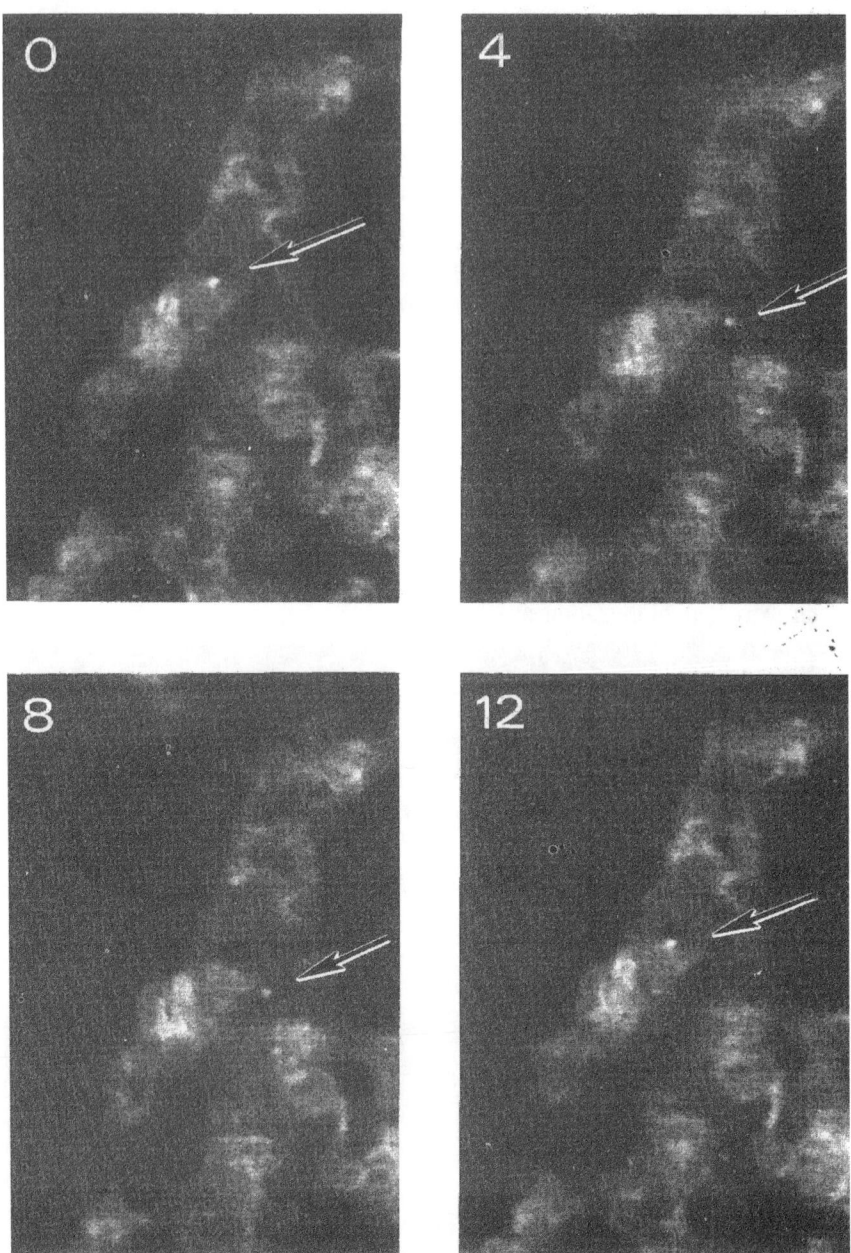

Figure 3-11: The arrow is pointing at a single mercury atom at different locations on the carbon film. The pictures were taken every 17 seconds over a 4-minute period, using a 43 keV electron beam. The horizontal field is 125 Å across. Again, notice the atoms of the carbon film are also moving, but to a much lesser extent.

in and out of our existence at rates of speed we call Planck's time (10^{-34} seconds). They do not actually move from Point A to Point B, but disappear at Point A, and reappear at Point B.

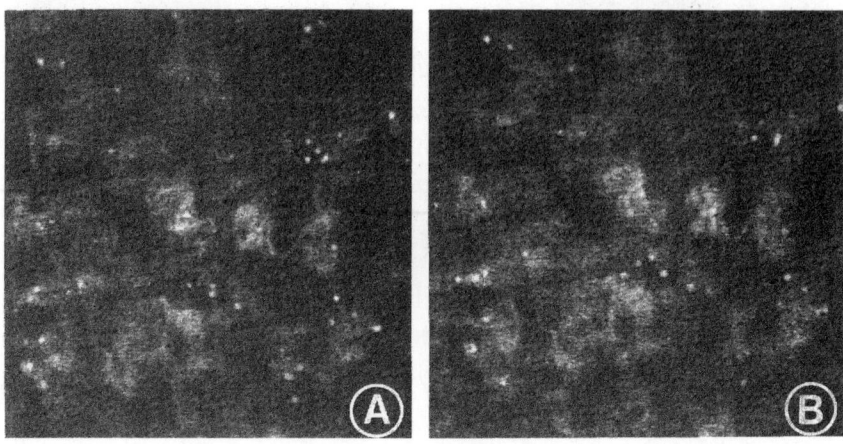

Figure 3-12: Two successive pictures, taken one-minute apart, of silver cyanide molecules on a thin-film of carbon about 20 Å using a 43 keV electron beam. The horizontal field is 195 Å across. Notice the carbon atoms are moving, as well as the larger molecules.

In normal energy states the atom assumes a standard pattern (Figure 3-13) as it blinks in and out of existence. The atom follows the pattern of a toroid (Figure 3-14) as it goes around in a complete circle. As it goes around forming the toroid pattern it also rotates vertically 360° as it flips around, with its equator pointing towards the middle axis.

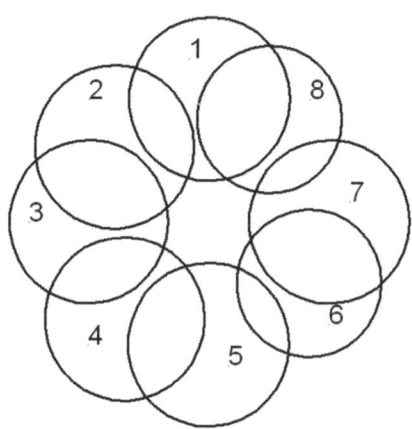

If you were to place a dot on the face of an atom on its equator, you would see this dot plotting the same modified square wave shown in Graph 3-4, except that the track would not be straight but would describe a complete circle. Figure 3-14 shows a wire model with a black

Figure 3-13: The toroid pattern produced by an atom as it "blinks" in and out of our existence. I only show eight positions, but in reality there must be millions of positions that make up one revolution.

Figure 3-14: Photograph of a model of how an atom revolves around 360° and also rotates 360°, forming a distinct pattern.

tube representing that track. This pattern is extremely important in understanding how the original Hebrew alphabet was formed.

The Diehold re-transmits the information for the same element, back to the same approximate coordinates, thereby causing the reappearance of the element, but at a slightly different location. The reason for the different location is that there is a tiny delay in the processing and propagation of information into the third dimension. This delay is the natural log e^x, which I will explain later.

The "electron" stream used to produce the images most likely cause the large movement during the process of demodulation and modulation of the atom's information. Since the electrons are really domains of energy they would be increasing the potential of these atoms enough to cause their greater movement.

Subatomic particles can be described as particle tracks of the individual frequencies of an element as it demodulates *out* of our dimension. The atoms demodulate or "come apart" because they are bombarded by another particle, which kinetically raises its potential, or energy, to a level where it can no longer exist in this dimension. The elements are made up of many different frequencies. Each frequency can be converted to a vector angle of energy. As a result, scientists are seeing multiple "particle tracks" for each frequency in their bubble chambers. The circular shapes of many of these tracks are due to the exponential decay of the information in this dimension.

Isotopes and Error in the Universe

Scientists have found and identified naturally occurring isotopes for every natural element we know of. They can create isotopes by bombarding atomic nuclei with high-voltage gamma rays, or other particles. Some of these isotopes last only a few milliseconds, and some for many thousands of years. Some of these isotopes release their energy in the form of radiation. Eventually, these isotopes decay down to one of the stable states for the element. The process of decaying isotopes occurs because the Diehold will eventually correct any unstable portions of the Universe.

One might conclude from my theory that the only elements that would appear would be stable, one atomic weight per element. For instance, all of the hydrogen found in our dimension would have an atomic weight of one, or all carbon found would have an atomic weight of 12. What is known today is that all natural elements have radioactive isotopes.

The definition of an isotope: *An atom of the same element with the same atomic number, but of a different atomic weight.* The isotope is identical in all physical and chemical properties. The only difference between them is their atomic weight.

Let us take the example of nickel. The vast majority of nickel has an atomic weight of 58.7. Why would we find 26.2 percentage of nickel with atomic weights of 61 and 3.6 percent at 63—and it will still be nickel? With an atomic weight of 63, it should be copper. But why is it still nickel? I don't know if anyone has ever thought about this phenomenon, or wondered why it occurs. The only way I can explain the occurrence of isotopes is by using the *Theory of Multidimensional Reality*. What I did was create a table with 83 natural elements. I listed all of the naturally occurring isotopes for each element. (There are also man-made isotopes which last only seconds, and some for many years, but I did not take them into account. I only wanted to deal with naturally occurring isotopes that the Diehold produces normally.) What I found was that some elements had isotopes that were as much as eight atomic weights greater than the stated average weight, and some were six atomic weights less. I plotted the results in Graph 3-3.

% at 6-	% at 5-	% at 4-	% at 3-	% at 2-	% at 1-	% normal
0.087%	0.109%	0.481%	0.618%	3.916%	5.868%	72.426%

% at 1+	% at 2+	% at 3+	% at 4+	% at 5+	% at 6+	% at 7+	% at 8+
4.472%	8.618%	1.751%	1.350%	0.114%	0.183%	0.000%	0.002%

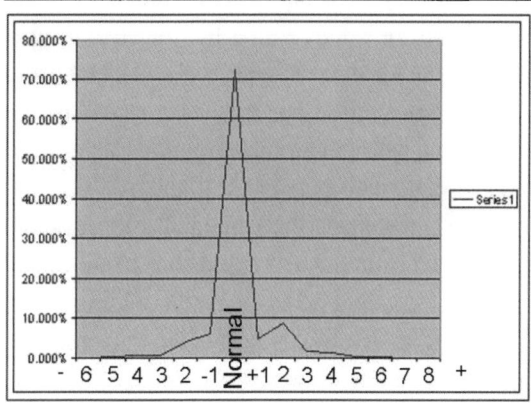

Graph 3-3: Chart plotting the percentage of isotopes for each natural element. The plot almost follows an exponential curve.

Why do Isotopes Exist in our Universe?

As you can see from the chart, the plot almost fits an exponential curve, such as the one shown in Graph 3-5. Now allow me to explain why isotopes show up in our existence. Let us assume that the Diehold is transmitting the information for nickel. The information consists of a variety of frequencies. The big question is: "How does the Diehold *know* that it is transmitting the majority of information for nickel at the correct frequencies?" It could be done utilizing the same method used in radio and television transmitters. The device used is called a *"phase detection circuit."* The circuit ensures that the transmitter is producing the correct frequencies. It does this by comparing the correct frequency with any higher or lower frequency produced by the circuit. If the frequencies produced go too high the phase detector senses it and corrects it by lowering the transmitted frequency. The same thing occurs if the transmitter frequency goes below its acceptable parameters.

The profound point to be made here is this: *Without the error, the transmitter does not know if it is producing the correct frequencies.* I think the same principle holds true for the Diehold. It knows that the majority of signal information is transmitted at the correct frequencies, which in turn modulates into the correct atomic weight for an element. The isotopes are the error factor found in our dimension. If this principle holds true for our Universe, then we should see this error factor show up in everything from childbirths to plant seeds. A certain percentage of the life forms will have natural defects, some worse then others, and of course these defects do in fact show up.

Explanation for Thermonuclear Reactions

If you raise the energy potential of atoms by kinetically forcing them into themselves, you eventually impart a very high-equivalent voltage to them. Eventually, the voltage, or potential, will be so great that the information that makes up the atom becomes so unstable it can no longer exist in our time and space. At this point, the atom demodulates out of our existence, resulting in a tear or void in our Universe. The Diehold will not allow a tear in the Universe, so it rushes information in to stabilize that small part of the Universe. The new information is usually of one or more other elements that can stabilize the location. The immense amount of energy released from nuclear explosions from a small quantity of atoms can be described and understood as the result of information from the Diehold rushing in to fill a tear in an unstable part of the Universe. The immense amount of energy does not come from within the atoms, but is the result of the process of matter modulating into our dimension from the Diehold. The energy is the force and power the Diehold uses to create this reality. This also explains why Quasars appear to produce more energy than can be explained, using traditional theories of nuclear Physics. As a result, it appears that more energy is produced than can be explained within our reality.

What is Light—Particles or Wave Forms?

We have all marveled at the color spectrum produced by a prism or rainbow. We have all turned on light bulbs, which produce every color imaginable. Even scientists still wonder what light really is, but so far light has defied understanding. Is it a particle, like Einstein envisioned and quantum mechanics describes, or is it an electromagnetic wave, as theorized by Maxwell? Using just one theory scientists cannot explain all of the phenomena exhibited by light. They have two to explain it, which should tell you that both are wrong! The old theory is called the *wave theory of light,* and the more modern theory is called the *photon theory of light.* This section will actually explain what light is.

Quantum Mechanics Viewpoint of Light

Max Planck first developed the theory of quantum mechanics in 1900. Later contributors included Albert Einstein. Planck was trying to discover what the energy distribution of light, emitted from a heated object, was. He concluded from his experiments that the energy emitted could only be radiated in bundles of energy, he called "quanta." Einstein later called these bundles "photons." Planck calculated that these quantums all had the same energy level. (Planck's constant is $h = 6.6261 \times 10^{-27}$ erg. sec., times the frequency of radiation.) Planck's conclusion was that solid matter can only radiate quantums of energy in the form of light.

Einstein added to Planck's conclusion by trying to explain the photoelectric effect. This is when electrons are produced from a metal after light strikes it. It was found through experimentation that the intensity of light has little to do with the velocity of the electrons "produced," but instead higher light intensity produced more electrons in direct proportion to the frequency of the light. It was also discovered that the "freed" electrons had a constant velocity. Einstein theorized that each element had a given number of electrons, held in place by magnetic forces. When light, with sufficient energy levels, strikes the element, the energy overcomes the attracting forces holding the electrons to the atom. If there is any excess energy left over, it is given to the electron as kinetic energy.

The quantity of energy needed to release an electron varies from element to element. The surface electrons on atoms receive the greatest amount of energy, so they therefore receive the greatest amount of kinetic energy. The Einstein-Planck theory considers light to be particles, called photons, with each photon having a certain amount of energy, depending upon its frequency (color). The momentum of each photon is equal to Planck's constant times the frequency of the light spectrum, divided by the speed of light. Planck's constant will be covered later.

$$h \times \frac{f}{c}$$

The quantum theory is not considered perfect because it cannot explain the phenomenon of *interference lines* and *diffraction spectrum* formed by a prism. Maxwell's electromagnetic theory of light explained those phenomena, but could not explain the photoelectric effect.

Multidimensional Reality Explanation

The next question is: Why does light have a constant speed and what is it? Light is the demodulated information of an element passing by us at the speed of approximately 2.997925×10^8 meters per second (186,000 miles per second). This speed is the rate at which the Diehold processes and transmits all of the information into our Universe. Using the videotape analogy, the speed of light reflects the speed at which the information passes over the head device. In other words, it is analogous to the speed of the video tape recorder.

Using a computer analogy, it would be the product of the clock speed of the computer, and the number of bits the CPU can process in one clock cycle (example: 32-bit or 64-bit). Since the matter world we live in is the central element that is being propagated into existence at the speed of light that means that light is really "motionless" information on the "tape." Our world (our dimension) is the *object* which being created at the speed of light.

A simple analogy: Imagine you are in a plane flying at 400 mph. You are looking out the back of the plane while it is releasing a contrail behind it. To your point of view, the contrail cloud is traveling away from you at 400 mph and you look as though you are standing still. The vantage point of someone on the ground shows the contrail speed to be zero, and your plane is flying at 400 mph. Light is like the contrail falling behind us. The only difference is that in our dimension the light, or information, leaves this dimension 360° around the object, in little packets reflecting the pulsing of the frequency creating the atoms. Graph 3-4 shows this. The spikes along the Z plan represent our matter world, and the light will match these spikes.

The term "black body" is a theoretical idea describing an object or area that would absorb all energy and reflect no radiation falling upon it. Its reflectivity would be zero, and its absorbability would be 100%. If light were aimed at it, the black body would appear perfectly black and be invisible. It could only be noticed by what it obscures behind it.

Combine this with the black body's temperature at absolute zero (−273.15°C or −459.67°F), which is theoretically where the thermal motion of the atoms stop. Now, let's assume we have a collection of black body atoms at this theoretical state of absolute zero. It will give off no light or other radiation nor reflect it. All of the information that is being directed from the Diehold is going to a point in time and space. The atom I theorize would no longer travel in a circle as shown in Figure 3-13. As the temperature increases and motion starts, the object starts to appear. First, it will start reflecting light. Then, as the temperature rises above freezing, it will start giving off infrared radiation. When the temperature starts approaching the melting point, the object starts to give off visible light. If the temperature continues to rise, the matter would melt and glow very brightly, and finally vaporize. Considerably above this level, the atoms would initiate either fusion or fission, depending upon whether the element was lighter or heavier then iron. At that point the atom would no longer travel in a circle, forming the toroid, but would fly apart and its individual frequencies go off in different directions as are seen in bubble chambers.

We see objects because light is reflected from them, and the light wavelengths seen are only frequencies not absorbed by the object. If the object is above the energy state of a "black box," and also above absolute zero, some information from the Diehold will demodulate and enable us to see it.

My explanation for the observations made by Max Planck and Albert Einstein: They are correct when viewing small packets of energy, which leave a heated object. We see the small packets of energy because the information, which encompasses the atom, is "blinking" in-and-out of our existence at frequencies of the information—including a carrier wave. Remember, the information that makes up the matter world appears to be digital in nature, so it would "blink" in-and-out of our reality. I created a diagram (Graph 3-4) which shows this.

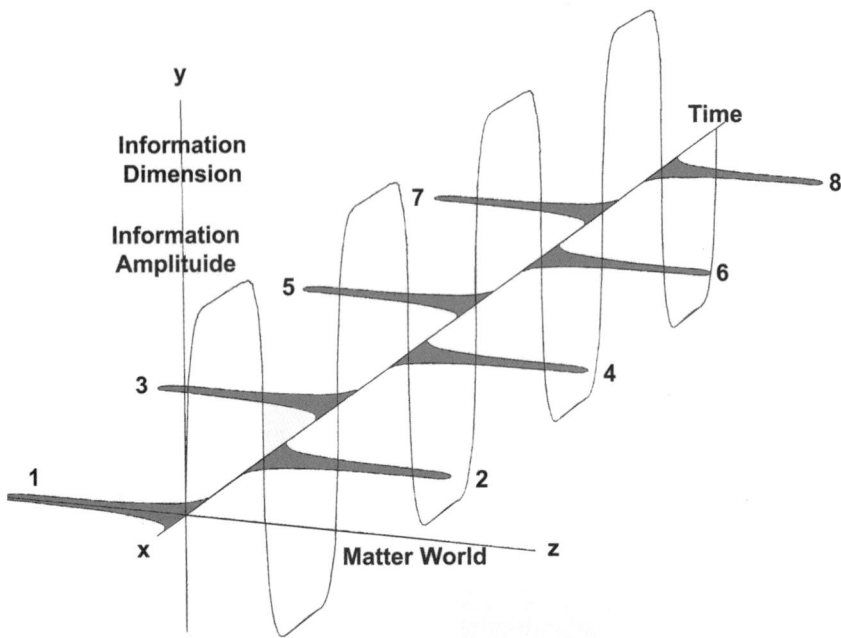

Graph 3-4: The diagram shows the theoretical idea of how a particle travels through space "blinking" in and out of existence in our dimension. The electromagnetic information of the atom is represented by the vertical square wave. The horizontal spikes represent the electrostatic waveform. They are drawn 90° out-of-phase from each other. The z plan represents our Matter World.

The figure above graphically explains five major phenomena within our existence. I will describe each phenomenon below, and what the diagram reveals. First, the vertically-oriented modified square wave represents the information Dimension as frequencies directed to a single point in time and space. Its amplitude and frequency vary with each element, and each element consists of many frequencies, but we are only showing one in Graph 3-4.

Fine Structure and the Fine Structure Constant

Using very sensitive spectral analysis equipment, it has been observed that many spectral lines are doublets (actually two lines very close together). This phenomenon is believed to be caused by the spin angular momentum of the electron, interacting with the orbital angular momentum of the atom. This phenomenon can be explained differently. The fine structure is the result of the two phases of the information coming into this dimension, as shown in Graph 3-4. The even and odd spikes would generate a light spectral line slightly different from the other, and therefore would show up as a split spectral line. As stated earlier, it also explains why we think there is a proton and neutron. It is just how we are seeing the same wave form but a different phasing of it.

Fine Structure Constant

The fine structure constant is a *dimensionless* constant that shows up as the ratio of many different things at the atomic scale. The electromagnetic force that holds atoms together is characterized by the dimensionless Fine Structure Constant = $e/\hbar c$. The equation incorporates the elementary unit of electric charge e, the speed of light c, and Planck's constant, $\hbar$. It incorporates electromagnetism, relativity, and quantum mechanics into one equation. All of the sciences are influenced by the fine structure constant. The following quote from the renowned physics professor, the late Robert P. Feynman, sums up the history—and the predicament scientists are faced with—regarding the discovery of the fine structure constant.

> There is a most profound and beautiful question associated with the observed coupling constant, e the amplitude for a real electron to emit or absorb a real photon. It is a simple number that has been experimentally determined to be close to −0.08542455. (My physicist friends won't recognize this number, because they like to remember it as the inverse of its square: about 137.03597, with about an uncertainty of about 2 in the last decimal place. It has been a mystery ever since it was discovered, more than fifty years ago, and all good theoretical physicists put this number up on their wall and worry about it.) Immediately you would like to know where this number for a coupling comes from: is it related to pi or perhaps to the base of natural logarithms [2.718281]? Nobody knows. It's one of the greatest damn mysteries of physics: a magic number that comes to us with no understanding by man. You might say the "hand of God" wrote that number, and "we don't know how He pushed his pencil." We know what kind of a dance to do experimentally to measure this number very accurately, but we don't know what kind of dance to do on the computer to make this number come out, without putting it in secretly!

The accepted value of the Fine Structure Constant today is 137. This is supposed to represent the ratio of the strength of the electromagnetic force of the electron, relative to the nuclear strong force which binds neutrons and protons together in the atomic nuclei. The problem is that their equation reflects their incorrect philosophy of the structure of matter and energy. If any one of their values changes, the value of the fine structure constant could well be 136, which is what it first was assumed to be.

Light Refraction and Dispersion

Refraction

A prism can be made of any clear, hard material. When light passes through a prism at an angle to the surface (Figure 3-15), the incoherent light immediately divides into separate color (frequency) lines. This is called dispersion. Electronically speaking this would be called demodulation. Each element in the Universe has its own unique spectral line frequencies. Another phenomenon happens when light passes through a prism—the speed of light slows down. In fact, light slows down when it passes through anything denser than a vacuum. This velocity-decrease is directly related to the refractive index of the materials of which the prism is made. Denser elements have higher refractive indices than do less-dense elements (Figure 3-16). The phenomena of diffraction, dispersion and the decrease in velocity are directly related. The traditional explanation for dispersion: short wave lengths are bent more than longer wavelengths of light, but this really is not a very satisfactory philosophical explanation of why light bends. For instance an obese person cannot fit through a narrow door because he or she is fat. This does not address the question of why he or she is fat.

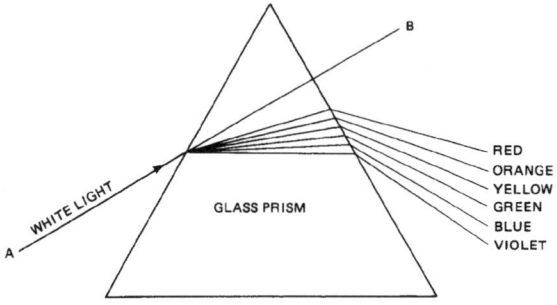

Figure 3-15: A glass prism demonstrating dispersion of white light. Direction AB is the straight direction a microwave signal would take.

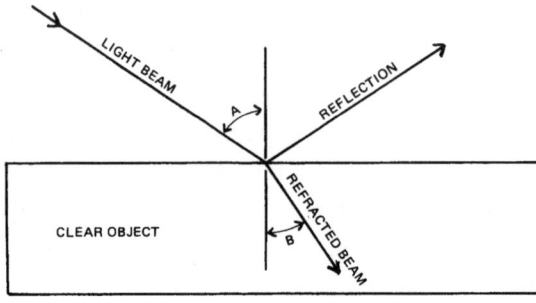

Figure 3-16: Example of refraction. ∠A is the angle of incidence; ∠B is the angle of refraction. ∠B is different for each element.

Why does Light Bend?

Why light bends and light dispersion is the second most difficult phenomenon in the Universe to explain, but the Theory of Multidimensional Reality explains it.

First, white incoherent light represents the information of one or more elements. When you raise the potential (energy) of an element high into the light spectrum, you have actually caused some of the information *not* to exist (to demodulate) out of this dimension, and thus we see it as light. To us it appears that the light is traveling faster, but in reality it is stopped: we are the ones who are actually moving faster in the stream of information from the Diehold. The light bends (refracts) in a prism, for exactly the same reason that it bends and appears to slow down while passing through a strong gravitational field. A gravitational field in really is a strong concentration of information directed towards a planet or star. The gravitational field is not as strong as the modulated information of a prism, as exemplified by its physical shape in this dimension. Also, a gravitational field does not have an immediate effect on the light beam, but rather its force follows the inverse square law. The result is that the light beam gradually curves in, towards the center of the gravitational field.

When the light beam enters the prism, it is immediately bent and changes velocity. This is because the light information, passing through the domains of information that make up the prism, is being increased, or "pushed" a little, by the modulated information of the prism, which is being modulated into existence at 2.997925×10^8 meters per second. The result is, in our dimension, the speed of light looks like it has decreased. Once the light beam passes through the prism's domains of information, the light resumes its zero velocity. To us in this dimension, the light appears to speed up again.

Light Dispersion

The immediate dispersion of light, when it enters a prism, is due to the different energy levels of the frequencies of light. Violet and ultraviolet light possess much more energy potential than infrared colors. To recap a principle mentioned earlier, the electromagnetic field dominates the storage dimension in the Diehold, and the electrostatic field dominates the matter world we live in. It is an established fact that the two fields are 90° out-of-phase from each other (Figure 3-15), as will be explained in the next section. Ultraviolet light bends more than infrared because the First Dimensional information is 90° out-of-phase from the Third Dimension. Thus, ultraviolet light bends closer to 90° from the angle-of-incidence. Higher frequencies where the primary information exists, bend more, approaching

90° from the direction of the incoherent light beam AB. This is directly related to the formula we derived, shown later, in the Pyramids and Angles section. The formula describes our matter world as it is formed 90° out-of-phase from the information dimension. Light dispersion exhibited from a prism illustrates this principle. Lower level frequencies, such as microwaves, would not be affected, as frequencies above it are. In Figure 3-15, track AB is the path microwave energy takes through the prism, more-or-less undeflected by the information of the prism.

The Natural Log e^x

The natural logarithm (e^x) shows up in everything. It grows faster than any power of x, as $x \to \infty$. Similarly, e^{-x} decays faster than any power of x^{-1} as $x \to$ $^-\infty$. This rapid exponential growth and decay is observed in all physical and chemical reactions—charging and discharging of capacitors, mechanical motion, atomic movement, biological reactions, and nuclear reactions. Radioactive elements will decay exponentially over time. Unchecked population growth tends to increase exponentially over time. It is a transcendental number used as the base for natural logarithms. Mathematically, it is defined by the equation:

$$e = \lim_{n \to \infty}\left(1+\frac{1}{n}\right)^n \cong 2.718281; \text{ or by } e = \lim_{x \to 0}(1+x)^{1/x} \cong 2.718281$$

It is represented by the infinite series:

$$e = 1 + \frac{1}{1!} + \frac{1}{2!} + \frac{1}{3!} + \frac{1}{4!} + \ldots + \frac{1}{n!} + \ldots$$

The numerical value for $e = 2.718281$. Graph 3-5 shows a graph plotting the values of the logarithmic function y = *ln* x. This gives you an idea of what a logarithmic function looks like. What is interesting is that there is no philosophical explanation as to why the natural log shows up in our reality.

I have not found a philosopher or scientist yet who has explained why e shows up in our reality! They know they have to account for it mathematically, but they cannot explain philosophically why it shows up in our existence. It is one of the most important numbers in science, and yet no one knows why it exists!

First you must ask: When does *e* show up? The answer is: When something is building up or decaying. The Multidimensional Reality explanation: The natural log shows up as a result of information first being processed, then entering our dimension. In the Basic Theory section, I wrote: An object exists in three different dimensions, *almost* at the same time." My computer-model of the Universe describes information in a storage dimension, a transmission dimension, and as created matter of the third dimension. For the information, time is irrelevant in the Diehold, and the transmission dimension (second dimension) is instantaneous. So where is the delay? It is in the central processing unit (CPU) of the Diehold. Remember, the Diehold is made up of matter and it is a computer, therefore the delay is due to one or more clock cycles used to process our information. We do not see the Diehold because it is in another time-space relationship, so do not ask: "Where in the sky can I look to see the Diehold?" I have actually been asked that question.

Since the CPU is made up of matter, and as information passes through it to get processed, the information slows down because of the few clock cycles necessary to process the information. This is why *e* shows up. It is a function of how the information is processed and transmitted into our existence. It is the delay caused by the *processing time* of the information from the First Dimension to the Third Dimension.

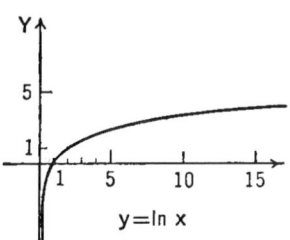

$y = \ln x$

Graph 3-5: Graph plotting the values of the logarithmic function $y = \ln x$.

In Graph 3-4, the vertical waveform along the *y* axis represents the information that makes up matter. Normally, this waveform would be a square wave, but I have depicted it as a slightly modified square-wave, modified for e^x, following the waveform shown in Graph 3-5. This waveform is important in understanding what the Hebrew alphabet design is derived from. The natural log *e* manifests itself as a delay in propagation of the signal, and that is why I show a modified square wave, rather then a classic square wave.

On a personal level, there is a phenomenon that you will relate to and understand concerning how the natural log reveals itself. All humans have experienced déjà vu multiple times. To define it, so that we all know what I am talking about, déjà vu happens when you are looking at some event, and all of a sudden you think you dreamt that same, exact event. Usually, the event lasts for less than a second. No action on your part can change the outcome of the event—

it just happens. Our usual reaction is we think we must have dreamt the event and did not recall the dream until now.

If our existence really emanates from the Diehold, in another dimension, then our thinking processes are also occurring there. The matter part of our brain is like a receiver for this information. What is happening with déjà vu is that the information that makes up what we are doing and seeing is perceived twice. The first signal is perceived directly from the Diehold, the second signal through our physical senses. It is "processed" by our brain, like audio feedback. The signal is sent through the circuit twice. I do not know why this phenomenon happens in the Diehold, or why some people have it more often than others. The delay in the signal is an example of e, the natural log.

Pyramids and Angles

The Great Pyramid of Giza is one of the Seven Wonders of the World. The angle of its sloped sides are believed by many to be 51°51'14.3" (51.854°), which is an angle resulting from a relationship of Pi π(3.14159). Now it is argued by archaeologists that the ancient Egyptians built the Great Pyramid, and they could have known the value of π. I agree they could have known the value but the problem is that when you take the actual base and height measurements of the Great Pyramid, you come up with an angle of 52.606°. You are a whole .755° off! The equation below calculates the correct angle within .054°. But since we can only guess at what the original dimensions were before the Arabs stripped off the white limestone fascia stones—to use for building mosques—I am most likely correct.

$$\text{Pyramid angle} = \text{arc cos}\left(\frac{1}{\sqrt{e}}\right) = 52.66093239°$$

I included the term "Pyramid angle" in the equation above because this angle is the correct exact angle of the Great Pyramid of Giza, Egypt. This simple equation gives us the *arc cos* of a frequency, which is an angle. An *arc cos* is shown in Figure 3-17. The *cos* of an angle can be interpreted as a vector being directed at a given angle θ in this case 37.34°. The *arc cos* of an angle is a plane drawn 90° from the direction of the vector. In our example, it is 52.66°. We would say that this plane is 90° out of phase from the 37.34° vectors. This formula is important because it states that our matter world is formed 90° out of phase from the information dimension.

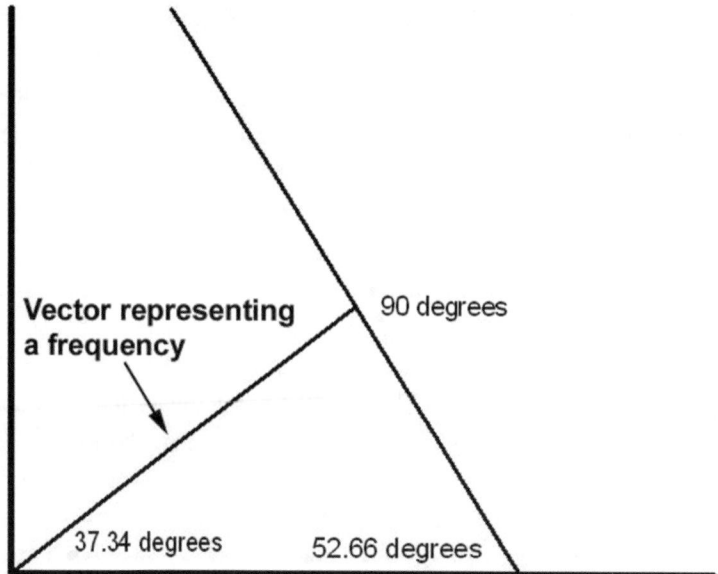

Figure 3-17: This graph demonstrating what an arc cos is. The Arc cos of 37.34° is 90° out-of-phase from it, resulting in a slope plane of 52.66°

The Great Pyramid is a monument modeled on the philosophy of how information comes into this dimension to create the matter world we live in. There is no proof the ancient Egyptians knew the natural log *e* and, therefore, we must conclude they did not build this Pyramid—someone else did!

Planck's Constant Clarified

The traditional explanation for Planck's constant: A physical constant can be seen as a conversion factor between frequency and energy, especially for photons (an explanation of light). It appears in all quantum mechanical equations. It is denoted by the letter *h* and named after the German physicist, Max Planck. Its value is approximately $h = 6.6261 \times 10\text{-}34$ Js (Joules). Energy is measured in Joules, often expressed as electron volts (eV) 1eV = 1.6x10-19Js. Energy of a photon (a *packet* of light) is given by: E = *hf*; *h* is Planck's constant = 6.6x10-34Js and f is frequency.

Graph 3-4 graphically reveals what Planck's constant is, using the Theory of Multidimensional Reality. The energy we call Planck's Constant is created when the information frequencies of an atom cross the x-axis. It creates a distinct amount of electrostatic energy, formed 90° out of phase to the collapsing information field. Some of you may have figured out what this spike-like curve is, depicted in Graph 3-4. **It is our Matter World!** What Planck had observed is that our matter world comes into existence in very short pulses. In essence, it blinks in and out of existence constantly, which I depict in Graph 3-4 as spikes 1-8. The odd ones can be looked at as appearing positively charged, and the even ones are analogous to the neutrons, but it is all the same atom, just two different states of its existence. The conclusion we must come to is that our matter world is made up of packets of energy restrained and directed by information.

The next challenge: What is the light we see in relation to the model, shown in Graph 3-4 The light is formed by these spikes of energy, as a result of some of the information leaving this dimension, assuming the atom is above absolute zero. As previously described, an atom is made up of many frequencies, so we have many of these frequencies going through the same process for the same atom. The resulting light includes the spectral lines of all of those frequencies. The resulting physical atom is the end result of the sums and differences of all its frequencies. The model clearly shows why light exhibits both a wave-nature and particle-nature. The light is the result of many packets of information coming into this dimension, creating matter. The matter "blinks" in and out of our existence and consequently the resulting light also appears as small packets of energy.

String Theory verses Quantum Mechanics

In standard Quantum Mechanics, particles are considered to be points moving through space, tracing out a line. Particles are said to have position, velocity, mass, electric charge, color (which is the "charge" associated with the strong interaction) or spin. This is part of a broader theory called Quantum Field Theory. It provides scientists a theory consistent both with Quantum Mechanics and the Special Theory of Relativity. Quantum Field Theory describes, with great success, three of the four known interactions in nature: electromagnetism, and the strong and weak nuclear forces at the atomic level. Unfortunately the fourth interaction, gravity, theorized in Einstein's General Theory of Relativity, does not work into Quantum Field Theory. To apply the rules of Quantum Field Theory to General Relativity you get results, which go to infinities, which make no logical sense using the "accepted" theories of existence. But does make sense using the Theory of Multidimensional Reality. "For instance, the force between two gravitons (the particles that mediate gravitational interactions), becomes infinite and we do not know how to get rid of these infinities to get physically sensible results."[10]

There are five different string theories (three superstrings and two heterotic strings). String Theory replaces the multitude of particles with a single fundamental building block called a 'string.' The current science philosophy envisions these strings to be either closed, in loops, or open, like a violin string. As it moves through time it traces out a tube if it is a closed ended string, or a flat sheet if it is an open string. In addition the string vibrates. The different vibrational modes supposedly create different sub-atomic particles, with different masses or spins. One mode of vibration, or 'note,' makes the string appear as an electron, and some as photons. Philosophically their theory does not explain where the frequencies, that make the strings unique, come from. There is a string mode called a graviton, which carries the properties of gravity. Allowing for gravity in the theory, is the reason String Theory has received so much attention. Unlike Quantum Field Theory, scientists can account for gravity using the interaction of two gravitons and there are no infinities to deal with. In Quantum Field Theory gravity is a factor artificially added to the equation. In String Theory it has to be there. The great hope of String Theory is that it would unify all known forces and particles in nature into a single 'Theory of Everything.'[11]

My theory of existence helps explain why both types of string theory are partially correct. Straight strings can be visualized within Graph 3-4. And since the atom travels in a circle and is made up of many different frequencies this would appear as a closed osculating string. The toroid model shown in Figure 3-14, looks like the close-end string theory model, if you cut the toroid and stretch it out.

I will elaborate further on string theory in Chapter 4 because the model of the Hebrew Alphabet most closely resembles closed-end string theory.

Quasars

Quasars are believed to signify the birth of a galaxy. Energy levels from quasars are greater than can be explained by traditional physics. Quasars emit huge amounts of energy (10^{58} ergs), which is the equivalent of consuming 500 million of our suns per second to produce the energies emitted throughout the electromagnetic spectrum.[12] A good summary of the problem traditional astrophysics faces is found in the following quote from the astronomer, Dr. George Abell:

> "We have then the perplexing picture of a quasar: an extremely luminous object of small size displaying enormous changes in energy output over intervals of months or less from regions less than a few light months across; 100 times the

luminosity of our entire galaxy is released from a volume more than 10^{17} times smaller than the galaxy."[13]

This contradiction between the philosophy that energy comes from matter, as exemplified by the famous equation $E = mc^2$, compared to actual observation on the cosmological level, appears to have a fatal problem. As the famous scientist-engineer Nikola Tesla said: "Nuclear energy is an illusion." If our galaxy, like all galaxies, went through the quasar stage of evolution some 15 billion years ago, and had consumed matter-for-energy to produce the power quantities mentioned before, the Milky Way in its quasar stage would have used up all of its matter (certainly all of its hydrogen and helium) billions of years ago. To put it simply, this galaxy should not exist today.

The energy produced by quasars can be compared to transient or bias voltages produced on the tape head of a recording device. The bias voltages on the head device far exceed the voltages produced by the magnetic domains of the tape passing over the head. If any of these voltages should leak through to the tape, more energy would appear available from the tape system than exists on the entire tape. Maybe the Diehold, when it creates a galaxy, opens up a portal and rushes a tremendous amount of information to that point in time and space thereby appearing to emit more energy than can be produced in the Third Dimension.

Kirilian Photography and the Phantom Leaf Effect

Not too often can an experimental piece of equipment, costing less than $100, prove a theory with one phenomenon. The phantom leaf effect, produced by Kirilian photography, is that experimental test. The phantom leaf is a phenomenon caused by passing high-voltage (low current) at a high frequency (ranging from 3,000 hertz to 5 MHz) through a leaf with a photographic plate on one side. It is not really photography using cameras or even lenses. It is a high-voltage, high-frequency effect. The electron discharge is what exposes the photographic film. Figure 3-18 shows a cross-section of how a unit is set up.

The phantom leaf effect is produced when 2% to 10% of a leaf is cut away and the remaining part placed between the plates. Remember the uncut leaf never touches the photographic film before the part is cut away. The reason I mention this is because some "scientists" have tried to explain away the phantom leaf effect by claiming that the whole leaf *must* have come in contact with the photographic film or paper. This is just not true. It does show how intimidated many scientists are when they see this phenomenon—and that is because some

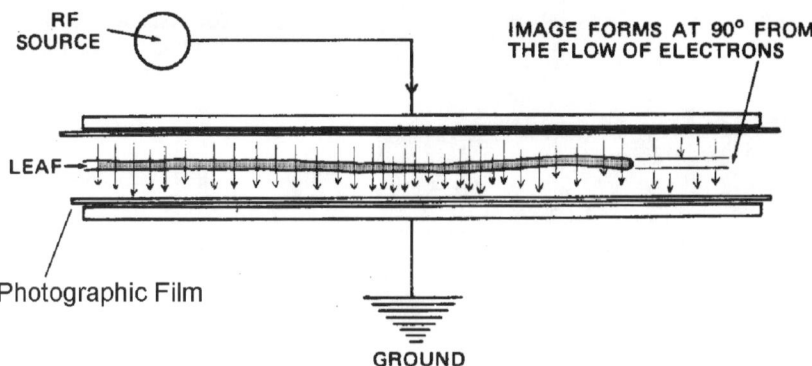

RF SOURCE

IMAGE FORMS AT 90° FROM THE FLOW OF ELECTRONS

LEAF

Photographic Film

GROUND

Figure 3-18: Cross-section of a Kirlian "photographic" unit, where the leaf or other living object would be placed.

know what it really means. When the leaf is properly "photographed," with the correct frequency, the cutaway section of the leaf will appear as if it is still there. It only works about 5% of the time, and that is because we just do not know all the variables involved when setting the frequency to produce an image correctly. The "photograph must be taken within one minute of cutting the leaf." Figure 3-19 and 3-20 are both black and white photos of two leaves, where part of the right side had been cut off.

I have indicated where the cuts are in both pictures. The surface of the leaf looks similar to small bubbles and streamers of light coming from the remaining part of each leaf, even from the areas that no longer physically exist! On the missing portion, the resulting light pattern outlines the superstructure of the leaf all the way to the edge. It is not often that this type of phantom appears. Figure 3-21 shows a photo where the top of the leaf was cut off. Only streamers of light outlining the former leaf shape are visible.

Some experimenters have reported a 5% success rate in "photographing" phantoms. The

Cut line of leaf

Figures 3-19: Kirlian photograph of a cut leaf, exhibiting the "Phantom Leaf" effect. Courtesy of T. Moss, UCLA

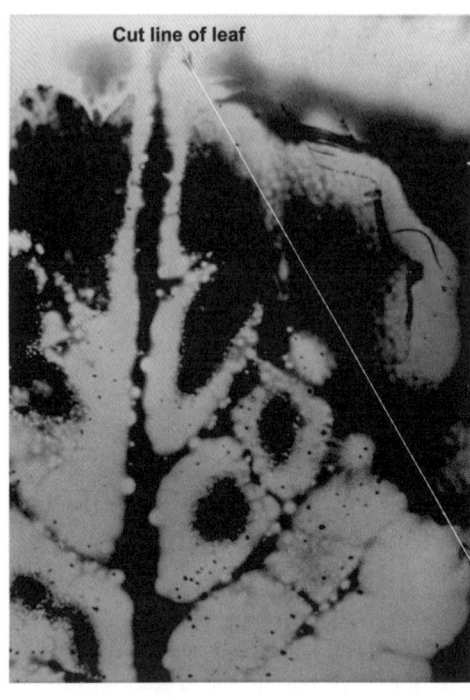

Cut line of leaf

Figures 3-20: Kirilian photograph of a cut leaf, exhibiting the "Phantom Leaf" effect. Courtesy of T. Moss, UCLA

phantom image is very amazing to observe in a real-time environment. When photo-graphed with a movie or television camera, it is seen to fade in and fade out of the picture for up to five to eight seconds before it "disappears" altogether![14] The phantom images are not a residue from the leaf before it was cut off. The first time it ever comes in contact with the film is after it was cut. The leaf was immediately photographed and the phantom could appear. The experiment was performed many times to show that it is a repeatable phenomenon. No one has ever understood why this could happen, or even thought about the consequences of what it means.

Multidimensional Reality Explanation

First you must ask yourself, "How does a leaf know how to recreate its original image?" If you consider what the leaf has going for it, in this dimension, the answer is, it does not know how to reconstruct its original shape. The only possible answer for the appearance of the phantom effect is that the information for the whole leaf exists in another dimension! The life and conscious energy part of its information, its "soul," exists in another time-space relationship. Since the only way to make this phantom appear is by using a high-voltage, high-frequency device, this implies that the signal that makes up the leaf is a modulated signal of some sort. Otherwise we could not possibly make the image appear. As we look at the phantom portion of the leaf, we notice that the bubbles of light and the aura around the edges are quite visible. The only thing that is not present is the light that makes up the actual matter in this Dimension. That means we are looking at the portion of the leaf's information that makes up its *conscious energy*. Another way of stating it is, only the information that makes up its consciousness

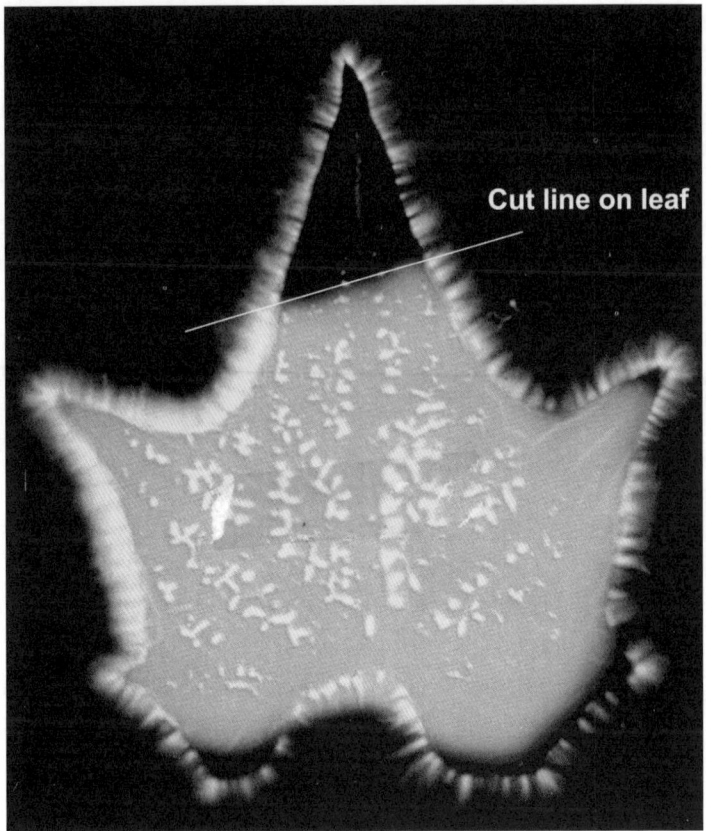

Cut line on leaf

Figure 3-21: "Phantom Leaf" effect from a cut leaf. Courtesy of Hernani Andrade, Brazilian Institute of Psychobiophysical Investigation.

is displayed. The pulsations observed from the phantom are the result of these different frequencies dynamically changing.

The phantom leaf effect proves two important principles: the first being that energy does not come from matter ($E = mc^2$ is wrong!). This is the reason why: Einstein and Planck theorized that light is a form of energy emitted when matter changes energy states. In all three pictures, we have light coming from *no matter*. If the *m* in Einstein's equation is equal to zero, then the equation is equal to zero. In other words, energy does not come from matter! It is the result of information coming into our dimension, forming and constantly creating matter. The second point it proves is that our consciousness exists in another time-space relationship. That the "template" or programming, which makes up the idea of a plant or man exists as information in the Diehold. Only God ultimately has the power to create or change life forms. That is all done from the information dimension.

Delayed Radio Signals, A Time Warp in Space

There is another high-voltage phenomenon that tends to prove my theory of existence. The phenomenon deals with delayed radio signals of unexplainable time duration. The phenomenon occurs as follows: A radio transmitter sends out an RF signal. A delayed signal is heard from a few seconds to ten, fifteen, or more minutes later. The delayed signal is usually pretty weak. Even a few seconds delay represents hundreds of thousands of miles. Reflection, which happens in the ionosphere, is highly remote. This is especially true when you are dealing with a five-minute long delay or more. In this case, you are dealing with tens-of-millions-of-miles of space. The phenomenon was first recognized in 1927 by a couple of Norwegian scientists.[15]

It is quite possible that the very high voltage produced on a resonant-type antenna produces a time warp. These high voltages can bleed energy from the transmitted signal to other areas of the Diehold. This can only be described by using the videotape analogy. For example, when a strong signal exists on a tape recording, the signal bleeds through the various layers of tape and produces a *cross talk* effect. Time and space for that signal is changed. The transmitter that produces this effect need not be high power. High-standing waves on the antenna system can raise the potential energy of the signal tens-of-thousands of volts, even millions of volts. It is important to note that when the two scientists adjusted the antenna to make it more efficient, the phenomenon disappeared. The standing waves had been reduced and, with them, the very high potentials.

Proof of the Carrier Wave of Existence

Earlier, I theorized several times that there is a carrier wave that information is strapped onto. This is the hardest phenomenon in the Universe to discover and figure out, but ironically, it was the first one I figured out, and it is the one that led me to write my first book, *Reality Revealed: The Theory of Multidimensional Reality*. The question is: How do you discover the carrier wave?

Mysteries of the Pyramid

During 1973 to 1975 I was experimenting with pyramids and studying them for their effects on living things. I am talking specifically about pyramids that were built in the same proportions and angles as the Great Pyramid of Giza. Earlier in this Chapter I covered the correct angle for the Pyramid, so I will not describe that now, other than to state that the angle of the pyramid one makes could be 1°

off and it will still "work." The unusual mysteries associated with such pyramids are listed below:

- The pyramid can be built out of any material. Some metal pyramids "work" better than non-metal ones.
- Food kept under the pyramid will stay fresh for two-to-three times longer than food not in a pyramid.
- Artificial flavorings in food will lose their taste, but natural flavors are enhanced.
- The tastes of foods change; they become less bitter and acidic. When we take a spectrographic reading of the treated item, it shows a change in the molecular structure.
- The pyramid will dehydrate and mummify organic objects, and it will not permit decay or mold to grow.
- There is a slowing or complete stopping of the growth of microorganisms.
- The Swedish Scientist, Dr. Carl Benedicks, discovered that the pyramid produced a resonance or frequency inside. Two German scientists, Born and Lertes, also discovered that this frequency was in the microwave range.[16]
- Researchers say that items placed under a pyramid stay "charged" for various lengths of time after being removed from the pyramid. It has been found that water keeps its "charge" longer than any other molecule; and water will lose its chlorine taste and generally taste better.[17]
- Plants grown under a pyramid grow about twice as fast, in their early life, than do plants that are not grown under it. The treated young plants look healthier and have less insect damage.
- It has been found that a copper pyramid has the best effects and intensifies the effect on organic materials.
- The pyramid has an effect on inorganic items. It is well known that razor blades keep sharper longer, if placed under a pyramid between uses.
- Bill Kerell[18] has done many experiments using brine shrimp, which usually live 6-to-7 weeks, but under a pyramid, Bill has kept them alive for over a year. He also noticed that pyramid-grown shrimp grew two-to-three times larger than normal.

- It has been found that the theta and alpha brain waves are increased. These frequencies are also higher and the signal strength is twice the normal amplitude.
- Inside a pyramid, a tingling sensation on the skin, similar to that of mild electricity, occurs, along with an increase in skin temperature, a tranquilizing effect on the nervous system, a deeper "dropping off" in the transcendental state, and finally, very graphic dreams in vivid color.

The Explanation of these mysteries

I am not going to go into a detailed explanation for all of these observations. That was covered in my previous book, and is not within the scope of this one, but the general explanation is this: The first thing we notice about a pyramid is that it puts things "back the way they are supposed to be;" it makes things more perfect. The first thing to ask is: "How does an object know how to change its condition to a more perfect state than before?" Living things use DNA as their template. What about inorganic objects?

The four phenomenal observations that tell us the most about what is going on in a pyramid are:

1.) The pyramid can be made out of any material and it will still work. Metal pyramids have a greater effect, probably because they are denser.
2.) The pyramid will preserve living objects and enable some to live longer.
3.) Inorganic objects are affected by the pyramid.
4.) The angle and alignment of the pyramid is critical to making it work.

When we build a pyramid with the correct angle we are building a tuned circuit that is a three dimensional representation of the carrier wave. The formula, that I presented earlier and list here is the three-dimensional representation of the carrier wave. When matter is positioned at this angle it oscillates at the carrier wave frequency. Any object you place inside the pyramid is, in essence, having its information amplified. It makes no difference what is put inside the pyramid. The razor blades will stay sharp because the information that makes up those microscopic metal crystals will "try" to remain in their original shapes by "trying" to move atoms to the areas worn away. The reason any material can be used in building the pyramid is because the one frequency, common to all elements, is the carrier wave frequency.

$$\text{Pyramid angle} = \text{arc cos}\left(\frac{1}{\sqrt{e}}\right) = 52.66093239°$$

The same explanation and conclusion can be applied to plants and foods placed under a pyramid. If the plant is growing under a pyramid, it receives amplified information about itself. The stronger the signal, the more energy the plant has. More of the information it receives about itself the less chance for imperfections in its DNA, which would cause disease, mold, or other organisms to attack it. The increased signals for humans show up as the theta and alpha brain waves, increasing in both frequency and amplification.

This I believe is the formula for the carrier wave and also ties in with e the natural log. The vector angle of energy formed by the carrier wave frequency is 52.66093°. By the way, as a response to those anthropologists/archaeologists who say that the Great Pyramid was built by the Egyptians, think about this: Our civilization did not discover e until after Gottfried Liebniz and Sir Isaac Newton invented calculus in the 18th century. There is no proof the Egyptians knew about the natural log. Besides we must ask ourselves: *Why* would they build such a huge building—with 2.5 million stones, based on this angle?

A final observation: The true builders of the Great Pyramid appear to have had the same philosophy of existence that is expressed in the "Theory of Multidimensional Reality."

The Carrier Wave and Magnetism

The average person may not realize that magnetism is still a phenomenon in physics. Nobody knows why iron can display something we call magnetism and only a few other natural elements display paramagnetic properties and the rest do not. There is something very special about the element Iron (atomic weight is 55.8, atomic number, 26). It is the only element that you cannot theoretically do fusion or fission to. It has 4,169 light spectral line frequencies that identify it, vastly more than any other element. I believe what makes iron so special is that some of its spectral lines are not used for its physical properties but some are the carrier-wave frequency (or frequencies, we just do not know). Each individual iron atom has this dipole (north-south poles) as a result of strong carrier-wave frequencies. You should think of this dipole as the different phasing of the carrier wave (see Chapter 4 for actual phasing). When the majority of these atoms are able to line up (their dipoles go in the same direction), the iron object becomes a magnet or has magnetic properties.

What is Time?

The concept of time is a very difficult concept to explain or understand. Some top scientists have tried, with varying degrees of success. The only theory, which can successfully explain time, is the Theory of Multidimensional Reality. The following example offers a very clear understanding of what time is, and its importance to living creatures.

Time is relative to the dimension you are in. In the first and second dimensions, time does not mean anything. In the first dimension of information time only means something when it is processed into some kind of reality. In the other dimensions, time means something. If you look at Figure 3-1 and Graph 3-4, you will come to the conclusion that time is the result of the constant rate that the Diehold transmits information into our reality. Put mathematically time = transmission speed c × information density, where c is the speed of light.

Time definitely has meaning to the matter world we live in. Life-forms in the fourth and fifth dimensions are linear beings. That means that they and we live and perceive reality over time—as in past, present and future. Sixth Dimensional beings may not be linear beings. They may be able to choose different time periods. The theory holds that all life forms cannot go ahead in time because they cannot go beyond the CPU, or "head device" that creates their existence. If God has not programmed it yet, you cannot "see" it.

Everything we or other living creatures do in time is recorded in the Diehold, but time is perceived differently by different life forms. Have you ever wondered about what the primary difference between man and animals are? My son, David, came up with the main difference: Animals do not perceive time as we do. To an animal, the past is perceived as one event. An animal does not differentiate between three hours in the past, or two years in the past. To the animal, it was just the past. For them, remembering the past is difficult, therefore learning is difficult. Actions are more instinct than learned behavior, due to environmental factors. People learn by comparing one event in the past with another. the mind analyses two or more events, then comes up with a conclusion. Learning is much easier for man because he can segment time and randomly retrieve parts of time, and learn from his experiences. I am going to present a striking example of this, but first I must offer you another computer analogy, so you will understand the example.

Computer hard drives are segmented in sectors and blocks. Sometimes the blocks are only 512 bytes but they could be as many as 32,000 bytes. The smaller the block sector the more efficient will be the storage, but the retrieval time will increase. At the root of a hard drive is a boot sector that contains the File Allocation Table (FAT table), which contains the address of the blocks and sectors where particular bytes of information are located. If the operating system wants to retrieve a file, it goes to the FAT table and the drive head retrieves the files from the correct sectors and blocks.

My example is from firsthand knowledge of two older men who had the beginnings of Alzheimer's disease. One was a friend of my in-laws, who was about 89 years old at the time. A group of five of them went to Las Vegas for a few days of fun. The group was sharing two hotel rooms with a shared door between them. In the middle of the night, the friend got up to go to the bathroom and, while walking by the glass mirror on a sliding closet door, he could not recognize himself in the mirror. He began yelling to the image in the mirror to get out of his room. Finally, others in the adjacent room came in and explained to him that he was looking at himself, but he still could not recognize himself.

The second person was my 77-year old father-in-law, who developed Alzheimer's in the beginning of 2001. In his bedroom were two wooden sliding doors to his closet. My mother-in-law changed the wooden doors to full-length mirrors. After they were put in, my father-in-law could not recognize himself in the new mirrors, but he could still recognize himself in the old mirrors in his room. This is just like their old friend seven years before. The baffling question is, why? What is the difference between the image in both mirrors? The answer is: None. The only difference was the age of the mirrors. It is a well-know observation that Alzheimer's patients can remember long-term past events, but not the recent or immediate past. Doctors know that parts of the brain show damage in such patients but it is not known why they have selective memory. The following explanation, I believe, is the only one, and it further proves that the Universe is the product of information.

I stated earlier that man is a program or process in the Diehold. Of course, our program is not affected by the aging of our frail bodies in the fourth dimension, but our brain is a reflection of this program. It also acts like a point of reception and transmission of our thoughts in this dimension. Let us examine the mirror. The only difference was whether the mirror was introduced to the person in his recent or distant past. The men could recognize themselves in the old mirrors, but not the new ones. Let us return to the earlier statement that man is able to segment time and retrieve specific events in his past, and know when they happened. The reason Alzheimer's patients have little or no short-term memory is because the part of the brain that performs the function of segmenting time is damaged or not working at all. It is like a computer hard drive that has a damaged FAT table. The operating system cannot find the file located in the proper sector. The sectors are analogous to the small segments of time containing the events in our lives.

The Diehold records everything in time but it is not part of our own program. Our program evidently stores our perspective (actually, time pointers) and events in our lives. The problem occurs with our ability or inability to store and retrieve the proper segment of time containing the information needed to come up with (retrieve) the proper answer. If you can not properly segment time you cannot store or retrieve the information, because you do not have an *address in time* to know where to go for the information. The reason the two old men could not recognize their reflections in the new mirrors was because their brains could not properly store or retrieve the correct segment of time which contained the new information concerning the new mirror. That processing had to be accomplished first, before their brains could go to the next process, which was analyzing the old information displayed on the mirror—namely their images.

Chapter Conclusion

A great deal of science-oriented information has been covered in a short amount of space, but this Chapter is intended to give you a brief knowledge of the Theory of Multidimensional Reality. Nevertheless, even with only a brief explanation of the phenomena described, you can grasp the idea that my information theory of existence explains concepts and phenomena with one model of the Universe whereas no other theory has ever been able to.

Einstein and other great men of science have sought the ultimate formula or theory that would explain everything. It has sometimes been described as a "unified field theory," but as you can tell from the title, they were still hopelessly stuck in a matter theory of existence. They will never figure it out if they continue on their current path, but that path will not last forever. There are fatal consequences to man and his society if he does not evolve to discover how the Universe really works—before it is to late to take action. The next Chapter and Chapter 8 reveal what they are, and how it ties in with the Torah and the prophets.

This Chapter was included in this book because without the reader having a working knowledge of the Theory of Multidimensional Reality there is absolutely no way to figure out and understand what the Torah really is, and what its message is. Most importantly, you will not know how the first Hebrew alphabet was created. You will also not discover what causes the polar reversal and ice age.

Chapter 4
The Hebrew Alphabet

The Orthodox Jewish tradition says that Moses brought the gift of writing to mankind, but the Hebrew priests had no way to prove this. The only place in the Torah referring to writing, is found in Exodus 32:15-16:

> "And Moses turned, and went down from the mount, with the two tablets of the testimony in his hand; tablets that were written on both their sides; on the one side and on the other were they written. [16] And the tablets were the work of God, and the writing was the writing of God, graven upon the tablets."

The quote clearly tells us the letter designs were unknown to these people before Moses brought the two tablets down from mount Sinai. It also states that the letter sequence was not created or written by Moses. Before 3406 B.C.E.[1] there were only two forms of written communication, the first being cuneiform writing (Figures 4-1 and 4-2), and the second Egyptian hieroglyphics (Figures 4-3 and 4). There was also an Egyptian script used as a kind of shorthand version of hieroglyphics (Figure 4-5). After the Exodus, miraculously somehow, writing came into being. The Jewish priests could never prove that Moses "brought down" the skill of writing because the only way it can be proven was through science

Figure 4-1: Examples of cuneiforms found at Tel El Amarna, Egypt.

Figure 4-2: Example of cuneiforms found at Tel El Amarna, Egypt.

and the Theory of Multidimensional Reality. Chapter 3 was presented first because without understanding how the Universe really works it is impossible to figure out and describe the Hebrew alphabet and understand what the Torah actually is. This Chapter will explain what the Torah really is and how the 22 Hebrew letters of the Hebrew alphabet were created—the very first alphabet in the world. I will also describe why the alphabet was originally credited to the Phoenicians.

What is Known about the Hebrew Alphabet and the Torah

The current-day Hebrew alphabet is not what the original alphabet looked like at the time of Moses. No copy of the original Torah writing style from before the destruction of the Temple (587 B.C.E.) exists today. During the Babylonian captivity, Baruch (the grandson of Jeremiah; Appendix D covers Baruch's lineage) redesigned the alphabet almost to the form we have today. We do not know the model he used to create the current alphabet, and some of the letter models may be wrong—we just do not know. This is a handicap we have and should be mindful of when analyzing these letters. With this in mind, I will proceed. Fortunately, whoever created these letter designs put them into three groups. These letter groupings form three distinct geometric shapes that are a philosophical statement by themselves. Two of the shapes incorporate e^x, of which neither the Hebrews nor Egyptians had knowledge of.

You cannot separate the Torah from the Hebrew alphabet, because you cannot figure out one without the other. There are 304,805 "letters" in the Torah and every one of them is important. Let us start with the Orthodox Jewish Tradition

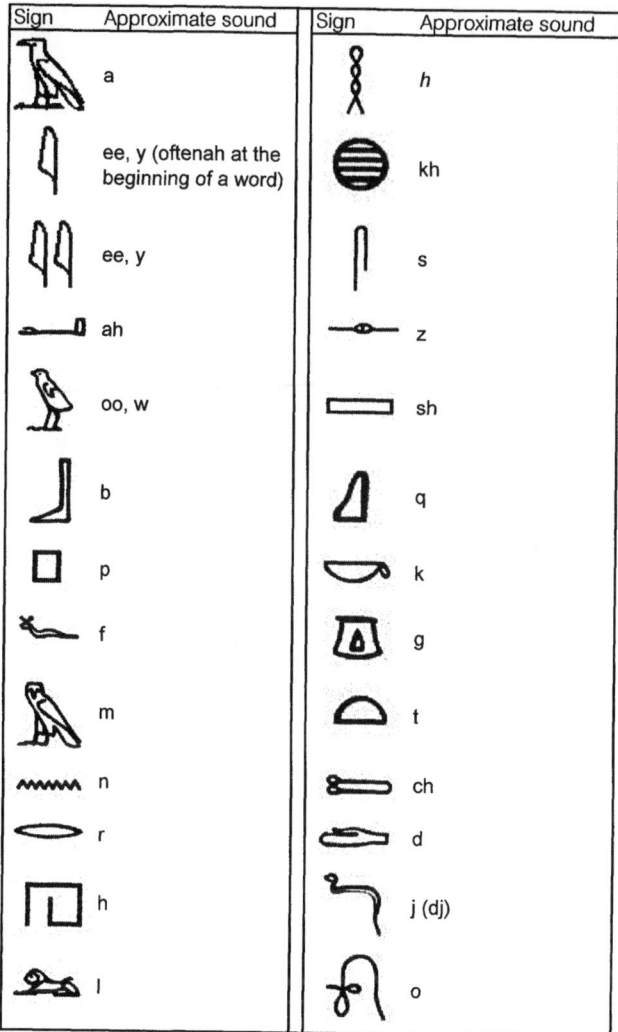

Sign	Approximate sound	Sign	Approximate sound
	a		h
	ee, y (oftenah at the beginning of a word)		kh
	ee, y		s
	ah		z
	oo, w		sh
	b		q
	p		k
	f		g
	m		t
	n		ch
	r		d
	h		j (dj)
	l		o

Figure 4-3: Egyptian Hieroglyphics (proto-alphabetical forms).

about the Torah, the alphabet, and what it says about them. It provides clues about the Torah. There are seven points taught about the Torah that come from very ancient oral traditions. They are:

1   The Torah was written before God created the Earth or the Universe. In fact the Torah was created seven days before the creation of the Universe. "God already had the Torah with Him, in the form of all the letters, before He created the world. But the letters were combined into words only afterwards, forming the Torah as we know it."[2]

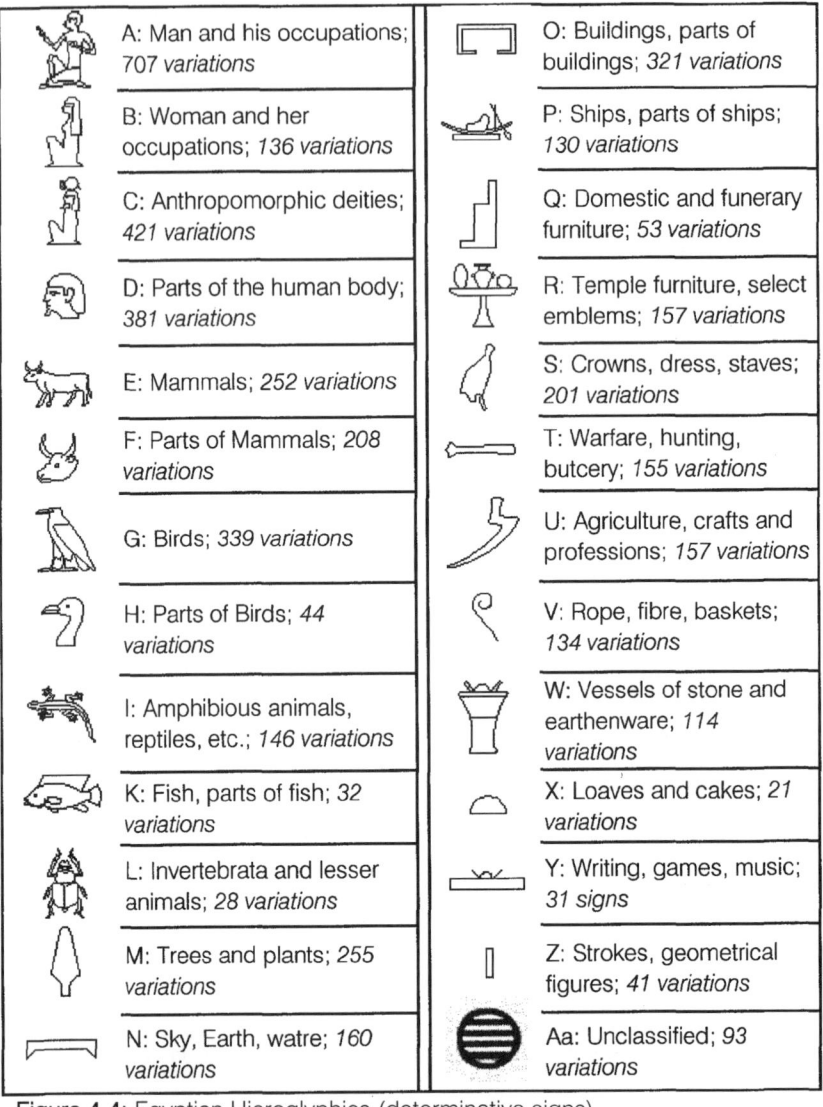

	A: Man and his occupations; 707 *variations*			O: Buildings, parts of buildings; *321 variations*
	B: Woman and her occupations; *136 variations*			P: Ships, parts of ships; *130 variations*
	C: Anthropomorphic deities; *421 variations*			Q: Domestic and funerary furniture; *53 variations*
	D: Parts of the human body; *381 variations*			R: Temple furniture, select emblems; *157 variations*
	E: Mammals; *252 variations*			S: Crowns, dress, staves; *201 variations*
	F: Parts of Mammals; *208 variations*			T: Warfare, hunting, butcery; *155 variations*
	G: Birds; *339 variations*			U: Agriculture, crafts and professions; *157 variations*
	H: Parts of Birds; *44 variations*			V: Rope, fibre, baskets; *134 variations*
	I: Amphibious animals, reptiles, etc.; *146 variations*			W: Vessels of stone and earthenware; *114 variations*
	K: Fish, parts of fish; *32 variations*			X: Loaves and cakes; *21 variations*
	L: Invertebrata and lesser animals; *28 variations*			Y: Writing, games, music; *31 signs*
	M: Trees and plants; *255 variations*			Z: Strokes, geometrical figures; *41 variations*
	N: Sky, Earth, watre; *160 variations*			Aa: Unclassified; *93 variations*

Figure 4-4: Egyptian Hieroglyphics (determinative signs).

2   "All the words and all the actions of all the worlds are contained in the Torah."[3] This declares that all of history is contained within the Torah.

3   The Torah is the Book of Life.

4   "The Torah is the cause of the world's Creation, and also the power that maintains its existence."[4] Another approach used to express this philosophy is that as long as God "thinks" the Universe it exists.

5   To understand who God is you must understand the Torah.

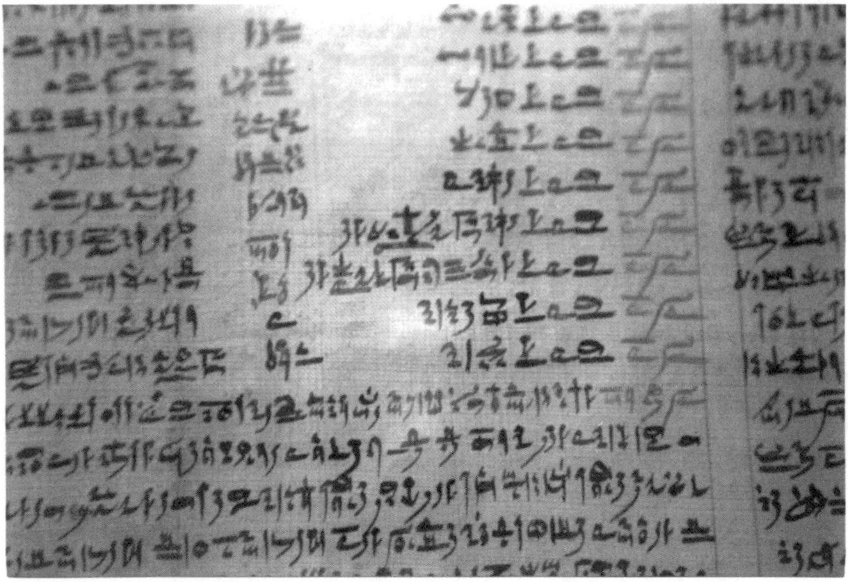

Figure 4-5: Example of a script type of Egyptian Hieroglyphic script.

6   The only book you can compare to the Torah is the Torah itself, because there is no other book like the Torah.

7   The 22 Hebrew letters came from the fiery pen of God, and ten of them fell from the fiery crown of God.

All of these statements are clues about the Torah and the Hebrew alphabet and give us an idea of what we are really dealing with. Some of these clues are true. This Chapter will explain all seven of them. I will cover the first clue because it opens the door for the reader and it explains why many of the Torah verses are out of chronological sequence in the book of Exodus.

The Torah was first written on two stone tablets, then transcribed onto parchment or skins by Moses. Then the two stone tablets were placed in the Ark. How can the traditions of the Torah say that the Torah was written before God created the Earth or the Universe? The stones came from the Earth which had to be formed first to create the stones the Torah was written on. So it is obvious that it is not the tablets themselves that are important, but rather the Information on them that is immortal! We must ask ourselves: "Are the words and Biblical surface stories the important message God is giving us?" Well, the second part of the first statement says that the Biblical surface story was put together after it was given to Moses in 1306-5 B.C.E. To help emphasize this point, there are Chapters in Exodus, which are out of chronological sequence, such as Chapters 6, 18, 21-23 and 32-36. Normally, a writer would sit down with pen and paper and write his story in some kind of sequence the reader will understand, so one can follow the story. The writers' tools are the alphabet, rules of grammar, and a vocabulary, but the Torah says, "the writing was the writing of God, graven upon the tables." According to the first and seventh traditions, the words were formed later from the sequence of "letters" which were already on the two tablets, and the 22 Hebrew letters were of God's design, not man's. So we must conclude that Moses did not have any of the tools a normal writer would have.

We must now conclude that the Biblical surface stories are not the real message of the Torah, but rather the sequence of 304,805 "letters" which make up the Torah and the 22 individual Hebrew "letters." This Chapter will explain what is so profound about the original design of the Hebrew alphabet and what the "letters" really are.

The Traditional Explanation for the Creation of the Alphabet

The traditional academic explanation of the origin of the alphabet is that it originated with a Semitic people sometime between 1600 to 1200 B.C.E. Examples of this early form of writing were found in numerous locations in the Sinai Desert. Figure 4-6 shows some of the proto-Canaanite alphabet.[5] Other written characters were found west of the Nile River, and west of Thebes in Southern Egypt. These symbols have been attributed to Semitic people living in Egypt sometime between 1900 to 1800 B.C.E.,[6] but this is mere guesswork by scholars, because there is no hard evidence to support their conclusions. As you can see from most of these symbols, they appear more like Egyptian hieroglyphics than an alphabet. The only character that looks similar to a Hebrew letter is number seven, below, which looks like a Hebrew Shin ש, but this could also look like a farmer's pitchfork, so we really do not know. Most of the letters seem to have an Egyptian origin, such as Numbers 5, 8, 14 and 16. Table 4-1 shows a comparison between the proto-Canaanite characters and the hieroglyphics listed in Figures 4-3 and 4-4. It seems to be a real stretch to conclude that these symbols were the actual original alphabet.

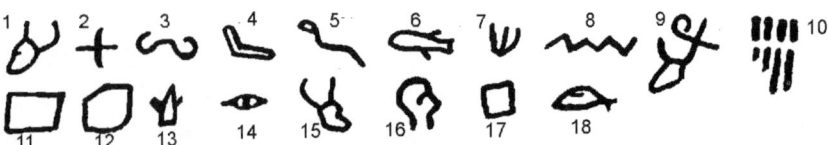

Figure 4-6: Proto-Canaanite characters found in a Sinai Mine.

Proto-Canaanite letter	Hieroglyphics Table 16-3	Hieroglyphics Table 16-4
1		F
2	?	
3	?	
4	f or v	
5	j(dj)	
6		K
7	?	
8	n	
9		F
10	?	
11	p or sh	Z
12	p or sh	Z
13		M
14	r	D
15		F
16		D
17	p or sh	Z
18		K

Table 4-1: Comparison of letters shown in Figure 4-6, and those of Figures 4-3 and 4-4.

The Creation of the Alphabet

The traditional Orthodox explanation of the Torah is that Moses received the entire Torah at Mount Sinai. Other traditional explanations state that Moses wrote the five books. Reformed Jewish and academic explanations state the Torah had multiple authors who wrote at separate times in history. This section will answer this age-old question, once and for all.

All writers have certain basic tools available to them when they write. They have an alphabet with vowels and consonants, proper grammar, spelling, and punctuation. Moses had none of these. He had to create the basic writing tools from scratch. In Moses' time there were only two forms of writing in the world—cuneiform and Egyptian hieroglyphics. So the question is: "How did he come up with this alphabet?" The answer is, he didn't! In the following chapter, I theorize that the family burial cave, called the Cave of Machpelah (not actually located at today's Hebron), contained highly advanced technology from a distant past civilization. The Genesis story of Adam and Eve in the Garden of Eden symbolizes this civilization. Later, I will try to date this civilization, in order to try to date the symbols/letters. To definitively prove there was once an advanced civilization on the Earth, one should have some of their technologies in hand as evidence. Unfortunately, Moses' Rod and the Ark have not been found yet, but the Hebrew alphabet is available and it is all the proof we need. For this reason, I will begin the Chapter by explaining what the Hebrew alphabet really is, because after you see and understand it, the Biblical stories, and everything else you thought you knew about the Torah, become subordinate.

The Basics

In Chapter 3, I presented a waveform (Graph 3-4) that is the model for how information forms the matter world (Graph 4-1). To reiterate, the Y-axis represents the Information Dimension. It creates a modified square wave, modified by e^x. The X-axis represents Time as the Information is propagated to create our matter world. The Z-axis represents the matter world we live in. Our matter world is created 90 degrees out-of-phase from the information dimension. Graph 4-1 illustrates the original model of what the Hebrew alphabet represents.

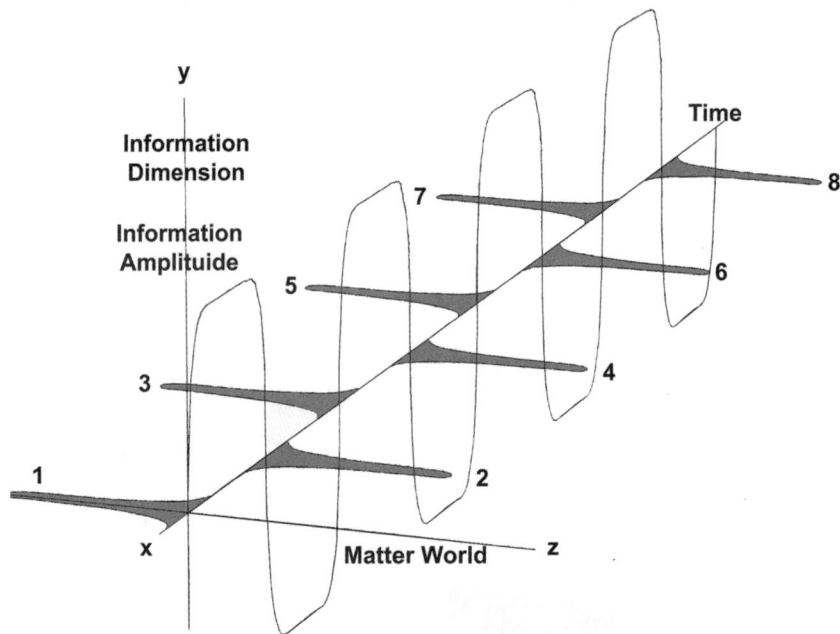

Graph 4-1: A graph depicting how the Information Dimension creates the matter world we live in.

I am going to place this model into a three-dimension shape so you can see what I mean. First, you must understand the matter I am referring to is a single atom as seen in Figure 4-7. I have drawn it so the pole is pointing towards the x-axis, which is what actually happens at the point where its Information creates the matter of the atom in this dimension. This is very important to understand, why the creator[7] of these symbols chose this waveform, and the angles or vectors they chose.

Now let us follow the pole of the atom as it traces the same waveform along the Y-axis (Figure 4-8) as shown in Graph 4-1. What you see is the pole moving

Figure 4-7: Representation of a single atom with its pole pointing along the X-axis.

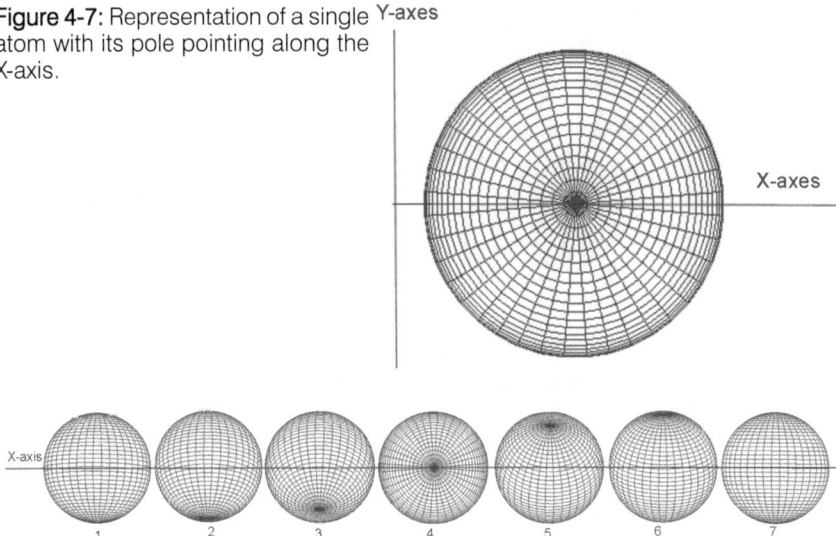

Figure 4-8: The path of the pole as it reverses polarity.

very quickly as it passes through the X-axis. We could rightfully say that it has a "polar reversal" as the waveform collapses and crosses the X-axis. The next point to understand is the atom does not stay in the exact same place, as the Diehold transmits the information to a specific point in time and space. There is a slight difference, as shown in Figures 4-9, with its eight positions—perhaps among thousands of positions as the atom makes one revolution. The shape the circular path forms is a toroid (Figure 4-10 and 4-11). The atom would have completed two polar reversals by the time it completes one revolution and circles back to position 1 shown in Figure 4-10.

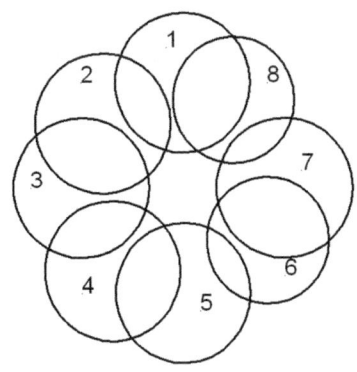

Figure 4-9: Eight positions of an atom as it modulates into the Third Dimension.

Now we are going to track the path of the polar axis, as shown in Figures 4-7 and 4-8, as the round atom turns over 360° and goes around 360° in a circle. That pattern looks like the dark line going around the toroid computer-generated model in Figure 4-11.

This model is what I used in order to decipher the 22 letters of the Hebrew alphabet. The Hebrew alphabet is the result of 22 views of this waveform! I used two models, because the computer-

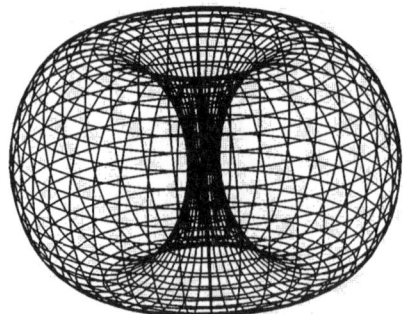

Figure 4-10: A classic toroid shape.

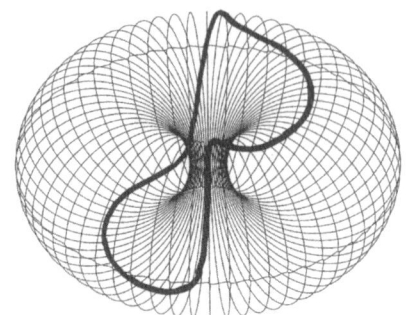

Figure 4-11: The track of the polar axis of the atom.

generated one (Figure 4-11) had a problem displaying perspective. So I built a brass wire model of the toroid with the waveform (Figure 4-12). Between the two models, I was able to figure out which view formed which letter. Another assumption I made was the diameter of the hole in the center of the toroid. We simply do not know if the atom rotates in a tight circle or not, so I made the assumption that it would have a fairly small hole in the center. The diameter of the hole is dependant on the energy state of the atom. At room temperature I assumed the hole would be small. Later, I will show a cross-section of a toroid with no hole in the center.

It took me over seven years to figure out which view produced each letter, but in September 2003, I had a breakthrough when I recognized the different views were fitting a pattern and, in fact, the 22 letters were grouped into three groups that formed three different shapes which were very recognizable. Once I figured that out the last few missing letters fit into place.

Figure 4-12: The brass wire toroid I built for discovering the Hebrew alphabet.

The first thing you immediately realize is that Moses, a late Bronze Age man, could not have invented these "letter" shapes. In order for me to figure them out,[8] I first had to discover the formula, shown in Chapter 3, which incorporates the natural logarithm (2.718281).

$$\text{Pyramid angle} = \text{ arc } \cos\left(\frac{1}{\sqrt{e}}\right) = 52.66093239°$$

Figure 4-13: The formula converting a vector to a plane, 90 degrees out-of-phase from the vector.

I also had to know what this means, and to incorporate it into a square wave. None of these things were known by any civilization at the time of the Exodus and Moses in 1306 B.C.E. There is no evidence the Egyptians used, or knew about, three-dimensional modeling or graphing. One must know all of these skills to create this waveform, and the Egyptians would have to have had the scientific philosophy to want to create an alphabet, or symbols representing 22 views of this waveform.

Once it becomes clear that we are now dealing with the work of a very highly advanced civilization the question arises: "Do the angles they chose convey a message universal to any intelligent society that may find them?" The following evidence will prove that a previous civilization was conveying a philosophy of science parallel to the Theory of Multidimensional Reality.

Closed-End String Theory

I covered the current basic ideas of String Theory in the previous chapter. The model for the Hebrew alphabet (the toroid model) represents a philosophy of existence that a very advanced previous society possessed. The following sections will clearly show that 22 views of a single waveform creates the 22 letters of the Hebrew alphabet. The previous section clearly shows the similarities with closed-end String Theory. I believe the waveform, in my model, represents the carrier wave. Instead of being only one string vibrating at different frequencies, I believe we have multiple frequencies making up an element, all spinning within the toroidal shape. The "string" is not really a string but the path of only one of these many frequencies that make up all the atoms.

In the Theory of Multidimensional Reality, gravity is the result of the force information exerts on other surrounding information and is not dependant on matter in this dimension. Remember matter in this dimension is the result of the information that makes it up. More matter in the third dimension results in greater weight, which is the result of more information in the first and second dimension. The problem dealing with infinities in Quantum Field Theory, becomes understandable and expected if you understand the equation listed before. The information dimension is 90-degrees out of phase from the matter world we live in. When their Quantum Field equations approach infinity that meant the matter went from this dimension back to the first or second dimension.

The best way to conclude this section is a quote from Dr. Lee Smolin, research physicist at the Perimeter Institute in Waterloo, Canada.

> A successful unification of quantum theory and relativity would necessarily be a theory of the Universe as a whole. It would tell us, as Aristotle and Newton did before, what space and time are, what the cosmos is, what things are made of, and what kind of laws those things obey. Such a theory will bring about a radical shift—a revolution—in our understanding of what nature is. It must also have wide repercussions, and will likely bring about, or contribute to, a shift in our understanding of ourselves and our relationship to the rest of the Universe.[9]

I sincerely believe that the Theory of Multidimensional Reality has fulfilled Dr. Smolin's requirements and that it shows that the Universe is the product of information.

The Letter Groupings

The following is a description of the letters as they were found in the three groupings mentioned earlier. I believe these are the correct letter matches. Some of the letters were very close but I believe I figured out how Baruch drew the new Hebrew letter designs, and I also believe he used a model very similar to what I have created as shown in Figure 4-12. The letters found in Tel el Arad dated back to 800 B.C.E. were very helpful. I also found the script version of the current alphabet also very helpful and provided clues which waveform was a match.

Ten Special Letters

The seventh Jewish legend say that 10 of the 22 letters "descended from the terrible and august crown of God whereon they were engraved with a pen of flaming fire."[10] There seems to be no references anywhere, which describe what 10 letters these are. So the question is: "Where did this legend come from and why does it say that?" First, you have to ask: "What is the shape of a crown?" It is round with a top, usually. Now let us look at a top view of the toroid, with the waveform (Figure 4-14 and 15). You will notice that I measured the degrees going counter-clockwise every 45 degrees.

We will start at the zero-degree position, along the X-axis, and look at the toroid from the side, as in Figure 4-16. I will then cover each letter-form as we go counter-clockwise around the toroid, every 45 degrees. Also shown is a table

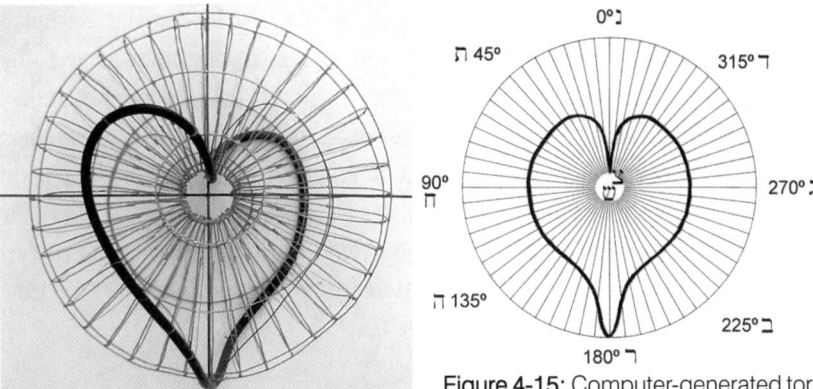

Figure 4-14: Wire-form model of the waveform, viewed from the top.

Figure 4-15: Computer-generated toroid with waveform, showing the familiar heart-shape and marking off every 45° counterclockwise around the toroid.

of ancient and present-day letter-forms for the same letter. The table will highlight, the name of the example that most closely fits the computer-generated waveform. I will then explain which letter-design compares closely or not to the waveform, and why. I measured from the top of the vertical Y-axis around to the bottom on both sides. These are the only angle directions one has to know in order to figure out the Hebrew alphabet. The first picture in the tables, located in the upper left-hand corner, is the total computer-generated waveform, with the portion used for the letter shape highlighted as a bolder line. The graphics program created these waveforms, so they differ only slightly with the actual brass model, shown in Figure 4-14. I have also shown seven ancient alphabet designs. I will be referring to the reference number, for reasons of brevity. The first two are of the modern-

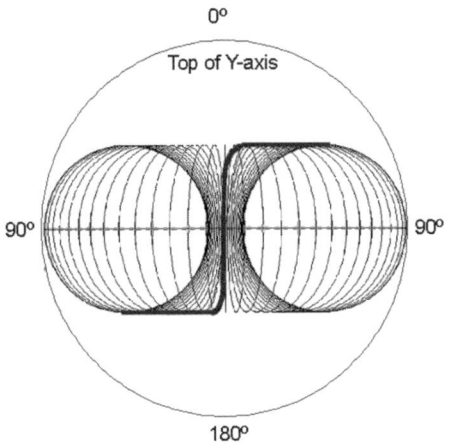

Figure 4-16: Looking at the waveform and the toroid from the side at the 0 degree position along the X-axis. I have measured the vertical Y-axis from the top, around to the bottom of the toroid.

day Hebrew alphabet, showing the standard and script styles. Baruch designed the standard letter shapes during the Babylonian captivity. Number three is the oldest common letter design.[11] It was found at Tel el Arad, located 22 miles northeast of Beer Sheva. Those letters date back to about 800 B.C.E., and their appearance was very helpful. Example numbers 4 through 7 cover the period from 520 B.C.E to about 1,300 C.E. The letter shapes have gone through many changes over the past 2,800 years, but the current-day Hebrew design still carries much of the same design that Baruch used when he redesigned the Hebrew alphabet while in Babylon (6 century B.C.E.).

Nun

The first letter we encounter at the zero-degree position on the X-axis is the letter *nun* (נ). I started from this viewpoint because the Torah mentions the word *nun* no less then 16 times. It was used as the name of Joshua's father, who was Moses' grandson.[12] Nun also meant *fish* (in Aramaic) and Moses was "fished" out of the Nile. So I am not surprised to see the letter *nun* show up first. An important point about the letter style used by Baruch and the earlier priests— and ultimately by the previous civilization—is that every single letter incorporates another view of the waveform, crossing the X-axis, because that is where our matter world is created! In essence, these "letters"—(really symbols)—are celebrating God creating the universe!

Nun

New Waveform	Viewing Angles	(1) Modern Hebrew	(2) Modern Hebrew Script	(3) Ariad Cursive, 800 BCE
	x = 0° y = 90°	נ	J	ל ﬥ ל
	(4) Early Book Cursive 520 BCE	(5) Rabbinic Script, 300 BCE	(6) Hebrew papyrus, Egypt 100 BCE	(7) Rabbinic Scripts Islamic Spain
	ﬥ ﬥ ﬥ	ƽ	ﬥ	ﬥ ﬥ

One of the clues I figured out, is that when they drew a loop on the top of a letter, it was intended to show the top of the toroid. Usually a straight down-stroke meant the waveform was crossing the axis, through the center, or outside of the toroid, such as with the letter *nun* ב.

The modern script (number two) is just like the waveform. Number three and four show the loop on top, indicating the top of the toroid. The down-stroke is similar to the actual waveform, except sloped. Number five and six look like an elongated S, which is what the whole waveform looks like, except it is not sloped like you would see with *italic* letters.

Tau

The *Tau* or *Tav* letter matches the waveform quite well. Numbers one, two, five, six, and seven match the waveform very closely. The right-hand down-stroke of the letter is not drawn as low as the left side so it would not be confused with other letters that look much like the same waveform, such as *He* (ה) and *chet* (ח). In other words, many times the original designer chose only parts of the whole waveform, in order to make it easier for people to write the letters and not cause confusion between letters. Unfortunately, these three letters were swapped many times, because of a scribe's error in not distinguishing between them. The reason numbers three and four were drawn as an X may be because of the similarity of *Tau, He* and *Chet*. The ancients solved the problem simply by creating a distinguishable character.

Tav

New Waveform	Viewing Angles	(1) Modern Hebrew	(2) Modern Hebrew Script	(3) Ariad Cursive, 800 BCE
(waveform drawing)	x = 45° y = 90°	ת	♪	X ✗ ✗
	(4) Early Book Cursive 520 BCE	(5) Rabbinic Script, 300 BCE	(6) Hebrew papyrus, Egypt 100 BCE	(7) Rabbinic Scripts Islamic Spain
	X ✗	ת	♪	ת ת

cHet

All of the examples are similar to the waveform. What is interesting is that numbers three and four are drawn with a thin line through the middle of the box. What I concluded was that the designer of these oldest versions placed a horizontal line through the middle, to indicate the X-axis—in other words, trying to show that the waveform crossed the axis. I found this characteristic incorporated into seven of the early letter designs. Again, the two straight vertical lines in all of these examples were to indicate the waveform crossing the X-axis. All of the computer-generated waveforms show a more curved line, depending on the observer's position relative to the center of the toroid and the waveform. I can only assume that it was felt that straight lines were easier than curved lines for people to write.

cHet

New Waveform	Viewing Angles	(1) Modern Hebrew	(2) Modern Hebrew Script	(3) Ariad Cursive, 800 BCE
	$x = 90°$ $y = 90°$	ח	ת	א אא
	(4) Early Book Cursive 520 BCE	(5) Rabbinic Script, 300 BCE	(6) Hebrew papyrus, Egypt 100 BCE	(7) Rabbinic Scripts Islamic Spain
	ध	ח	ת	ת ק

He

Number one, two and five were the closest to the correct waveform. The left vertical line does not touch the top to indicate that the waveform at the center of the toroid is farther away from the observer than the right vertical line. In other words, it is behind the other line. The script in number two is very important because, for the first time, it is correctly depicting the curviness of the waveform, including the center wave crossing the X-axis, which is the smaller curved line under the larger one. The older design types, as shown in numbers three and four, are somewhat confused. They do have the line in the middle, indicating crossing the X-axis, but the top line is a poor representation of the top of the waveform. This is one of those three letters which must have entailed a redesign, so the three similar letters (Tav, cHet, and He) would not confuse anyone.

He

New Waveform	Viewing Angles	(1) Modern Hebrew	(2) Modern Hebrew Script	(3) Ariad Cursive, 800 BCE
	x = 135° y = 90°	ה	ך	ﻷﻷﻷ
	(4) Early Book Cursive 520 BCE	(5) Rabbinic Script, 300 BCE	(6) Hebrew papyrus, Egypt 100 BCE	(7) Rabbinic Scripts Islamic Spain
	ﻷ ﻷ	ת	ﻷ	ה ק

Resh

All of the examples reflect the correct waveform. The oldest samples, numbers three and four, showed something a little different. They look like a number nine, or a stylized four. I think the designers were trying to show, with the loop

Resh

New Waveform	Viewing Angles	(1) Modern Hebrew	(2) Modern Hebrew Script	(3) Ariad Cursive, 800 BCE
	x = 180° y = 90°	ר	ﺭ	9ﻷ9
	(4) Early Book Cursive 520 BCE	(5) Rabbinic Script, 300 BCE	(6) Hebrew papyrus, Egypt 100 BCE	(7) Rabbinic Scripts Islamic Spain
	9 9	ﻷ	ﻷ	ﻷ ﻷ

that joins with the vertical line, that the line is crossing the X-axis. It is like the other short horizontal lines drawn on some of these letters.

Bet

Numbers one, two, five and seven are good likenesses of the actual waveform. Our current day standard Bet was an exact match. The base line of the Bet extends beyond the vertical to the right because the bottom of the waveform extends further right as you look at the toroid.

Bet

New Waveform	Viewing Angles	(1) Modern Hebrew	(2) Modern Hebrew Script	(3) Ariad Cursive, 800 BCE
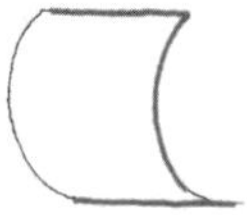	x = 225° y = 90°	ב		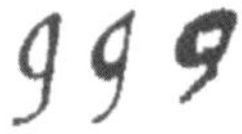 9 9 9
	(4) Early Book Cursive 520 BCE	(5) Rabbinic Script, 300 BCE	(6) Hebrew papyrus, Egypt 100 BCE	(7) Rabbinic Scripts Islamic Spain
	 9 9 9	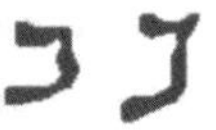 ב ב	 y	 ב ב

Vov

The *Vov* could have been formed from several different positions but the 2/0 degree mark was the best because it produced the best likeness to the known design forms. Needless to say the modern day Hebrew standard letter was a very good likeness. Number three displayed the familiar U-shaped design on the top of a vertical line, depicting the top of the toroid. The vertical line depicts the waveform crossing the X-axis. Number four even added the short horizontal line across the vertical line, to show that it crossed the X-axis.

Dalet

Again, the current-day design in number one is the closest. Numbers three and four look very close to the Bet and Resh letters, which must have caused a

Vov

New Waveform	Viewing Angles	(1) Modern Hebrew	(2) Modern Hebrew Script	(3) Ariad Cursive, 800 BCE
x = 270° y = 90°		ן	/	ϒϒϒ
	(4) Early Book Cursive 520 BCE	(5) Rabbinic Script, 300 BCE	(6) Hebrew papyrus, Egypt 100 BCE	(7) Rabbinic Scripts Islamic Spain
	ϓ ϒϒ	ι	nothing	ן ן

great deal of confusion. That is most likely why Baruch had to redesign the letters based upon the original model. Numbers five and seven are all based on the same general shape, and are similar. Number six has a curve on top to indicate the top of the toroid. The reason the top of the *Dalet* hangs over the vertical, is because the top of the waveform curves to the right, as it goes over the top of the toroid.

Dalet

New Waveform	Viewing Angles	(1) Modern Hebrew	(2) Modern Hebrew Script	(3) Ariad Cursive, 800 BCE
x = 315° y = 90°		ד	ꟼ	ᐊ ᐊ ᐊ
	(4) Early Book Cursive 520 BCE	(5) Rabbinic Script, 300 BCE	(6) Hebrew papyrus, Egypt 100 BCE	(7) Rabbinic Scripts Islamic Spain
	ᐊᐊᐊ	ך	५	ㇴㇴ

Tzadi, *The Top of the toroid*

The *Tzadi* was not an easy letter to figure out. When viewed from the top of the toroid down one sees the same waveform shape, as if seen from underneath. The letter on the bottom is Shin. The modern standard letter (number one) was no help. But the modern script and number three, the oldest, was much more helpful. I discovered some of the waveforms looked identical if viewed from 180° on the other side of the toroid. The original designers of the letters simply rotated the waveform on three of the letters ninety degrees to avoid confusion.

Zadi

New Waveform	Viewing Angles	(1) Modern Hebrew	(2) Modern Hebrew Script	(3) Ariad Cursive, 800 BCE
	x = 0° y = 0° rotated 90°	צ	3	צ ז ﬡ
	(4) Early Book Cursive 520 BCE	(5) Rabbinic Script, 300 BCE	(6) Hebrew papyrus, Egypt 100 BCE	(7) Rabbinic Scripts Islamic Spain
	tℏ	nothing	ﬣ	5 ﬃ

Shin, *The Bottom of the toroid*

The letter *Shin* was one of the easiest to figure out because it was so obvious when you looked at the toroid from the bottom through the hole. This view, rotated 180°, is obviously the origin of the heart shape. Numbers one, three, four, five and six are all the same basic shape and design. The letter is drawn with one half smaller than the other, because if you look at the brass model in Figure 4-14, the curve farther away appears smaller than the top curve.

The seventh legend mentioned ten letters that were part of the fiery crown of God. They are the following: Nun (נ), Tav (ת), cHet (ח), He (ה), Resh (ר), Bet (ב), Vov (ו), Dalet (ד), Shin (ש), and Tzadi (צ). The reason why the legend talks about a crown becomes obvious. Ask yourself, "What is the shape of a crown?" A crown is round because it has to fit over a head and it sometimes has a top. When you look at the top of the toroid (Figure 4-15), eight of the letters are every 45 degrees around the toroid and the remaining two are the top and bottom. After you have acquired the correct angles for all these letters, you have to remember that these are not just angles, but represent *vectors* of information. If you go back to the formula presented in Chapter 3 and earlier in this Chapter (Figure 4-13), you have to create a plane perpendicular to the vector. You then have the sides of a geometric shape (see figure 4-27). The 10 letters form an eight-sided crystal with a top and bottom. This is the first geometric shape.

Shin

New Waveform	Viewing Angles	(1) Modern Hebrew	(2) Modern Hebrew Script	(3) Ariad Cursive, 800 BCE
	x = 0° y = 180° Bottom view			
	(4) Early Book Cursive 520 BCE	(5) Rabbinic Script, 300 BCE	(6) Hebrew papyrus, Egypt 100 BCE	(7) Rabbinic Scripts Islamic Spain

The Eight Letter Grouping

The next group of eight letters forms a very recognizable geometric shape. In fact, a very large structure incorporates the same shape and angles that this grouping forms. This shape and the next are very profound. You will see that the original creators of these waveforms were expressing a very specific philosophy of science when they chose these positions and angles. It is the same philosophy you read in the previous chapter. The following eight letters are viewed at the zero, 90, 180, and 270-degree marks along the X-axis. There are two letters seen at each location. One is located **52.66** degrees along the Y-axis (down from the top), and the other is at 127.4 degrees along the Y-axis. Figures 4-18, 19, and 20 will illustrate what I am describing, along with the sides the vectors created, but I will present these Figures after describing the eight letters.

Koof or Qof

Qof was simple to spot because it looks just like most of the alphabet examples. Modern examples one and two are obviously very close to the actual waveform.

Koof or Qof

New Waveform	Viewing Angles	(1) Modern Hebrew	(2) Modern Hebrew Script	(3) Ariad Cursive, 800 BCE
	x = 0° y = 52.6°	ק	ק	
	(4) Early Book Cursive 520 BCE	(5) Rabbinic Script, 300 BCE	(6) Hebrew papyrus, Egypt 100 BCE	(7) Rabbinic Scripts Islamic Spain

Kaf or Chaf

New Waveform	Viewing Angles	(1) Modern Hebrew	(2) Modern Hebrew Script	(3) Ariad Cursive, 800 BCE
	x = 0° y = 127.4°	כ	כ	
	(4) Early Book Cursive 520 BCE	(5) Rabbinic Script, 300 BCE	(6) Hebrew papyrus, Egypt 100 BCE	(7) Rabbinic Scripts Islamic Spain

Kaf or Chaf

This letter was not easy to figure out because it is one of only two letters that does not appear, going through the X-axis. The older designs did show it, such as in numbers three through six. They all look like the letter *y* but what I think they were trying to show was the top of the toroid and the down thrust through the X-axis. The modern-day standard letter and script designs and script look just like part of the waveform.

Ayin

The *Ayin* went through several evolutionary stages in its design history. The modern-day design takes a part of the waveform that goes through the X-axis, but the script shows it better. The script design is the loop at the right-hand side, but turned 90-degrees clockwise. Number three seems to have focused on the left-hand side of the waveform, which looks like a circle. It is actually a partial circle. I completed the circle so you can better see it. The right-hand side loop may have been utilized, which turned it into a circle. Number four could have been a modification of number three. Number seven looks more like the *y*-design of the original waveform.

Ayin

New Waveform	Viewing Angles	(1) Modern Hebrew	(2) Modern Hebrew Script	(3) Ariad Cursive, 800 BCE
	$x = 90°$ $y = 52.6°$	ע	ð	o o o
	(4) Early Book Cursive 520 BCE	(5) Rabbinic Script, 300 BCE	(6) Hebrew papyrus, Egypt 100 BCE	(7) Rabbinic Scripts Islamic Spain
	◁	yᴜ	nothing	ע ע

Pe

Pe could be seen at several different positions, but this one fits the best. The modern-day standard design is the best match. The modern script depicts the coiled circular form of the waveform at this angle. Number five seems to be closer to the modern-day design, but is missing the indent line. Number seven is more like the modern letter design.

Pe

New Waveform	Viewing Angles	(1) Modern Hebrew	(2) Modern Hebrew Script	(3) Ariad Cursive, 800 BCE
	$x = 90°$ $y = 127.4°$	ב	⊙	)))
	(4) Early Book Cursive 520 BCE	(5) Rabbinic Script, 300 BCE	(6) Hebrew papyrus, Egypt 100 BCE	(7) Rabbinic Scripts Islamic Spain
	⅃	ฉ	)	כ כ

Yod

The *Yod* could arguably have been located at more than three positions. Numbers three and four were the most helpful. I concluded that when a letter had a line drawn through the vertical line that usually meant crossing the X-axis. I show a second drawing, highlighting a different part of the waveform which corresponds with the designs in numbers three and four. I have even added a horizontal line indicating where the waveform crossed the X-axis. As you can see, they match very well with numbers three and four. The reason *Yod* was redesigned to the current small apostrophe shape, is because the original design was too close to the letter *Vov,* and must have caused a great deal of confusion.

Lamed

Lamed was not difficult to figure out, because its top vertical line narrowed down the possibilities. Number one, the modern-design standard letter was the most helpful because it looked so close to the correct waveform. Number seven is also a take-off on that. The oldest designs, as shown in numbers three and four, appear focused on the vertical line and then, as it turned to the right, making it look like our letter *L*.

Yod

New Waveform	Viewing Angles	(1) Modern Hebrew	(2) Modern Hebrew Script	(3) Ariad Cursive, 800 BCE
	x = 180° y = 127.4°	׳	∕	٦٦Ζ
	(4) Early Book Cursive 520 BCE	(5) Rabbinic Script, 300 BCE	(6) Hebrew papyrus, Egypt 100 BCE	(7) Rabbinic Scripts Islamic Spain
	ζ϶ι	▲ゝ.	nothing	׳ ׳

Lamed

New Waveform	Viewing Angles	(1) Modern Hebrew	(2) Modern Hebrew Script	(3) Ariad Cursive, 800 BCE
	x = 180° y = 52.6°	ל	∫	lll
	(4) Early Book Cursive 520 BCE	(5) Rabbinic Script, 300 BCE	(6) Hebrew papyrus, Egypt 100 BCE	(7) Rabbinic Scripts Islamic Spain
	ϩll	c	nothing	ϧϧ

Alef

Alef could not be discovered by looking at the modern Hebrew standard design, but the script is obviously a remnant of the ancient design in number three. Again, the oldest design was the best clue as to what letter matched the waveform. Number three, seems to be trying to depict the wave crossing and looping towards the vertical line, as it crosses the X-axis. The ancient design looks a little like our modern-day *k*. The letter design seems to have changed by 520 B.C.E. to something that looks like our letter F. This is one of the few letters about which I would say the designer of the modern-day version lacked a good picture of what the original model looked like and came up with something else.

Alif

New Waveform	Viewing Angles	(1) Modern Hebrew	(2) Modern Hebrew Script	(3) Ariad Cursive, 800 BCE
	x = 270° y = 127.4°	א	k	⊬ɬɬ
	(4) Early Book Cursive 520 BCE	(5) Rabbinic Script, 300 BCE	(6) Hebrew papyrus, Egypt 100 BCE	(7) Rabbinic Scripts Islamic Spain
	⊬ ⊬⊬	⋏	✗	ክክ

Mem

Mem is one of the three letters that are rotated 90-degrees, once one finds the correct viewing position. This was apparently done to delineate it from *Pe*. The oldest designs shown in number three and four do not do this rotation. They use the double-loop on top to show the top of the toroid, and the curved vertical line depicting the waveform crossing the X-axis from the side of the toroid. The modern standard design shows the general curved arch design on top. The straight line at the bottom depicts the waveform crossing the X-axis through the center of the toroid.

Mem

New Waveform	Viewing Angles	(1) Modern Hebrew	(2) Modern Hebrew Script	(3) Ariad Cursive, 800 BCE
(waveform diagram)	x = 270° y = 52.6° rotated 90°	מ	א	(cursive forms)
	(4) Early Book Cursive 520 BCE	(5) Rabbinic Script, 300 BCE	(6) Hebrew papyrus, Egypt 100 BCE	(7) Rabbinic Scripts Islamic Spain
	(cursive forms)	(cursive form)	(cursive form)	(cursive forms)

Summary of the Eight-Letter Grouping

The eight-letter grouping consists of Qof (ק), Kaf/Chaf (כ), Ayin (ע), Pe (פ), Yod (י), Lamed (ל), Mem (מ), and Alef (א). Figure 4-18 shows a top-view where these letters are located, and Figures 4-19 and 20 show the viewing angle of each letter.

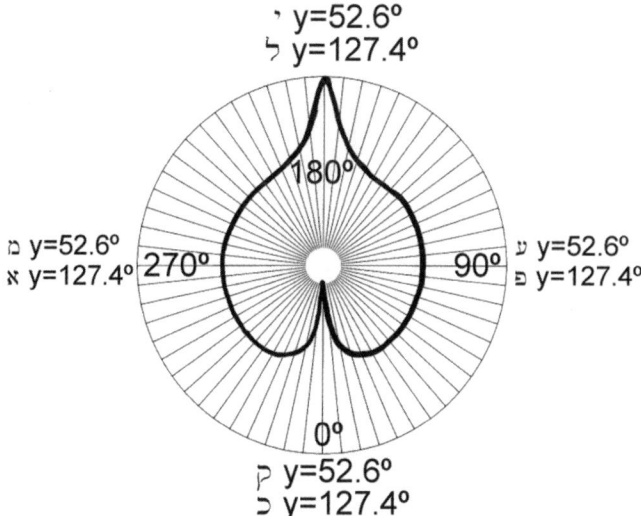

Figure 4-18: Top view of the toroid showing the location of the eight-letter grouping.

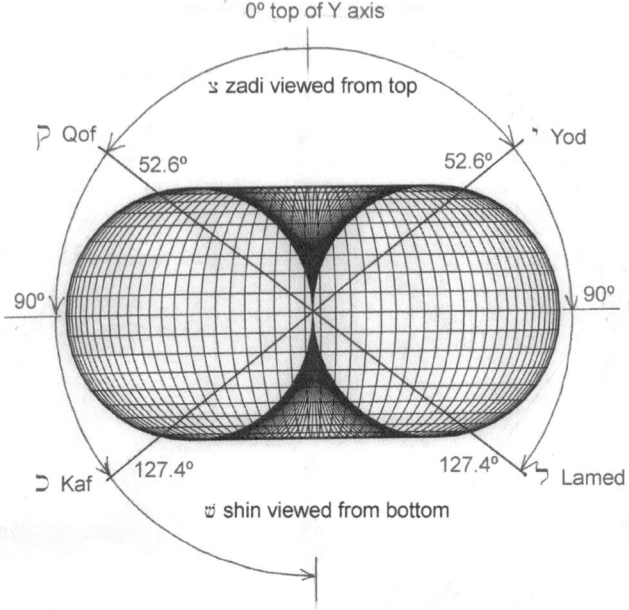

Figure 4-19: Cross-section along the 0° to 180° plane, showing views of the toroid which reveal the Hebrew letters Qof, Kaf, Yod, and lamed.

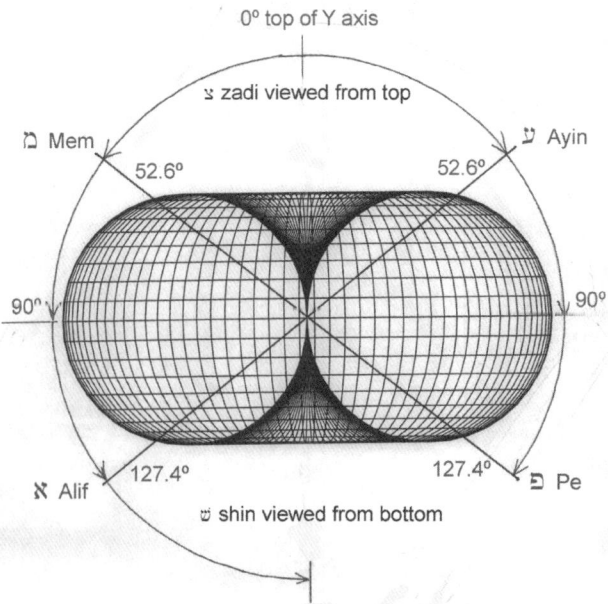

Figure 4-20: Cross-section along the 90° to 270° plane, showing views of the toroid which reveal the Hebrew letters Mem, Alef, Ayin, and Pe.

The only two Hebrew letters that have no sound values, but which acquire a value only from the vowels associated with it, are the *alef* (א) and the *ayin* (ע). Notice that they are exactly opposite each other.

The next point to remember is that these views are really vectors and use the formula in Figure 4-13. The matter world is formed 90 degrees out-of-phase from the Information Dimension. So we draw lines which represent planes perpendicular to these four or eight vectors, and discover what shape we come up with. Figure 4-21 is the result of both planes, shown in Figures 4-19 and 20.

Figures 4-23 shows how you can clearly see the Star of David in the octahedron form. No one previously knew how king David came up with this symbol for the Jewish religion. But now you know. This is the first time in history

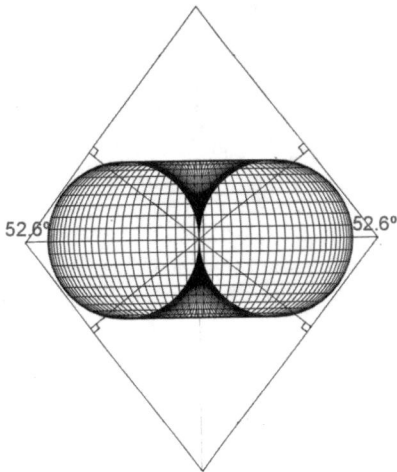

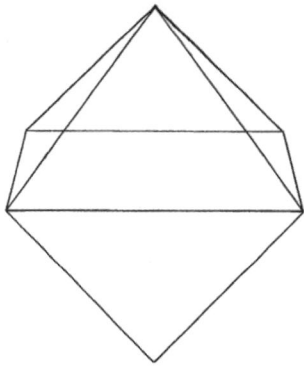

Figure 4-21: The resulting planes formed from the vectors along either the 0°-180° or 90°-270° cross sections.

Figure 4-22: A three-dimensional illustration of the eight-sided shape formed from the eight vectors. The vectors form a perfect octahedron, with the slope of the sides being 52.66°.

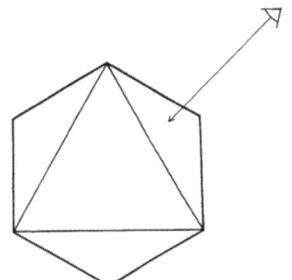

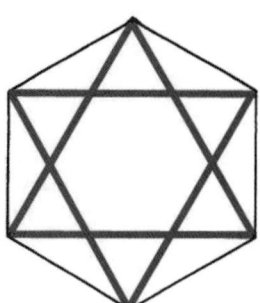

Figure 4-23: Viewing the octahedron shape perpendicular to one of its sides. The Star of David is clearly seen.

anyone has definitively connected the Hebrew alphabet with one of the symbols of Judaism, the Star of David. The fact that he chose this design proves that the priestly class knew the correct model, which created the 22 Hebrew letters, as well as the scientific philosophy, which created the sides of the three geometric shapes. When David chose this symbol, he was making a philosophical statement about who God is, and the alphabet created by God. What you are looking at is profound, and should be savored.

The Four-Letter Group:

The last four letters also form a similar shape. The four letters are viewed at the 45, 135, 225 and 315-degree marks, along the X-axes around the toroid. Figure 4-24 shows their angles from the top down.

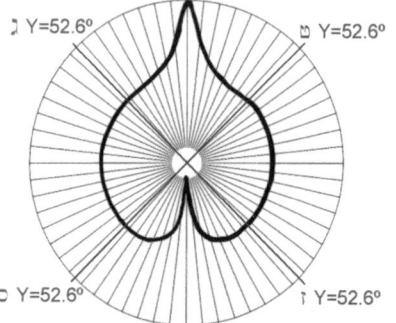

Figure 4-24: Position of the last four Hebrew letters around the toroid.

Zayin

Zayin is located at the 45-degree mark on the toroid. The modern Hebrew standard design looks just like the waveform. The top small loop was designed as a sloped line on top of the vertical line, crossing the X-axis. The script design (number two) seems to have accentuated the top loop and redrawn

Zayin

New Waveform	Viewing Angles	(1) Modern Hebrew	(2) Modern Hebrew Script	(3) Ariad Cursive, 800 BCE
	x = 45° y = 52.6°	ז	ﬞ	nothing
	(4) Early Book Cursive 520 BCE	(5) Rabbinic Script, 300 BCE	(6) Hebrew papyrus, Egypt 100 BCE	(7) Rabbinic Scripts Islamic Spain
	ﬞﬞ	ﬞ	nothing	ﬞ ﬞ

it pointing upward, but it is still very close to the actual waveform. Number four seems to have accentuated the top and bottom lines, with the vertical line either in the middle or the right edge. The design with the vertical line in the middle seems to be close to the waveform. Finally, number seven seems to be close to the modern-day script. All of the examples are close representations of the actual waveform.

Tet

Tet is viewed from the 135-degree location and 52.66-degrees down from the top, or Y-axis. It is the third letter the designers chose to rotate 90-degrees, in this case to prevent confusion with the letter *Zayin*. The only number letter that fit was the modern Hebrew capital-letter design. It appears that the other designers had no idea what the original shape should be. Baruch must have had a correct model of the waveform.

Tet

New Waveform	Viewing Angles	(1) Modern Hebrew	(2) Modern Hebrew Script	(3) Ariad Cursive, 800 BCE
	x = 135° y = 52.6° Rotated 90°			
	(4) Early Book Cursive 520 BCE	(5) Rabbinic Script, 300 BCE	(6) Hebrew papyrus, Egypt 100 BCE	(7) Rabbinic Scripts Islamic Spain
	nothing			

Gimel

Numbers one and seven are the closest to the actual waveform. In fact, once I realized that the vertical line was usually one of the wave parts which crossed the X-axis and that they were drawn straight, it was easy to figure out. The current-day standard *gimel* is almost an exact match to that portion of the waveform. Number three was close, and I understood what the designer was trying to show. We can see that the vertical line and the top short line was intended to depict the top of the waveform as it goes over the top of the toroid. This letter was close to *vov,* but the reason I decided that *gimel* belonged here was because these four letters are the least used letters in the Torah, whereas the *vov* is the second-most used letter. It seemed to me that the designers of these letters would not have combined the second-most used letter with three of the least used-letters. It is more logical that they would group the four least used letters together.

Gimel

New Waveform	Viewing Angles	(1) Modern Hebrew	(2) Modern Hebrew Script	(3) Ariad Cursive, 800 BCE
	x = 270° y = 90°	ג		
	(4) Early Book Cursive 520 BCE	(5) Rabbinic Script, 300 BCE	(6) Hebrew papyrus, Egypt 100 BCE	(7) Rabbinic Scripts Islamic Spain
	nothing			

Samek

The last letter in this group is *samek* and I figured it out by default after most of the other letters were discovered. The only ancient letter design that resembles

Samek

New Waveform	Viewing Angles	(1) Modern Hebrew	(2) Modern Hebrew Script	(3) Ariad Cursive, 800 BCE
	x = 315° y = 52.6°	ס	O	
	(4) Early Book Cursive 520 BCE	(5) Rabbinic Script, 300 BCE	(6) Hebrew papyrus, Egypt 100 BCE	(7) Rabbinic Scripts Islamic Spain
Flipped	nothing			

samek is number three, the oldest letter design known. The downward vertical line depicts the wave crossing the X-axis, while the top loop is shown as a loop with a line through the middle. This may indicate the people and priests did not know what the original waveform looked like, and only had a vague idea. Number five and six looked like the design, but flipped 180 degrees for some reason.

Summary of the Four-Letter Grouping

The four letters in this final grouping are: *zayin* (ז), *tet* (ט), *gimel* (ג), and *samek* (ס). The letters form a pyramid, with the same angles as the eight-letter grouping. They form a pyramid similar to Figure 4-22, except that there is no bottom to form an octahedron. Figure 4-25 shows the geometric shape the four letters form. What is interesting about these four letters is they are the least-used letters in the Torah.

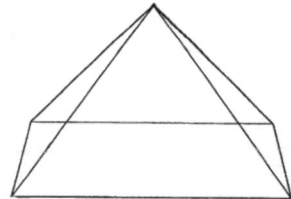

Figure 4-25: The classic pyramid-shape created by the last four Hebrew letters. The pyramid-shape has the same slope of 52.66 degrees as the octahedron shape shown earlier.

Section Conclusion

What I have presented is the evidence to irrefutably prove that Moses did not create or design the Hebrew alphabet. It also proves that Moses told us the truth when he wrote: "the tables were the work of God, and the writing was the writing of God, graven upon the tables." This is sufficient proof that the "letters" of the Torah were the work of a highly advanced previous civilization that once lived on Earth, and which left a storehouse of highly advanced technology in a cave in the area of the world we now call the Sinai desert. I have also shown that the philosophy of science that produced these "letters" parallels my Theory of Multidimensional Reality. The principle that the Universe is the product of information is the only method one can use to figure out how these letters were formed and understand the significance of the pyramid or octahedron-shape, with a slope of 52.66 degrees. It is the only explanation why the Great Pyramid of Giza was built with the same angles! Now you can understand why I presented Chapter Three first and explained the Theory of Multidimensional Reality, in order for people to understand what the letters really showed.

The Hebrew alphabet was the very FIRST alphabet and the Phoenician and all the others were merely adaptations of the original Hebrew alphabet. More profoundly, God took a direct hand in giving man the very first alphabet, which was of invaluable help in writing down ideas and stories throughout history. I doubt if man would have evolved very far without developing an alphabet. What no one realized before is that God, in essence, gave man the alphabet!

The Atom and the Toroid Model.

I mentioned earlier that the toroid represents a single atom entering our existence or dimension. The next question is, "What atom is it?" You might think that there isn't enough information to tell us, but I think there is. Earlier in this Chapter I referred to the Torah as the *Book of Life*. But why do they say that? There are multiple reasons why, but this is one of the reasons: I believe what we are looking at is a model of the carbon atom as it is modulated into our existence, and here is why. The following are the formulas for DNA (deoxyribonucleic acid) and RNA (ribonucleic acid).

DNA: CHO·CHOH·CHOH·COCH·CH$_2$OH
RNA: CHO· CH$_2$·CHOH·COCH· CH$_2$OH.

All life forms are carbon-based. These two formulas are common for all life forms we know of. They are the building blocks of our DNA strand. The formulas are limited to combinations of carbon, oxygen and hydrogen atoms. Oxygen and hydrogen are gases in their natural state. Carbon is normally a solid within the temperature ranges of living creatures. There are 22,000 genes in our DNA strand, and there are only 22 spectral line frequencies of the carbon atom in the visible infrared spectrum. I will add a condition to this statement, and that is that the visible light spectrum does not represent the primary frequencies of the elements, but rather a much lower harmonic of the primary frequency above the ultraviolet frequencies. My assertion is that if we knew the actual frequencies of carbon, there would be only 22 of them. The atomic weight of carbon is 12, the same as the number of tribes of the kingdom of Israel. Finally, the crystal shape of carbon (a diamond) is that of an octahedron, just like Figure 4-22.

I believe the designers of this waveform chose the carbon atom because they were celebrating the creation of life. That is why there are 22 letters in the Hebrew alphabet. Every original letter was a different view of the waveform as it crossed the X-axis, and that is why I have made repeated mention of it. The Torah is said to be the *Book of Life* because of the letters that make it up. There is also another reason, and I will cover that in the Chapter conclusion.

Creating the Biblical Surface Story

Forty years after receiving the two Torah tablets, Moses had finished most of the writing of the Biblical story of the five books. He had created a written working-copy of the same text that appeared on the two tablets. This account appears in Deuteronomy, Chapter 31:16:

> [31:24] And it came to pass, when Moses had made an end of writing the words of this law in a book, until they were finished, [25] That Moses commanded the Levites, which bear the ark of the covenant of the Lord, saying, [26] Take this book of the law, and put it in the side of the ark of the covenant of the Lord your God, that it may be there for a witness against thee.

After Moses' departure, Aaron's two sons finished writing the Biblical story of the Torah including Moses' death and the people entering the Promised Land. The two questions now to be answered are: "Why are some chapters in the Torah out of chronological order, and why are there skip patterns found in it?" These will be covered next.

Moses Creates the Hebrew Vocabulary

I stated earlier that Exodus Chapters 6, 18, 21-23 and 32-36 are out of sequence. Besides these there are other Torah stories out of order. The big question is why? Once you understand how Moses received the Torah and how he wrote the Biblical story you begin to understand what he was up against and the task at hand. Moses did not have the normal tools a writer has when one writes a story. What he received was 304,805 symbols carved on the front and back of the two stone tablets. He could not change the sequence of the symbols, or add any. The Hebrews most likely spoke a form of Canaanite and Hittite language's as well as some Egyptian words incorporated into their vocabulary. The first thing Moses must have done was to assign sound values to 20 of the 22 symbols. He then came up with the device of adding vowels to the letters by adding dots and dashes to the bottom, sides, and top of some letters. They are

. ָ ֹ וֹ . ֻ . . With these vowels applied to the 22 letters, you can have over five billion permutations and combinations. That means you could create almost any kind of word or Biblical story you wanted to. Adding the vowels to the letters made it much easier to create words to fit the symbol patterns on the stone tablets. Once you think about the fact that Moses had to create the spelling of the vocabulary, and sentence structure at the same time, and create coherent Biblical stories, you realize that he must have had help. This task must have been very difficult. I can only come to the conclusion that Moses must have spent a lot of time in front of the Ark getting God's help writing the stories, as well as the vocabulary. Some chapters are out of chronological order because, after the grammar and vocabulary were finalized, Moses must have had some blocks of symbols that were impossible to compose into a coherent consecutive story line, so he had to put another story in its place. Of paramount importance was that he could not change any of the symbols on the two tablets. The overriding question is: "Why, and what, is the Torah?" I will answer that later.

Authorship

Some academics believe that the Torah was the compilation of many authors, written over 700 years. Some say that Jeremiah wrote Deuteronomy because the current acceptable translation includes words that did not exist until after 800 B.C.E. I believe Jeremiah translated Deuteronomy from the original copy which had no word breaks, so he obviously grouped letters together making words that sounded best to him in the 7th century B.C.E. That does not mean he wrote Deuteronomy. It does mean he did the best translation job he could, and I accept his word groupings.

There is a way of testing a book to see if only one author wrote it. The method is not easy, but it does indicate whether a document has more than one author. The method is to graph the number of times each letter appears in a document. Each author has their own vocabulary; therefore they use the same mix of words throughout their writing, except for proper nouns. The average author uses the same combinations of words (writing style) throughout their book. The primary investigatory tool used by code breakers to discover who wrote a document, or if there is a code, is to graph the letter distribution found in the document. I used the same technique in analyzing the Torah against English and American novels. To test the hypothesis, I selected a few popular English-language novels and used the code breakers' website[13] to graph the letter distribution. I did not count all of the vowels in the English books because there are no vowel letters in Hebrew. The vowels I did not count were: *e, i* and *o*. There are some Hebrew letters represented by English vowels, such as *a* for alef א, and I counted those.

Table 4-2 shows the aggregate for three novels. I chose: *Moby Dick,* by Herman Melville (697,484 letters); *Animal Farm,* by George Orwell (98,068 letters), and *A Brave New World,* by Aldous Huxley (213,642 letters). The table shows the average letter distribution for all three books. As you can see they are all very similar.

	M. Dick	An. Farm	BNW	Average
a	11.10%	11.35%	10.72%	11.06%
b	2.41%	2.35%	2.15%	2.31%
c	3.21%	2.87%	3.22%	3.10%
d	5.47%	6.44%	6.31%	6.08%
f	2.99%	3.13%	2.90%	3.00%
g	2.98%	2.83%	3.01%	2.94%
h	9.00%	8.79%	8.66%	8.82%
j	0.15%	0.22%	0.14%	0.17%
k	1.15%	1.01%	1.18%	1.11%
l	6.12%	6.00%	5.92%	6.01%
m	3.34%	3.76%	3.42%	3.51%
n	9.40%	9.68%	9.68%	9.59%
p	2.45%	2.23%	2.37%	2.35%
q	0.22%	0.16%	0.13%	0.17%
r	7.44%	7.83%	7.78%	7.68%
s	9.22%	8.18%	8.59%	8.66%
t	12.60%	12.17%	12.37%	12.38%
u	3.81%	3.51%	3.96%	3.76%
v	1.23%	1.22%	1.26%	1.23%
w	3.17%	3.73%	3.11%	3.34%
x	0.15%	0.28%	0.24%	0.22%
y	2.29%	2.24%	2.72%	2.41%
z	0.09%	0.05%	0.17%	0.10%

Table 4-2: Graph of the percentages of each letter in the three books. Key: **M. Dick** = Moby Dick; **An. Farm** = Animal Farm; **BNW** = A Brave New World.

My analysis of the three books included graphing how many letters were one percent or more over or under the average letter usage (the standard deviation of each letter). The graphing of these statistics revealed much about the authors' writing style. Graph 4-2 shows the standard deviation for Moby Dick. The book has 135 chapters and I grouped them into 48 sections. As you can see, the beginning of his book has more letters with a deviation of one percent (+/–) than later chapters. He only had six chapters where one letter had two percent (+/–) deviation from the norm.

Graph 4-3 shows the letter distribution for the ten chapters in *Animal Farm*. Notice that only Chapter One has five letters, one percent (+/–). There were no letters that were two-or-more percent different. The rest of the book is very even. The Author's vocabulary or writing style is very predictable.

Graph 4-4 shows the letter distribution for the 18 chapters in *A Brave New World*. Notice that only two chapters have four or five letters that are one percent (+/–) different from the norm and they are mostly in the front of the book. The rest of the book is very even and no letters had a deviation of two percent. The Author's vocabulary or writing style is again predictable.

The next five tables are for each one of the five books of the Torah. You will see an obvious difference. All five books are totally different from a normal novel written by one author. Genesis and Exodus have somewhat the same patterns, but the other three books are totally different from one another. All five books have at least some letters that show five percent (+/–) difference from the norm, and the book *Numbers* has one letter with a six-percent difference. It is very common to have a substantial number of letters that are two-to-three percent (+/–) off. This pattern does not exhibit multiple authors, but something totally different. No author or authors write like this in any language.

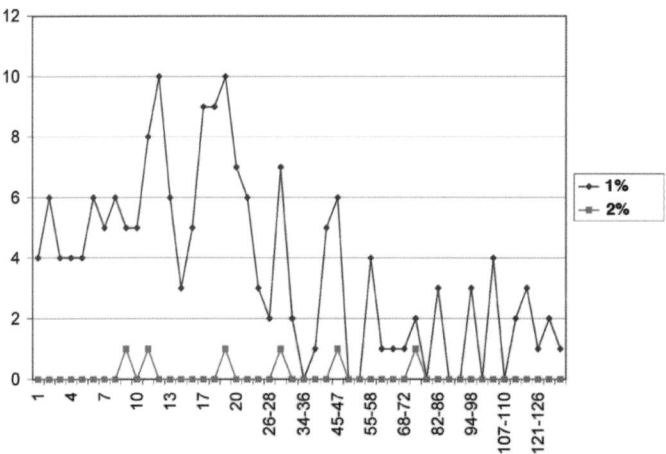

Graph 4-2: Plot of the Standard deviation of the letters in Moby Dick.

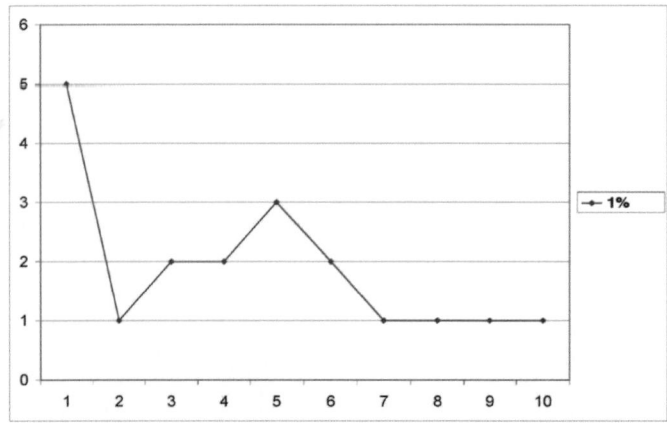

Graph 4-3: Plot of the Standard deviation of the letters in Animal Farm.

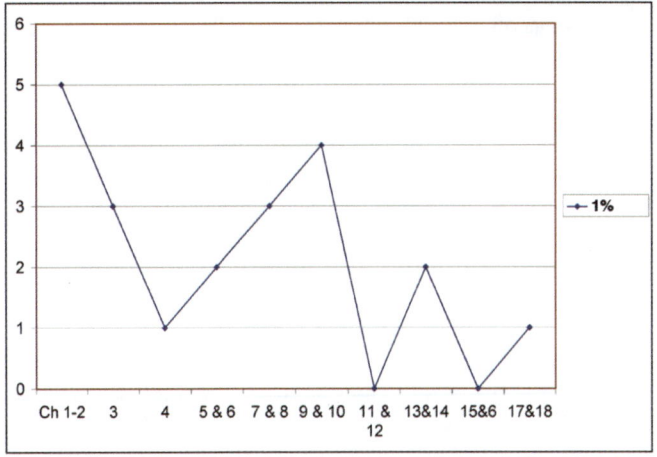

Graph 4-4: Plot of the Standard deviation of the letters in A Brave New World.

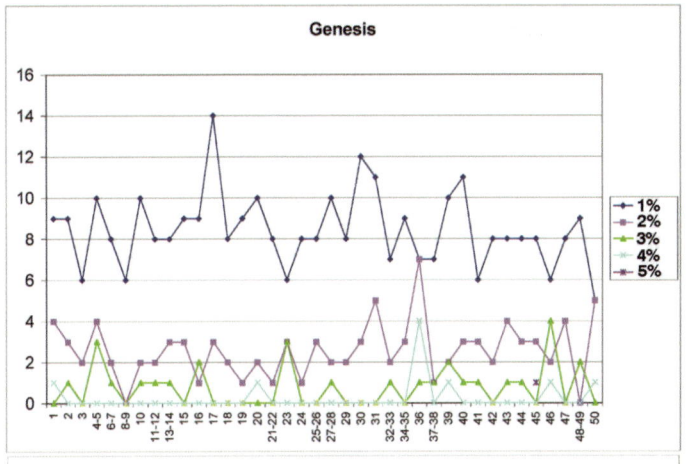

Graph 4-5: Plot of the Standard deviation of the letters in Genesis.

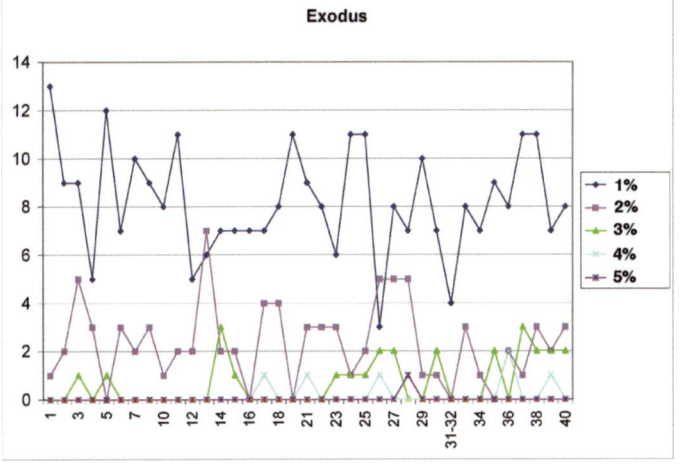

Graph 4-6: Plot of the Standard deviation of the letters in Exodus.

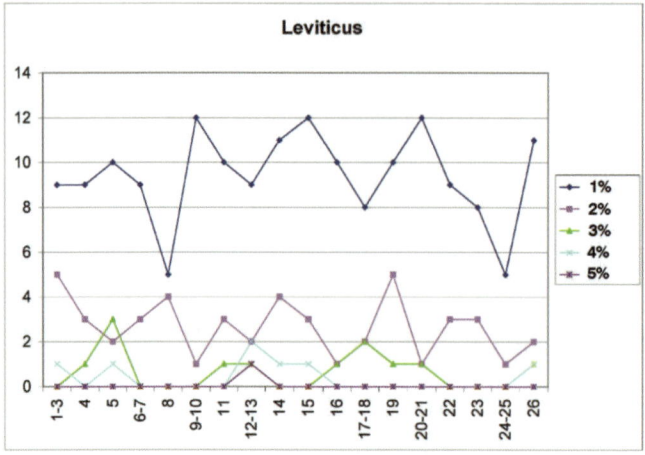

Graph 4-7: Plot of the Standard deviation of the letters in Leviticus.

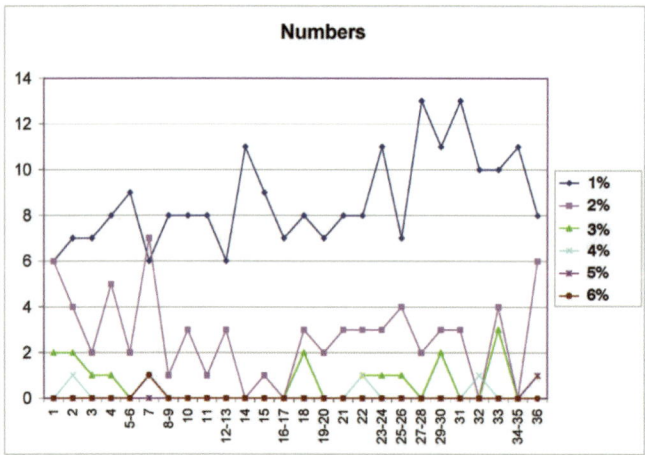

Graph 4-8: Plot of the Standard deviation of the letters in Numbers.

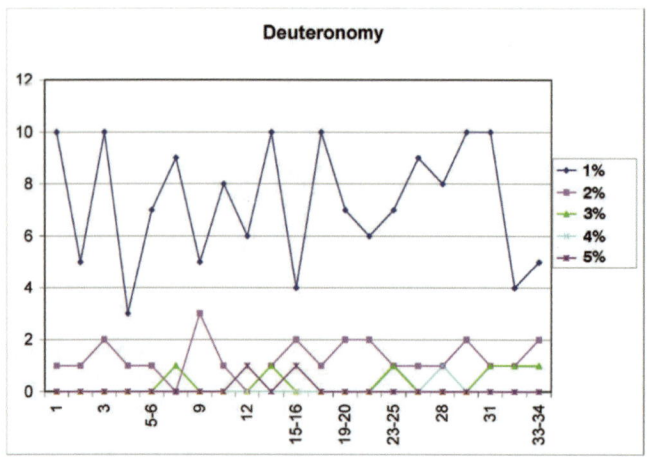

Graph 4-9: Plot of the Standard deviation of the letters in Deuteronomy.

The Book of Ezekiel

Next, I compared the book of Ezekiel with the Torah to see if Hebrew books have the same standard deviation over small groupings of chapters. I chose Ezekiel because I knew that Baruch wrote it and was the only author. I wanted to see if there was an unusual quark about the Hebrew vocabulary that would cause such wide swings in letter distribution. The book of Ezekiel has 74,505 letters, or 24.44% of the size of the Torah, which is a substantial sampling. I grouped the Ezekiel chapters into mostly two-chapter groups, which had similar letter totals in each group, as compared to what I tested in the Torah. Ezekiel had 192 letters in the 23 chapter groupings that were a 1% or more deviation from the norm. As you can see in Graph 4-10, it does have three groups that had five letters that were 3% deviation (+/−), or 2.6% of the 192-letter deviation population. The book of Ezekiel is closer to resembling Moby Dick, than it is to any one of the Torah books.

The Torah had a total of 1,644 letters in chapter groupings that were over 1% deviation from the norm. It had 7.8% of the letters over 3% deviation, which is a very high percentage. Ezekiel had 12.4% deviation that were 2% (+/−), compared to the Torah, which had 20.3% at 2% deviation. That is a very big difference. When you compare the Torah to the three English novels, there is really no comparison. In summary, it is obvious the Torah is a totally different kind of document.

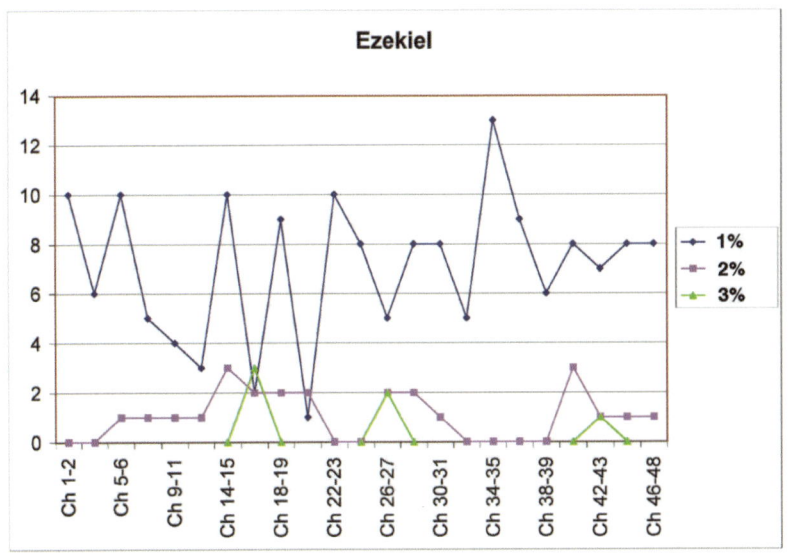

Graph 4-10: Graph of the standard deviation of the book of Ezekiel.

Two options for the Torah

We have two alternative theories about the Torah. First, there is some kind of code within the story. Second, the Torah is not a story or communication at all, but something else and the surface story was forced over the letter sequence, leaving us with a disjointed story sequence.

The Torah exhibits very large swings from chapter to chapter, but it does not abruptly change at the end of a chapter. Most of the time, the writing styles change a few verses into the next chapter. Such is the case with chapters one to three. This is because the chapter and verse breaks are artificial. When you compare the Torah letter distribution with a normal novel written by one author in his or her native language, you can clearly see a major difference between the two. Regular fictional literature has letter deviations, from chapter to chapter, of only one-to-six letters, with only a one percent-deviation on the average for a whole book. A two-to-four percent letter-deviation is very common in the Torah. Biblical scholars who say the Torah was written by more than one author over many hundreds of years are wrong. There is another logical explanation, which I will present later, and you will see what the Torah most closely resembles.

The Layout of the Torah

The letters of the Torah were arranged on four sides of the two stone tablets. Rabbinical scholars think the letters were arranged in 49 or 50-letter rows per column. I estimate there were 50 letters per row. The next task was to see how 304,805 symbols would fit on the four sides. I already knew the outside dimensions of the Ark, which are 60.34 inches by 36.204 inches. I made several assumptions about the thickness of the sides of the Ark, and decided upon three inches. That leaves an interior space of 30.204 inches by 54.34 inches to fit the maximum size of the two tablets, which were placed inside the Ark. My next step was to figure out how many columns of symbols were on each side. I tried two-to-five column arrangements, and settled on five as the best fit because the others resulted in columns over the 54-inch requirement. The calculations turned out as follows: 304,805 ÷ 20 columns = 15,240.25 symbols ÷ 50 symbols per row = 304.805 rows (lines) per column. This is also assuming a leading or spacing between lines. In typographic terms, we have a document with eleven or twelve-point type with a spacing (leading) of 12-point (six lines to the inch). I think this is a good assumption as to how the letters or symbols were arranged on the tablets because of four considerations: 1. The number of rows per column of 304.805 is a factor of the 304,805 symbols in the whole document. 2. Five columns matches the number of books in the Torah. It may have had column headings. 3. The total number of columns, 20, matches the age of adulthood for Hebrew men. 4. The final physical size of the tablets fits within the physical constraints of the interior of the Ark. I would also conclude that the two tablets were most likely rectangular and not curved on top as the popular renditions like to show.

The Skip patterns in the Torah

One of the many mysteries of the Torah are the skip patterns that have been found by many researchers.[14] Some authors have attempted to show the skip patterns foretell the future. I do not subscribe to this theory because of the extreme flexibility of the letters and vowel structure in the language. There are too many permutations and combinations one can create with the letters for someone to say for certain that a proper name or other words are in close proximity to each other. Those authors may have thought they could foretell the future, because one of the Jewish legends states that "all the words and all the actions of all the worlds... are in the Torah," but they did not realize the legend is only a clue as to what the Torah really is.

What is important and significant are words that are found with short-to-medium skips in the same area as the Biblical surface stories that talk about the same subject. For instance, in the Garden of Eden section, researchers have found the names of over 16 tree and plant species in the skip patterns. In the story, Adam is naming the plants and animals. Moses may have chosen to place the plant names in this story because he was, in essence, doing the same thing. Figure 4-26 shows seven of these words.

What is the Torah?

The main question is: "What is the Torah?" It is obvious that a highly advanced previous civilization created the two tablets and the writing on them. Why was it so important for this highly advanced people to immortalize these symbols on stone so the message could last forever? Why was it important that none of the "letters" be changed, deleted or added? Why do the legends of the Jews say things about the Torah that the average modern person would say are "far-fetched," such as: "The Torah is the cause of the world's creation, and also the power that maintains its existence." "The Torah was written before God created the Earth or the Universe." "All the words, and all the actions of all the worlds are contained in the Torah." And finally, and most importantly: "To understand who God is, you must understand the Torah."

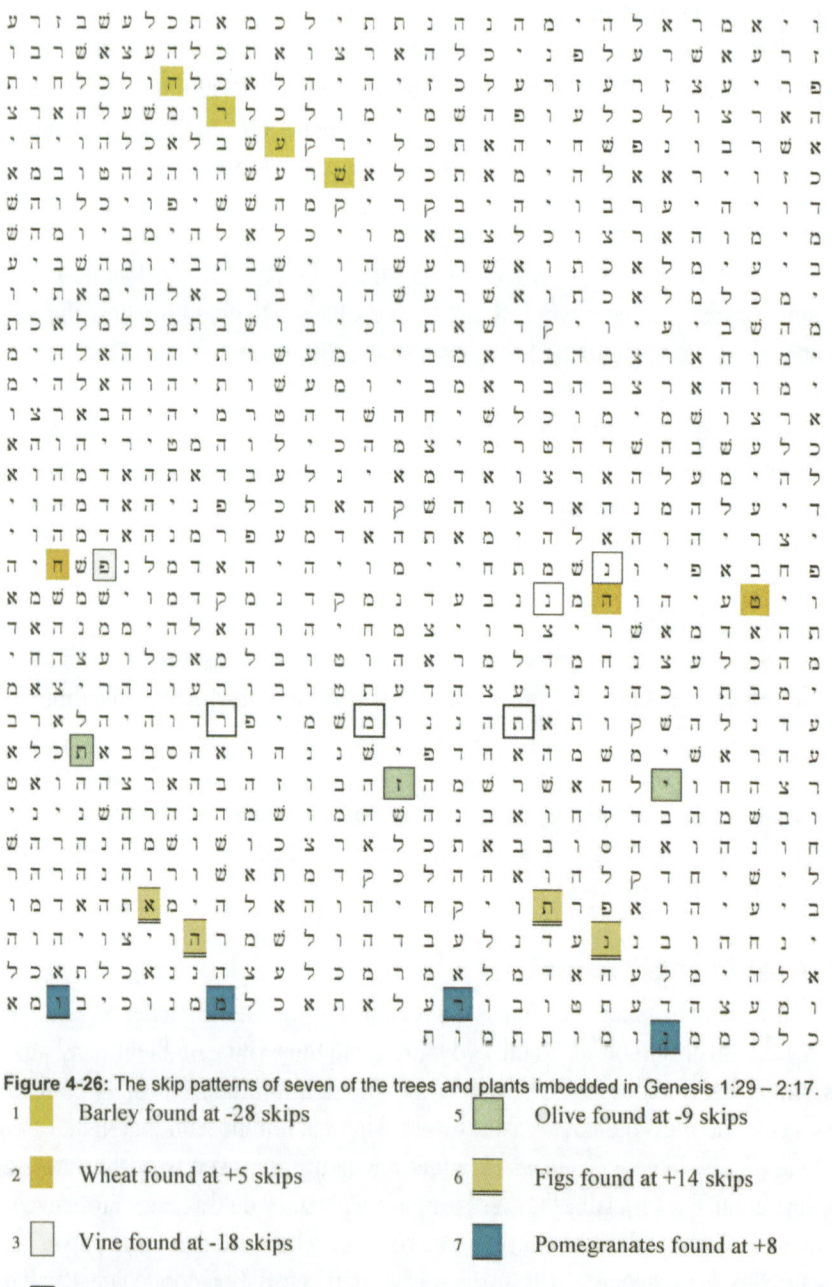

Figure 4-26: The skip patterns of seven of the trees and plants imbedded in Genesis 1:29 – 2:17.

1. Barley found at -28 skips
2. Wheat found at +5 skips
3. Vine found at -18 skips
4. Dates found at +5 skips
5. Olive found at -9 skips
6. Figs found at +14 skips
7. Pomegranates found at +8

Figure 4-26: The skip patterns of seven of the trees and plants imbedded in Genesis 1:29 – 2:17.

I will now explain what I believe the Torah really is. I have already explained what the letters are. Next, I will describe what they are for, and what the overall document is. Later, I will attempt to date this technology.

Analysis of the Letters

I now ask you to think about what computer technology would be like in another 10,000,000 years in the future. I know this is not easy to imagine. After all, we have only been building and programming computers since 1945, and I am asking you to reject the current matter-oriented theory of existence, and use an information-based theory—and all that this will imply.

After I discovered all of the correct letter shapes, and was reasonably comfortable with the vectors for each letter, I then applied the vector information for the 22 letters to the formula in Figure 4-13, which creates a flat plane perpendicular to the vector. Figures 4-27 through 4-31 show what each letter-form represents. The drawings are based on Figures 4-15, 4-19, 4-20 and 4-24. Remember the vectors are only the first stage in finding out what the letters represent.

The next step was to substitute the letters for the corresponding planes and word groupings in the Torah. I sampled many random chapters and verses. What I found was that there is a pattern to the letters. Repeatedly, I found three-to-seven letters (planes) were adjacent to each other. Some sets were opposite each other. If you used mirrors for the planes, a laser beam could reflect off of the adjacent sides and return. Some beams would reflect off and go in different directions. Most of the time the letters form planes next to each other. I sampled hundreds of letter combinations throughout the Torah. I chose both sequential letters, not corresponding to Hebrew word breaks, and also common Hebrew words, thinking that if a word appears multiple

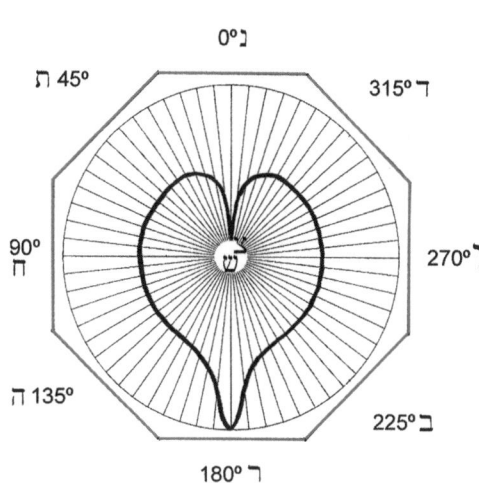

Figure 4-27: Top view of the planes formed from the first group of ten letters.

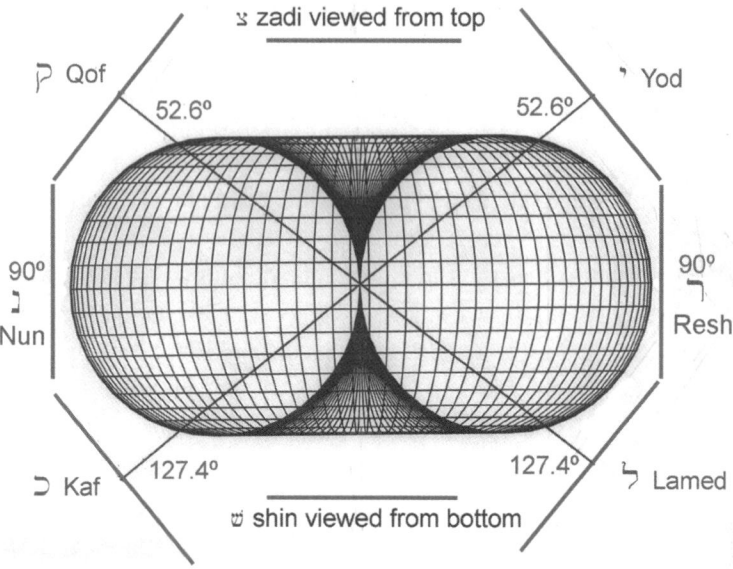

Figure 4-28: Side view of the second group of eight letters along the 0 - 180-degree plane.

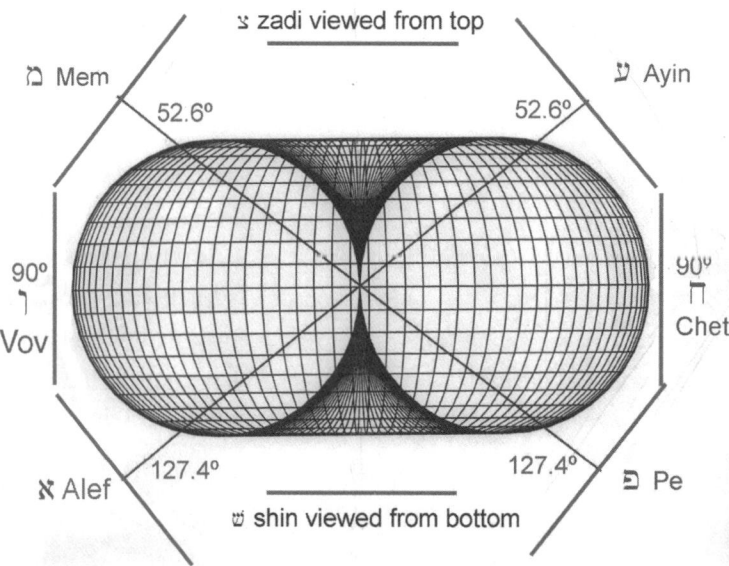

Figure 4-29: Side view of the second group of eight letters along the 90 - 270-degree plane.

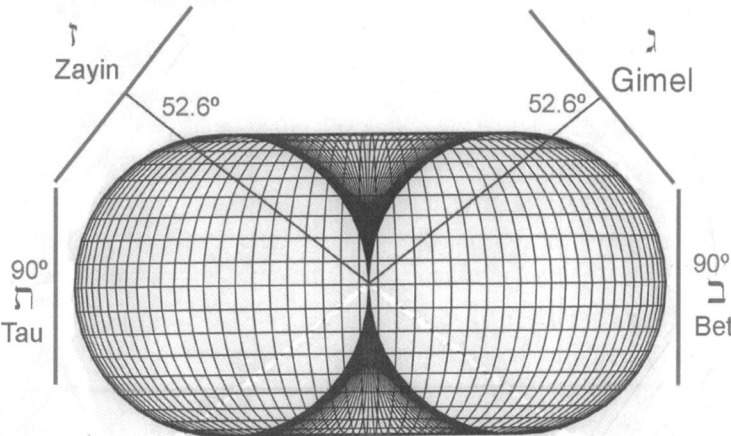

Figure 4-30: Side view of the third group of four letters along the 45 - 225-degree plane.

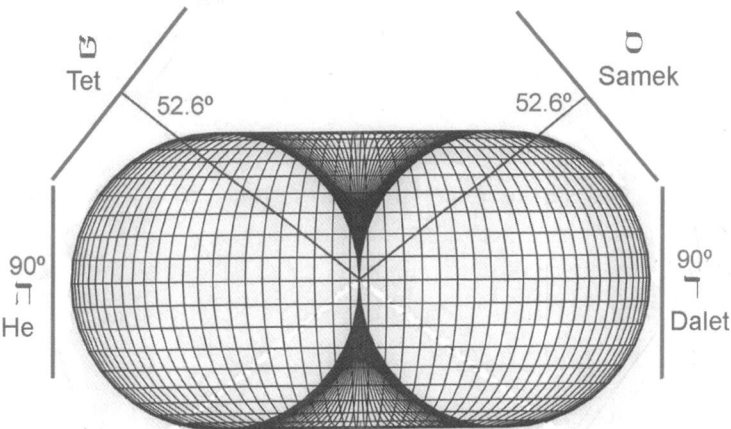

Figure 4-31: Side view of the third group of four letters along the 135 - 315-degree plane.

times, it may be a command that would form a functional pattern. I wanted to show what these words or letter sequences created. At first I tried using a two-dimensional representation of the Toroid but that really did not show the three-dimensional shape very well. So I built a model with removable sides that could be attached to the frame to show a real three-dimensional representation of the letters. The following photos are samples of what the letters form. I have also added lines with arrows representing the coherent light information passing through a light-based crystal computer. To show the letters that form an upward or downward slope, I used tinted translucent plastic. It is my opinion that these letters or symbols form functional logical patterns.

You will remember that I grouped the 22 letters into three general categories. Ten of the letters form a 10-sided prism with a top and bottom. Eight of the letters form a perfect octahedron prism with slopes of 52.6 degrees, and the final four letters form a perfect pyramid with the same upwardly sloped angles. After studying these shapes for years, I came to the conclusion that the first category represents the information that makes up our created Universe. The eight letters represent the different phasing of the carrier wave that our information is strapped onto. The final four letters are another type of phasing of the carrier wave. What is interesting is that in Chapter 3 I theorized that there are eight dimensions. There are also eight upwardly sloped pyramid sides in my model and only four downwardly sloped sides. I believe that the top eight sides are used for the eight dimensions I believe our reality is composed of. The downwardly sloped sides are for information within the Diehold, controlled by clocking and resynchronizing frequencies.

Most of the following Figures display designs that have direct electronic and photo-electronic counterparts. Some of the shapes are currently beyond our comprehension because we do not know how these crystals are put together and what elements they are made from.

All of the examples shown are of words or sequences of letters that appear many times in the Torah, except for Figure 4-32.

The first grouping is the most important because the literal translation of the two words philosophically corresponds to the shape they form—which proves that Moses created words that matched shapes he was looking at! These two words are the most important of all. The words appear in Genesis 1:8, and are commonly translated as "firmament of Heaven," but the first word רקיע means *vault of Heaven,* and the second word שמים (I only used the first two letters) means *Heaven* or *sky.* The two words can mean *vault of the universe* or *vault of God.* In the Theory of Multidimensional Reality, the term would be synonymous with the term Diehold, and the Diehold transmits our existence into this dimension on a carrier wave represented by a three-dimensional shape, called a pyramid or octahedron. The letters displayed in Figure 4-32 form the same shape as in Figure 4-25, a pyramid! In summary, the shape of the planes of the six letters form the four sides and bottom of a pyramid, with the R (ר) forming the vertical side representing our information. The angles of this pyramid are the same as the Great Pyramid in Egypt! The second example, Figure 4-33, is what the name Jacob forms with an (מ) *mem* and (ו) vav, both of which appear many times together.

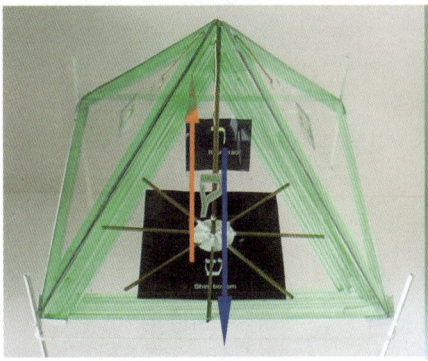

Figure 4-32: רקיע שמ; Vault of heaven. Information being directed 90 degrees from each other within a full four-sided pyramid. In Jewish Tradition, the letter (ר) *resh* is the first letter because it is the second letter in the beginning of the Torah after (ב) *bet.* There is no good rabbinical explanation why, but I believe I found out why. It looks like information starts from this point and travels in two directions, 90 degrees out of phase, from each other.

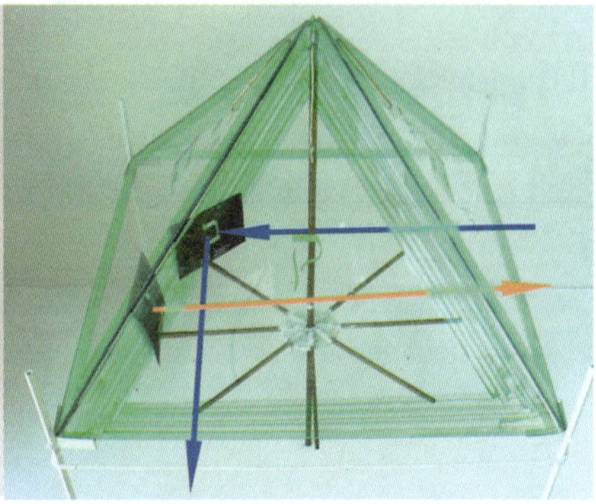

Figure 4-33: מו יעקב; Jacob with an added *mem* and *vov*

The very important name Jacob, with two additional letters, occurring over 28 times both before and after the word Jacob. The shape forms a full pyramid with information changing direction by 90 degrees, and also inducing information back to its origin. The letter shin (ש) also appears many times, which would make it look even more like Figure 4-32.

Figure 4-34: יעקב; The name Jacob. The name Jacob appears 216 times and forms a three-sided pyramid, changing the direc-tion of a beam of light. This could be analogous to a diode, which lets information pass in only one direction.

The following two examples (Figure 4-35 and 36) are significant because they involve one of the names used for God and the other is Jacob's changed name of Israel. You can plainly see the similarity between the two shapes. The name Israel (ישׂראל) shows up 548 times in the Torah. The name adjacent to an (מ) *mem,* shows up 54 times. The example shown displays it with the *mem*.

Figure 4-37 is the name Moses (משה),[15] which shows up 723 times. I found it 193 times with an *el* (אל) in front of it.

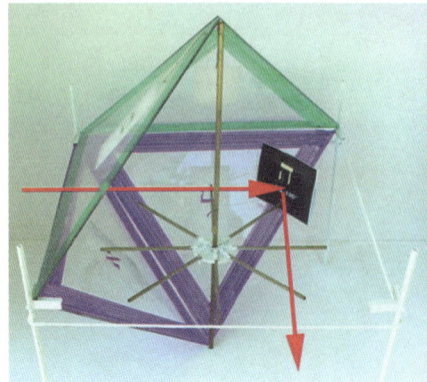

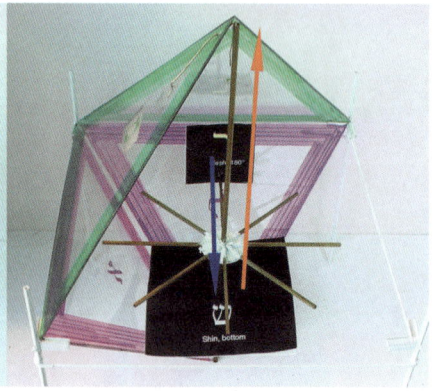

Figure 4-35: אלהים; God, *Elohim*. Information changing direction 90 degrees. The half of an octahedron shape is similar to the full pyramid shape of Figure 4-32.

Figure 4-36: ישׂראל מ; The word "Israel" with an added *mem*.

This is the same shape as *Elohim*, except for the information plains inside. The word appears 548 times in the Torah. The *mem* appears 54 times adjacent to it.

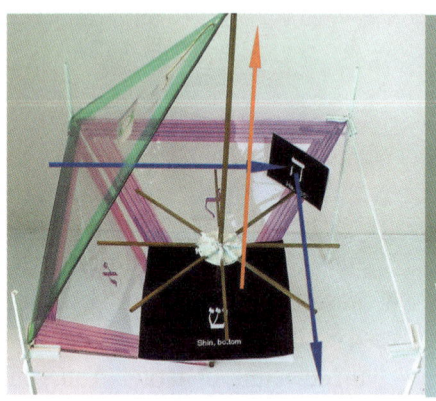

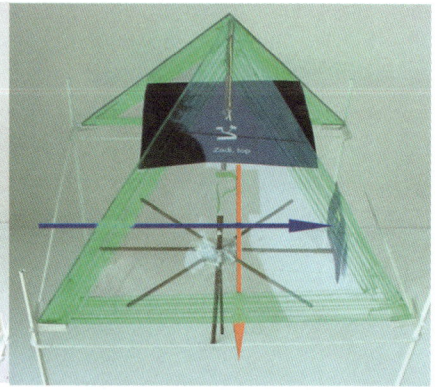

Figure 4-37: אל משׂה; The name Moses with an *el*.

The shape is only missing one side closely resembling *Eloheim*. The shape is also very close to the shape for Israel. The word Moses appears 723 times and 193 times with *el*.

Figure 4-38: יצחק; Isaac. Information changing direction 90 degrees and maybe into another dimension. The sides appear to be the reverse of Isaac's son, Israel Figure 4-36.

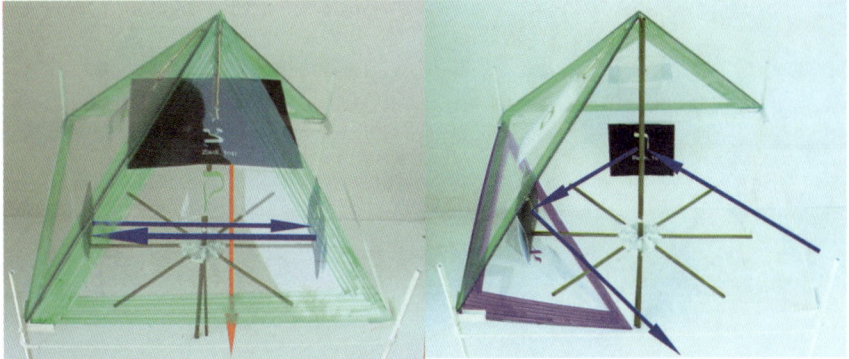

Figure 4-39: יצחק מו ; "Isaac" with an added *mem* and *vav*.

The name Isaac appears 101 times and over 51 times with *mem* and *vav*. The design functionality appears almost like a transistor, which amplifies information or creates a signal in a different direction.

Figure 4-40: ויאמר ; The word for "saying" or "speak."

This combination of letters can be "saying," "speak" or "command." The design functionality is that of a circuit that changes the informations direction.

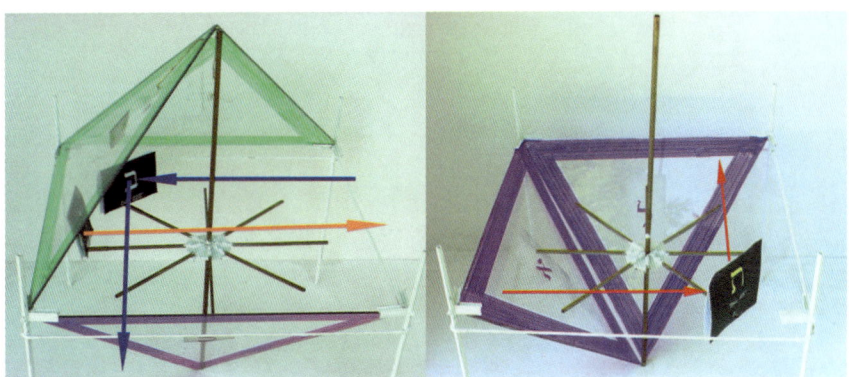

Figure 4-41: כוכבים; The word "Stars."

Information changing directed 90 degrees and inducing different information back to the origin.

Figure 4-42: לאת; The word "not."

Information changing directed 90 degrees. Notice it is opposite the shape for "stars." This combination of letters shows up together no less than 500 times in the Torah.

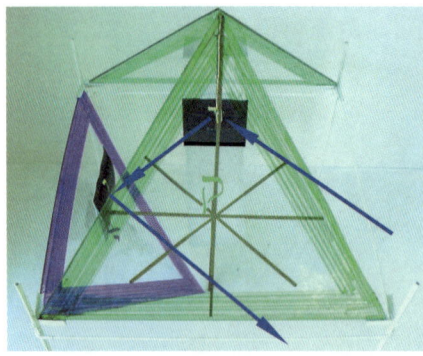

Figure 4-43: ויקרא; The word "and the named" or "called."

The design functionality is that of a circuit that changes information direction and adds information to the stream of information.

The following four examples appear to have the same functionality as our NPN or PNP transistors. A transistor is used to switch signals or amplify information. Some of the following letter patterns are prevalent in the Torah. The letters I searched for sometimes did not correspond to a single word, but may have been part of one word and part of the adjacent word. Remember that the sequence of letters is important, not the arbitrary word breaks Moses created to form the surface story.

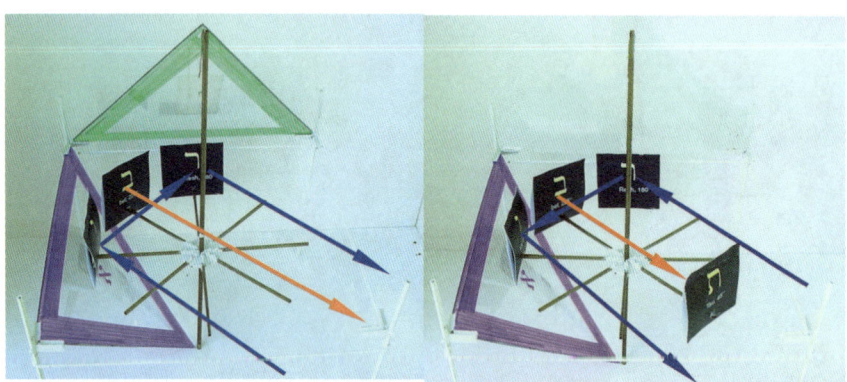

Figure 4-44: ויברא, The words "create, form, or make" with the added letter *bet*.

The design functionality is that of a transistor, switching on some information and amplifying another stream of inform-ation.

Figure 4-45: ב ראות, The words "sight" or "seeing" with a *bet* added.

The design functionality is that of a transistor, switching on some information but also amplifying that information.

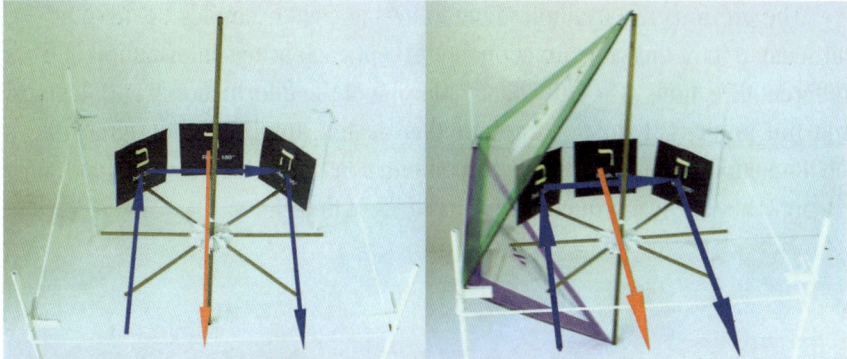

Figure 4-46: בהר, The words "in the mount."

The design functionality is that of a transistor, switching on some information. The word is important and appears 70 times in the Torah.

Figure 4-47: אברהם, The name Abraham.

Abraham appears 151 times in the Torah. The design functionality is that of a transistor, switching on some information. Notice the similarity to "in the mount" in Figure 4-46. This may be why his name was changed from Abram to Abraham, so the mount he purchased was within his name.

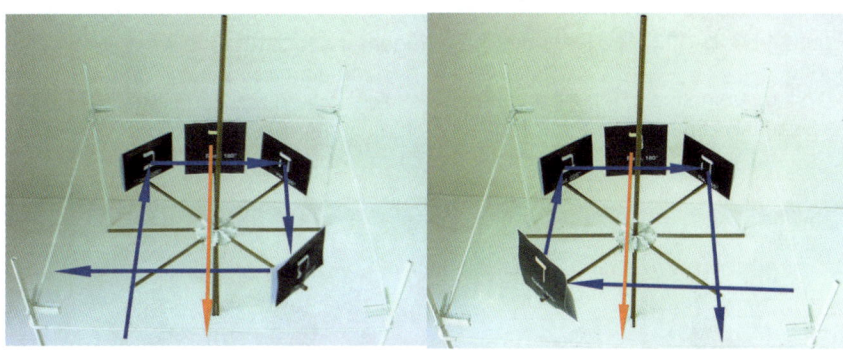

Figure 4-48: רתהב, The words "sight" or "seeing."

The design functionality appears to change infor-mation 90 degrees without changing polarity. It also looks like a transistor, switching on some information, but also amplifying that information. Combinations of תהב appear no less than 340 times, sometimes spelling "Ark" and another combination spelling "some-thing cut off."

Figure 4-49: ר + דבה, The word "talk or report" with an added *resh*.

The design functionality is that of information changing 90 degrees without changing polarity. It also looks like a transistor, switching on some information but also amplifying that information. The הדב appears no less than 296 times in the Torah and with the *resh,* 132 times.

The previous two examples, and following eight examples (4-48 to 4-57), all seem to have the same functionality, except each brings information in from different directions. They all change phasing of the information by 90 degrees, but not polarity. I do not know if this design amplifies the signal like a photomultiplier. Some of these letter combinations form Hebrew words and some do not. They are just sequential occurrences of the letters.

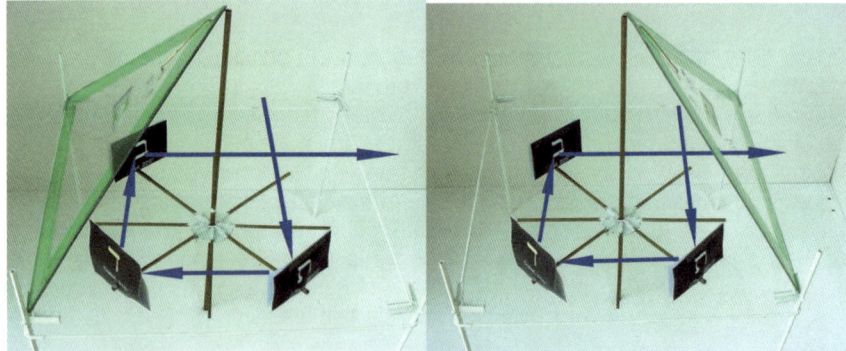

Figure 4-50: בדת מ, The letter *mem* added.

This combination of letters appears no less than 215 times in the Torah.

Figure 4-51: רתהב, The word for "bondsman."

This design functionality appears to change information 90 degrees, without changing polarity.

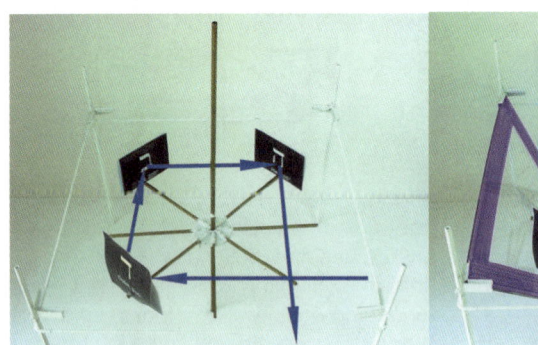

Figure 4-52: דבה, The word "talk" or "report."

The design functionality changes information 90 degrees, without changing polarity. The combination of הדב appears no less than 296 times in the Torah.

Figure 4-53: דאבה, The word "fear."

The combination of rearranged letters הדב appears no less than 296 times in the Torah.

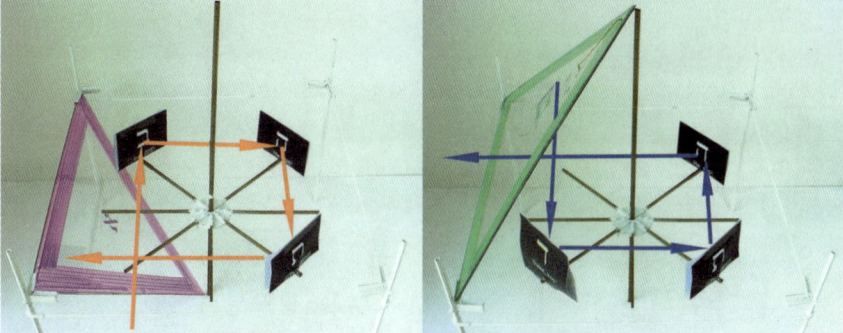

Figure 4-54: א תבה, The word "ark" or "box" with an *alef*.

The design functionality changes information 90 degrees without changing polarity. The combination of תבה appears no less than 340 times in the Torah..

Figure 4-55: תהד, The letters *dalet, he, and tav* with an added *mem*.

This combination of letters appears no less than 112 times in the Torah.

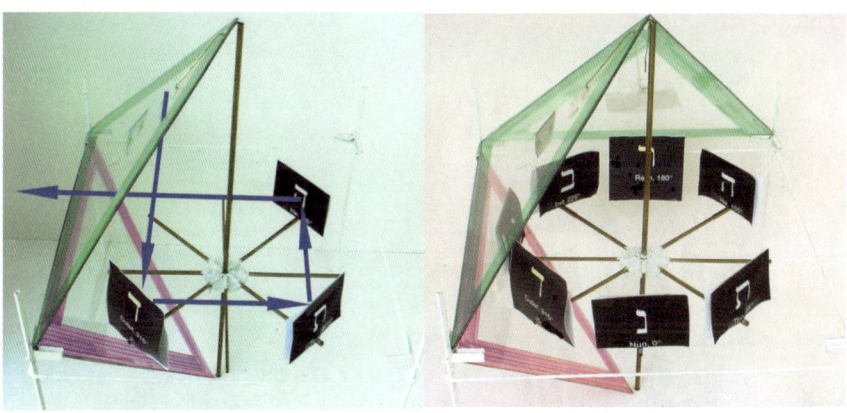

Figure 4-56: אתההדמ, The letters *alef, tav, he, delet,* and *mem*.

The design functionality appears to change infor-mation 90 degrees without changing polarity. This combination of letters appears 17 times in the Torah.

Figure 4-57: הרנאתידובמ, Part of the phrase "Aaron held out the rod."

The Verse comes from Exodus 8:13. The functionality of the shape most closely resembles a photomultiplier, or a microwave tube.

I wanted to find out if there were any letter combinations that resembled photomultipliers or microwave tunnel-amplifiers. Figures 4-57, 4-59 and 4-60 most closely resemble functionality of these types of devices. This type of device is used to amplify signals. Figure 4-59 almost forms a complete octahedron and it incorporates the bottom shape of Figure 4-58, which suggests to me that this is

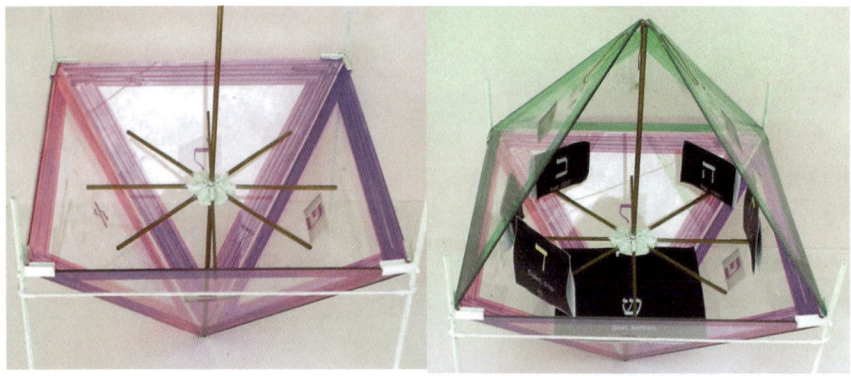

Figure 4-58: פכאל, These four letters are parts of two words.

These four letters appear 6 times in the Torah and in the same clockwise order and one time counter clockwise in reverse order, but in no other arrangements. The pattern is the bottom of an octahedron.

Figure 4-59: שפכאליסודמזבחהע, The letters are part of the phrase "at the base of the altar."

The design functionality appears to be like a photo-multiplier. This combination of letters appears 5 times in the Torah (Lev. 4:7, 4:18, 4:25, 4:30 and 4:34), with some minor variations of letter sequence. The letters form almost a complete Octahedron.

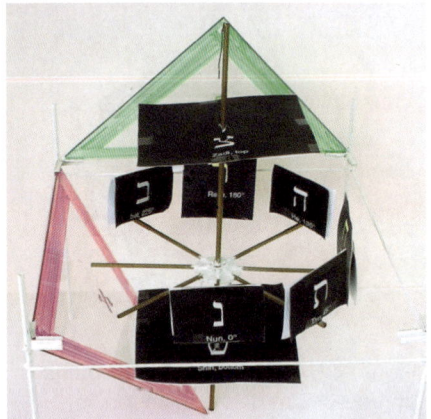

Figure 4-60: בנצחררהחחתיאש, Part of the phrase "son of Zohar the Hittite."

The functionality of the shape most closely resembles a photomultiplior, or a microwave tube.

all happening in the storage dimension. The sequence of letters was in clockwise order starting from *lamed* (ל) and one in counter-clockwise order, but always with the same sequence of letters. Figure 4-59 is interesting because it shows information (a signal) coming from the Diehold (first dimension), then being ampliphied and transmitted into the third dimension. Examples 4-56 to 4-60 do not represent a single Hebrew letter, but are the sequence of adjacent letters which create a non-random organized pattern. Obviously it is impossible to know with certainty what function these shapes are performing. In Figures 4-57 through 4-60, I did not try to guess how the information went through the crystals.

The next step is to explain why we have different planes and slopes (52.66° sloping up, 90° vertical, and 52.66° sloping down). In Chapter 3, I described the carrier wave idea and how it reveals itself in this dimension. A pyramid with sides that slope up 52.66 degrees, like the Great Pyramid of Giza, is a three-dimensional representation of the carrier wave that all information is modulated onto. In the 1970s, pyramid research and resulting products became a popular fad in the United States. I knew several serious researchers in the field and I have already included a summary of their research findings in the previous chapter. Electronically speaking, the carrier wave is modulated by the information one wishes to transmit. The transmitter creates the carrier wave and, after the receiver detects the signal, it strips off the carrier wave and amplifies the information. The first ten-letter grouping (Figure 4-27) which represents the information part of the signal which makes up our physical Universe, including our consciousness. The top and bottom letters (ש א) may be information produced as a byproduct of the other information, passing through the crystal but proceeding 90 degrees out of phase from the other information. I believe the eight sloping planes of 52.66 degrees represent the different phasing of the carrier wave that creates our Universe. The downward sloping planes of 52.66 degrees may represent the carrier wave back in the Diehold, but of course this is all conjecture because we do not have the computer this program was used in. This is all part of what is called reverse engineering. One develops a working theory about how software or a device works, then applies the theory to what one knows about the device or program.

I believe these planes are used in crystals, which are grown and programmed to perform various tasks. The angles are produced within the crystals as growth planes, or molecular lattices, within the crystals. The angles, combined with doping of the crystals with various elements, produce the desired processing of the light-information passing through the crystals. I realize this is somewhat technical and theoretical, but the patterns seen in the letters strongly suggest an organized pattern for processing information. Some of the shapes are similar to NPN or PNP transistors (Figures 4-44 to 4-49) and others are phase-change designs (Figure 4-50 to 4-57). My conclusion is that many of the letter sequences form recognizable, organized patterns that are not random. The forms are functional designs.

Dating the Technology

There is a way of dating this technology, and Moses gave us the clues within the Garden of Eden story. First, I must repeat my definition of dimensions from Chapter Three.

The first dimension is the storage dimension as information in the Diehold. The second dimension is the transmission dimension. The third dimension is the realm of inanimate objects of the matter world, down to the atoms.

The fourth dimension consists of living forms of all types including plants, animals, insects, and everything that walks, crawls and talks on the face of the Earth. We humans are fourth dimensional life forms. Fourth dimensional beings must consume food to sustain their physical bodies, breathe oxygen, feel heat and cold and pain. They are dependent on the matter world around them for all forms of communication and substances. Their physical senses are the only way they receive information about their environment. Thoughts and emotions are closely connected to their physical bodies.

A fifth dimensional being would be something like Uri Geller, the psychic. He can bend keys and other metal objects just by using his mind. Fifth dimensional beings begin to perceive and interact with the Universe from their consciousness in the Diehold. They do not physically have to touch something to move it. They can communicate with others by using their minds, instead of talking. They can effect the matter world around them by using their minds, instead of physically touching an object. This is accomplished by thinking commands in the Diehold, resulting in the action performed in the third dimension. Fifth dimensional beings are still intimately connected to their physical bodies. They still have to eat, they get sick, suffer pain, grow old and die. Physically they should be very similar to fourth dimensional beings, but I belive they live much longer: Because of a strong mind-body connection, they can "will" themselves healthy.

Sixth dimensional beings are much different from the previous two. They may not have a physical body at all. They perceive most of their surroundings from the Diehold. The matter world may appear as translucent domains of information, something like Kirilian photography, because they perceive the structure of the thought that created the object, not necessarily the physical object itself. They may not see color as fourth dimensional beings see it. They can manipulate the matter world by using their minds. Pain, heat, cold, food, emotions and sickness have no relevance to them, because they are not really part of the matter world. Emotions such as hate, love and desire may not be as enhanced or exist at all, because emotions have a physiological component associated with them. If you do not have a physical body then emotions take on a different meaning. And finally, the main difference and maybe the most important is that sixth dimensional being do not grow old and die. They literally "live" or exist forever, and that is their key feature.

In summary, the major difference, which separates the different dimensions, is the amount of potential (energy or voltage) which each dimension has available to them to control and interact with information in the Diehold. More potential translates to more information, which an entity can access, manipulate and process. This is a much different definition of dimensions than what is taught today, but after all, we are dealing with a computer that is creating our reality. Length, width, and depth are only a small factor in reality. There is so much more that divides the evolutionary levels in our reality. It comes down to this: All of our thinking is done in the Diehold. Our physical brain is merely a reflection of what is going on in the Diehold. The structure of our brain reflects the holographic memory structure that exists in the Diehold.[16]

Now let us get back to the clues for dating this technology. The Garden of Eden story is very important because it gives the foundation of everything that follows, namely *who* put the highly advanced technology in that cave and why they did it.

I have been able to figure out the total years from the time of Adam to the present day by totaling up the years in the generations listed in Genesis (Figure 2-1). The total came to 24,060 years, or about two polar reversals ago, but this is not the real age of the technology because it goes back much further in time. To figure that out, we must carefully read the Adam and Eve story and turn the sentences around. When God tells Adam that if he "eats" from the "Tree of knowledge of good and evil," he would surely die. That means that when God created Adam, he was not designed to die. Of course the "tree of knowledge" was not really a tree but a structure with a special computer inside of it. I put "eats" in quotes because the Hebrew word could also mean *consume,* and if you break apart the word האכל used in the verse, you can get *power* or *strength* from איל by swapping the letter adjacent to כ. If we use the first two letters תא, we get *chamber* or *room,* and if we take the last two letters כל, we get *the whole* or *totality.* The words *whole* or *totality* may be why some translations say, "all knowledge." After analyzing the word האכל, you now get the impression that they did not literally eat anything, but went into a room (structure) that could give them *total power* or *total information.* After Adam and Eve came out of this structure, Genesis states "their eyes were opened." But what does that mean?

After all, they had been "living" there for some time and everything was familiar to them. So why does it say, "Their eyes were opened?" This seems to be telling us that the world they saw before they went into the structure looked very different when they came out. They saw the same things we see in a beautiful woods or garden, such as vibrant colors, the smell of the flowers and green trees. The very next sentence says, "They saw they were naked." This could imply that they had the emotion of shame or embarrassment for the first time. The question is why didn't they have this feeling before? They had seen themselves before without clothing. Why is Moses telling us "they were naked"? Either we can conclude that such a feeling was unknown to them before or it had nothing to do with emotion at all, but had to do with the next part of the sentence, which says they started to "sew leaves together" to make clothing. Making clothing could be connected to the first part of the sentence, but I do not think so, because the next sentence says, "God was walking in the Garden in the *cool of the day*. [emphasis added]" Ask yourself: "Why do we wear clothing? Don't we put them on to keep warm?" That means Adam and Eve were not affected by cold before they went into that structure, but after they came out, they felt the sensation of cold. After a short discussion, God tells them that they will surely die, implying they were not designed to die before they did what they did inside the structure. God tells Eve that she will be emotionally bound to her husband, implying this emotion did not exist before, and she will suffer pain in childbirth, again implying they did not experience pain before. God tells Adam that he will be forced to till the fields and eat the herbs of the earth. This statement implies that they did not have to eat before. A few verses later, there is a strange discussion implying multiple gods, fearing that Adam would reach his hand out and sample of the "Tree of Life" and live forever, like them. Again this implies that a state of immortality existed for Adam before he went into the structure and changed to a lower life form. It also implies there were other sixth dimensional beings who decided not to follow Adam, and chose to remain sixth dimensional beings.

We can certainly say that Adam and Eve were nothing like *us* before they went into the structure, but they came out very much like us. We can now come to this conclusion: The individuals we identify as Adam and Eve were sixth dimensional entities who did something in a structure that converted them back into fifth or fourth dimensional beings. There is a very important question we must ask ourselves: "What did they do and why did they do it?" Sixth Dimensional beings do not make mistakes like this. I can only conclude that a select group of Six Dimensional beings made a conscious decision to de-evolve and become

Fifth or Fourth Dimensional beings. This event is dated by the Torah to have occurred about 24,060 years ago, or just before two polar reversals ago. The question we must ask ourselves is, Why? I will try to answer this question in the conclusion of this book, but for now, we can try to date the technology they created and used.

It has been estimated that it took about 4.8 million years for the earliest form of human-like animals, called australopithecines—or australopiths, for short—to evolve into homo sapiens, and another 130,000 years to become like man today. The question is, "How long would it take us to evolve to fifth dimensional beings? It could be another two or three million years. Then how many years for a fifth dimensional being to evolve and become a sixth dimensional being? The differences between the two are huge. It could take over 50-to-100 million years.

The society Adam and Eve came from had incredibly advanced technology. I believe that when Adam and Eve left the area we call "the Garden of Eden" they took this advanced technology and put it deep in a cave in the Sinai desert. It seems they made a conscious decision to de-evolve, but they evidently did not want to give up the ability to communicate with God, the Operating System of the Universe. So, in conclusion, we get a very rough estimate for this technology: at least ten million years more advanced than ours.

The Torah and what it is.

Generally, advanced societies do less written communication and more electronic communication and computer storage of information. If we conclude that Adam came from a sixth dimensional society, that society did very little, if any, written communication. Ask yourself, "Why does a society chisel words in stone?" The answer is obvious. What they are chiseling is so important to them they want it to last forever. Keeping that in mind, I want you to recall the impossible things the legends of the Jews say about the Torah. There are only two things the Torah could be, but only one that fits the criteria perfectly. Both alternatives are a computer program, but there is one program that fits the legends perfectly, and that is an operating system. The other program alternative would be a terminal program, or BIOS program, that would enable man to communicate with God and everything in the Diehold. The legends point to an operating system, but I would expect such a program to be much longer and have a repetitive section that would program the 89 natural elements in our Universe. But I found no such long, repetitive sequence.

A BIOS program would be a different situation. A BIOS program stands for Basic Input Output System, and this is how it works: When you turn on a computer, the first thing that happens when the voltage hits the CPU (Central Processing Unit), is a command string says to go to address zero. At address zero, there are instructions which say to go to another address that the CPU recognizes as an EPROM address (Erasable Program Read Only Memory) that holds the BIOS program. The BIOS program is a basic computer program that tells the system how to generate the words on the screen when you first start up. It tells the computer the basic instructions and parameters of the keyboard, mouse, video board, hard drive and RAM memory locations.

The reason BIOS programs are put onto EPROM's is so they can be upgraded as improvements occur in the operating system and CPU. But we are dealing with the Operating System of the Universe. This does not change, so when a society finally fully evolves to understanding how the Universe really works, and has figured out how to communicate with the Operating System, they would immortalize the program by engraving it on stone, so future generations can have it long after they are gone. The program becomes Sacred and Holy, something worth saving for all eternity.

That brings us to the next important point, "How the Torah was found." It was a string of symbols with no word breaks. No society would create written communications with no divisions between ideas and concepts (words), but the CPU sees a computer program precisely as an unbroken string of computer commands. This alone is a strong indication of what the Torah symbols are.

The next question is: "Do we have the entire program?" Moses brought down two sets of tablets. The first set he supposedly broke when he saw what his brother had done. The second set was a "replacement" of the first set, but it was "given" or produced only about one and a half days later. We really do not know if the same information was on both sets. There is a possibility that the program was chiseled on more then two tablets, and Moses only brought two out. The only way we will know for sure is to go into the cave and compare the two tablets in the Ark with the first two that should still be in the cave.

I believe that some of the legends attribute too much to the Torah. Considering the size of the Torah, it is most likely a terminal program, because this is all man needs to communicate with God.

A Computer Program

The following explanation is not going to be easy for everyone to understand, because I have to get somewhat technical in order to help prove my theory about the Torah. If you can follow along, you will understand why the legend says: "The only book you can compare to the Torah is the Torah itself because there is no other book like the Torah."

The Ultimate Computer

The kind of computer I theorize this program would be used in is a light-based computer known today as *optical computing,* or *quantum computing,* but still very different. Optical computing technology is in its infancy, but some of the basic principles can be applied to the type of computer this advanced civilization built. There would be multiple CPUs made of crystals, all connected together by some kind of fiber-optic connections. Each crystal would behave like a CPU, but would have special program functions imbedded in them. The crystal shape would also indicate the function of the crystals' programming. One major difference between this futuristic computer and ours today is that it would not have storage memory to hold massive amounts of information. It does not need one. The user would not upload any information into memory, as in today's computers, and I will explain why later.

The next issue is time standards in this computer. Some of the issues facing computer scientists in developing light-based computers are: counting of bits, pulsing a laser fast enough, and the power needed to perform the operations with minimum error correction. What I think this advanced civilization did was use the light as the time standard. The crystals would generate no clock cycle themselves, but rather the light passing through them would introduce the pulsing, or clocking, because of what light is—information leaving this dimension. Every element in the universe has its own unique set of frequencies which uniquely identify it. The spectral light frequencies of each element are the time standard in this computer. Besides these frequencies, we also have the carrier wave.

Let us say this advanced civilization solved the problem of pulsing and separation of frequencies by using the natural spectral line frequencies of the elements. We would not have to worry about a time standard because the Diehold would provide that, and considering the elements would each have their own unique frequency, they would have the ability to build a massively parallel processing computer and separate the information by using the elements' unique frequencies. The clock frequencies would be within the light frequencies, resulting in a computer so fast it could emulate our reality. There are tens-of-thousands of different frequencies available for the task.

The next question is, "How do you get the information in and out of the system?" Perhaps the only things programmed into these CPU crystals are processes, formulas and routines that perform actions to light information as it passes through the crystals. We know from Chapter Three, that if atoms are made very unstable they leave this dimension and the Diehold rushes replacement information to the same coordinates in time and space to stabilize that portion of the Universe (our created reality). What if this type of computer makes the information itself unstable, so that it "opens up" an access port in the Diehold, enabling the computer to access the information in the Diehold? If this is the case, then the computer becomes a terminal to the Diehold, and no data needs to be stored in the computer, but rather retrieves information already existing in the Diehold. In essence, everything in the Universe "the Tree of *all knowledge* good and evil. [emphasis added]"

How does mankind interact with such a computer? This comes under the broad heading called input/output functions (I/O). These are not performed with a keyboard and monitor, even though such a computer could display what is processed. The interface goes directly *into* man. Chapter 3 only partially covered what man is in relation to the Diehold, but to understand the I/O functions of such a computer, I will have to explain it.

First, let me state that all life forms are a task or computer program in the Diehold. This may shock you, and you may not want to accept it, but that is exactly what all life forms are—including you! If I have proven to you that the Universe is the product of information, then you may draw no other conclusion than we are a program in the Diehold. Man is a special program to which God has given special functions and parameters. He has not given these to other animals or to plants. In Genesis 1:26-27, it reads:

> And God said: 'Let *us* make man in *our* image, after *our* likeness; and let them have dominion over the fish of the sea, and over the fowl of the air, and over the cattle, and over all the earth, and over every creeping thing that creepeth upon the earth.' [27] And God created man in *His* own image, in the image of god created *He* him; male and female created *He* them. [emphasis added]

What has never been fully explained is why Moses first uses the plural in creating man, then the singular terms of *He* and *His* or masculine forms of *it* or *its*. There is a reason, and it is a clue to the real nature of the Universe, the Diehold, and why it exists, but that will be left for the conclusion of the book. For now we will say that Moses is telling us that God had an idea of what man should look like and how he should behave, including his unique intellectual properties. That is what God programmed into us, and when I say that all life forms in the

Universe are individual programs in the Diehold, I mean that literally, not as a metaphor. You can think of this program as being your soul—which it is!

The next question is, "What tool is used as the I/O device?" How does man interact with such a computer? We now have to refer to the Exodus story of Moses, and the Rod he used. A full explanation of the Rod will be covered in Chapter 6. In the Exodus story, God told Moses when he was about to divide the Red Sea, or Sea of Reeds (Bay of Suez), he was to command the sea to divide by using the Rod, and pointing it to the water and the opposite shore, and the sea would divide. In the Garden of Eden story, Eve used the Rod (allegorically described as the snake, a long round shape) to communicate with the "Tree of all knowledge good and evil." I believe there is a Rod which is held in the user's hand that acts as the interface between the user and the computer/terminal to the Diehold. The Rod would generate a strong laser light which also has your thought patterns modulated onto it. Remember light is information in the Diehold, so therefore it is receiving your thought-forms from the Diehold, where your information is, as well as everything else. In our dimension, we see light transmitted from the Rod to the crystals, then through the crystals, as it processes the information and returns the information back to the one holding the Rod. All of this is happening in the first dimension, within the Diehold. We perceive it in this dimension, but it is really in the Diehold where we do our thinking and perceive our reality. The Rod in Moses' example was the computer and interface into the Diehold, so whatever he thought was manifest in the Third Dimension— and the waters were divided.

The Torah as a Program

My hypothesis is that the Torah is actually a computer program like our BIOS programs. It is a terminal program to the Diehold. The programming used should be analogous to our assembly language programs. This section will explore this hypothesis and examine the similarities.

Assembly language translates directly into machine code by a computer. Higher computer languages, such as C, C++, Fortran or Cobol, need an interpreter to convert the code into assembly code so the computer can use it. Assembly languages are very fast because they do not have to go through that extra step. In addition, assembly language gives the programmer much more control over the computer. Every assembly language is CPU and operating system dependant. Next, I will present an example of assembly language so you know what it looks like. The version I chose was for the Intel 80x86 CPU, running Linux, which is a type of UNIX.

Table of Linux Assembly language instruction sets for Intel 80x86

Syntax	Syntax	Syntax	Syntax
Control Instructions:	*Control Instructions:*	*Control Instructions:*	*Control Instructions:*
aad()	repe.cmpsb( )	movsb( )	shr(imm, reg)
aam()	cmpsw( )	movsx(reg, reg)	shrd(cl, reg, mem)
adc(imm, reg)	cmpxchg(reg, mem)	mul(mem)	stc( )
adc(mem, reg)	cwd( )	neg(mem)	sti( )
add(mem, reg)	cwde( )	nop( )	stosb( )
bound(mem, reg)	daa( )	not( )	stosw( )
bsf(mem, reg)	das( )	or(reg, mem)	sub(imm, reg)
bsr(mem, reg)	div(mem)	out(al, dx)	test(imm, reg)
bswap(reg32)	enter(imm16, imm8)	pop(reg)	xadd(imm, reg)
bt(reg, mem)	idiv (reg)	pop(mem)	xchg(reg, mem)
btc(reg, mem)	in (dx, al)	popa( )	xlat( )
btr(reg, mem)	int (imm8)	popad( )	xor(imm, reg)
bts(reg, mem)	intmul(imm, reg)	push(reg)	
call label	into( )	pusha( )	*Floating Point Inst:*
call(mem32)	iret( )	rcl(imm, reg)	f2xml( )
cbw( )	ja *label*	rcr (imm, reg)	fabs( )
cdq( )	lbe *label*	rdtsc( )	faddp( )
clc( )	jge *label*	ret( )	fbld(mem80)
cld( )	jle *label*	rol(cl, mem)	fchs( )
cmc( )	jne *label*	ror(imm, reg)	fcmovae(sti, st0)
cmova(reg, mem)	jmp *label*	sahf( )	fcmovnb(sti, st0)
cmovae(reg, mem)	lahf( )	sal(imm, mem)	fcom( )
cmovl(reg, mem)	lea (reg32, mem)	sar(imm, reg)	fist(mem)
cmovna(reg, mem)	leave( )	scasb( )	fld1( )
cmovnb(reg, mem)	lock prefix	scasw( )	fptan( )
cmovnle(reg, mem)	lodsb( )	repe.scasd( )	frstor(mem108)
cmovns(reg, mem)	lodsw( )	seta(reg)	fscale( )
cmovv(reg, mem)	loop *label*	setle(mem)	fsin( )
cmp(reg, mem)	loopne *label*	shl(imm, reg)	fwait( )
cmpsb(reg, mem)	mov(reg, mem)	shid(cl, reg, reg)	fsubrp( )

Table 4-3: Example of the instruction codes in an assembly language. The parentheses contain the different operands and flags.

Table 4-3 shows an example of some of the instruction codes. There are actually 351 instruction types, covering integer and control instructions, and floating-point instructions. Among these 351 instructions, there are over 660 operands, which are values placed within parentheses "( )" and follow the instruction. The operands are usually memory addresses, register locations and flags—all programmed as either numbers or letters. So you can have an operand which is a string of letters, or numbers after a command. The CPU will know what to do with them because it knows which and how many operands can follow a given assembly instruction. The average number of operands that can follow an instruction is usually from two to three, but there are some which can have as many as eleven. The point being there will be a known pattern of letters with a known set of other letters/numbers after it, and the CPU knows what to do with them. The result is that an assembly program can have a very rich diversity of commands and codes, which result in a large kind of "vocabulary."

I plotted the letter frequency of the Torah against assembly language programs to see if there was a closer match than what I found while comparing the Torah with English novels.

The next step was to get an assembly program, so that I could plot the letters as I did with the Torah and the other books. I received an assembly program from a friend[17] who is a professional programmer. I was also able to get a copy of a real BIOS program which goes into an EPROM on a computer motherboard. The BIOS program had over 87,000 letters that I plotted. I did not plot the numbers because the Code Breakers website did not count numbers. They probably should be counted, because the Hebrew alphabet also uses letters for numbers. Therefore we should not discount the possibility that some of the letters that follow repeated patterns (common words) could be integers.

You will recall that the letter distribution of the Torah text did not even remotely resemble that of a novel or even the book of Ezekiel. Graph 4-10 shows the letter deviation for the BIOS program.

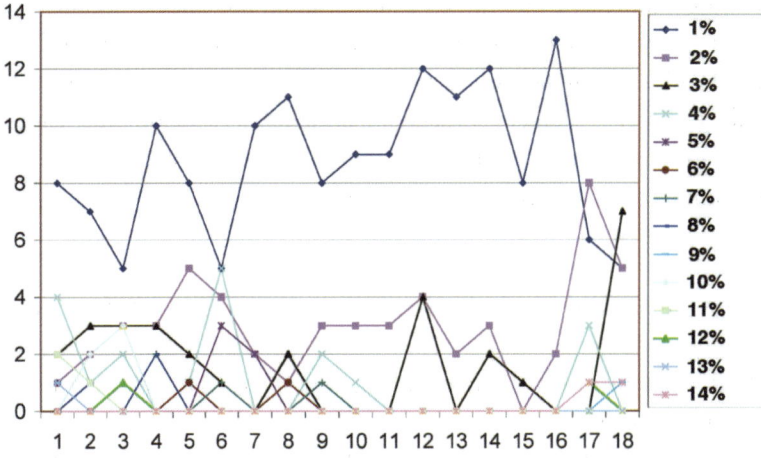

BIOS Assembly Language Program for a 8086 Chip

Graph 4-11: Letter deviation from the norm for an assembly language program.

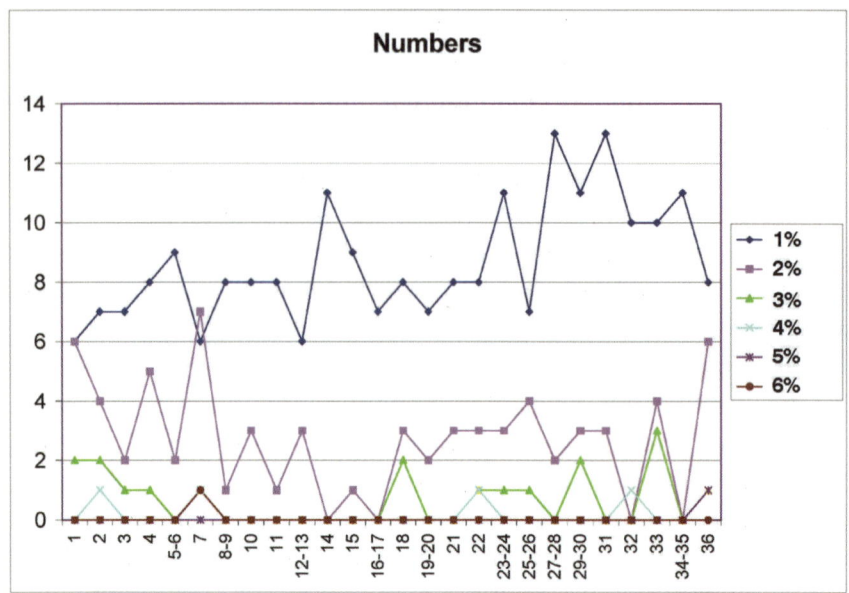

Graph 4-8: Plot of the standard deviation of the letters in the book of Numbers.

The letter deviation of the assembly program looks very similar to the books of the Torah. I have displayed Graph 4-8 of Numbers again because it closely resembles the BIOS program. There are, of course, differences between the two, such as in the BIOS program which had some letters in text groups that were 7% to 14% (+/−) of the norm. That can easily be explained because our programming language must perform different kinds of operations, such as loops, subroutines, moving memory addresses, etc., which I do not think this ancient technology had to perform. It was programmed for a totally different type of computer than ours, with a totally different purpose in mind, so there will be major differences. Even with those differences taken into account, the letter distribution of the Torah compared closely to assembly language programs. It is a closer match than what was found compared with English novels or Ezekiel.

Chapter Conclusion

I hope I have presented enough information to prove the points I covered. They are:

1.) God had a direct hand in manipulating the Hebrews to bring the art of writing to man.

2.) Moses brought the art of writing to mankind by mastering the 22 alphabet symbols in the Torah.

3.) Moses could not possibly have designed the Hebrew alphabet himself, because the model was mathematically and philosophically beyond his abilities and those of his civilization.

4.) The Garden of Eden story of Genesis hints that a very advanced older civilization created those tablets and placed them in the cave which Abraham eventually found and purchased.

5.) The Biblical surface stories of the Torah only give clues and hints as to the true nature of the Torah.

6.) The Torah is the *Book of Life* because the 22 Hebrew letters are 22 views of the carbon atom as it comes into this dimension. All life forms are carbon based. The atomic weight of carbon is 12, the same as the number of Hebrew tribes. The Torah is also the *Book of Life* because by understanding what it really is, one understands that our very existence is an emanation from the Diehold.

7.) The Torah is a program language which enables man to communicate with the Creator.

8.) The Torah and its letters are the first physical evidence of a very advanced civilization living on our Earth a very long time ago.

I stated in Chapter One that to understand what the Torah really is, you have to know the Theory of Multidimensional Reality. You have to know that the Universe is the product of information. After reading this Chapter, you can understand my reasons for stating that.

There is still one more Great Secret embedded in the Torah that I have not yet covered. I will give you a few clues. Ask yourself: "Why has no one found any evidence of the 41-year sojourn of the Hebrews in the Sinai desert after they left Mount Sinai? No one has ever found mass graves, or even the debris that we would expect from hundreds of thousands, or even tens of thousands of people. Out of so many thousands of Hebrews who left Egypt, why did none of them over 20 years of age, from the original group, enter the Promised Land? Why did God send an "angel," who He gives a male gender,[18] to lead Moses and the Hebrews to a "place, which I have prepared."[19] These should be sufficient clues to get your mind thinking. I tell you this: The secret of redemption—how to save yourselves during the polar reversal—is embedded within the Torah. I have shown you that we are dealing with very high technology, way beyond ours. You must think way beyond what you were taught in school in order to understand how the Torah can save mankind. I will explain this further in Chapter 10.

Chapter 5:
The Cave

The second most important event in Abraham's life was the purchase of the burial cave in the field of Machpelah, the mount above it (the real Mount Sinai), and the surrounding land. The first most important event was, without question, the contract with God, which occurred because of Abraham's willingness to sacrifice his son, Isaac. This also occurred at the same location. Today, the cave is called the Cave of Machpelah, commonly thought to be in the city of Hebron in central Israel. The cave is not located there, and the city was only renamed Hebron after the Hebrew invasion of Canaan, sometime after 1266 B.C.E. The cave in Hebron was explored in the 19th century, and also after the 1967 middle-east war, by Moshe Dayan. He had a slender, young girl, the daughter of a friend, lowered down into the cave. She found only three burial locations, not nearly the number there should have been. The true location of the cave had long-since been lost in time. Now, only tradition remains and unfortunately it is incorrect.

Clues referring to the true location of the cave of Machpelah and mount Sinai permeate much of the Torah. It is most likely why we have seven days to the week, why the number seven shows up 205 times in the Torah and why many of the names in the Torah contain words which refer to the mount or cave. Almost no one understands the importance of the cave to the Jewish religion, and what the cave contains. Common knowledge claims that the cave is the burial place of Abraham, Isaac, Jacob, 11 of the 12 brothers, and other members of the immediate family. That is far from what the cave's true importance really is, and the three secrets it possesses. The first is the cave contains technology from a highly advanced previous civilization, which Moses hints at in the Garden of Eden story. The second secret is what Joseph placed in the family cave when he was Prime Minister of Egypt, and the third what Baruch placed in it after the Temple was destroyed in 587 B.C.E. This book will not go into Joseph's life and his identity in Egyptian history, nor what he placed inside the family cave and how his actions brought shame and slavery to the other 11 tribes. That will be revealed in another book.

I can write about these subjects from a very unique position: I am the first person since Baruch (over 2,590 years ago) who has figured out where the real Mount Sinai is located and the first person to see Abraham's altar since Moses buried it 3,310 years ago. This book will not disclose the location of Abraham's altar, the other altars there, Mount Sinai, or the cave, so that they will be protected from anti-Semitic vandals. When that part of the Sinai desert again changes hands, and is rightfully part of Israel again, I will reveal the location of Mount Sinai and all of its treasures. To prove these claims, in Chapter 7 on the Ark I will show you the altars found on top of mount Sinai. Here I will show you a picture of Abraham's altar, after I uncovered it. After the photo was taken, I reburied the altar so that it would remain safe.

The following Jewish legend states the same thing: That Mount Sinai is mount Moriah (more accurately "in the land of Moriah"), where Abraham was supposed to have sacrificed his son, Isaac.

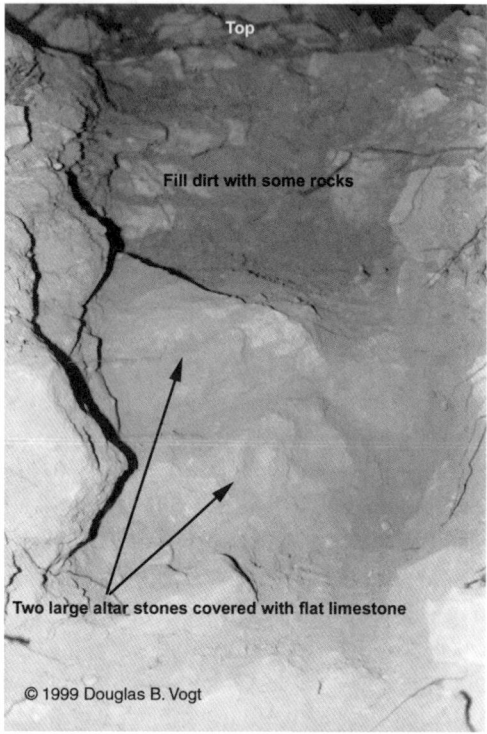

Figure 5.1: The real Abraham's altar buried by Moses in 1305 B.C.E.

The following legend is one of the most beautiful I have read because it gives us some insight into how God thinks and works. It should also settle the question of whether mount Sinai was a tall mountain, or a small hill, and where Abraham's altar was located.

THE CONTEST OF THE MOUNTAINS

While the nations and peoples were refusing to accept the Torah, the mountains among themselves were fighting for the honor of being chosen as the spot for the revelation. One said: "Upon me shall the Shekinah[1] of God rest, and mine shall be this glory," whereupon the other mountain replied: "Upon me shall the Shekinah rest, and mine shall be this glory." The mountain Tabor said to the mountain Hermon: "Upon me shall the Shekinah rest, mine shall be this glory, for in times of old, when in the days of Noah, the flood came over the earth, all the mountains that are under the heavens were covered with water, whereas it did not reach my head, nay, not even my shoulder. All the earth was sunk under water, but I, the highest of the mountains, towered high above the waters, hence I am called upon to bear the Shekinah." Mount Hermon replied to Mount Tabor: "Upon me shall the Shekinah rest, I am the destined one, for when Israel wished to pass through the Red Sea, it was I who enabled them to do so, for I settled down between the two shores of the sea, and they moved from one side to the other, through my aid, so that not even their clothes became wet." Mount Carmel was quite silent, but settled down on the shore of the sea, thinking: "If the Shekinah is to repose on the sea, it will rest upon me, and if it is to repose on the mainland, it will rest upon me." Then a voice out of the high heavens rang out and said: "The Shekinah shall not rest upon these high mountains that are so proud, for it is not God's will that the Shekinah should rest upon high mountains that quarrel among themselves and look upon one another with disdain. He prefers the low mountains, and Sinai among these, because it is *the smallest and most insignificant of all.* Upon it will He let the Shekinah rest." The other mountains hereupon said to God, "Is it possible that Thou art partial, and wilt give us no reward for our good intention?" God replied: "Because ye have striven in My honor will I reward ye. Upon Tabor will I grant aid to Israel at the time of Deborah, and upon Carmel will I give aid to Elijah."

Mount Sinai was given the preference, not for its humility alone, but also because upon it there had been *no worshipping of idols*; whereas the other mountains, owing to their height, had been employed as sanctuaries by the idolaters. Mount Sinai has a further significance, too, for it had been originally a part of *Mount Moriah*, on which Isaac was to have been sacrificed; but Sinai separated itself from it, and came to the desert. Then God said: "Because their father Isaac lay upon this mountain, bound as a sacrifice, it is fitting that upon it his children receive the Torah." Hence God now chose this mountain for a brief stay during the revelation, for after the Torah had been bestowed, He withdrew again to heaven. In the future world, *Sinai will return to its original place, Mount Moriah,* when "the mountain of the Lord's house shall be established in the top of the mountains, and shall be exalted above the hills."[2] [Emphasis added.]

The legend clearly says Mount Sinai is also Mount Moriah, since mountains do not separate and fly to the desert, no matter what size. What the author is saying is that Mount Sinai was the same hill on which Abraham was going to sacrifice Isaac. The prophecy stated that in the future everyone will know Mount

Sinai as the same mount. Finding Abraham's altar inside another altar seems to settle the question. There is much more to the story, including why the world was lead to believe it was in Jerusalem, but that story will be for another book.

Biblical Background

The first Biblical reference to a cave appears in the story of Abraham, who purchased the Cave of Machpelah from Ephron, the son of Zohar, for 400 shekels of silver.[3] This section will reveal some of the literary proofs the cave was not located in Hebron, and it was actually near Mount Sinai. Also revealed is that mount Moriah was actually mount Sinai, and not a hill in Jerusalem. Also shown is that Abraham, Isaac and Jacob lived near there at one time or another, because it was the family's property. The following subsections will describe additional clues that Moses created to hint at a cave at Mount Sinai.

Purchasing the Cave of Machpelah

Abraham purchased the cave of Machpelah, as the burial place for his wife. This event is the second most important event of his life and the Jewish people.[4] The story first appears in Genesis 23:8-10. The pertinent facts: Abraham purchased the cave of Machpelah from Ephron, the son of Zohar, a Hittite, for 400 shekels of silver. The purchase included the land and the trees around it, up to the gate of the city. The cave is described as being before Mamre,[5] said to have been Hebron in the land of Canaan. I should also note that the name Hebron appears earlier in verse 23:2, where it says: "And Sarah died in Kiriath-arba—the same is Hebron—the land of Canaan." The border of Egypt was the Brook of Egypt, today known as Wadi El Arish. So when this verse tells us that Mamre, or Kiriath-arba, is within the borders of Canaan, the town must be east of Wadi El Arish. It just does not tell us how far east.

Follow me as I decode the five key names: Machpelah, Ephron, Zohar, Mamre, and Hebron, and find out what Moses is really telling us, but first you should read Appendix C for Moses' code systems:

1.   The Hebrew spelling of Machpelah is מכפלה, which is made up of two smaller words. The first word מכ is short for the word מכון, which means "place" or "site." The second word פלה means "wonder," "wonderful" or "miracle." The resulting meaning is "place of wonder" or "a miracle place." This does not sound like the name of a burial cave.

2.   The word Ephron בעפרן is made up of two smaller words. The ב (b) before a word means "in." The עפרן means: "dry earth" or "gold dust." The final ן (nun) is equal to 700 in large numbering and 7 in small numbering. Since

the number 7 represents the Mount Sinai then the final meaning is "gold in the hill." Moses is writing this as a clue to indicate what Joseph put into the family burial cave. You will find out that "gold in the hill" fits the story best.

3.   The word Zohar is spelled צהר in Hebrew. The whole word actually means "opening for light," which you could interpret as a cave entrance. You can also get two separate words by swapping the ה (ha) for the next successive letter, ו (vov) giving you the word צו, which means "law." The end word, הר (har) means mount or mountain. The resulting phrase is "mountain of the law," which is a clear reference to Mount Sinai, where Moses received the Law or Torah.

4.   The word Mamre is spelled ממרא. There are likewise two interpretations of this word. The first one is, if you take the first three letters, ממר, they mean "sorrow" or "bitterness." If you take the last letter, א (alef), to represent "one" you get the meaning "the first sorrow or bitterness." This can be an allusion to the first death, that of Sarah, Abrahams wife.

5.   The last word, "Hebron," is the most familiar, because today there is a city in Israel called Hebron. It was the first city David ruled from before he moved his capital north, to Jerusalem. However, the city was not called Hebron at the time of Abraham. There was a son of Kohath, whose name was Hebron[6] and his name appears before the Exodus. The word is spelled חברון in Hebrew and is made up of two smaller words, and a final number. The word is חבר means "to be united, to be bound," or "joined." The final letter, ן (nun) is equal to 7 in small numbering. Since the number 7 represents Mount Sinai the final meaning is "bound to Mount Sinai" or "bound to the mountain." Moses must be saying that Abraham had bound his family to the cave and the mount throughout history. The name for the current-day city of Hebron at the time of Abraham, even through the time of Moses, was Kirjath-arba, because Arba means four. It was the home of Goliath and his three brothers,[7] so the town was called "Town of the four."

Linking the locations

Moses links the names Mamre, Hebron, Cave of Machpelah, Canaan and Gerar together, as if they were in the same location. He links Mamre and Hebron twice;[8] Mamre, Ephron, and the cave of Machpelah three times;[9] and Mamre, Hebron, Canaan and the cave of Machpelah once.[10]

Summary

The evidence I have discovered, both physical evidence discovered on two expeditions, as well as references from the Hebrew Scriptures, and the *Legends of the Jews*, are overwhelming. Mount Sinai is where Abraham built his altar. It is the same place where he periodically lived. It was in essence the family homestead. Moses gave us enough clues so that we can conclude that Sinai was the same place as the cave of Machpelah, and this cave was a special place of "wonder" to the family, besides being the family burial place. The names associated with the purchase of the cave, including the name "Machpelah," are all coded names, which further amplify the description, or aspects, of the location. The Biblical surface story of the Torah is merely a shadow of the real story, concealed within.

Traditions dies hard. Two Temples were built on the Temple Mount in Jerusalem, and both were destroyed. Everyone assumed that mount Moriah is in Jerusalem, only because Solomon instructed that it be inserted the Hebrew Scriptures, to give him the Biblical authority for the construction of his Temple.[11] As quoted earlier, "In the future world, *Sinai will return to its original place, Mount Moriah . . .*" [emphasis added], this is *like* a prophecy. A high priest in the past was told that one day everyone would know the truth. Well, this may be that day.

In the mountain

The word בהר (*be'har*) means "in the mount," is seen repeatedly in the Torah, implying that someone went into the mount, not just on top of it. A few of these references follow:

[Genesis 22:14] And Abraham called the name of that place Adonai-jireh: as it is said to this day: '*In the mount* where the Lord is seen.'

[Exodus 4:27] And the Lord said to Aaron: 'Go into the wilderness to meet Moses.' And he went, and met him *in the mount of God*, and kissed him.[12]

[Exodus 25:40] And see that thou (Moses) make them after their pattern, which is being shown thee *in the mount*.

[Exodus 26:30] And thou shalt rear up the tabernacle according to the fashion thereof, which hath been shown thee *in the mount*.

[Exodus 27:8] Hollow with planks shalt thou make it: as it hath been shown thee *in the mount*, so shall they make it.

[Exodus 34:32] And afterward all the children of Israel came nigh, and he gave them in commandment all that the Lord had spoken with him *in mount Sinai*.

[Numbers 3:1] Now these also are the generations of Aaron and Moses in the day that the Lord spoke with Moses *in mount Sinai.*

[Numbers 28:6] It is a continual burnt offering, which was ordained *in mount Sinai* for a sweet savor, a sacrifice made by fire unto the Lord.

[Deuteronomy 9:9] When I was gone up *into the mount* to receive the tables of stone, even the tables of the covenant which the Lord made with you, then I abode *in the mount* forty days and forty nights; I neither did eat bread nor drink water: 9:10 And the Lord delivered unto me two tables of stone written with the finger of God; and on them was written according to all the words, which the Lord spoke with you in the mount out of the midst of the fire in the day of the assembly. [Emphasis added].

What is important about these quotes from Genesis to Numbers is that all, except the first, took place at Mount Sinai before the Ark, the Tabernacle and the Tent of Meeting were built and placed on top. There was no reason for Moses to go to the top of the mount. In Exodus 25:40 to 27:8, Moses was instructed by God about how to build the items that would later be assembled on top of Mount Sinai. All of His instructions took place inside of the cave. The verses from Deuteronomy came after the Tabernacle and Ark were built, but the Torah clearly says that Moses received the two tablets "in the mount." In other words, from within the cave. None of these verses exclude the alternative explanation, that Moses went into a cave to communicate with God and, in fact, the text reinforces this because it repeatedly says, *in the mount*.

The proper Hebrew for *on* would be על, not a ב (bet) in front of the word.

Mount Horeb

Another name used for Mount Sinai was mount Horeb. The Hebrew spelling is חֹרֵב or חֹורֵב.[13] The first letter is pronounced with an "H," even though it is spelled with a "Ch" sound. When you take the Hebrew word apart, the first two letters mean "only" or "whole." And the last two letters are an abbreviation of the word רבצ, which means "resting place." The word Horeb really means "only resting place," which fits the description of the cave of Machpelah, which was the family's only burial place. The first time the name Horeb appears, it is described in Exodus 3:1 as the place where Moses brought his sheep ". . . to the farthest end of the wilderness, and came to the mountain of God, unto Horeb." This tells us that the mountain of God is also known as Horeb. So the resulting formula would be, Mount Sinai = mountain of God = Horeb = the family cave. The next time the word Horeb appears, God was standing on a rock in Horeb, implying a cave, because there are no noticeable large boulders on top of Mount Sinai unless you consider the whole hill one big rock.[14]

Other references imply that God spoke to Moses from within Mount Sinai, aka Horeb, as in Deuteronomy 1:6: "The Lord our God spoke unto us *in* Horeb, saying . . ." [emphasis added.] I believe the correct interpretation would then be that God spoke to them from within the family burial cave. Continuing this train of thought, additional references are also found in Deuteronomy 4:10-12:

". . . the day that thou stoodest before the Lord thy God *in* Horeb, when the Lord said unto me . . ."

[4:11] "And ye came near and stood *under the mountain*; and the mountain burned with fire unto the heart of heaven, with darkness, clouds, and thick darkness. [4:12] And the Lord spoke unto you out of the midst of the fire; ye heard the voice of the words, but saw no form; only a voice." [Emphasis added].

The expression "under the mountain" can have two meanings, "inside a cave located at the base of the mountain" or "at the foot of the mountain." Most of us have been taught that the Hebrew Tribes were waiting for Moses at the base of mount Sinai until he returned from the top. Another important point to remember from verse 4:12 is that Moses heard a voice originating from a flame just like in the burning bush story. This is important to remember when it comes to explaining how the Ark worked, which I will cover in Chapter 7.

Another example is found in Deuteronomy 5:2 to 5:5. Moses revealed that God was inside of Mount Sinai when He announced His holy Covenant with the Jewish people:

The Lord our God made a covenant with us *in* Horeb. [5:3] The Lord made not this covenant with our fathers, but with us, even us, *who are all of us here alive this day*. [5:4] The Lord spoke with you face to face *in the mount* out of the midst of the fire, [5:5] —I stood between the Lord and you at that time, to declare unto you the word of the Lord; for ye were afraid because of the fire, and went not up *into the mount*—saying, [5:6] I am the Lord thy God, who brought thee out of the land of Egypt, out of the house of bondage. [Emphasis added].

This quote plainly states that when Moses first arrived at mount Sinai, with the whole congregation, they were all in the mount when they received God's instructions.[15] The verses clearly say *in the mount*, not on top of it. What is also very interesting about the quote is that Moses spoke to the adults who left Egypt— but this speech was presented at the end of the 40 years in the Sinai desert. The whole generation was supposed to have been entirely "consumed by the desert" by then. So why is Moses saying, "who are all of us here alive this day"?[16] The next several verses further reinforce my theory that God spoke to Moses, or the congregation, *in* the mount in a cave, the cave of Machpelah.

[Leviticus 7:38] . . . which the Lord commanded Moses *in mount Sinai*, in the day that he commanded the children of Israel to present their offerings unto the Lord, in the wilderness of Sinai.

[Leviticus 25:1] And the Lord spoke unto Moses *in mount Sinai*, saying: [2] Speak unto the children of Israel, and say unto them: When ye come into the land which I give you, then shall the land keep a Sabbath unto the Lord.

[Leviticus 26:46] These are the statutes and ordinances and laws, which the Lord made between Him and the children of Israel *in mount Sinai* by the hand of Moses.

[Leviticus 27:34] These are the commandments, which the Lord commanded Moses for the children of Israel *in mount Sinai*.

[Deuteronomy 4:15] Take ye therefore good heed unto yourselves—for ye saw no manner of form on the day that the Lord spoke unto you *in Horeb* out of the midst of the fire—[16] Lest ye deal corruptly, and make you a graven image, even the form of any figure, the likeness of male or female, ."

[Deuteronomy 18:16] According to all that thou didst desire of the Lord thy God in Horeb in the day of the assembly, saying: "Let me not hear again the voice of the Lord my God, neither let me see this great fire any more, that I die not." [Emphasis added].

All of the previous references state either *in mount Sinai* or *in Horeb*, which means *in* a cave. This will all make more sense as the chapter unfolds, but for now, we can say the references in the Torah reinforce the presence of a cave in Mount Sinai, where Moses "spoke" and heard God's voice.

The Irrefutable Proof

All the necessary proof is found in Genesis, verse 22:14, because of the man who wrote it—Moses. "And Abraham called the name of that place Adonai-jireh; *as it is said to this day*: '*In the mount where the Lord is seen*.'" [Emphasis added.] This quote referred to Abraham traveling to the mount and calling it, "the Lord is seen." Then Moses reiterates and says, "it is called that to this day." At that point, the writer, Moses, exits the story and speaks to the reader. It is the only place in the Torah where this expression is used. Ask yourself: "What day is he writing this?" The answer is, between 1306 to 1266 B.C.E., and "Where did Moses and Aaron see God?" At Mount Sinai.

The evidence presented above should be irrefutable proof that the mountain Abraham traveled to was not in Jerusalem, but mount Sinai, where Moses communicated with God and received the Torah.

Elijah the Prophet

This diminutive mount is also the same mount, which Elijah went to when he spoke to God. He also started out from Beer Sheva and went into the desert for three days. I know that First Kings states that Elijah went one day into the desert, then rested, and traveled another 40 days and 40 nights in the desert. First Kings was also written as a codebook by the high priest. The number "40" does not literally mean "40." The number 40 is an allusion to Gen. 37:3, which adds up to 40 (37 + 3). There are three verses in Genesis Chapter 37 that the chapter and verse adds up to 40, 60 and 70. All three verses refer to Joseph's coat of many colors. Its real meaning is explained in Appendix C. That is why the number 40 appears 42 times in the Torah. The number 40 is mentioned twice here, therefore it really means two days. Added to the first day, the grand total comes to only three days. The following is an excerpt from the Elijah story, found in First Kings, Chapter 19.

> [19:8] And he arose, and did eat and drink, and went in the strength of that meal forty days and forty nights unto *Horeb the mount of God*.

> [19:9] And he came thither unto a *cave*, and lodged there; and, behold, the word of the Lord came to him, and He said unto him: 'What doest thou here, Elijah?' [10] And he said: "I have been very jealous for the Lord God of hosts; for the children of Israel have forsaken Thy covenant, thrown down Thine altars, and slain Thy prophets with the sword; and I, even I only, am left; and they seek my life, to take it away.' [11] And He said: 'Go forth, and *stand upon the mount* before the Lord."

> [19:13] And it was so, when Elijah heard it, that he wrapped his face in his mantle, and went out, and stood in the entrance of the cave. And, behold, there came a voice unto him, and said: 'What doest thou here, Elijah?" [Emphases added.]

The story of Elijah tells us the "Mount of God" is in an area called Horeb. Elijah was able to go from the secret cave, at the base, to the top of the mount in a short period of time, so therefore it could not have been a tall mountain. And finally, we see another allusion to a cave at the same location. It states that he knew where the real mount Sinai was. In verse 19:3 Elijah makes a point of saying, "and left his servant there." In other words, he wanted to travel to Sinai alone, implying that this place was to be kept secret.

The *Legends of the Jews* say the same thing:

> "The cave in which Moses concealed himself while God passed in review before him with His celestial retinue, was the same in which Elijah lodged when God revealed Himself."[17]

Who was buried in the Cave of Machpelah?

Abraham purchased the cave after his wife, Sarah, died. He could have used it and had access to the cave without paying Ephron, but he insisted on purchasing it for 400 shekels of silver. Abraham's act of purchasing the land shows us he wanted to make sure his family had good title to the cave and land around it FOREVER. He wanted to make sure that the people in the area knew that the cave and the land belonged to him and his descendants. The first person buried in the cave was Sarah. The second was Abraham, followed by Isaac and his wife, and then Jacob and his wives, as stated in these verses:

The Death of Abraham:

> [Genesis 25:9] And Isaac and Ishmael his sons buried him in the cave of Machpelah, in the field of Ephron the son of Zohar the Hittite, which is before Mamre; [25:10] The field which Abraham purchased of the sons of Heth: there was Abraham buried, and Sarah his wife.

The Deaths of Isaac and Jacob:

> [Genesis 49:29] And he [Jacob] charged them, and said unto them: "I am to be gathered unto my people; bury me with my fathers in the cave that is in the field of Ephron the Hittite, [49:30] In the cave that is in the field of Machpelah, which is before Mamre, in the land of Canaan, which Abraham bought with the field of Ephron the Hittite for a possession of a burying place. [49:31] There they buried Abraham and Sarah his wife; there they buried Isaac and Rebekah his wife; and there I buried Leah. [49:32] The field and the cave that is therein, which was purchased from the children of Heth."

> [50:12] And his sons did unto him according as he commanded them: [50:13] For his sons carried him into the land of Canaan, and buried him in the cave of the field of Machpelah, which Abraham bought with the field for a possession of a burying place of Ephron the Hittite, before Mamre.

Notice the repeated reference to the history of the land and cave purchase. Moses mentioned all of the names involved in the transaction; just to make it clear to the reader that something very important had been stated. Normally, a writer would only mention it once, very rarely twice, but Moses mentions the cave of Machpelah no less then six times, just to make sure we notice. That is how important this cave was to the Hebrews—as you will find out soon.

Moses Sees the Image of God in the Cave

There is a very unusual story found in Exodus, Chapter 33:12. The story states that Moses asked God to go with the congregation as they journeyed through the Sinai desert. Moses requested, "For wherein now shall it be known that I have found grace in Thy sight, I and Thy people? Is it not in that Thou goest with us, so that we are distinguished, I and Thy people, from all the people that are upon the face of the earth?"[18] God accepts the request, but that was going to happen anyway, because the Ark was already built and the Ark was how God communicated with them. In essence, God's presence was being displayed before them as they journeyed. What happened next was even more unique, and it could not have happened on top of the mount, but only inside a cave. Moses asked God, "Show me, I pray Thee, Thy glory." And God replied in the affirmative!

> [33:19] And He said: "I will make all My goodness pass before thee, and will proclaim the name of the Lord before thee; . . ." [33:20] And he said: "Thou canst not see My face, for man shall not see Me and live." [33:21] And the Lord said: "Behold, there is a place by Me, and thou shalt stand upon a rock. [33:22] And it shall come to pass, while My glory passeth by, that I will put thee in a cleft of the rock, and will cover thee with My hand until I have passed by. [33:23] And I will take away My hand, and thou shalt see my back; but My face shall not be seen."

The key word in that verse is "cleft." The definition of a cleft is a space, or fissure, formed by the splitting apart of a rock. It could be shaped like a "V" or just an opening. There is only one small area on the east side of the mount that could resemble a cleft, but there is no cave or opening there. This physical problem disappears if we speak about a large underground cave system under Mount Sinai.

In 2003 I want on a family trip to explore the Carlsbad Caverns in Carlsbad, New Mexico. The limestone that forms the caves is dated from the Cretaceous and Jurassic periods, similar to the Sinai Peninsula. The U.S. Parks Department literature declared that the Carlsbad caves extend to over 20 miles, and reach a depth of over 1,200 feet below the surface. As we walked from the surface to the main rooms, at the 800-foot depth, we saw many small and large caverns interconnected by a tunnel system. What I also observed were hundreds of small-room sized pockets, or clefts, in the rock that were formed by acidic water dissolving the limestone away as it flowed through the openings in the rock. Their literature notes that the caves were formed 5-6-million years ago. The top opening was formed later by a pond located above it. A large opening was eventually created when the limestone roof of the cavern below collapsed during an earthquake.

Back to Mount Sinai: let us say that if the Cave of Machpelah is similar to Carlsbad Caverns, then explaining the cleft in the rock becomes easy, because there should be hundreds, if not thousands, of clefts in the rock throughout the cave system. The only question is, which cleft was it, and how long is the cave system? In the following section on mount Sinai, Ginzberg's *Legends of the Jews* states that Moses was inside of a cave when he saw the back of God's presents.

The Legends of the Jews

The legends are rich with references to the Cave of Machpelah. Let us start with a legend about Abraham first purchasing the cave.

> Abraham first of all gave thanks to God for the friendly feeling shown to him by the children of Heth, and then he continued his negotiations for the Cave of Machpelah. He had long known the peculiar value of this spot. *Adam* had chosen it as a burial place for himself. He had feared his body might be used for idolatrous purposes after his death; he therefore designated the Cave of Machpelah as the place of his burial, and in the depths his corpse was laid, so that none might find it. When he interred Eve there, he wanted to dig deeper, because he scented the *sweet fragrance of Paradise*, near the entrance to which it lay, but a heavenly voice called to him, Enough! Adam himself was buried there by Seth, and until the time of Abraham the place was guarded by angels, who kept a fire burning near it perpetually, so that none dared approach it and bury his dead therein. Now, it happened on the day when Abraham received the angels in his house, and he wanted to slaughter an ox for their entertainment, that the ox ran away, and in his pursuit of him *Abraham entered the Cave of Machpelah*. There he saw Adam and Eve stretched out upon couches, candles burning at the head of their resting-places, while a sweet scent pervaded the cave.[19] [Emphasis added.]

The priest who wrote this legend knew that Abraham lived near the cave. In the reference to Adam and Eve and Paradise, this is hinting to the Garden of Eden story. You will see this theme in the next couple of legends, because Mount Sinai is near the Biblical Garden of Eden. The next reference also links Adam with the Cave of Machpelah, Paradise and also includes the number seven, to hint at the location of Mount Sinai:

> The Garden of Eden was the abode of the first man and woman, and the souls of all men must pass through it after death, before they reach their final destination. For the souls of the departed must go through *seven* portals before they arrive in the heaven Arabot. There the souls of the pious are transformed into angels, and there they remain forever, praising God and feasting their sight upon the glory of the Shekinah. The first portal is the *Cave of Machpelah*, in the vicinity of Paradise, which is under the care and supervision of Adam.

If the soul that presents herself at the portal is worthy, he calls out, "Make room! Thou art welcome!"[20] [Emphasis added.]

We now jump ahead to Noah and a legend written about a great book of knowledge he received that was originally possessed by Adam.

Samael[21] departed, but Adam was sore grieved, and he put on sackcloth and ashes, and he fasted many, many days, until God appeared unto him, and said: "My son, have no fear of Samael. I will give thee a remedy that will help thee against him, for it was at My instance that he went to thee." Adam asked, "And what is this remedy?" God: "The Torah." Adam: "And where is the Torah?" God then gave him the book of the angel Raziel, which he studied day and night. After some time had passed, the angels visited Adam, and, envious of the *wisdom he had drawn from the book*, they sought to destroy him cunningly by calling him a god and prostrating themselves before him, in spite of his remonstrance, "Do not prostrate yourselves before me, but magnify the Lord with me, and let us exalt His Name together." However, the envy of the angels was so great that they stole the book God had given Adam from him, and threw it in the sea. Adam searched for it everywhere in vain, and the loss distressed him sorely. Again he fasted many days, until God appeared unto him, and said: "Fear not! I will give the book back to thee," and He called Rahab, the Angel of the Sea, and ordered him to recover the book from the sea and restore it to Adam. And so he did.

Upon the death of Adam, the holy book disappeared, but later the *cave* in which it was hidden was revealed to Enoch in a dream. It was from this book that Enoch drew his *knowledge of nature*, of the earth and of the heavens, and he became so wise through it that his wisdom exceeded the wisdom of Adam. Once he had committed it to memory, Enoch hid the book again.

Now, when God resolved upon bringing the flood on the earth, He sent the archangel Raphael to Noah, as the bearer of the following message: "I give thee herewith the holy book, that all the secrets and mysteries written therein may be made manifest unto thee, and that thou mayest know how to fulfill its injunction in holiness, purity, modesty, and humbleness. Thou wilt learn from it how to build an ark of the wood of the gopher tree, wherein thou, and thy sons, and thy wife shall find protection."

Noah took the book, and when he studied it, the Holy Spirit came upon him, and he knew all things needful for the building of the ark and the gathering together of the animals. The book, which was *made of sapphires*, he took with him into the ark, having first enclosed it in a *golden casket*. All the time he spent in the ark it served him as a *timepiece*, to distinguish night from day. Before his death, he entrusted it to Shem, and he in turn to *Abraham*. From Abraham it descended through *Jacob*, Levi, *Moses*, and Joshua to Solomon, who learnt all his wisdom from it, and his skill in the healing art, and also his mastery over the demons.[22] [Emphasis added.]

What is so interesting about this legend is that it spans so much time. It covers from Adam possessing a great book of knowledge, something akin to the story of the tree of all knowledge good and evil, to Solomon. It covers the main personalities who lived at or knew about the Cave of Machpelah. I personally do not think that Solomon knew about any such book other than the Torah because there is no evidence that he did. What is important is that it describes the book being hidden in a cave more then once. It tells us Abraham and his descendents knew about the book and it being hidden in a cave. Finally the book is very unusual because it was made out of sapphires, in other words crystals and it told time for Noah. This book sound more like a computer. It was also placed in a golden casket reminiscent to the Ark of the Covenant, which was a golden box. The author is most likely showing the parallels between the two, so the reader could form the conclusion that they are one and the same.

Next, we will jump ahead in time and describe Moses in the cave when he requested to see God's illumination, or presence. Notice that a cave is clearly described, not as a cleft in a rock outside of the mountain. Notice it also states the cave Moses went into was the same cave that Elijah found, where he heard God speak. These legends emphasize my point that some of the high priests knew about the cave, and they also knew that it was located *in* Mount Sinai.

> Although God had now granted all of his wishes, still Moses received the following answer to his prayer, "I beseech Thee, show me Thy glory": "Thou mayest not behold My glory, or else thou wouldst perish, but in consideration of My vow to grant thee all thy wishes, and in view of the fact that thou art in possession of the secret of My name, I will meet thee so far as to satisfy thy desire in part. *Lift the opening of the cave*, and I will bid all the angels that serve Me pass in review before thee; but as soon as thou hearest the Name, which I have revealed to thee, know then that I am there, and bear thyself bravely and without fear."[23]

> The cave in which Moses concealed himself while God passed in review before him with His celestial retinue, was the same in which *Elijah lodged when God revealed Himself to him on Horeb*. If there had been in it an opening even as tiny as a needle's point, both Moses and Elijah would have been consumed by the passing Divine light, which was of an intensity so great that Moses, although quite shut off *in the cave*, nevertheless caught the reflection of it, so that from its radiance his face began to shine.[24] [Emphasis added.]

The following Jewish legend is part of a discussion about the levels a soul goes through, but it mentions and cave.

> "The first portal is the Cave of Machpelah, in the vicinity of Paradise, which is under the care and *supervision of Adam*. If the soul that presents herself at the portal is worthy, he calls out, 'Make room! Thou art welcome!' The soul then proceeds until she arrives at the gate of Paradise guarded by *the cherubim and the flaming sword*. If she is not found worthy, she is consumed by the sword; otherwise she receives a pass-bill, which admits her to the *terrestrial Paradise*. Therein is a pillar of smoke and light extending from Paradise to the gate of heaven, and it depends upon the character of the soul whether she can climb upward on it and reach heaven."[25] [Emphasis added.]

The Garden of Eden

Adam appears in Genesis as the first occupant of the Garden of Eden. As described in Chapter 4, Adam represents a highly advanced previous civilization, and the story links them together with the cave. The "Cherubim and the flaming sword" are mentioned in Genesis (Chapter 4:24) in the Garden of Eden, "to keep the way to the tree of life," and to keep man out. The reference to a "terrestrial Paradise" is again telling us that there was an earthly place called "Paradise." The "pillar of smoke and light" describe what happened at Mount Sinai when the Tabernacle was finished, assembled and put into use.

> "In Paradise stand the tree of life and the tree of knowledge, the latter forming a hedge about the former. Only he who has cleared a path for himself through the tree of knowledge can come close to the tree of life, which is so huge that it would take a man five hundred years to traverse a distance equal to the diameter of the trunk, . . ."[26]

The reason 500 is used is because $500 \times 24.136 = 12,068$. This number represents the real message within the paragraph. It says that once you understand what the number 12,068 means, then you will discover the secret of the Universe, and how it works—It is that simple.

Other Legends About an Advanced Civilization

The Following two legends pre-date the Hebrew religion. Both hint at the existence of an advanced civilization located somewhere in the Sinai Peninsula. Both stories write about the previous civilization as a paradise, similar to the Garden of Eden story.

The Land of Tilmun

The legend of Tilmun (sometimes called Dilmun) first appears in the mythologies of the city of Sumer, prior to 3000 B.C.E. The tablets describing Tilmun are the oldest evidence of writing yet found.[27] Tilmun's description most closely resembles the Garden of Eden and the Noah stories. Other scholars have noticed the similarities, since the publication of Samuel N. Kramer's translation of the cuneiform tablets in 1945.[28] It is also reminiscent of repeated ancient references to a Golden Age of man in our very distant past.

The reader should also remember that the oldest and earliest forms of writing were found in Erich, Samaria, dated from 3500 B.C.E., which means that all previous legends were communicated orally. There most likely were changes to the Sumerian writings and the meanings of the stories, due to alterations to fit the teller's inclination. We have to examine the common threads that run through these legends. As you read them, note their similarities with the Genesis story. Moses wrote Genesis between 1305 B.C.E. to 1266 B.C.E. The story of Gilgamesh dates prior to 3000 B.C.E., as does the Tilmun tale. In truth, no one knows how old these stories are, or which civilization handed them down to the Sumerians, or how far back the Sumerians go.

We know the Sumerians had a great influence on the Babylonians and the Semites, including the Canaanites. Whether Moses merely wrote down the Hebrew version of this creation-and-paradise legend, cannot be stated for certain. The reason that I say this is because the phrases, "And God said to Moses" or, "And the Lord said unto Moses," does not show up until Exodus, Chapter 4:2, after Moses received the Rod of God and later after the Ark and the Tabernacle were built. Before then, Abraham and Jacob used the two phrases. So we do not know for sure if God told him what to write in Genesis or if he was writing what he thought their history was. It is obvious the origins of these stories date back further then the Hebrews, or anyone else we know of.

The Sumerian story of Tilmun does not tell us its location, but it does parallel the Gilgamesh story through the character Uruk, or Utnapishtim, the Babylonian Noah. There are strong clues, in the Gilgamesh story, where the location is of the "tree of life"—and for this reason I present them here.

The Story of Tilmun

The word "Tilmun" is interchangeable with the word "Dilmun" because the D and the T are interchangeable. Also keep in mind that the spelling of the words was done phonetically, because they used a phonetic type of notation when writing. The story of Tilmun first shows up in the flood legend of the Sumerians. Ziusudra is the Sumerian equivalent of the Biblical Noah.

The following excerpt was translated from the cuneiform tablets, line-by-line. I have <u>underlined</u> the words I wished to emphasis because the italicized words were like that in the original translation.

After, for <u>seven</u> days and <u>seven</u> nights,
The deluge had raged in the land,
And the huge boat had been tossed about on the great waters,
Utu came forth, who sheds light on heaven and earth.
<u>Ziusudra</u> opened a *window* of the <u>huge boat</u>,
Ziusudra, the king,
Before Utu prostrated himself,
The king kills an ox, slaughters a sheep.
[A long break because of lines lost, or unintelligible.]
Ziusudra, the king,
Before An and Enlil prostrated himself;
<u>Life like a god</u> they give him,
Breath eternal like a god they *bring down* for him.
In those days, Ziusudra, the king,
The preserver of the name of . . . and man,
In the mountain *of crossing,* the <u>mountain of Dilmun</u>, the place <u>where the sun rises</u>,
They <u>(An and Enlil)</u> caused to dwell.[29]

The previous mythology links Ziusudra (the Noah character) with "life like a god," because he lives eternally. In Genesis, Noah lives to the ripe old age of 950.[30] Ziusudra became king in a land *of crossing*. This would properly describe the Sinai Peninsula, because it bridges the African continent with Asia Minor. They mention "the mountain of Dilmun," not as a city or a country, but as a mountain, and the names of their gods, *An* and *Enlil*, live there. Thus is the source for the reference that Tilmun is the home of the gods, just as the Garden of Eden is where God dwelt. The phrase, "where the sun rises" is not literally speaking about the sun. It is an allusion to another kind of light, such as the light of knowledge, or a holy light from the Creator.

The following legend is of Enki, the Sumerian water god, and Ninhursag, their mother Earth. The legend is a mirror of the Garden of Eden story.

The land Dilmun is a clean place; the land Dilmun is a clean place,
The land Dilmun is a clean place, the land Dilmun is a bright place;
He who is all alone laid himself down in Dilmun,
The place, after Enki had laid himself by his wife,
That place is clean, that place is bright;
He who is all alone laid himself down in Dilmun,
The place, after Enki had laid himself by Ninsikil,
That place is clean, that place is bright.

In Dilmun the raven uttered no cries,
The *kite* uttered not the cry of the *kite,*
The lion killed not,
The wolf snatched not the lamb,
Unknown was the kid-killing dog,
Unknown was the grain-devouring *boar,*
The bird on high . . . not its *young,*
The dove . . . not the head,
The sick-eyed says not "I am sick-eyed,"
The sick-headed says not "I am sick-headed,"
Its (Dilmun's) old woman says not "I am an old woman,"
Its old man says not "I am an old man,"
Its unwashed maid is not . . . in the city,
He who *crosses* the river utters no . . . ,
The *overseer* does not . . . ,
The singer utters no wail,
By the side of the city he utters no lament.[31]

The above describes a paradise where there is no sickness, old age, death
or predatory animals. All of these attributes are common in the Garden of Eden
story. They describe people who live forever, just as I concluded from the Garden
of Eden story. The legend continues by saying that the only thing Tilmun is lacking
is water. So the Enki, the water god, brings forth water from underneath. In the
Garden of Eden story, water comes in the form of a mist.

Her city drinks the water of abundance,
Dilmun drinks the water of abundance,
Her wells of bitter water, behold they are become wells of good water,
Her fields and *farms* produced *crops* and grain,
Her city, behold it is become the house of the *banks and* quays of the land,
Dilmun, behold it is become the house of the *banks and* quays of the land.[32]

After Dilmun has water, the plant goddess Uttu (the great-grandmother of
the Earth goddess, Ninhursag, gives birth to eight types of plants. The plants are
named and eaten by the water god Enki.

His messenger Isimud, answers:
"My king, the tree-plant," he says to him;
He cuts it down for him, he (Enki) eats it.
"My king, the honey-plant," he says to him;
He plucks it for him, he eats it.
"My king, the roadweed (?)-plant," he says to him;
He cuts it down for him, he eats it.

"My king, the water-plant," he says to him;
He plucks it for him, he eats it.
"My king, the thorn-plant," he says to him;
He cuts it down for him, he eats it.
"My king, the caper-plant," he says to him;
He plucks it for him, he eats it.

"My king, the . . . -plant," he says to him;
He cuts it down for him, he eats it.
"My king, the cassia-plant," he says to him;
He plucks it for him, he eats it.[33]

The creation of plants and the naming of them was yet another parallel to the Garden of Eden story. However, in this story, Enki eats the plant, whereas Adam and Eve eat only the fruit. After comparing the two stories, I would say the Genesis story is more correct and gives more detail about why God did what He did. There is a greater coherence in the Genesis story.

The Assyrians

The name "Tilmun" shows up later in history on cuneiform tablets from the time of the Assyrian ruler, Sargon I (2334 – 2279 B.C.E.). On the first tablet was written: "(Sargon) moo[red] the shi[ps of Meluhha Magan, and Tilmun] a[t the quarry of] Ag[ade]."[34] The next inscription helps us find the location of Tilmun: "He (the god Enlil) gave to him the Upper Sea and the Low[er] (Sea)."[35] Another reference to the upper sea and lower sea, says: "Ana . . . and Kaptara, lands beyond the upper sea, <u>Dilmun and Magan, lands beyond the lower sea</u>, . . ."[36]

The upper sea is the Mediterranean and the lower sea is the Red Sea.

The final inscription from Sargon I: is one that reveals a distance from the Euphrates River to Tilmun. "120 *biru* (about 800 mile) miles from the tail of the Euphrates, to the border of Meluhha is Bit-Sin . . ."[37] That is about 800 miles from the mouth of the Euphrates to Eilat and to the Sinai desert. The wilderness of Sin (Sinn in Egypt) mentioned in Exodus is located at the mountain pass leading into the interior of the Sinai peninsula, just south of the Port of Suez.

The next Assyrian king who mentions Tilmun or Dilmun is Sargon II (722 – 705 B.C.E.). He evidently conquered a great deal of territory, just as did Sargon I. His boastful descriptions of the kingdoms he conquered were very helpful in providing clues about where Tilmun was located. The following quotations are from a catalog of cuneiform translations. The numbers in brackets are the tablet numbers where the verses are found.

"And Upêri, king of Dilmun, who lives (whose camp is situated), <u>like a fish</u>, 30 bêru ("double-hours") away in the <u>midst of the sea</u> of the rising sun

[Indian Ocean], heard of my lordly might and brought his gifts. [inscriptions: 41, 43, 70, 92, 99, &185][38]

The description of Tilmun that looks "like a fish," describes an aerial view of the Sinai Peninsula because it looks like an upside-down triangle thus like a fish head (Figure 5-2). The reason they could write that it is in the "midst of the sea," is because the Gulf of Suez is on its west side, and the Gulf of Aqaba is on its eastern shore. When a traveler approaches the Peninsula from the south, via the Red Sea, the Sinai abruptly sticks out in the middle of the Red Sea, with its tall southern mountains.

Figure 5-2: Aerial photo of the Sinai Peninsula.

(King Sargon) " . . . all of the Chaldea, as much as there was; Bît-Iakin and the shore of the Bitter Sea (Dead Sea) as far as the border of Dilmun;—all of these I brought under my sway, over them I sat my officers and governors, the yoke of my sovereignty I placed upon them." [Inscriptions: 54, 96, 102].[39]

" . . . Bît-Iakin, which is on the shore of the Bitter Sea (Dead Sea), up to the Dilmun border, I brought under one rule and added them to the territory of Assyria. My officials I sat [mine] over them as governors, the yoke of my rule I imposed upon them." [inscription 82].[40]

Assyria was north of the Bitter Sea, or what is currently called the Dead Sea. Both of the previous inscriptions indicate Tilmun is further away from Assyria than the Dead Sea. We can assume that the writer is also giving us directions to Tilmun, thus it must be south of the Dead Sea. The obvious reason is that the Dead Sea is over 1326-feet below sea level. The writer is saying, "up to the Dilmun border," because the traveler has to walk uphill to get to the south, and into the Sinai to get to Dilmun.

"After I (Sennacherib) had destroyed Babylon . . . I removed its ground and had it carried to the Euphrates (and on) to the sea. Its earth (dust) reached unto Dilmun (more figurative then actual, more like the tales of the account). The Dilmunites saw it, and the terror of the fear of Assur fell upon them and they brought their treasures. With their treasures they sent artisans, mustered from their land, carriers of the headpad, a copper chariot, copper tools, vessels of the workmanship of their land, —at the destruction of Babylon." [Inscription 438].[41]

What is important about this inscription is that it describes the natural resources located in Tilmun, mainly copper. The copper mines of Sinai had been worked for thousands of years. The Egyptian Pharaohs mentioned them and archaeologists have found copper mines and other mining activities in the southern and southeastern parts of the Sinai, including the Eilat area.

"The city of Arzani (Arza), which is on the border of the Brook of Egypt, I plundered. Asuhili, its king, I cast into fetters and brought to Assyria."[inscription 515][42] "Upon Qanayah, king of Tilmun, I imposed tribute." " . . . which is on the border of Egypt [toward] Magan, I spent the night. . ." [inscription 558][43]

These three quotations provide us a more detailed description of its location. The first provides the name of a city near the Brook of Egypt. The Brook of Egypt is the seasonal river that flows through Wadi El Arish and empties out into the Mediterranean Sea. The next inscription gives the name Qanayah as the king of Tilmun, and sets its location near the border of Egypt and on the way to Magan, which was a land and people known to live in the Sinai desert.

The next inscription was during the reign of Esarhaddon, king of Assyria. It lists the four regions of his kingdom. The last inscription lists, "the kings of Dilmun, Magan (and) Meluhha, king of the four regions of the world, . ."[insc. 668][44]

During the reign of King Assurbanipal (668-626 B.C.E.), the following quotation describes a general location of Tilmun, by the lower sea, which is identified as the Red Sea.

"Assurbanipal, the great king, . . . king of the four regions, king of kings, unrivaled prince, who from the Upper to the Lower Sea has brought under his sway all princes and has made them to submit at his feet; who established the yoke of his rules over Tyre, which is in the midst of the Upper Sea, and Dilmun, which is in the midst of the Lower Sea." [Inscription 970][45]

I am not the only writer who identified Tilmun, located in the Sinai. Zecharia Sitchin's book, *The Stairway to Heaven*,[46] came to the same conclusion. He also couples Tilmun as the origin of the Paradise stories, and being the home of the gods. However, he arrives at a different conclusion then I do regarding who the gods were.[47] My conclusion is they were from the previous advanced civilization that lived in the Sinai desert. Their origin is yet unknown.

The Story of Gilgamesh

The story of Gilgamesh goes back past 2700 B.C.E. The story seems to be a combination of the Garden of Eden and the Cain and Abel stories. The story was discovered on stone tablets in the ancient city of Nineveh in the 1860's. The entire epic poem includes twelve songs (cantos) of about 300 lines each. Each canto is inscribed on a separate tablet. The eleventh tablet primarily describes the previous Deluge. The tablets found in Babylon date to about 1700 B.C.E. Some parts of Gilgamesh's journey is jumbled around, and out-of-sequence. I will attempt to reveal the correct sequence, so you can understand Gilgamesh's travels and where he was heading.

Gilgamesh was very wise: "He saw secret things and revealed hidden things; he brought intelligence of the *days before the flood*; he went on a long journey." To *seek eternal life*. (Tablet I, Column 1).[48] Gilgamesh was two-thirds divine and one-third mortal. His father was Uruk or Utnapishtim, the Babylonian Noah [tablet 9, column 3], and his mother was Ninsunna, a "goddess," as well. He was said to be a bad ruler of the city of Erech. He had a competitor named Enkidu, who lived in the fields like an animal with long hair all over his body, and he was very strong. After a test of strength (a fight), Enkidu became his friend and, later Gilgamesh called him his brother (like Cain and Abel) [Tablet 7].[49] On tablet 8, he called Enkidu his "younger brother"[50]—similar to the Biblical, Cane and Abel story.

The two of them went on a long journey to a large cedar forest (the Cedars of Lebanon), to find a large tree in the middle, called Huwawa (Humbaba). Gilgamesh wanted to climb the tree and cut it down. The gate to enter into the forest was *irrational* and had *no understanding* [Tablet 7].[51] The tree was the product of Shamash (the sun god, who later became the "patron of travelers"), which represents death and all the evil that Shamash had brought to man. [Tablet 3]. This is an allusion to the Tree of Life in the middle of the Garden of Eden, and the Sun that brought destruction upon the Earth.

At the mountain, Enkidu dreamt: "The heavens roared, the earth resounded; Daylight failed, darkness came; Lightning flashed, fire blazed; [the clouds] thickened, raining death. The brightness vanished, the fire went out; and that which fell down, turned to ashes [Tablet V].[52]

The next event in their journey started as they entered the gates of the cedar forest. They killed the bull which came down from heaven, and it was decreed that Enkidu must die because he killed the bull of Heaven. He died from a disease a few days later and Gilgamesh mourned for him, and then went on another journey to find immortality. Before Enkidu died he cursed the prostitute who led him into civilization to meet Gilgamesh and convince him to go on his journey—an allusion to Eve causing the downfall of Adam, and Abel, who died because of Cain.

Gilgamesh said he "would roam over the desert" to find the land where he could receive immortality [Tablet 8].[53] He arrived at the towering mountain range of Mashu in the southern Sinai. Gilgamesh said he was on a journey to the mountains of Mashu, which may have been the original name for the southwestern Sinai mountain range. The legend continues by describing a mountain pass guarded by two "*scorpion-people*". They permitted him to enter the pass. The "two scorpion-people" were most likely the two naturally carved stones that resemble a man and woman located near the entrance to the pass that leads to the interior of the Sinai. Wadi Sudr flows through it and the description of the two natural stone statues that look like a man and a woman, are in the folklore of the Bedouins in the area. Most of the Bedouins who live in the Wadi Sudr area know that there are two such statues but we did not meet anyone who had been there and knew where they were. I am confident the stone statues do exist there, but their discovery we will have to wait for another expedition. Even the legends of the Jews describe this place, but they mixed up the location of Baal-zephon with the two stone statues, which are located further south.

> ". . . two rectangular rocks form an opening, within which the great sanctuary of <u>Baal-zephon</u> was situated. The rocks are shaped like human figures, the one a man and the other a woman, and they were not chiseled by human hands, but by the Creator Himself.[54]

The mountain pass of Gabriel Raha was the same one which Moses traveled through to cross the coastal mountain range, to access the interior of the Sinai Peninsula, en route to Mount Sinai. That is the only pass with ample water and is the easiest route to get into the interior of the Sinai. It must have been the most common route to travel into the interior, and that is why the Gilgamesh legend alludes to it.

The number seven is also written a number of times in the Gilgamesh legend. When Gilgamesh goes into the desert, he meets the "scorpion-people" who keep watch at the gate of the forest. [Tablet 9 column 2]. He asked the Scorpion-man to open the gate to let him cross the mountains, which he does and Gilgamesh continues on his quest. The "Scorpion-man" was most likely an allusion to one of the stone statutes near the entrance to the mountain pass. Gilgamesh came across a "marvelous garden of precious stones" with trees and shrubs, fruit and vines, all glittering with stones [tablet 9, column 6]. This is another allusion to the many plants of the Garden of Eden, with the addition of multi-colored stones, mentioned in the construction of the Ark, and the priest's breast plate.

At this point in the story, we find it out of sequence. What follows should have come after Gilgamesh left the large cedar forest, because he obviously found the Cedars of Lebanon near the Mediterranean coast. The legend continues with Gilgamesh reaching the seacoast where he came to a tavern with a barmaid named Siduri, who told him the way to Utnapishtim (Noah), where the Tree of Life was located. She told him to travel to the Sea of Death, which was very deep. This is an excellent description of the Dead Sea, which is 1300-feet below sea level, with the high salt content, that is damaging to skin over long exposures. This also meant that Gilgamesh was traveling south from Lebanon, through the Dead Sea valley, to the Gulf of Aqaba. She told him that at the sea he would meet "Urshanabi, the boatman of Utnaposhtim" [Table 10, col. 2].[55] He took Gilgamesh to the "stone images," to locate the pass to cross (the mountains). The boatman told Gilgamesh to cut down 120 trees, and after that, he would take him on the journey. They sailed three days, but it was the equivalent of 45 days. The distance from Eilat to the mountain pass of Gabriel Raha, is 340-miles by water, and would take about that time to sail in a small boat.

Gilgamesh arrived at Utnaposhtim and spoke to Utnaposhtim. Utnaposhtim told him that nothing was permanent, and death was the fate of man [tablet 10, column 6].[56] Utnaposhtim looked like an ordinary man. He told Gilgamesh of a "hidden thing," or "secret of the gods." [Tablet 11] He described a city by the rived Euphrates, which existed before the Great Flood.

Utnaposhtim then told him the story of the Great Flood, and what he did to build a ship to save a number of people and animals (the Noah story, more or less). We learn from Tablet 11 that Utnaposhtim or Atrahasis a.k.a., "the exceedingly wise," learned the secrets of the gods from a dream. Gilgamesh returns to his land with the boatman, but first, Utnaposhtim told him of a plant in the sea, which he had to eat that would keep him young. On the trip home Gilgamesh, took a bath in a pond, and than a serpent ate the plant. This was similar to the serpent in the Garden of Eden story, that was responsible for Adam's expulsion from the Garden of Eden.

The Gilgamesh story had some key components, which reinforces my conclusion—that the Sinai was once the home of a very advanced civilization that lived in a true paradise absent of death and want.

Developing a Working Theory

The literary evidence presented indicates there is a cave in Mount Sinai, and this cave is the family burial cave described as the cave in the field of Machpelah.

The question that came to me after I proved to myself that this is the real Mount Sinai, was: "Why did these people come back to this insignificant mount after escaping from Egypt? Since returning from the November 1999 expedition this question has been answered. Only after March 2001 did I realize how important the cave really was. What Abraham had discovered was not just an interesting cave systems, but he found so much more. What he discovered was a repository of very advanced technology, left by a civilization that might be over ten million years more advanced than ours!

The Secret Cave's Past

A cave like this must have begun forming at least ten million years ago. The cave entrance may have formed some 300,000 years ago. Other civilizations must have known about the cave, and most likely used it. The Torah hints at this in the Garden of Eden story, where Adam and Eve represent a highly advanced civilization, able to communicate with God, the Operating System, and visa-versa! Their civilization and paradise came to an abrupt end during a polar reversal.

The Gilgamesh story hints at the same thing. The next civilization, represented by Noah tries to warn his people about the cataclysm that was about to happen, but most of the people who heard his warning did not listen, and they perished. The legends of the Jews seem to hint that the character Noah also found the cave, and used what was inside of it to gain knowledge on ways to save themselves from the Cataclysm.

In Chapter 11 of my first book, "*Reality Revealed; The Theory of Multidimensional Reality,*" I covered most of the cataclysm mythologies from around the world. They all have the same theme—that there was a previous advanced civilization which was destroyed by a cataclysm.

Let us assume that a highly advanced civilization lived in the Sinai over 24,000 years ago, and they knew of an impending cataclysm that was about to destroy the Earth.

Let us further assume that some of them wanted to save their lives, technology, knowledge, history, and their art. So they found a deep cave and placed their most-prized and important possessions in it, with hopes of retrieving them after the cataclysm. Let us also add the possibility that a few of them put themselves into the cave, in suspended animation, with the hope of "waking up" a number of years after the cataclysm, to continue their civilization. But something happened, and they did not "wake up," but died in their chambers. After the cataclysm, the few survivors were not those who buried these valued items, so the knowledge of their whereabouts was lost.

The next cycle, God intervened and gave a man (Noah) the knowledge of where this cave was located. Noah explored the cave and found this treasure trove of ancient knowledge. After the next cataclysm the "book" (if that's what it was) was hidden once again back into the cave.

Then finally, Abraham came along. First, God lures him to the mount in order to sacrifice his son, Isaac. He may have gained knowledge of the existence of the cave at that time, and perhaps explored it. The natives in the area may never have explored the cave systems, because they may not have had oil lamps for long-term portable lighting. This is not unusual for native peoples, because at Carlsbad Caverns the native Indians never explored the caves, due to superstitions, fear, and the lack of portable lighting. An American cowboy did the main exploration work at Carlsbad, with his kerosene lantern and lots of courage and curiosity. We know that the Canaanites and the Hebrews, had oil lamps at the time of Abraham, therefore they had portable lighting. Let us also assume that Abraham was curious and brave and explored the cave system. We do not know how far into the cave he had to explore (it may have been days), but he came across a treasure trove of ancient and highly advanced technology from a previous civilization. He accidentally activated something that played out something that gave him great knowledge, but it also frightened him, because he did not understand how it worked. To him, it was miracle that God spoke to him. I wish it to be clearly understood that I believe that this advanced civilization had the ability to communicate with God, just as the Garden of Eden Story tells us.

The reason the cave and the location of Mount Sinai were kept secret, hidden, and coded within Torah, is what Abraham discovered inside. It is not just the fact that all of the Patriarchs were buried there, but that it is also because of the repository of highly advanced technology left there by a long-lost civilization. What is concealed inside there must have frightened them, but also amazed them as well and taught them great knowledge. Now you know why the name Machpelah means *place of wonder* or *a miracle place*—because it truly is.

Chapter 6:
The Rod of God

Twentieth-century man has been molded by the rationalist philosophy of the 17th and 18th centuries. The scientific rationalist of the 20th century would reject the Exodus story because it cannot be explained using any scientific theories of existence now available. The whole idea that an 80-year-old man who stuttered would be the spokesman for God and leader of the Hebrew people and would use a wooden staff to bring Egypt to its knees, is unbelievable. Today, our sciences are based on a matter-oriented theory of existence. All theories and unusual phenomena are filtered through this matter-oriented viewpoint of existence. The problem we are faced with is to determine if our scientific philosophies are correct, or if the Exodus even happened. This chapter will pose the answer to the question: "What was the Rod of God that Moses held and how did it work?"

An Historical Analogy

During World War II in the South Pacific, aircraft flew over many islands inhabited by primitive peoples who had never encountered modern, industrial societies. Their cultures developed without any outside influences. They had seen modern airplanes fly overhead but did not know what they were. After the war, an anthropology team visited one of these islands.[1] The anthropologists found that the pre-industrial natives had constructed a wooden and palm-leaf model of an airplane. They were worshiping it as one of their gods.

The lesson for us is this: The technology of these islanders was so primitive that they could not understand what they were looking at, so they put it into the category of "magic" and attributed it to "the gods." In our society we are much more sophisticated in how we deal with an unknown phenomenon. Academia hides it, denies it exists, forgets it, and government will not fund it!

This introduction will help frame a clear explanation of what I think Moses' Rod[2] really is. I will present an educated theory, but there are enough descriptive references of the Rod in the Hebrew Scriptures and in the Legends of the Jews to lead me to believe that it is much more than the wooden walking stick of a shepherd. We know from Chapter 4 on the Hebrew alphabet that we are dealing with the product of a very highly advanced previous civilization. Like other topics in this book, we require a variety of other subjects and principles in order to explain this one. We must refer back to Chapter 3 on the *Theory of Multidimensional Reality*. The technical and philosophical explanations of how the Rod would have worked will be described at the end of this chapter. We have been given a glimpse of a technology far beyond ours. It is a certainty that our technology is closer to that of Neanderthal man, than to the technology that created Moses' Rod. If we were able to go to a planet where a civilization had an uninterrupted history and technological evolution of tens-of-millions of years, we would NOT find a more advanced technology than Moses' Rod. I do not mean this figuratively, nor am I attempting to be outrageous for shock value. I mean this literally.

Multiple Appearances of the Rod of God

The Rod makes three appearances in the Hebrew Scriptures. It shows up in the Garden of Eden story, the Exodus story, and the Gideon story in the book of Judges. It is described as the *Rod of God*, the *Rod* or the *right hand*, because Moses held the Rod in his right hand. The Rod is mentioned cryptically by many of the prophets, writing that the Rod will appear in the future. I will examine each of these prophecies in this chapter.

The Garden of Eden Story

Literary references have been presented in Chapter 5, which describe the Garden of Eden location as being in the same area as Mount Sinai. There is a Jewish tradition about the Torah: "The first time the Torah mentions anything, it is the most important." The Rod first appears in the first story in Genesis after creation. The story is, of course, about the Garden of Eden, with Adam and Eve. The story is an allegory of something that has to do with the evolution of man, and a highly advanced civilization some time in the far distant past. This society had evolved to build a technical device which could communicate with the Operating System of the Universe (the Lord of Hosts).

Here is a brief summary of the Garden of Eden story: After the creation of the Universe, God created a garden in an area called Eden, where there were all types of food-bearing trees and plants. Located in this garden was "the Tree of Life also, in the midst of the garden, and the *tree of all knowledge of good and evil.* [2:9, Emphasis added]." God tells Adam—representing man—that he can eat of all the trees in the garden "but of the tree of the knowledge of good and evil, thou shalt *not eat* of it; for in the day that thou eatest thereof thou shalt surely die [2:17, emphasis added]." At the beginning of Chapter 3 of Genesis, the serpent is introduced as the seducer of Eve—representing woman and the materialist nature of mankind. The serpent tells her that she would not die if she "eats" the "fruit" from the "tree of the knowledge of good and evil." The serpent continues by saying:

'. . . for God does know that in the day ye eat thereof, then your eyes shall be opened, and ye *shall be as God, knowing* good *and evil.*' And when the woman saw that the tree was good for food, and that the tree was to be desired *to make one wise*, she took of the fruit thereof, and did eat; and she gave also unto her husband with her, and he did eat. . ." [3:5-7], [Emphasis added].

First of all, food does not make you wise. Only information does! So you know right from the start we are not talking about food or a tree at all. Tradition tells us that the serpent temped Eve with an apple. As explained in Chapter 4, the apple represented the matter world, in this case the knowledge of how our matter world is created. The story continues with God discovering what had happened while He was "walking" through the garden. God ends man's stay in the Garden of Eden with the following decision:

"And the Lord God said: 'Behold, the man is become as one of us, *to know good and evil*; and now, lest he *put forth his hand*, and take also of the *tree of life*, and eat, and live for ever.' Therefore the Lord God sent him forth from the Garden of Eden, to till the ground from whence he was taken. [3:22-23, emphasis added.]"

Let us start the analysis by stating that the "Tree of Life" and the "Tree of the Knowledge of good and evil" are one and the same object because it is revealed in Chapter 2:17, when God says, "but of the *tree* of knowledge . . ." it is presented as singular. Also, in Chapter 3:3, when Eve says to the serpent, "but of the fruit of the *tree*, which is in the midst of the garden . . ." it is also described as singular. Therefore we must conclude that both qualities or abilities are in one "tree." The next concept to consider is that the "tree" is not really a tree, but a structure. "Eating" does not necessarily mean taking in food into your mouth, but can also mean consuming or absorbing information, as in knowledge. Since the fruit of this tree is not food, then the apple is not really an apple: It is a symbol of something else. The phrase "put forth his hand" indicates some device or tool held in Adam's hand. Finally, the serpent is not really a snake, but represents a tool.

The Tree

Clues that tell us what the tree really is include: "the tree of all knowledge of good and evil," and, the tree will "make one wise." Information is not good or bad by itself—it is just information. How the information is used determines whether it is used for good or evil. Earlier I stated the Garden of Eden story was really an allegorical account of a previous advanced civilization which was able to communicate with the God of the Universe. This was accomplished with the use of a computer. The "tree of knowledge of good and evil" is the allegory for that computer. The Torah and the legends of the Jews do not elaborate on the phrase "the tree of all knowledge of good and evil." This is likely because they did not know what it could possibly have been. After all, in order to conceive what it is, you first must have a relatively high technological level of understanding. The reason why a tree is used in this allegorical story is that trees bear fruit which contains seeds, which reproduce themselves. In Genesis it states: "fruit tree bearing fruit after its kind, whose seed is in itself, upon the earth: and it was so."[3] This is an allegory for the cycles of matter shown in the model of the Hebrew alphabet.

The Fruit of the Tree of Good and Evil

The fruit was described as an apple. The apple or fruit represents the model of how matter is transmitted into this dimension. It is also an analogy for the Hebrew alphabet, as explained in Chapter 4. To summarize its meaning, the fruit represents the philosophical model of how information becomes matter in this dimension. A cross-section of an apple looks like the cross-section of the toroid model of the atom (Figure 6-1).

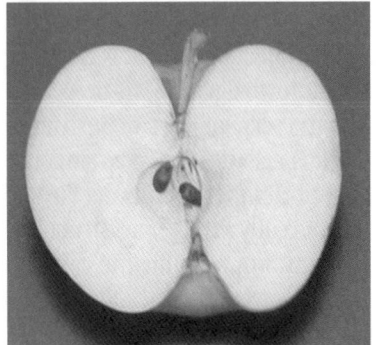 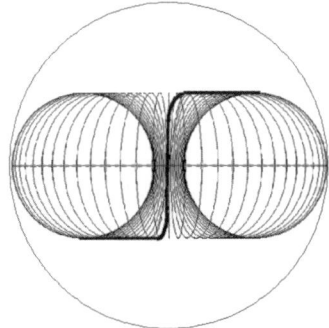

Figure 6-1: Cross-section of an apple compared to the cross-section of a toroid

The Snake and the Rod

The snake, or serpent, and the Rod are one and the same. Twice in the Torah the Rod turned into a serpent and twice more the two appeared close together in the story. The first time, in Exodus, it appeared as the staff God placed into Moses' hand.

[Exodus 4:3] And He said: 'Cast it on the ground.' And he cast it on the ground, and it became a serpent; and Moses fled from before it.

[4:4] And the Lord said unto Moses, put forth thine hand, and take it by the tail. And he put forth his hand, and laid hold of it, and it became a Rod in his hand.

The Rod changed into a serpent and then back again, after it had been held a certain way. The next time it appeared, Moses and Aaron were standing in front of Pharaoh asking him to let the Hebrews go. "And Moses and Aaron went in unto Pharaoh, and they did so, as the Lord had commanded: and Aaron cast down his Rod before Pharaoh, and before his servants, and it became a serpent."[4] You must realize that the snake allegory was used by the writer as a literary tool to tell the reader that they are one and the same.

The Rod and the serpent show up together, or in close proximity, in Isaiah 14:29, and Micah 7:14-17. These references are sufficient to prove my point that the Rod and the serpent are one and the same.

Let us return to my original discussion of the Garden of Eden, when Eve came in contact with the serpent, which you now know was the rod. The Rod *is* a tool, a very special tool, but nevertheless, a tool. What the Garden of Eden story tells us, between the lines, is that a previous highly advanced civilization possessed the technology to access all information directly from the Diehold. The way Adam and Eve interacted or accessed this computer (the tree of all knowledge of good and evil) was through the Rod. The civilization hinted at in Genesis was far beyond anything we can imagine. What I believe Adam and Eve used the Rod for was to de-evolve to fifth dimensional beings. They chose not to live forever. I will explain why in Chapter 12.

Verse 22 in Genesis Chapter 3 says: "And the Lord God said: 'Behold, the man is become as one of us to know good and evil; and now, lest he *put forth his hand*, and take also of the tree of life, and eat, and live for ever.' " [Emphasis added.] This verse suggests that there was a civilization that lived forever.

What is the Rod?

Chapter 3 and 4 of Exodus are very significant chapters because they contain some very vital and revealing information necessary for us to understand what really occurred between God, Moses, the Hebrews and Egypt. It is in Chapter 4 that the Rod makes its second appearance in the Torah. The following section covers the reappearance of the Rod and a partial description of how it worked. As you read this, ask yourself, "Why is the Torah so obscure about how the Rod worked, and when it disappeared?"

How did Moses Obtain the Rod?

Most people have thought that the Rod was Moses' wooden staff, which he used to herd sheep. But it was not so simple. As you read the following quote, notice that Moses does not use the possessive when identifying the Rod. He says: "A Rod" not "my Rod." [Exodus 4:2] And the Lord said unto him: 'What is that in thy hand?' And he said: 'A Rod.'

The first thing you should notice is the question God asked was a rhetorical question, because He already knew what was in Moses' hand—He's God! God really asked Moses, "what have I put into your hand?" And Moses answered "A Rod." He didn't say it was *his* rod or staff, just a Rod.

To further prove this point, I found three other Exodus verses stating the same thing. Notice that Moses did not call the Rod his, but rather the "Rod of God," indicating he did not own it. He was merely possessing it for a time.

> [Exodus 4:17] "And thou shalt *take in thy hand this Rod*, wherewith thou shalt do the signs."

> [4:20] And Moses took his wife and his sons, and set them upon an ass, and he returned to the land of Egypt: and Moses took the *Rod of God* in his hand. [21] And the Lord said unto Moses: "When thou goest back into Egypt, see that thou do before Pharaoh all the *wonders* which *I have put in thy hand* . . ."

> [17:9] And Moses said unto Joshua:[5] "Choose us out men, and go out, fight with Amalek; tomorrow I will stand on the top of the hill with the *Rod of God* in mine hand." [Emphasis added.]

The next verse in Chapter 3 is revealing because it gives us some insight into how God used this Rod and how he was going to use Moses for His purpose to force Egypt to do what He wanted.

> [3:19] And I know that the king of Egypt will not give you leave to go, except by a *mighty hand*. [20] And I will put forth *My hand*, and smite Egypt with all *My wonders*, which I will do in the midst of them.

What is revealing about these two verses is that they ignore Moses holding the rod and performing the "wonders." In verse 20, it states "My hand." God was speaking about His hand holding the Rod, even though it was Moses who was physically holding the Rod, and administering the plagues upon Egypt. The word "wonders" is a reference to the power of the Rod, just as it also states in Chapter 4:20. It seems that God wanted a man physically to hold the Rod and go through the actions required, but it was really God who orchestrated what happened. As you can see, this was a complicated scenario that played itself out. It is like saying that God was the one placing the thoughts into Moses' head to perform the action in this dimension.

How did God Communicate with Moses?

God communicated with Moses and Aaron through the Rod before the Ark were built. The way we know this originates from two sources. The first source is a verse in Exodus 4:8. In order to place this verse into the context of the story, I will describe what preceded it. Towards the end of Chapter 3, Moses needed a way to prove to the Hebrews in Egypt that he was a messenger from God. In Chapter 4:2, Moses received the Rod. Next, God taught Moses how to perform two wonders with the Rod in front of the people. God told him to cast the Rod on the ground, where it turned into a serpent. The next wonder, God told Moses to put his hand into his shirt (chest covering). Moses pulled it out and it had turned white. He then put his hand back into his shirt, then pulled it out again, and it was restored to its natural flesh color. Verse 8 is the next verse, and is the key clue:

> [4:8] And it shall come to pass, if they will not believe thee, neither hearken to the <u>voice</u> of the first sign, that they will believe the <u>voice</u> of the latter sign.

The word "voice," לקל (Lu-Kol in Hebrew), is the word used in the sentence. The problem is that it does not fit into the context of what is presented in verses 3 to 7. Some modern translations[6] omit the word "voice," which destroys the true meaning of the sentence. The word "voice" does not fit because it is the Rod that was necessary to perform the wonders, and neither of the wonders include a voice being heard. The reason the word "voice" is used, is because the voice of God came from the Rod—it is another clue how the Rod may have worked. Notice how subtle the clues are in the Torah. A scholar must literally take every sentence apart to understand what is meant, and they must step back to see what the broader picture should be.

Another very important clue pertaining to the Rod is revealed in verse 6, when Moses' hand turned white. There is a good scientific reason why his arm turned white, and this will be explained in the next section.

The second source was discovered in a much more obscure way. The phrase, "And the Lord said unto Moses," shows up only after the Rod was in Moses' possession.[7] Since the Ark was not built and put into service until after September 1305 B.C.E, fifteen months later, Moses had to have been receiving communications from God through the Rod. The phrase shows up again only after the Ark was built and Moses or Aaron were in the Tabernacle in front of it. After the Ark was built the Rod of God drops out of the story, but where did it go?

The Rod in Legends

The word Rod was sometimes translated as "a staff." This brings to mind a wooden shepherds' staff, which is the common interpretation, but this is not the complete story. The Jewish oral tradition included in *Legends of the Jews,* describes an object that does not sound like a wooden staff at all. It describes an object much more technologically modern than that culture could have produced. After reading all the legends, it became obvious to me the Jews did not fully believe the story told in Exodus about Moses receiving the Rod from God. So they created a story and attempted to show that the lineage of the Rod came from Adam to Moses. The Jewish religious leadership must have felt they had to somehow explain how this magical Rod came into Moses' possession.

It was a *staff made of sapphire*, which the Almighty had created in the twilight of the first Sabbath eve. When Adam was driven out of the Garden of Eden, he carried this staff with him, as one of the gifts he had received from the Creator. He handed it to Enoch, who transmitted it to Noah, who again handed it to Sheth. The staff reached Abraham, who transmitted it to his son Isaac. The latter gave it to Jacob, who brought it with him to Egypt and handed it to his son Joseph. When the Viceroy (Joseph) died, the Egyptians pillaged his house and took away this sapphire rod, which they brought to Pharaoh. Reuel, who was one of the counselors of Pharaoh, saw this Rod and made up his mind to possess it. His desire was so great that he did not hesitate to steal it and carry it away when he left Egypt. He planted the Rod in his garden, and no one could uproot it or even approach it. [Emphasis added.]

Now Jethro, or Reuel, made it known all over the country of Midian that he who could pluck up this staff would take Zipporah to wife.[8]

Moses came along and pulled out the Rod and then married Zipporah. In another part of the legends the Rod was cut from the Tree of Knowledge of good and evil, from the center of the Garden of Eden. This legend was saying just about the same thing I am.

There is no evidence for the truth of this legend from the viewpoint of the Torah or the rest of the Bible. For this reason it is likely that the author of the legend was tying up a few loose ends. It also suggested that the author did not wholeheartedly believe the Exodus version of how Moses received the Rod.

The beginning of the previous excerpt referred to the Rod as "a staff made of sapphire." I found several other references of this type in the legends.

"In his hand Moses held the wonderful *sapphire rod*, and the speech of the sons of Amram was like fiery flame."[9]

"Upon God's bidding, Moses told the people to choose from which rock they wished water to flow, and hardly had Moses touched with his *sapphire Rod* the rock, which they had chosen, when plenteous water flowed from it. The spot where this occurred, God called Massah, and Meribah, because Israel had there tried their God . . ."[10]

I have an easier time accepting the "Rod of God" looking similar to a sapphire Rod than accepting it being constructed of wood. If it were made out of a crystalline-blue material, then it would bring to mind a device that might resemble a large hand-held laser. If it were some kind of laser then that would also explain the earlier clue, when Moses' arm turned white. Extremely high-intensity light will kill and bleach the epidermal layer of your skin white. The Rod might have produced a very bright light when in use, similar to a laser. In the Exodus story Moses puts his arm back into his tunic and pulls it out appearing normal. This miraculous recovery was probably a literary device to show one of the side effects of using the Rod without explaining how the device really worked. The only way to remove the white epidermal layer of skin would be washing it off and the story does not tell us Moses did.

Aaron's Staff

By now you may have come to the conclusion that the Torah was written so it would appear to be confusing, written out of sequence in many portions, and obscure to the point of being secretive. This conclusion is absolutely correct. The present section on Aaron's staff will reinforce that conclusion. Moses did not have the Rod after they left Mount Sinai. Aaron's staff first appears in Numbers 17:16. To properly frame the following quotation, I will briefly describe what happened just before. The time period was about 4 to 33 days after the congregation left Mount Sinai (about November 1305 B.C.E). Korah and his men rebelled against Moses and the ground swallowed 250 of them up.[11] The next day the congregation assembled in front of the Tabernacle against Moses and Aaron. God then brought a quick plague upon them that killed 14,700. God then decided to settle the issue of who He had chosen to lead the Hebrews. God asked all the tribes to place their staffs in front of the Ark within the Tabernacle.

[Numbers 17:16] And the Lord spoke unto Moses, saying: [17] "Speak unto the children of Israel, and take of them rods, one for each fathers house, of all their princes according to the house of their fathers' houses, twelve rods; thou shalt write every man's name upon his rod. [18] And thou shalt write Aaron's name upon the Rod of Levi, for there shall be one Rod for the head of their father's house. [19] And thou shalt lay them up in the tent of meeting before the testimony [Ark], where I meet with you.

The next day, Aaron's staff miraculously "bloomed blossoms, and bore ripe almonds." [17:23]: It is said that this was the end of the "murmuring against Me."

This story is primarily the surface reason for the creation, or anointing, of Aaron's rod. Aaron's rod resembled a staff more than Moses' Rod. Korah's rebellion occurred only after Moses' Rod disappeared. The reason? No one in their right mind would have challenged Moses while he possessed the Rod. Everyone saw what it did to the Egyptians. No sane person would want to challenge Moses while he still had the Rod of God.

God had to do something after the entire congregation challenged Moses' leadership. What He did was to make Aaron's staff appear to be as miraculous as the Rod of God. The congregation may have thought it was the same one. God may have chosen a plague to strike the congregation, to force them to remember the plagues Moses' Rod had brought upon Egypt. God wanted the 12 tribes to get the message. This was not a democracy. God made His choice and it was not put up for a vote.

There was another reason for the creation of Aaron's staff. As long as the Egyptians thought that Moses and the Hebrews had the Rod of God, they were not going to harass or attack them in the desert or later in the Promised Land. That is why the vassal kings in the land of Canaan did not receive any military help from the Pharaohs when the Hebrews invaded Canaan. The Egyptians had learned their lesson some 41 years earlier and they were not about to repeat the experience, even if it meant losing their northern territories.

The Last Appearance of the Rod

The last appearance of the Rod is found in the Story of Gideon in the book of Judges 6:21. The Rod performed another miracle at the time of Passover, and then disappeared.

> "Then the angel of the Lord put forth the end of *the staff that was in his hand*, and touched the flesh and the unleavened cakes; and there rose up fire out of the rock, and consumed the flesh and the unleavened cakes. Then the angel of the Lord departed out of his sight." [Emphasis added.]

How was the Rod Used and How Did it Work?

Let us review the literary clues about what we know about the Rod.
1. The Rod showed up in the Garden of Eden story as an "interface device" which increased one's knowledge.
2. The Rod showed up again in Mount Sinai when God gave Moses the Rod and placed it in his hand.

3. The Rod may have had a sapphire appearance.
4. The Rod had the capability of changing its appearance to the observer.
5. God "Spoke" through the Rod.
6. God revealed that He was going to use the Rod to redeem the Hebrews and change Egyptian society.
7. Moses used the Rod to bring the plagues upon Egypt. Whatever Moses thought, or was told by God to think, was manifest in this dimension. Whether it was to multiply frogs, or locusts, or bring hailstones, or plagues, all of these were the direct result of a thought-form. The splitting of the Red Sea at the Bay of Suez was the same thing.

God's Mighty Right Hand

The device God loaned Moses was vastly more powerful than anything you can imagine. The Egyptians really did not have a chance. With this in mind, after reading what the plagues wrought upon Egypt over the span of three months, you come to the realization that the Exodus could have taken place one day after Moses first met the Pharaoh.[12] If Moses knew what he really had in his hand, he could have destroyed the entire country with one thought-form. Thus it became obvious that Moses did not know how this Rod really worked, and that was fortunate for the Egyptians.

God told Moses only what he had to know to perform the task at hand. Moses had the Rod in his possession long enough to do what God wanted performed, and then it was taken away. No one can "own" the Rod, nor can they accidentally "possess" it. If you have it, it is because God wants you to have it—and for a good reason.

How Will the Rod be used in Prophesy

A number of Prophets have written that the Rod is going to make another appearance during the "End of Days." I have calculated that this period of time being 50 years before the polar reversal, also known as "God's Day of Judgment," or the Cataclysm ("Kataclymos" in Greek literature).

[Isaiah 30:30] And the Lord shall cause his glorious voice to be heard, and shall show the *lighting down of his arm*, with the indignation of his anger, and with the flame of a devouring fire, with scattering, and tempest, and hailstones. [31] For through the voice of the Lord shall the Assyrian be beaten down, which smote *with a rod*.

[Psalms 74:2] Remember thy congregation, which thou hast purchased of old; *the Rod of thine inheritance*, which thou hast redeemed; this mount Zion, wherein thou hast dwelt.

[Psalms 89:25] I will set his *hand* also in the sea, and his *right hand* in the rivers.

[Psalms 89:31] If they break my statutes, and keep not my commandments; [32] Then will I visit their transgression with *the Rod*, and their iniquity with stripes.

[Psalms 110:2] The Lord shall send the *Rod of thy strength* out of Zion: rule thou in the midst of thine enemies.

[Psalms 110:5] The Lord at thy *right hand* shall strike through kings in the day of his wrath.

[Micah 6:9] Hark! The Lord crieth unto the city—and the man of wisdom shall see thy name—hear ye *the Rod*, and who hath appointed it. [Emphasis added.]

I also found one remarkable legend which reveals what happened to the Rod, and that it will make another appearance in the End of Days. The legend states that just before Moses' death he asked God to show him the future of His people. God would not describe it to him. The conversation continued:

Moses, at the same time, begged God that in the future world He might restore to Israel the heavenly weapon that He had taken from them after the worship of the Golden Calf. God said, "I swear that I shall restore it to them."[13]

This is the only reference I found that describes the Rod as a "heavenly weapon"—but that was one of its attributes. The quote also reinforces my conclusion that the Rod disappeared, or, more accurately, that God instructed Moses to leave it in the cave where he received it, at the time they left Mount Sinai and started on their travels through the Sinai desert (1305 B.C.E., second Jewish month, 20th day).

All of these quotations are prophecies for the future, which suggest the Rod will be an integral part of the fulfillment of these prophecies. I do not know if the future is fixed in time, but I do believe God has a definite influence on its outcome and you will see the proof of that statement in Chapter 11. There are many circular stories in the Hebrew Scriptures. The Rod has shown up three times already in history. It would not surprise me if the Rod of God shows up one more time, before the End of Days is completed.

Chapter 7:
The Ark of the Covenant

The average person would say that Moses and the Hebrews made and used the Ark of the Covenant as a way to talk and listen to God. This is not the viewpoint of the vast majority of academia. Their views are that Moses was a fictional character, the Exodus was a myth, Mount Sinai was a myth, the Red Sea crossing never happened, and—since all of the Biblical characters, places, and events were fictional—therefore everything that happened at Mount Sinai must have been fictional because Mount Sinai was fictional. They also say that the Ark, if it ever existed, was nothing more then a piece of statuary that could not speak to anyone. Their reasoning seems to be that because no academician[1] has ever found Mount Sinai therefore it does not exist. The Ark could not possibly have spoken to Moses or the Hebrews because their civilization had not evolved to building or understanding electricity, transmitters, radios or anything technologically advanced. Therefore, the Ark could not have existed. Since the Bible story relates that God "spoke" through the Ark in a supernatural way, and academia cannot see how the Hebrews could have had such technology, the only logical conclusion to them would be that the Ark did not exist *as described,* or it was just a myth like everything else in the Bible.

Fortunately for all of us, I know more then they do about the location of the true Mount Sinai and I have an unshakable belief that the Torah is basically a true story that had to be decoded before anyone could understand what the real message was that Moses left us. This Chapter will explain how the Ark of the Covenant worked and what the priests did when they knelt in front of it. Finally, I will explain who took the Ark out of Jerusalem in 587 B.C.E. just before the fall of Jerusalem and hid it safely away where it will one day be found again and used to speak and listen to God.

All the following dimensions are presented in sacred cubits (sc), with standard measurement equivalents in parentheses. As an interesting addition, all of the measurements or descriptions of all the objects are written twice in the book of Exodus.

How did Moses Know How to Build the Ark?

Moses received all of his instructions of how to build the Ark and the other Temple articles when he was in the cave at Mount Sinai. Moses did not build the holy objects himself, but rather God had somehow given the knowledge of how to build these objects to a few people, especially to one man, named Bezalel (also spelled Bezeleel), in Hebrew בצלאל.

> [Exodus31:1] And the Lord spoke unto Moses, saying: [2] "See, I have called by name Bezalel the son of Uri, the son of Hur, of the tribe of Judah; [3] and I have filled him with the spirit of God, in wisdom, and in understanding, and in knowledge, and in all manner of workmanship, [4] to devise skilful works, to work in gold, and in silver, and in brass, [5] and in cutting of stones, to setting, and in carving of wood, to work in all manner of workmanship. [6] And I, behold, I have appointed with him Aholiab, the son of Ahisamach, of the tribe of Dan; and in the hearts of all that are wise hearted I have put wisdom, that they may make all that I have commanded thee: . . ."

It is very unusual for anyone to acquire so much knowledge instantly. I felt there was something strange about this person. Therefore, I looked at what his name meant in Hebrew. There are two interpretations to his name: The Rabbis have used the translation, *in the shadow of God.* The first letter ב means *in.* The second part of the word צל, means *shadow,* and finally the last two Hebrew letters אל are another way of expressing the name of God. The other interpretation for his name is, *Only God,* because the first three letters בצל means "only," and the last two mean *God.* Either Moses was trying to tell us that Bezalel received his instruction from God while he was in the shadows of the dimly lit cave, or Moses was telling us that God was the real designer of the Ark, and man was merely following instructions. My feeling is that it is most likely both.

After the Ark was built, Moses communicated with God through the Ark, and not from within the cave or from the Rod. The following verse describes a picture of the finished Ark, shown to the workers so they would know precisely what to build. This could be explained if they were looking at a hologram of the object, just as when Moses heard God's voice coming from a burning bush that was not consumed by fire.

> [Exodus 25:8] And let them make Me a sanctuary; that I may dwell among them. [9] According to all that I show thee, after the pattern of the tabernacle, and the pattern of all the furniture thereof, even so shall ye make it.

> [25:40] And look that thou make them after their pattern, which was showed thee in the mount. [Emphasis added.]

Overall Positioning of all the altars and platforms found on top of Mount Sinai

I am the first person since Baruch (587 B.C.E.) who has figured out and discovered where the real mount Sinai is located. My father did not tell me where it is nor did I have to spend years trekking all over the Sinai to find it. I figured out its location before I ever left my home in Washington State for Egypt. How I found Sinai is an interesting story by itself, but I will leave that for another book.

I also do not want to reveal the location of the mount, at this time, so vandals and souvenir hunters do not strip the place clean. I am also concerned that the Muslim-dominated Egyptian government would destroy the altars and the Mount to spite the Jews. Their hatred towards Jews is only surpassed by their desire to breathe.

Because my claims are extraordinary, I will show you photographs of the altars, so you will know the evidence I illustrate is overwhelming proof that this is the real Mount Sinai!

The Site Layout

During my last expedition to Mount Sinai in November 1999, I had a chance to measure the location and direction of all the altars and platforms found on top of the real mount Sinai. Figure 7-1 is the resulting site map of what I found on top.

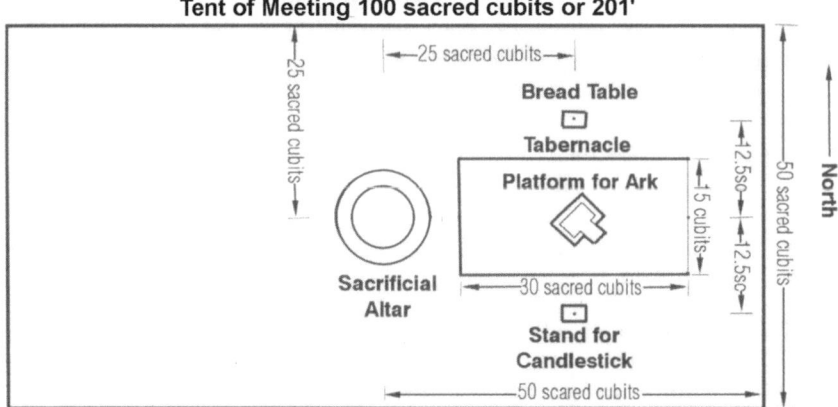

Figure 7-1: Site map of the top of Mount Sinai.

The Sacrificial altar was 48′ to 50′ due west of the center of the stone Ark altar. There is a problem with this measurement, because the center of the Sacrificial altar is difficult to ascertain. The altar has been greatly worn down, but I believe the measurement was originally close to 50 feet, because it means the layout of the Sacrificial altar is in position and symmetrical to its placement within the Tent of Meeting, and to the Tabernacle. In fact, one can analyze the placement of all the objects within the Tent of Meeting as a microcosm of the numbers used in the Torah. For example, the length of the Tent of Meeting is 100 sc (100 sc × 24.136″ = 2,413.6″). The distance from the Sacrificial altar to the east or west end of the Tent of Meeting is 50 sc (50 sc × 24.136″ = 1,206.8″). The Ark altar is equidistant between the sacrificial altar and the eastern wall of the Tent of Meeting, or 25 sc (25 sc × 24.136″ = 603.4″). And finally, the distance of the bread stand and the candle stand are both equidistant between the center of the Ark altar and the north and south walls of the Tent of Meeting, which is 12.5 sc (12.5 sc × 24.136″ = 301.7″). All of these numbers (24,136; 12,068; 6,034 and 3,017) are used, below the surface, within the Torah. For this reason I am confident that these measurements are correct.

The Sacrificial Altar

The reference and dimensions for the Sacrificial altar are found in Exodus 27:1: "And thou shalt make the altar of acacia-wood, five cubits long, and five cubits broad; the altar shall be four-square; and the height thereof shall be three cubits." The altar I found is about four-and-one-half feet high (Figure 7-2) and I am sure it was higher but, over the 3,300-plus years in the desert, it has been worn down and also damaged by the Bedouins in the area.[2] The brazen altar was described as five sacred cubits (sc), or ten feet square by three cubits, or six-feet high. The table was described as square. That means the supporting altar had to be round, so the table legs would fit underneath it. A complete description and drawing is covered later.

Figure 7-2: The sacrificial altar.

The Ark Altar

The Ark was placed on a stepped platform inside a prop-up portable building which the Torah calls the "Tabernacle." My friend and assistant, Vic Ardelean and I fully measured the Ark altar and took a full set of pictures, including the bread stand and the candle stand. Moses, in Exodus, describes the construction of the Ark altar as follows:

> "And if thou make *Me* an altar of stone, thou shalt not build it of hewn stone; for if thou lift up thy tool upon it, thou hast profaned it."[3]

"Neither shalt thou go up by steps unto *Mine* altar, that thy nakedness be not uncovered theron."[4] [Emphasis added.]

These verses were describing the stone altar or platform where the Ark was placed, because it is the only occurrence in the Torah where it reads, "make *Me* an altar" or "*Mine* altar." Since God's voice came from the Ark itself, that is where He resided. So that is why the possessive term is used. My next task was calculating the minimum size of the Ark altar. That was determined by the size of the Ark, which was 1.5 sc wide (36.204″) by 2.5 sc long (60.34″). So the platform altar had to be wider and longer than the Ark. We also measured the height of the altar, which we estimated to be about 2.5-feet. Figures 7-3 and 7-4 shows the altar where the Ark was placed. Figure 7-5 is a side view of the ramp on the south side, which enabled Moses and Aaron to ascend the altar without uncovering their legs, as required in verse 20:23. The length of the ramp is six feet, located on the southeast side of the altar. The center of the altar was exactly due west of the Sacrificial altar as it is described in the Torah. What was a surprise to me was the altar was not aligned in an east-west direction, but rather it was rotated about 30 degrees to the southeast, as the drawing in Figure 7-6 shows.

At first, this was a puzzle to me, but once I drew the layout I realized why it had to be aligned at an angle. The reason was so the incense table could be placed at the foot of the ramp. It measured one-sacred cubit square (24.136″). If the altar was aligned north-to-south, then the ramp would have faced us, with the incense table in front. That layout would have been logical, but it was impractical because the two staves used to carry the Ark were too long to rotate 90° inside of the Tabernacle. That means the staves were longer then 30-feet. The legends of the Jews imply the staves were 40 feet, but there is no reference in the Torah stating how long they were, so it is only guesswork, but there is a strong argument that the poles of the staves were 20- to 30-feet long. If the altar was aligned in an east-west direction, then the Ark could have been placed on the Altar, while holding the staves but the incense table would be too close to the south wall of the Tabernacle, or the Ark would have had to been placed closer to the north wall. By rotating the Ark altar 30 degrees, they placed the Ark in the center of the Tabernacle building and the incense table in front of the ramp, with enough room to go around it. I will explain later how the Ark worked and why the Ark had to be placed in the center of the Tabernacle.

After Vic and I completed measuring the Altar, I was able to come up with a drawing of the altar (Figure 7-7). You will notice that there was a step up to the platform where the Ark rested. You are able to see the steps when you look above them even though so many of the rocks were taken off the altars and used by the Bedouins for their nearby graves.

Figure 7-3: Northwest side of the Ark altar.

Figure 7-4: Southwest side of the Ark altar.

Figure 7-5: Side view of the ramp picture, taken with the camera on the ground. The ramp is clearly visible.

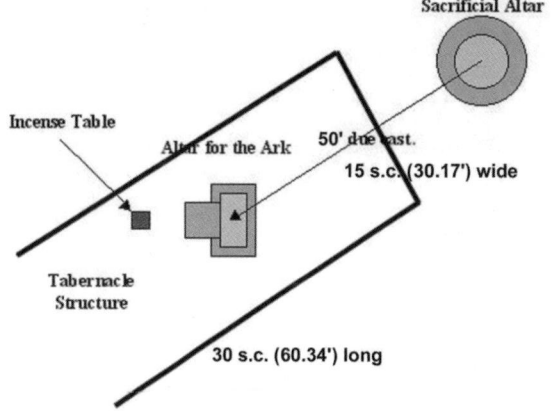

Figure 7-6: Layout of the Ark altar in relation to the sacrificial altar and the Tabernacle.

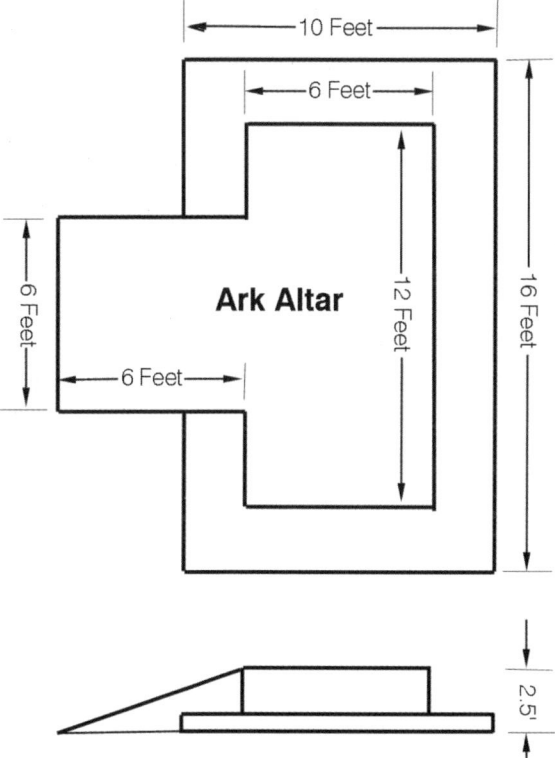

Figure 7-7: Drawing of the Ark altar derived from my detailed field measurements and photographs. The Ark measured 5.028' feet long by 3.017' wide, which means there was enough room on top of the Altar for the ark to be placed and for someone to walk around it.

After we came back from Egypt I wanted to find drawings of other altars which archaeologists strongly felt were also used for the Ark. Figure 7-8 is an artist's conception of the reconstruction of an altar found on Mount Ebal in Israel. The archaeologist, Adam Zertal, found it on April 6, 1980. He estimated it was built about 1260 B.C.E. by Joshua. It was located just north of Shechem. Similarities between this altar and the altar I found on Mount Sinai, are obvious.

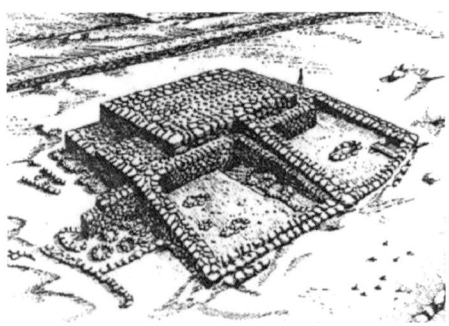

Figure 7-8: An artists conceptions of another stone altar near Shechem. Credit: Shechem organization, Israel.

They both have two levels to them, they both have ramps in the center leading up to the platform, and they are both made from unhewn stone. The only disagreement I would have with the picture of the altar is that I think the width of the ramp should be wider.

The missing keys—The Candle Stand and the Bread Stand

The last two platforms I found on top of Mount Sinai were crucial because they helped prove that my calculation of the length of the sacred cubit was correct, and that this mount was, in fact, Mount Sinai! These are the candle stand and the bread stand. In the following verses, Moses refers to the Tabernacle by saying, "without the veil." The Tabernacle had a veil, or curtain, placed over the eastern opening.[5] He is telling us that these two holy articles were not located inside the Tabernacle. Everything inside the veil was within the Tabernacle, and everything "without the veil" was outside of the Tabernacle. The book of Exodus informs us that both objects were not within the Tabernacle. The following scripture proves the point.

> [26:35] And thou shalt set the table [bread table] *without the veil*, and the candlestick over against the table on the side of the tabernacle toward the *south*: and thou shalt put the table [the bread table] on the *north side*.

> [27:21] In the tent of meeting, *without the veil* which is before the testimony [the Ark], Aaron and his sons shall set it in order, to burn from evening to morning before the Lord; it shall be a statute for ever throughout their generations on the behalf of the children of Israel.

> [40:22] And he put the table [bread table] in the tent of meeting, upon the side of the tabernacle *northward, without the veil*. [23] And he set a row of bread in order upon it before the Lord; as the Lord had commanded Moses.

[40:24] And he put the candlestick in the tent of meeting, over against the table, on the side of the tabernacle *southward*. [25] And he lighted the lamps before the Lord; as the Lord commanded Moses. [Emphasis added.]

It is evident from the three verses that the bread stand and the candle stand were not inside the Tabernacle.

You will find some confusion in the *Legends of the Jews* about whether or not the candlestick and the bread stand were placed within the Tabernacle. They were not, for two reasons. First, I discovered the platforms for both holy objects and they were located outside of the walls of the Tabernacle. They were equidistant between the outside curtain of the Tent of Meeting, and the center of the Ark altar—a distance of 25 sc (50.28′), so both objects were located about 12.5 sc (25′) from the center of the Ark altar. They were both placed exactly north and south of the Ark altar, exactly as the Torah describes. Figures 7-9 and 7-10 show the remains of the candle stand and bread stand. The Bedouins removed the majority of the stones for use on their graves.

Figure 7-9: Close-up top view of the candle stand platform, 24.5± feet from the Ark Altar.

The locations, directions and dimensions of the sacrificial altar, the Ark altar, Abraham's altar, the bread stand and the candle stand prove that this is the real Mount Sinai, because they all fit the Biblical story perfectly. The location and direction of the bread and candle stands fit the story perfectly. They are the keys that could fit no other lock. I discovered much more evidence which proves this hill is the real Mount Sinai, but the nature of the proof would also give away its location. I also found the first altar Moses built at the western base of the hill, I found Aaron's altar and the platform built for the golden calf just west of it. I found the location where the 12 standards are located. All that is left of them is 12 small piles of rocks in three rows of four each. I believe the physical evidence I have found irrefutably proves this is the real Mount Sinai.[6]

Figure 7-10: Top view of what is left of the bread stand, located 25.4 feet north of the Ark altar.

Description of the Ark, Tabernacle, and Sacrificial Altar

The Ark did not work by itself. The four components that were necessary for it to work were the Tabernacle (a tilt-up building), the Incense table in front of the Ark (placed inside the Tabernacle), the smoke produced on that table, and the Sacrificial Altar. The Sacrificial altar in front of the Tabernacle on the west side produced the charcoal that was placed onto the incense table, but any altar could have produced the charcoal. I will first describe each component, then later present a detailed description of how the objects worked together, what the Ark really was, and how it worked.

The Tabernacle

The first holy object that must be described is the Tabernacle. There is historical-literary confusion that still exists about the Tabernacle and the Tent of Meeting. In the King James Version of the Bible, the Hebrew words אהל מועד are rendered *Tabernacle of the Congregation* instead the correct translation, *Tent of Meeting*. The Hebrew word for Tabernacle is spelled משכן, and the word for tent is אהל. As anyone can see, the two words are very different. I believe the translation problem was due to confusion as to what the function of the Tabernacle was and what was placed inside of it. The English translations of the Hebrew Scriptures display no such problem.

The Tabernacle Roof

The Tabernacle was basically a tilt-up wooden building which measured in length 30 sc (60.34′) by 15 sc (30.17′) wide, and 10 sc (20.11′) high. The Tabernacle had a four-layer roofing system. The underside was of linen with a woven pattern of Cherub's.[7] A Cherub looked like young baby boy. There were ten long strips, each measuring 4 sc × 28 sc (8.045′ × 56.317′).

The ten strips were clasped together into two separate sections, of five strips each. The resulting two sections were looped together to make the final covering, measuring 36 sc × 28 sc (72.4′ × 56.137′).[8] The next layer was of goat's hair, comprised of 11 strips, measuring 4 sc × 30 sc (8.045′ × 60.34′). The 11 strips were clasped together into two separate sections, one of five strips, the other of six strips. One strip overlapped the other panel. The two resulting sections making up the final covering measured 40 sc × 30 sc (80.45′ × 60.34′).[9] The final two top layers were of ram's skins, with the outer layer made from badger[10] skins. The dimensions of these two outer layers are not given, but we can safely assume they were the same measurements as the goat's hair layer, because the outer skin layers were used as waterproofing. The top three layers overlapped the Tabernacle by about 10 feet on the eastern and western ends, and 15 feet on the north and south sides. That is what formed the top half of the veil on the eastern side.[11] If you were to go inside of the Tabernacle, you would be inside the Veil. In addition to this, there was a veil or curtain supported by four gold-covered posts placed at the eastern opening as the door.[12] If you were outside of the Tabernacle, you were *without the veil*.

The most important observation about the roof system is that there was no opening for the smoke to rise and escape from the Tabernacle. You will understand *why* later. Some artists' drawings show the Tabernacle with an opening through the roof, and smoke rising out, but there is absolutely no Biblical proof for this interpretation. There was no opening in the roof.

The Walls of the Tabernacle

The Tabernacle had three prop-up walls made of acacia wood (*shittim wood*), 10 sc long by 1.5 sc wide (20.11′ × 3.017′), except for two corner-boards on the shorter western side, that measured 3 sc × 10 sc[13] The final measurements of the Tabernacle were: 30 sc × 15 sc (60.34′ × 30.17′). Five horizontal bars were attached to the vertical boards by gold rings on the boards. The Torah gives no description of any wall on the eastern side of the building.

The reader is supposed to deduce that the door is on the eastern side of the building. The practical reason for the door being on the eastern side is because the Sacrificial altar was at the western end and when the wind blew from west to east it would blow smoke into the Tabernacle, and that was obviously not desired.

To make the Tabernacle even more spectacular and to protect the wood from the elements, the boards were all covered with gold, even the bars—the same gold that Joseph had taken out of Egypt. The gold coating also had a practical application, which I will explain later.

The Sacrificial Altar

The sacrificial altar (Figure 7-2 and 7-11) was placed on the western side of the Tabernacle. The Torah describes it as constructed of acacia wood and brass. The dimensions were 5 sc × 5 sc, and 3 sc high (10.056' square, height 6.034'). This is a very large table for use as a sacrificial altar. The height bothered me because six feet is too high for anyone, without a ladder, to work on an animal sacrifice, because it is burned on top. Also, a full-grown cow or ox was too heavy to be supported by a brass table, with a fire again on top. If it were made with wooden beams running underneath the middle, the brass top would get hot enough to burn the wood supports below it. So the question is: What was supporting the table from the center?

When I arrived at Mount Sinai the actual shape and construction of the stone platform answered the question. The height mentioned in the Torah meant the final height of the altar was designed to be six feet, not the actual height of the table itself. A round stone altar was created with one step to it (Figure 7-11). The first layer was a platform and walk-around area about 3-feet wide. The center portion was a round stone altar 4-feet high. The overall diameter was 16 feet. The center part supported the square brass table placed above it. To further adduce, the Torah also describes the use of a round grating: "And thou shalt make for it a grating of network of brass; and upon the net shalt thou make four brazen rings in the four corners thereof. And thou shalt put it under the ledge *round the altar beneath,* that the net may reach halfway up the altar" [Emphasis added].[14] All of the translations I have studied are imprecise, but I believe Moses was trying to describe a round altar beneath the square brass table. The legs of the table would easily had fit around the center of the stone altar. The stone platform underneath helped support the weight of the sacrificial animal and dissipated heat from the fire in the middle of the table.[15]

Besides the table, other objects were constructed for this purpose. Moses was told to construct fire pans to collect the charcoal, plus shovels, basins, flesh hooks, and a grating, all made out of brass.[16] The Sacrificial altar also had four rings mounted on opposite sides, so the two brass-covered acacia wood staves could be put through the rings to lift up and move the table.[17]

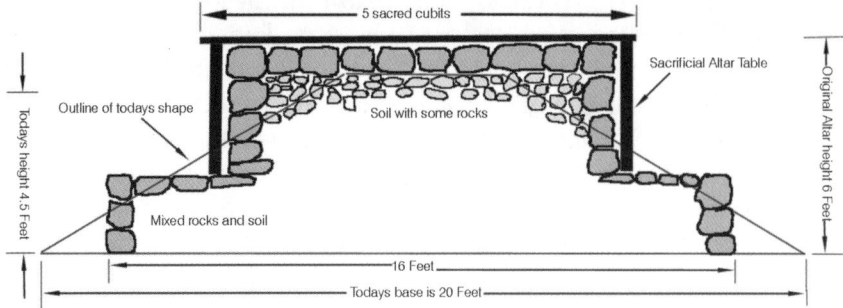

Figure 7-11: Cross-section of the stone altar on which the brass sacrificial altar was placed. It was located on the west side of the Tabernacle.

The Fat

Besides the cattle, goats, lambs and other animals sacrificed on the altar, God made it very clear that He wanted the fat burned also, and He made it a law.

[Exodus 29:13] And thou shalt take all the *fat* that covereth the inwards, and the lobe above the liver, and the two kidneys, and the fat that is upon them, and *make them smoke* upon the altar.

[Exodus 29:22] Also thou shalt take of the ram the *fat*, and the rump, and the *fat* that covereth the inwards, and the lobe of the liver, and the two kidneys, and the *fat* that is upon them, and the right thigh; for it is a ram of consecration; [23] and one loaf of bread, and one cake of oiled bread, and one wafer, out of the basket of unleavened bread that is before the Lord. [24] And thou shalt put the whole upon the hands of Aaron, and upon the hands of his sons; and shalt wave them for a wave-offering before the Lord. [25] And thou shalt take them from their hands, and *make them smoke*, on the altar upon the burnt offering, for a sweet savour before the Lord; it is an offering *made by fire* unto the Lord.

[Leviticus 3:14] And he shall present thereof his offering, even an offering made by fire unto the Lord: the *fat* that covereth the inwards, and all the *fat* that is upon the inwards, [15] and the two kidneys, and the *fat* that is upon them, which is by the loins, and the lobe above the liver, which he shall take away by the kidneys. [16] And the priest shall *make them smoke upon the altar*; it is the food of the offering made by fire for a sweet savour; all the *fat* is the Lord's. [17] It shall be a perpetual statute throughout your generations in all your dwellings, that ye eat neither *fat* nor blood.

The three previous quotations are all instructional and similar in meaning. God told Moses that He wanted the fat from these animals put on the fire, so it would smoke. The references to fat fall into two categories: The first was instructional, the others asking that the fat be burned on the fire and made to smoke. There were 10 instructional verses, all in Exodus and Leviticus. The other references were mentioned 19 times in Leviticus, Exodus and once in Numbers. The other Hebrew Scriptures mention the fat burning on the altar a total of 10 times (Isaiah, First Kings, First Samuel, Ezekiel, Psalms and Second Chronicles). One would think that a writer would only mention it two or three times to make the point. But the Torah mentions it 29 times! Moses is really trying to tell us something very important about a smoky fat fire! The fat is very important, as you will soon learn.

Priestly Procedures that had to be followed

The procedures the priests followed were first, to start a wood fire on top of the brazen altar. After the fire began, they placed the cut-up animal on the fire. As the animal and the wood burned and turned to charcoal, they then placed the fat on the red-hot charcoal. The fat would then start to produce the typical smoke which fat generates. The priests then removed the smoldering charcoal and placed it onto brass pans that were brought into the Tabernacle, and set atop the gold Incense Altar.

The priest would then place ground-up sweet incense on top of the red-hot coals. So the fat and incense would smoke and fill up the top of the Tabernacle with smoke. Moses was instructed by God to fill the Tabernacle with smoke until the smoke reached the Ark cover—sometimes translated as the "Seat of Mercy." Remember, there was no venting in the roof of the Tabernacle. When the Tabernacle was first assembled in the first week of the second year, Moses did not want to enter the smoke filled Tabernacle. "Then the cloud covered the tent of meeting, and the glory of the Lord filled the Tabernacle. And Moses was not able to enter into the tent of meeting, because the cloud abode thereon, and the glory of the Lord filled the tabernacle."[18] Most people would not enter a smoked-filled room, for obvious reasons.

Moses was described in Exodus Chapter 34:33 as placing a veil over his face. The story says that Moses took off the veil when he went into the Tabernacle, and put it back on when he came out. But I have a feeling that the order was reversed! He put the veil on over his nose and mouth so he would not get smoke in his lungs. I do not see any other way a person could stay in a smoke-filled room for a long period of time without getting smoke in their lungs. The fact that the Torah states Moses wore a veil is a significant clue.

The Ark of the Covenant and the Incense Table

The Ark of the Covenant was the central and most important holy object to the Hebrews, from its creation in 1305 B.C.E., up until the destruction of the

Temple, when it was hidden from the invading Babylonians in 587 B.C.E. Over the years, the Hebrews replaced some of the objects they made at Mount Sinai such as the Sacrificial Altar, fire pans and tables, but not the Ark or its related objects. The Ark of the Covenant was totally unique. This does not rule out the distinct possibility that the Hebrews made a copy of the Ark, so that if an invading army, such as the Egyptians, were to capture Jerusalem, the real Ark would not be stolen from them. This I believe happened when Pharaoh Shishak captured Jerusalem in the fifth year of king Rehoboam (917 B.C.E.). In First Kings the Pharaoh "took away the treasures of the house of the Lord, and the treasures of the King's house; he even took away all;"[19]

It is not clear whether or not the Egyptians took away the Ark. I know they did not get the real one, because after this event, the High Priests, who were the prophets of Israel, were still in front of the Ark, receiving messages from God. And they continued doing so until the destruction of the Temple. This implies that the priests had made a duplicate copy of the Ark, just in case something like this might happen. If they had made a replica, it is most likely they left off the five items the real Ark had. I will explain this later.

The Table for the Ark

The Ark was placed on a small table constructed of acacia wood, measuring 2 sc (48.272″) in length by 1 sc (24.136″) in width and, 1.5 sc (36.204″) high. It was also overlaid with gold, with a "crown" around the top to form an edge. Two gold rings were attached to both long sides of the table, for transporting it.[20] Two acacia wood staves, overlaid with gold, were also built to transport the table.

The Incense Table

The incense table was a key component, for the Ark to work as a communications tool. Its main purpose was to create smoke to fill up the top of the Tabernacle, until the smoke reached down to the level of the Ark cover. The purpose was to make sure the two cherubim were in the smoke cloud. The table measured 1 sc square (24.136″) by 2 sc high (48.272″) and with a gold crown around the top edge. The purpose of the raised edge was to prevent the hot coals from rolling off the table, and falling on the floor. The instructions for building it first appear in Exodus 30:1, and were repeated in Chapter 37:25. All of the Torah stories about the objects in the Tent of Meeting and the Tabernacle are repeated twice. The key phrase that tells us how they used the table is found in Leviticus 16:12:

> And he [Aaron] shall take a censer full of burning coals of fire from off the altar before the Lord, and his hands full of sweet incense beaten small, and bring it within the veil. [13] And he shall put the incense upon the fire before

the Lord [the Ark], that the *cloud of the incense may cover the ark-cover that is upon the testimony*, that he die not: [14] And he shall take of the blood of the bullock, and sprinkle it with his finger upon the ark-cover on the east; and before the ark-cover shall he sprinkle of the blood with his finger seven times. [Emphasis added.]

The *testimony* referred to is the Torah, engraved on the two stone tables placed inside the Ark. The role of the High Priest was to make sure there were hot coals on the incense table every morning so if God wanted to communicate with the priest the Ark would work. A description of the kind of incense is given in Exodus 30:34.

And the Lord said unto Moses: Take unto thee sweet spices, stacte, and onycha, and galbanum; sweet spices with pure frankincense; of each shall there be a like weight: [35] And thou shalt make it incense, a perfume after the art of the perfumer, blended together, pure and holy. [36] And thou shalt beat some of it very small, and put of it before the testimony in the tent of meeting, where I will meet with thee: it shall be unto you most holy.

The incense was placed on the incense table which was placed inside the Tabernacle that was, in turn, located within the Tent of Meeting. I believe it was placed in front of the ramp on the south side of the Ark altar.

The Ark

The Ark was comprised of five parts: They were the rectangular acacia wood box that was covered inside and out with gold, the gold ark cover, and the two solid gold cherubim facing each other. Between them there was a Seat of Mercy, which I believe was a blue-crystal which resembled a stone chair (throne seat) placed in the middle of the cover directly between the two Cherubim. Finally, there were four round crystalline disks placed on top of the Ark cover on either side of the cherubim. Only the first three items did the Hebrews construct. The crystalline Throne seat and the four round disks they did not. I believe Moses retrieved them from within the cave, and they were part of the advanced technology left by the previous civilization. The following will describe each item and how they worked.

The Ark Box

The Ark box measured 2.5 sc (60.34″) in length, by 1.5 sc (36.204‴) width, and a height of 1.5 sc (36.204″). It was made from acacia wood found in the area, and covered inside-and-out with gold. We found acacia trees nearby, and there was ample evidence the area resembled a savanna 3,300 years ago. Combined with the gold-covered lid, the finished object was a box. The Ark box and lid immediately reminded me of a large electrical capacitor that could hold a huge electric charge. What is also interesting is the volume of the Ark is exactly one-half the volume of the stone sarcophagus of the Great Pyramid of Giza.

The Ark Cover

The Hebrew word for the Ark cover is כפרת, and its English translation is *cover.* In many non-Jewish Bibles it was incorrectly translated it as "Mercy Seat." When the phrase "Mercy Seat" is used, it becomes confusing because most of the time when the word shows up, it means just the cover of the Ark—not what is mounted on top of it. One of the holy objects was a seat, or throne, but the Torah does not use another word for it, so the reader is required to figure out that there was a model of a throne seat in the center of the cover. I will describe that in the next section.

Placed at opposite ends of the cover were two golden cherubim facing each other.

> [Exodus 37:7] And he made two cherubim of gold, of beaten work made he them, at the two ends of the Ark cover: [8] One cherub at the one end, and one cherub at the other end; of one piece with the Ark cover made he the cherubim's at the two ends thereof. [9] And the cherubim's spread out their wings, and sheltered with their wings over the Ark cover, with their faces one to another; even towards the cover were the faces of the cherubim's.

What is strange about the last sentence is that you are led to believe that the cherubim are sitting on the Ark cover and also looking at each other. But it also says they are looking down towards the Ark cover. They were looking down at something placed between them. According to the *Legends of the Jews*, the Cherub's heads were slightly tilted back. In other words, their lips were the closest things to each other, rather then their noses. This is a somewhat un-natural pose. That is why I began to realize there was something else placed between the Cherubim. But first, let's examine additional information about what the Cherub resembled.

In the *Legends of the Jews,* there is an account of what they looked like and a few measurements:

> "On the Ark were the cherubim with their faces of *boys* and their wings. Their number was two, corresponding to the two tables, and to the two sacred names of God, Adonai and Elohim, which characterized Him as benevolent and as powerful. The face of each Cherub measured *one span*, and the wings extended each ten spans, making twenty-two spans in all, corresponding to the twenty-two letters of the Hebrew alphabet." It was "from between the two Cherubim" that God communed with Moses, for the Shekinah [presence of God] never wholly descended to earth any more than any mortal ever quite mounted into the heaven, even Moses and Elijah stood a slight distance

from heaven; for, "The heaven, even the heavens, are the Lord's: but the earth hath He given to the children of men." Therefore God chose the Cherubim that were *ten spans* above the earth as the place where the Shekinah betook itself to commune with Moses. The heads of the Cherubim were slightly *turned back*, like that of a scholar bidding his master farewell; but as a token of God's delight in His people Israel, the faces of the Cherubim, by a miracle, "looked one to another" whenever Israel were devoted to their Lord, yea, even clasped one another like a loving couple.[21]

Please notice that the cherubim had the faces of young boys, probably under the age of two. The next step is to figure out what is the length of a *span*? Before this quote, the legend stated the length of the Ark was 10 spans. Since we know that the length of the Ark was 2.5 sc or 5-feet, that means the length of a span is six inches. With that information, we can now get a measurement for the cherubim. The head would be six inches wide; the body would be one-foot wide. We can now even calculate the height of these statues. The legend reveals the Cherub stood 10 spans above the ground and the ground, they are referring to is measured from the base of the Ark, which is three-feet high, or six spans, then the remainder is 4 spans, or two feet above the cover of the Ark.

The Ark box, the Ark cover and the Cherubim are the only objects the Torah says the Hebrews built, which are part of the total device we call the Ark of the Covenant. However, there are two other objects on the Ark cover which they did not build, and are the most important. The following describes these items and possibly how they worked.

The Throne

After carefully studying all of the Biblical references and the legends about the Ark and the cherubim, it became obvious that there were more objects on the Ark cover. One of them was a model of a Throne Seat. The priests mention a Throne, but only refer to it in very cryptic terms. The following are some of these references, scattered throughout Hebrew Scriptures:

> [I Samuel 2:8] He [God] raiseth up the poor out of the dust, He lifteth up the needy from the dunghill, to make them sit with princes, and inherit the *throne of glory*: for the pillars of the earth are the Lord's, and He hath set the world upon them.

> [Jeremiah 14:20] We acknowledge, O Lord, our wickedness, even the iniquity of our fathers; for we have sinned against thee. [21] Do not contemn us, for Thy name's sake, do not disgrace *the throne of thy glory*; remember, break not Thy covenant with us.

[Jeremiah 49:38] And I will set My *throne* in Elam, and will destroy from thence kings and princes, saith the Lord.

[Psalms 9:5] For thou hast maintained my right and my cause; thou sattest in *the throne* judging right.

[Psalms 9:7] But the *Lord is enthroned* forever; He hath established His *throne for judgment.*

[Psalms 45:7] Thy *throne* given by God is forever and ever; a *scepter* of equity is the *scepter* of thy kingdom.

[Psalms 47:8] God reigneth over the nations; God sitteth upon *His holy throne.*

[Psalms 89:14] Righteousness and justice are the foundations of *thy throne*; mercy and truth shall go before Thee. [Emphasis added.]

The following paragraph refers to such a Throne Seat, placed in the middle of the Ark cover, that the Cherubim were facing. The most important reference to a throne on the Ark cover comes from Baruch, who was the author of the book of Ezekiel. The faithful scribe and grandson of Jeremiah, he later became the high priest in exile. He was the last priest to kneel in front of the Ark, at the time of the fall of Jerusalem in 587 B.C.E. His description was the most detailed and revealing. In Ezekiel Chapter 10:1, he wrote:

Then I looked, and, behold, upon the firmament that was over the head of the cherubim, there appeared above them as it were a *sapphire stone*, as the appearance of the *likeness of a throne*. [Emphasis added.]

The Throne Seat apparently was not made of gold, but of some blue crystalline material. We do not know the size of the Throne but I estimate it to have been about 6″ to 8″ wide, and about 6″ to 8″ high. The above quotation may have described a hologram that formed above the heads of the Cherubim and not the actual Throne Seat on top of the Ark. Baruch may also have used this description as a literary device to describe the Throne Seat, without drawing too much attention to its specifics. Other references are described in the book of Psalms, which I also believe was primarily written by Baruch.

[Psalms 80:2] Give ear, O Shepherd of Israel, thou that leadest Joseph like a flock; thou that art *enthroned* between the cherubim's, shine forth.

[99:1] The Lord reigneth; let the people tremble; he *enthroned* them between the cherubim's; let the earth quake. [Emphasis added.]

Finally, from the *Legends of the Jews,* the following quote also states a Throne Seat was placed on top of the Ark:

> The Ark contained the two tables of the Ten Commandments as well as the Ineffable Name, and all His other epithets. The Ark was an image of the *Celestial Throne*, and was therefore the *most essential part of the Tabernacle*, so that even during the march it was spread over with a cloth wholly of *blue*, because this color is similar to the color of the *Celestial Throne*.[22] [Emphasis added.]

Replicas of the Ark Built

All five previous references imply or directly state that an object, appearing like a throne, was on the Ark cover between the cherubim. The mental picture one sees at the top of the Ark cover are of two golden Cherubim with their heads slightly tilted back, so their lips are the closest points to each other. Their eyes were looking down at a blue crystalline Throne in the middle. It will become obvious what symbolism God was illustrating here. The question is: "How was the voice of God heard from above the Ark cover?" The next subsection describes in detail how the sound was formed.

I believe the Throne and the four crystal disks are objects not present on any replicas of the Ark. I conclude this because there was so little description of them within the Hebrew Scriptures, almost nothing except the writings of Baruch. The reason Baruch felt it was safe to write about the Throne and the four crystal disks was because the Ark was already safely hidden away, and would not be found for a long time—so it did not matter whether he gave more details. The lack of details implies there were copies of the Ark made at some time in the past in order to trick any invading army into thinking they had stolen the real Ark. If key components were missing from the Ark, then a knowledgeable priest would know immediately which was the real Ark of the Covenant.

Biblical References to High Voltage

A high-voltage spark was formed between the Cherubim lips, and passed over the Throne. The only Torah clue for such a spark is found in Leviticus 10:1-2, detailing the strange deaths of Aaron's two older sons, Nadab and Abihu.

> [10:1] And Nadab and Abihu, the sons of Aaron, took each of them his censer, and put fire therein, and laid incense thereon, and offered strange fire before the Lord, which He had not commanded. [2] And there came forth *fire* from before the Lord, and devoured them, and they died before the Lord. [Emphasis added.]

The *Legends of the Jews* reveals even more details of the same event:

"But how soon was her [Aaron's wife] joy turned to grief! Her two sons, Nadab and Abihu, carried away by the universal rejoicing at the heavenly fire, approached the sanctuary with the censers in their hands, to increase God's love for Israel through this act of sacrifice, but paid with their lives for this offering. From the Holy of Holies issued *two flames of fire*, as *thin as threads*, then parted into four, and two each pierced the *nostrils* of Nadab and Abihu, whose souls were burnt, although no external injury was visible."[23] [Emphasis added.]

The first reference from Leviticus describes it as *"fire"*, but there was no fire seen or put on top of the Ark cover, so it must be something that looked like fire to a Bronze Age people but was not fire. A spark of high voltage fits the description precisely. The second reference is even easier to visualize, because it describes the *"fire"* being as "thin as threads." This is a very accurate description of lightning, or high-voltage, because voltage will split up into multiple threads over a long distance. The two sons of Aaron were electrocuted.

There were two "flames of fire" because the two Cherubim produced two high-voltage sparks. Aaron's sons' noses were hit because the top of the Ark was eight feet above the ground: two feet for the stone altar base, three-feet for the Ark table, and three-feet for the Ark itself, and the cherubim were another two-feet above the Ark cover. So Aaron's sons would had to have looked up at the top of the Ark to see the Cherubim. When they looked up, the closest part of their body to the Cherubim would have been their noses. Lightning or high voltage hits the closest point to ground. Their noses became the ground connection.

Another reference from *Legends of the Jews* reveals that electrical sparks issued forth from the Cherubim:

It was through the Ark, also, that all the miracles on the way through the desert had been wrought. *Two sparks* issued from the Cherubim that shaded the Ark, and these killed all the serpents and scorpions that crossed the path of the Israelites, and furthermore burned all thorns that threatened to injure the wanderers on their march through the desert.[24] [Emphasis added.]

The next quote is very important because it describes what the sparks looked like, from where they originated, and the sound the spark produced.

The voice that called Moses came from heaven in the form of a *tube of fire* and rested over the *two Cherubim*, whence Moses perceived its sound. This voice was as powerful as at the revelation at Sinai when the souls of all Israel escaped in terror, still it was audible to none but Moses.[25] [Emphasis added.]

This legend links the spark (tube of fire) between the Cherubim, with a voice from the Lord of Hosts. The only mistake the legend made was that the spark did not form over the Cherubim heads but from between their lips.

A Voice of the Lord of Hosts was Heard

The symbolism of a voice forming from a spark over an empty celestial Throne brilliantly illustrates that God is not anyone you can see in this dimension. He is not a celestial body, like a sun, nor can He be described as a person. An excellent quotation from the Book of Numbers, Chapter 22:19, provides us with a good idea of the Hebrew philosophy of God: "God is not a man, that He should lie; neither the son of man, that He should repent."

The following quotes refer to a voice emanating from between the Cherubim. Considering the number of times it is repeated in these books, one gets the impression that it was important for the High Priests to inform us that these words were the words of God, not theirs. The Priests only reported what God told them while they were in front of the Ark. The following quotes clearly state that the voice came from between the two cherubim, not from above them.

> [Exodus 25:21] And thou shalt put the Ark-cover above upon the Ark; and in the Ark thou shalt put the testimony that I shall give thee. [22] And there *I will meet with thee, and I will speak with thee from above the Ark-cover*, from *between the two cherubim*, which are upon the ark of the testimony, of all things, which I will give thee in commandment unto the children of Israel.

> [Numbers 7:89] And when Moses went into the tent of meeting that He [God] might *speak with him*, then he *heard the Voice speaking unto him from above the Ark-cover* that was upon the ark of testimony, *from between the two cherubim*; and *He spoke* unto him.

> [I Samuel 4:4] So the people sent to Shiloh, and they brought from thence the Ark of the Covenant of the Lord of Hosts, *who sitteth between the cherubim*; and the two sons of Eli, Hophni and Phinehas, were there with the Ark of the Covenant of God.

> [II Kings 19:15] And Hezekiah prayed before the Lord, and said, 'O Lord God of Israel, which *sitteth between the cherubim*, Thou art the God, even Thou alone, of all the kingdoms of the earth; Thou hast made heaven and earth.

> [Isaiah 37:15] [This is the exact same quote, word for word, as in II Kings 19:15. See above.]

[Ezekiel 9:3] And the glory of the God of Israel was *gone up from the cherub, where upon He was*, to the threshold of the house; and He called to the man clothed with linen, which had the writer's inkhorn on his side. [Emphasis added.]

All of the previous quotations state the same thing: God spoke to Moses and later to the High Priests who sat or knelt in front of the Ark. God spoke to them from between the cherubim. They all believed that the one speaking to them was God, because many times He told them about events that were going to happen in the near future, and they occurred exactly as the Voice said.

The Round Crystalline Disks

The last objects that were a key part of the Ark were four round crystalline disks. Only Baruch, writing as Ezekiel, describes them several times. The reason we have so few descriptions of the Throne and the disks was explained earlier, but I will add another, more practical, reason. Since both the Throne and the disks were on top of the Ark cover, and the Ark cover was at least eight feet above the temple floor, someone sitting or kneeling on the floor, in front of the Ark, would not have seen the disks at all. They may not have been able to see the Throne except for the very top of it. In summary, the ultimate explanation for so little description is a combination of all of these previously mentioned reasons.

The Book of Ezekiel has Baruch in Babylon informing us that he was with the first group of exiles when the Babylonians first came down to Jerusalem in 598 B.C.E. The Babylonians forced king Jehoiachin and the skilled and educated population into exile. By Chapter 1:4, he describes seeing an object with four wheels:

> "The appearance of the wheels and their work was like unto the color of a beryl; and they four had one likeness, and their appearance and their work was as it were a wheel within a wheel."

Even though the surface story says he was in Babylon, I knew he was sitting in front of the Ark in Jerusalem. The way I figured this out was as follows: there is a date given at the beginning of Ezekiel and it is a coded date representing a number which is important to what he is really saying. When he gave the date as the "thirtieth year, in the fourth month, in the fifth day of the month . . ." he was giving us the number of years and months since the beginning of the Babylonian empire, started by Nabopolassar. He then reveals it was also the fifth year of king Jehoiachin's captivity. By verse 1:4 to Chapter 3:11, he said he saw a vision and spoke to God. I knew that Baruch could not actually be in Babylon at that time, because the only place this could happen would be in front of the Ark, and that was still in Jerusalem. So the next question was, "What code did Baruch use to tell us he was really in Jerusalem, sitting in front of the Ark?" I first tested to see if the number of days in 30 years, three months and five days added up to one of the numbers, most specifically 6,034, which represented the Torah and the Ark that contained it. That did not work because I came up with too many days. Next, I tried the number of months, and that produced exactly 300 months. Remember we are dealing with the lunar calendar, which means you have to add three leap months every eight years to the total. The 300 months produced the proof that I needed: $(300 \times 24.136'' = 7,240.8 \div 12 = 603.4)$. So Baruch was secretly telling the reader he was in front of the Ark in Jerusalem, when he received each of his visions and messages from God.

Baruch later repeated the same description in Chapter 10:9-10, but there he was in front of the Ark in Jerusalem. I believe the first time he mentioned seeing something in Babylon he was just using a literary device to describe what the Ark looked like. This most likely was written to fool the Babylonians, who also were interested in finding the Ark. So Baruch threw them off by making it sound like he had just seen a vision, and nothing else.

He also wrote that the sound of high voltage was associated with the four disks. In Ezekiel, Chapter 3:13 he described:

> [Ezekiel 3:12] Then a spirit lifted me up, and I heard behind me a voice of a great rushing, saying: "Blessed be the glory of the Lord from His place." [13] I heard also the noise of the wings of the living creatures that touched one another, and the *noise of the wheels* over against them, and a noise of a *great rushing*. [Emphasis added.]

The sound of high-voltage static sounds similar to that of rushing water. Baruch tried to give us a more detailed description of the wheels, by reporting what they looked like, and what they were doing, as he watched them. The following are his descriptions:

> [Ezekiel 10:9] And I looked, behold four wheels beside the cherubims, one wheel beside one cherub, and another wheel beside another cherub; and the appearance of the wheels was as the *color of a beryl stone*. [10] And as for their appearances, they four had one likeness, as if a wheel had been within a wheel. [11] When they went, they went towards their four sides; *they turned not as they went, but to the place whither the head looked they followed it; they turned not as they went.* [12] And their whole body, and their backs, and their hands, and their wings, and the wheels, were *full of eyes round about, even the wheels that they four had.* [13] As for the wheels, they were called in my hearing, The wheelwork. [14] And every one had four faces: the first face was the face of a cherub, and the second face was the face of a man, and the third the face of a lion, and the fourth the face of an eagle. [15] And the cherubims mounted up—this is the living creature that I saw

by the river of Chebar. [16] And when the cherubim went, the wheels went by them; and when the cherubims lifted up their wings to mount up from the earth, the same wheels also turned not from beside them. [Emphasis added.]

Most readers have a difficult time understanding what Baruch described here. The important verses are 11 and 12. What Baruch is describing is a hologram. A hologram is the only technology that could produce these visual effects. The effect I am referring to occurs when you rotate around a hologram—it follows you. The scene Baruch describes in Chapters 9 and 10 is a hologram that was projected above the Ark, and seen in the smoke cloud above. The following verse is a very good description of the hologram. The phrase "were full of eyes round about" sounds similar to a string of laser diodes around the disk that produced the laser image.

[Ezekiel 10:4] And the glory of the Lord mounted up from the cherub, to the threshold of the house; and the house was *filled with the cloud, and the court was full of the brightness of the Lord's glory.* [5] And the sound of the cherubims was heard even to the outer court, as the voice of God Almighty when he speaketh. [Emphasis added.]

Another description written in Leviticus states:

[Leviticus 16:2] And the Lord said unto Moses: "Speak unto Aaron thy brother, that he come not at all times into the holy place within the veil before the Ark-cover, which is upon the ark; that he die not; for *I appear in the cloud upon the Ark-cover.*" [Emphasis added.]

So we have two distinct instructions from God that He is heard and seen above the Ark cover.

What was the Ark of the Covenant?

I have presented sufficient background information on the ways the Hebrews described the Ark and what they saw. But is the Ark real, or merely the vivid imaginings of a group of recently-freed slaves, returning to their homeland. Or was it merely the elaborations of their descendants?

To recap what we have learned from the Hebrew writings, the only components of the Ark which they built were the Ark table, Ark box, Ark-cover, wooden staves, and two Cherubim. All of these components were either covered with gold or solid gold, such as the Cherubim. The Ark was placed inside of the Tabernacle, which was a gold-plated wooden tilt-up building with a four-layer roofing system, with no venting in the roof. The gold-covered incense table was also placed inside the Tabernacle. Its function was to fill the top half of the 20-foot high Tabernacle with a certain type of smoke, down to the level of the Ark cover. The Voice would be heard from a spark that originated between the Cherubim. Moses, or the High Priest, went into the Tabernacle and waited for God to speak. If God spoke, the priest would write down every word and later announce it to the congregation or the king. There were two parts of the Ark which the Hebrews did not build—the blue-crystal Throne, located in the middle of the Ark cover, and the four crystal disks that were on either side of the Cherubim. Both objects were most likely acquired from within the Cave.

How did the Ark Work?

I will refer to the Ark in the present tense because it still exists. It may be hidden right now—but someday it will be found and used once again.

The Ark is an actual device that works and is not a figment of Moses' imagination. It seems to have two functions—the first, a device, which enables man to hear God speak, and the second, to create a hologram seen in the smoke cloud above the Ark. I previously described the formation of the holograms, and now I will explain how His voice is created.

If you have any electronics knowledge, especially of early radio and speaker design, you should have deduced what the Ark actually is. For the rest of you, here is my description: the Ark is, in essence, a crystal receiver, with a flame speaker above that created the sound. The crystal Throne was most likely the receiver and power supply all in one. The Ark box was used as a capacitor, which had the capability of holding a very large electrical charge. The Torah does not describe whether priests placed oil inside of the Ark, but if they did, the oil would be the dielectric for the capacitor. The two Cherubim served as positive and negative emitters (anode and cathode) of the spark gap. The Throne beneath somehow induced a modulation in the spark above it, forming the sound.

How Flame Speakers Work

Flame speakers were invented in the late 1800s. They are fairly easy to construct, and the principle is fairly simple, but the results are spectacular, especially appreciated if you are a sound or music connoisseur. When a flame or particles are ionized by an electrostatic modulated charge, the flame produces perfect sound. The ionized particles in the flame oscillate at frequencies from a few hertz to well above the audio range.[26] There are three different ways to produce sound: compression drivers, direct radiators, and ionic charges, with a sub-type called plasma speakers. They all convert a variable electrical signal into varying air pressure changes, also called sound. I will not cover the first two or the last, because they are not the method used by the Ark, just the third—ionic charges—because it is the method used by the Ark.

Logos Foundation[27] builds a speaker which most closely resembles how the Ark produced sound. Their speaker uses ionized air and an electric arc, which is modulated with an audio signal. The ionized air is supplied from a gas flame between the spark gap. The spark gap ionizes the air around it, thus creating a sound-radiator which does not use any moving parts, such as diaphragms, cones or coils. The spark from the spark gap forces the molecules in the ionized air into immediate vibrations creating perfect sound. This type of system also can convert digital-audio information into sound, using the ionized air around the electrostatic arc as a digital-to-analog converter. To the observer, it would look as though flames are talking—reminiscent of God's voice coming from the burning bush. Figure 7-12 shows a picture of such a speaker. Notice the two metal rods on top, where the spark forms across the gap, arc similar in function to the two Cherubim on top of the Ark cover.

Figure 7-12: Flame speaker system built by Logos Foundation in Belgium.

How the Ark Works

The Ark works like the flame speaker. The Ark box itself acts as the capacitor, which holds a large electrical charge. Smoke from the Incense Altar provided the ionized particles suspended in the air. The Torah instructed Moses that the smoke had to come down to below the Ark-cover, to make sure the Cherubim were in the smoke. The question is, "What modulated (vibrated) the spark that formed across the Cherubim lips?" I believe the Crystal Throne provided that function, because it was in the correct position, and we know that crystals can be made to oscillate. As the spark formed across the Cherubim the Throne must have modulated the spark and therefore created the sound. The Throne must be the key component, acting as the receiver, transmitter and power source—and was most likely a computer.

There must have been wiring between the Cherubim, the Ark-cover and the Throne seat, but the Torah provides us with no such description. This does not surprise me because the technical components of such a device would have had no frame of reference to Bronze age people, such as the Hebrews. Therefore, no description was deemed necessary.

The Ark produced a high voltage. We know this because two of Aaron's older sons were electrocuted when they came too close, and they also probably approached it from the wrong direction. The sound the Ark produces must be nearly deafening when it was activated, because the speaker size would have been measured by the volume of smoke within the Tabernacle, which was 30′ × 60′ × 13′.

The reason why the boards of the Tabernacle were covered with gold was so that any stray sparks would ground themselves after striking the boards.

The four round crystal disks on both sides of the two Cherubim created the holographic image in the smoke above the Ark. That is why several prophets reported seeing images above the Ark. They watched a hologram complete with sound.

The description Moses gave us perfectly describes a flame-speaker device, even down to how the sound was created. That there were holograms formed above the Ark is a good educated assumption, but regarding the flame speakers, there are enough detailed descriptions given so that we can safely conclude that the Ark was a flame speaker connected to a device which served as a receiver, modulating the spark gap. We are faced with the challenge of attempting to explain how the ancient Hebrews could have known how to build a crystal set, and connect it to a high-voltage flame speaker. How could they have known about RF signals, modulation and tuned crystals? The answers to these questions come next.

Where did Moses Find the Crystalline Throne Seat and the Four Disks?

There are enough literary clues provided us to enable us to arrive at a conclusion. These "modern" components were acquired from within the cave at Sinai. I have proven that the Hebrew alphabet is the product of a very highly advanced civilization. In addition to the Throne and the four disks, Moses' Rod and Aaron's staff were also acquired from within the Cave. We are told in Exodus that Moses was shown how to make everything, within the Cave. The four crystal disks explain how he saw the images which revealed how to build the objects. The same was also true for the Rod he was given by God. The burning bush was most likely a holographic projection, which Moses saw when inside the Cave. That is why the flame did not consume the bush. It did not really exist except as information. The hologram instructed Moses to take the Rod and on how to use it.

As explained in Chapter 5, Abraham insisted on purchasing the Cave of Machpelah because he wanted to possess good title to the hill, the cave, and the land around it. His reason was obvious. It was incorporated into the name itself. Machpelah means *place of wonder* or *miracle place*. The land of Moriah means *teaching* or *instruction of God*. It was God's plan to lure Abraham to Mount Sinai, not for sacrificing his son Isaac, but rather to have him discover the Cave located at the hill. The circumstances were more a test of loyalty, to see if Abraham was worthy to receive God's gift of knowledge. What was, and still is, in the Cave is a repository of highly advanced technologies from a very ancient civilization. I suspect this because that civilization seems to have had the ability to communicate directly with the Lord of Hosts and that is whom we are told spoke to all of the Patriarchs, including Moses, and the Priests who came after him.

What seems obvious to me is that the description of how the Ark worked implies a technology that the Hebrews could not possibly have known. In essence, the Ark was a crystal receiver connected to a flame speaker and a laser hologram projector. It communicated in real-time to Moses and all the Priests after him who sat before the Ark. There seemed to be no question about Who they were communicating with.

In my opinion, we are dealing with the technology of an extremely advanced civilization, that God decided to reveal to a specific Bronze Age people, the Hebrews. It is not for us to question why He did this, but it is obvious to me that this was part of a larger long-term plan that may be unfolding before us even now. We must ask ourselves, "Why did it take so long to figure out all of these secrets and codes within the Torah and the Hebrew Scriptures? Why did it take so long for someone to discover the exact location of Mount Sinai?"

What did the Priests do in Front of the Ark of the Covenant?

The Torah prescribes a continuous routine the priests were to perform every day. Additional services were to be performed on the Sabbath, on the first day of each month, and on other designated holidays. The underlying reasons for all of these traditions, routines and animal sacrifices were so God could communicate with man when He decided. That meant the High Priest had to keep the smoke level in the Tabernacle at the correct level so God's voice could be heard when He wished to communicate. The High Priest or other priest would sit in front of the Ark and wait until he heard a voice, then he would write down what he heard and report it. The main point here is that the priests, who were also the prophets, did not report their dreams or psychic impressions. They were only supposed to report what God told them, or showed them, when they sat in front of the Ark. That is why the Hebrews made it a capital offense for anyone to profess having visions from God, but who were only lying to make themselves seem important. A discussion about what a prophecy is will be discussed in Chapter 10.

Where is the Ark and who hid it?

This has been the greatest question since the fall of the Temple over 2,594 years ago. Some have theorized it was in Egypt; other theories placed it in Babylon. Kings and countries have searched for the Ark, believing it would bring them power and glory—but no one was worthy enough in the eyes of God. What none of them understood was that no one could *own* the Ark or the Rod. One must first understand how the Universe really works in order to begin to understand what the Torah really is, and who is speaking from it. You must first understand and accept man's relationship with the Universe and the Creator, plus what Gods' relationship is with the Universe. Only after that will you have the slightest chance of discovering its secrets. I also firmly believe that it is a question of whether you are worthy, in the eyes of God, to receive such a gift.

This section presents the clues about where the Ark and other Temple furnishings are, and who hid them away. I will then cover the prophecies that predict that the Holy Ark will be found again.

Where is the Ark of the Covenant?

There are many clues about where the Ark was hidden. Primary Biblical clues are found in the Book of Ezekiel. In Chapters 8 through 11 Baruch wrote that God showed him that the City of Jerusalem would be destroyed because of idolatrous priests who were worshiping other gods. One such priest was Jaazaniah, the son of Jeremiah. Now you can understand why Jeremiah confided in Baruch, his grandson, and not in his own sons.

Who was Baruch?

First, it is essential to present you with a brief background history of Baruch so you will have a more complete understanding of how important he became to the Jewish religion. A more complete explaination is in Appendix D. His given name was Seraiah[28] and his priestly name was Baruch. He wrote under three separate prophets' names, which I will cover in Chapter 10 and Appendix D. His father was Neriah,[29] the son of Jeremiah, aka Maaseiah (also spelled Mahseiah), aka Shaphan.[30] His Uncle was also Seraiah[31] who was the High Priest at the time of the fall of the Temple. The Babylonians murdered his uncle, Seraiah, after the fall of Jerusalem. Even though his uncle was High Priest, Baruch was loyal to Jeremiah, who had been High Priest before he was removed from the office by the last Judean king, Zedekiah. The king prevented Jeremiah from accessing the Ark, so he had his servant, and grandson, Baruch sit in front of the Ark and write down whatever God spoke, and then gave those messages to Jeremiah. What occurred just before the fall of the Temple was that God would not speak to anyone other than Jeremiah or Baruch. The other priests were under intense pressure from the king to come up with a favorable prophecy and when they could not do so they would lie.[32] The following quotation from Ezekiel, Chapter 12, is the key because it directly reveals to us that Baruch was the person who moved the Ark out of the city.

> [Ezekiel 12:1] The word of the Lord also came unto me, saying: [2] "Son of man, thou dwellest in the midst of a rebellious house, that have eyes to see, and see not, they have ears to hear, and hear not; for they are a rebellious house. [3] Therefore, thou son of man, prepare *thee stuff for exile*, and *remove as though for exile by day in their sight*; and thou shalt remove from *thy place to another place in their sight*; it may be they will perceive, for they are a rebellious house. [4] And thou shalt bring forth thy stuff by day in their sight, as *stuff for exile*; and thou shalt go forth thyself at even in their sight, as when men go forth into exile. [5] Dig thou through the wall in their sight, and carry out thereby. [6] In their sight shalt *thou bear it upon thy shoulders, and carry it forth in the darkness*; thou shalt cover thy face, that thou see not the ground; for I have set thee for a sign unto the house of Israel." [7] And I did so as I was commanded: I brought forth *my stuff by day*, as stuff for exile, and in the even I digged through the wall with mine hand; I carried out in the darkness, and *bore it upon my shoulder* in their sight. [Emphasis added.]

The quote does not directly mention the Ark, but the reader must use common sense and figure out that it is the Ark and the other furniture in the Temple that is referred to as his "stuff." You are supposed to ask yourself, "Would anyone care if a low-level priest took his personal belongings out of a city that was besieged for over a year?" The answer is obvious—No. But if he were taking the Ark of the Covenant and other Temple objects out and preparing them for exile, that would signal to everyone that the city was going to fall—and that is exactly what happened. The day after Baruch and some well-chosen priests and scribes removed the Ark from the city and headed south, Zedekiah the king also left, but he went northeast towards Jericho, and was captured by the Babylonians. Baruch wrote "born it upon my shoulder" because the Ark was carried on ones' shoulders using the staves. What throws the reader off guard is that the verses sound like he escaped all by himself. He was not alone. It takes more than one person to carry all of those objects by cart or by hand.

There is a legend that offers us a clue as to where Baruch traveled after the fall of Jerusalem, and it reconfirms that he was not alone:

> At the time of the destruction of the Temple, one of the prominent figures was Baruch, the faithful attendant of Jeremiah. God commanded him to leave

the city one day before the enemy was to enter it, in order that His presence might not render it impregnable. On the following day, he and *all other pious men* having abandoned Jerusalem, he saw from a distance how the angels descended, set fire to the city walls, and concealed the sacred vessels of the Temple. At first his mourning over the misfortunes of Jerusalem and the people knew no bounds. But he was in a measure consoled at the end of a *seven days fast*, when God made known to him that the day of reckoning would come for the heathen, too. Other Divine visions were vouchsafed him. The whole future of mankind was unrolled before his eyes, especially the history of Israel, and he learned that the coming of the Messiah would put an end to all sorrow and misery, and usher in the reign of peace and joy among men. As for him, he would be removed from the earth, he was told, but not through death, and only in order to be kept safe against the coming of the end of all time.[33] [Emphasis added.]

The "seven days" are inserted into the narrative to inform us where Baruch was headed with the Ark. These are the same seven-days that Samuel took to meet Saul at "the hill of God."[34] It required four days to travel from Jerusalem to Beer-Sheva and three days to go to Mount Sinai. The reason why the legend states that they were fasting was to show that they were in a hurry to get to Mount Sinai and hide the Ark away. The allusion to "The whole future of mankind was unrolled before his eyes . . ." is a good description of his books of Ezekiel and Joel, because that is exactly what is in those books.

The next two clues he gives us are a lot more abstract, but if you remember what these two objects were you should know what he is writing about. The two objects are Aaron's staff and the High Priests' breastplate.

[Ezekiel 19:11] "And she had strong *rods* for *the scepters of them that bare rule*; And her stature was exalted among the thick branches, and she was seen in her height with the multitude of her branches. [12] But she was plucked up in fury, she was cast down to the ground, and the east wind dried up her *fruit*; Her strong rods were broken off and withered, the fire consumed them. [13] And now she is planted *in the wilderness, in a dry and thirsty ground.* [14] And fire is gone out of the *rod* of her branches, it hath devoured her *fruit*, so that there is in her no strong *rod* to be a scepter to rule." This is a lamentation, and shall be for a lamentation. [Emphasis added.]

The phrase "the scepters of them that bare rule" refers to Aaron's staff, which was used by the Judean kings and High Priests. The word "fruit" was an allusion to the story in the Torah which states that Aaron's staff bore almond blossoms. The words "in the wilderness, in a dry and thirsty ground" tell us that the staff was hidden in the wilderness of Sinai, which is a dry desert. From the *Legends of the Jews*, we have another reference to Aaron's staff:

It was this rod, which never lost its blossoms or almonds, that the Judean kings used until the time of the destruction of the Temple, when, in miraculous fashion, it disappeared.[35]

Another reference to Aaron's staff is found in Ezekiel, 21:33 and 35:

[33] O sword, O sword keen-edged, furbished for the slaughter . . .

[35] Cause it to return into its sheath! *In the place where thou wast created*, in the land of *thine origin*, Will I judge thee. [Emphasis added.]

The place of its origin, where it was created and found, was inside of the cave in Mount Sinai. Baruch wrote, between the lines, that Aaron's staff was returned to the cave in Sinai. The next quotation concerns the High Priests' breastplate, which they wore in front of the Ark. The breastplate had 12 stones imbedded in the front of it. This reference is in Chapter 28:11. The surface story appears to be a prophecy about the king of Tyre, but it is not. All of the stones mentioned are the same type that were imbedded in the Priests' breastplate. Verse 14 was the key verse. "Thou wast the far-covering cherub; and I set thee so that thou was upon the holy mountain of God;" This makes it very clear that Baruch was hinting about the breast plate the High Priests wore while in front of the Ark.

Zin, Zion, and Mount Sinai

Moses used seven names to refer to Mount Sinai. One was *"The Wilderness of Zin."* The reason he created the name Zin was that in Hebrew each letter has a meaning by itself. The word Zin meant *to hunt for God.* The hunt Moses and the Hebrews were on to find God led them directly to Mount Sinai. Therefore, the word Zin is a substitute for the word Sinai.

The word "Zion" (צִיּוֹן) first appears in First Kings 8:1 and Second Chronicles 5:2 (most likely that part was written at the same time by the same priest), where it was used to describe the name of the City of David, a hill just south of the Temple Mount in Jerusalem, after he conquered it from the Jebusites and made it his capital. The name Zion describes the same place by adding the extra יו (yud vov). We get the same meaning with just two more adjectives. The י (yod) means *"choice of direction"* and ו (vov) means *"unfolding."* So the word Zion can translate into *"personal choice to hunt for God."* The word shows up again some 220 years later, written by the prophet Isaiah, then later by Jeremiah and Baruch. The differences are they use the word to describe four separate places. A few times Isaiah used the word to describe the Temple Mount, where the Ark was located. At times he used it to describe the people, *e.g.,* "daughters of Zion." But mostly he wrote it to describe Mount Sinai, written in the following verse. I have added the bracketed [Sinai] so you should substitute the word *"Zion"* for *"Sinai,"* and you will see that Sinai fits the subject of the verse:

> [28:16] Therefore thus saith the Lord God: 'Behold, I lay in Zion [Sinai] for a foundation a stone, a tried stone, a costly corner-stone, a sure foundation; He that believeth shall not make haste.

Let's jump ahead in time to Baruch. Besides writing Ezekiel, it appears he also wrote many of the chapters of Psalms. I believe this is because he included many prophesies in Psalms, as well as including events that occurred after the fall of the Temple. Since he was the last person to have access to the Ark, Baruch must be the writer of much of Psalms. In the *Legends of the Jews*, it is similarly implied that he wrote Psalms:

> Five years after the great catastrophe, he composed a book in Babylonia, which contained penitential prayers and hymns of consolation, exhorting Israel and urging the people to return to God and His law. This book Baruch read to King Jeconiah and the whole people on a day of prayer and penitence.

> . . . Baruch sent his book also to the residents of Jerusalem, and they read it in the Temple on distinguished days, and recited the prayers it contains.[36]

Within Psalms, Baruch uses the word Zion 15 times, many of them hinting where he hid the Ark and Aaron's staff. The following are some of them:

> [9:12] Sing praise to the Lord, who dwelleth in Zion [Sinai]: Declare among the people His doings. [He could say this because he put the Ark in the cave in Sinai.]

> [14:7 & 53:7] Oh that the salvation of Israel were come out of Zion [Sinai]! When the Lord bringeth back the captivity of his people, Let Jacob rejoice, let Israel be glad. [The Ark was already in the Cave and Baruch was hoping God would end the exile soon. What is interesting is the verse is repeated twice in Psalms.]

The following are references to Aaron's staff and Moses' Rod. Moses' Rod disappeared after Moses left Mount Sinai and headed south, and Aaron's staff was stored in the Temple until its destruction when Baruch hid it.

> [23:4] Yea, though I walk through the valley of the shadow of death, I will fear no evil, For Thou art with me; *Thy Rod and thy staff*, they comfort me. . . [The staff is Aaron's staff and the Rod is Moses' rod given to him by God.]

> [110:2] The *Rod* of thy strength the Lord will send out of Zion [Sinai]: 'Rule thou in the midst of thine enemies.' [This Psalm is a prophesy indicating that

Moses' Rod will make a new appearance in the future and also indicating that the Rod is currently in the cave.] [Emphasis added.]

Baruch used the phrase, "I walk through the valley of the shadow of death" because you should visualize that he was walking through a long dark cave, dimly lit by oil lamps, and the graves of his entire family, from Abraham to the eleven sons and two wives of Jacob are on both sides of him. The following legend also tells about Aaron's staff and predicts that it will be found again.

> Aaron's rod was then laid up before the Holy Ark by Moses. It was this rod, which never lost its blossoms or almonds, that the Judean kings used until the time of the destruction of the Temple, when, in miraculous fashion, it disappeared. Elijah will in the future fetch it forth and hand it over to the Messiah.[37]

Following are additional legends, which write about the whereabouts of the Ark:

> Now God commanded Kenaz to deposit twelve stones in the Holy Ark, and there they were to remain until such time as Solomon should build the Temple, and attach them to the Cherubim. Furthermore, this Divine communication was made to Kenaz: "And it shall come to pass, when the sin of the children of men shall have been completed by defiling My Temple, the Temple they themselves shall build, that I will take these stones, together with the tables of the law, and put them *in the place whence they were removed of old*, and there they shall remain until the *end of all time*, when *I will visit the inhabitants of the earth*. Then *I will take them up*, and they shall be an everlasting light to those who love Me and keep My commandments.[38] [Emphasis added.]

The key sentence from the above quotation is "in the place whence they were removed of old," which refers directly to the two stone tablets, the twelve stones on the Priests' breastplate, and the Ark components which came from within the Cave at Mount Sinai. When he wrote, "*I will take them up,*" he means they will be discovered in the future and taken out of the cave once again. The legend further predicts that the Ark will be found again in the End of Days.

The next legend begins by writing about the candlestick, but it also revealed some of the other holy objects concealed by Baruch.

> On account of its sacredness the candlestick was one of the five sacred objects that God concealed at the destruction of the Temple by Nebuchadnezzar, and that *He will restore* when in His loving-kindness He will erect His house and Temple. These sacred objects are: the Ark, the candlestick, the fire of the altar, the Holy Spirit of prophecy, and the Cherubim.[39] [Emphasis added.]

This legend predicts the Ark, the candlestick, and the Cherubim (top part of the Ark) are the physical objects which will be restored by God in the future.

What is also vital to note, the writer linked the Ark with the ability to restore the "spirit of prophecy," and that was because the Priest had to sit in front of the Ark and write down whatever God "spoke." Without the Ark, no prophecy!

The next two legends mention that Jeremiah and Baruch were the priests who were responsible for hiding the Ark. The only error the legend makes is that Jeremiah was in prison at the time of the fall of Jerusalem. Since Baruch was his servant, perhaps the author gave the ultimate responsibility and credit for the Ark's safety to Jeremiah. I would disagree. I think that God commanded Baruch directly, just as his book of Ezekiel tells us. I do think that Jeremiah must have told Baruch where Mount Sinai was and where the Cave entrance was located because that information was only handed down from father to son, who would then become the next High Priest. Jeremiah did not trust his own sons so he passed the responsibility to his grandson, Baruch.

> The task laid upon Jeremiah had been twofold. Besides giving him charge over the people in the land of their exile, God had entrusted to him the care of the sanctuary [the cave] and all it contained. The *Holy Ark*, the *altar of incense*, and the *holy tent* were carried by an angel to the mount whence Moses before his death had viewed the land divinely assigned to Israel. There Jeremiah found a *spacious cave*, in which he concealed these sacred utensils. Some of his companions had gone with him to note the way to the cave, but yet they could not find it. When Jeremiah heard of their purpose, he censured them, for it was the wish of God that the place of hiding should remain a secret until the redemption, and then God Himself will make the hidden things visible. [Emphasis added.]

> . . . The sacred musical instruments were taken charge of and hidden by Baruch and Zedekiah until the advent of the [someone in the future], who will reveal all treasures. In his time a stream will break forth from under the place of the Holy of Holies, and flow through the lands to the Euphrates, and, as it flows, it will uncover all the treasures buried in the earth.[40]

This was the first legend which directly states that the Ark was concealed inside a cave. It also correctly mentioned that Baruch took part in the concealment of the Temple vessels. It referred to "all the treasures buried in the earth" which include Joseph's stolen riches concealed within the cave system. And again, the legend predicted that God would reveal the location of the cave and its contents at the appointed time.

I found a reference in a legend that spoke of Baruch's visit to Mount Sinai, but the writer does not say Sinai directly. He called the place "Paradise," which, of course, is Mount Sinai, as explained earlier. Note that the number seven is used as a clue to the real location the writer was describing:

Baruch is one of the few mortals who have been privileged to visit Paradise and know its secrets. An angel of the Lord appeared to him while he was lamenting over the destruction of Jerusalem and took him to the *seven* heavens, to the place of judgment where the doom of the godless is pronounced, and to the abodes of the blessed.[41] [Emphasis added.]

Chapter Conclusion

The Ark is not the figment of the priests' imaginations, nor is it just a cold piece of gold statuary. It is a working electronic device, portions of it built by the Hebrews, and parts borrowed from advanced technology found in the cave, which contained the artifacts of a very advanced civilization.

The Hebrew Scriptures and the legends of the Jews, all predict the Ark of the Covenant will be found again in the "End of Days." When it is found will anyone believe that it is real? If someone gets it to work, and God speaks through it, will anyone believe that it is God speaking? Will the atheists in and out of academia merely ridicule the Ark as an electronic trick, created by religious Jews to rejuvenate their faith and increase their numbers? I am sure a majority of anthropologists and archaeologists will insist that they should have the Ark so it "can be studied in a *controlled* environment," which means *they* control it and no one but them would have access to it.[42] I am sure government would love that solution also. Fortunately for the human race, they will not have their way. The Ark is not intended for a museum sideshow.

I have asked the following question of many people over the years: "How would it affect your life if someone found the Ark of the Covenant and got it to work and then God spoke?" The answers I received range from, "It wouldn't have any affect on my life, but I think it would be interesting," to, "It would be very scary to me and I wouldn't know what to think." Over the years I have given considerable thought to the very real possibility that the Ark will be rediscovered and made to work. I believe that when God speaks to us, it will be very frightening for everyone, and that no one will be able to ignore what He has to say. He is, after all, the Operating System of the Universe, and He does not like to be ignored or have people think He does *not* exist. When this event happens, it will be such a huge wakeup call, that it will change mankind forever! At first there will be great religious conflicts over what God is, but once that is settled, there will most likely be only *one* religion on the planet and everything else will be relegated to the dustbin of history.

Was Abraham directed by God to find this cave? I would say yes! Did God choose the descendants of Jacob to be His people? Again, I would say yes! Should we criticize God's judgment why He chose the Hebrews? No—it is not our function or place to criticize why God does what He does! But that is not to say that we should not try to understand why He chose the Hebrew tribes. The big question is: "What does God want from His relationship with man? What are His long-term goals for us, and what are we supposed to evolve to? These weighty questions I will attempt to answer in Chapter 10 and the conclusion of this book.

Chapter 8:
The Polar Reversal and the Ice Age

Chapter 3 presented a science philosophy, which explains that the Universe is the product of information and behaves like a synchronistic system—a computer. I have named this system The Diehold. Chapter 4 reveals that the Hebrew alphabet was the creation of a highly advanced previous civilization, which had the same philosophy of the Universe that I developed in my fist book, *Reality Revealed: The Theory of Multidimensional Reality.* Similar to all computers, the Diehold utilizes clocking, synchronizing, and resynchronizing frequencies. This chapter will reveal the most important of these cycles—and what its affects are on the planets and the rest of the Universe.

I believe the main clock cycle is what we call the polar reversal, which occurs every 12,068 years. This is the same number that I discovered encoded within the Hebrew Scriptures. The clock cycle causes the polar reversal and the subsequent ice age. This event is the most important event in human history. It is without question the biggest story in history. Some may respond to this statement by saying that receiving the Torah or the whole Bible was the greatest event in human history. But remember, in Chapter 2 clearly I showed that the Torah and some other Hebrew books had the 12,068-year number repeatedly imbedded within its pages. You will clearly see in Chapter 10, that one of the main purposes of the Torah and the Prophets was to save mankind from this terrible event. God wants mankind to survive it—and that is why He left us these clues within the Hebrew Scriptures and no other place!

Because the subject of this Chapter is so important, I will describe every event, which occurs on the Earth and Moon, during and after the polar reversal. The Chapter is heavily footnoted to allow academia the opportunity to examine my research conclusions and to decide for themselves. Hopefully they will put aside their pre-notions, biases and PhD. thesis and examine the subject with an open mind.

The Basics

Geologists divide great expanses of time into four general eras. Each era is in turn, divided into sub-periods. Table 8-1 describes these eras and periods, and the estimated years in each period.

The current assumption is that there were three previous ice age periods. Each period will include multiple polar reversals and ice ages. The most recent occurred during the Pleistocene period starting about 1,800,00 years ago and ending 11,000 years ago. Geologists can identify at least seven major ice field advances within the Pleistocene period, in addition to some minor ones, including some intermittent warmer periods.[1,2] The second ice age period is thought to have happened during the beginning of the Permian period 275 million years ago. The oldest is thought to have occurred at the end of the Precambrian period, 600 million years ago, occurring in Africa, India and Australia. I have doubts about their methods of determining and detecting the two older ice ages. I do not believe it is possible for the Earth to have an ice age that long ago, which I will explain later in the Chapter.

ERA	Approx. Age (millions of years)	Periods	Approx. Age in Millions of years	Sub-Periods
Cenozoic	11,000 yrs. to present	Holocene		
	11,000 to 1.8 million yrs.	Pleistocene	(man)	
	1.8 to 65 million	Tertiary	1.8 to 13 million yrs.	Pliocene
			13 ot 25	Miocene
			25 to 36	Oligocene
			36 to 58	Eocene
		(Mammals)	58 to 65	Paleocene
Mesozoic	65 to 135	Cretaceous	(flying reptiles and first birds)	
	135 to 181	Jurassic	(Dinosaurs)	
	181 to 220	Triassic	(First reptiles)	
Paleozoic	220 to 280	Permian		
	280 to 345	Carboniferous		Pennsylvanian
				Mississippian
	345 to 405	Devonian	(first amphibians)	
	405 to 425	Silurian	(first land plant fossils and insects)	
	425 to 500	Ordovician	(first fish)	
	500 to 570	Cambrian		
Precambrian	570 to 700	(First multi-celled organisms)		
	700 to 3400	(First one-celled organism)		
	4000	Approx. age of oldest rocks discovered		

Table 8-1: Geological eras.

Other Theories of the Ice Ages

The great insurmountable mystery for the science community has been: What mechanism causes ice ages? Why does a polar reversal immediately precede an ice age? How do they relate? What caused the mass extinction of species and the creation of similar new species? What caused an ice field 4,000 to 10,000 feet thick in less then 40 years? What caused the increase in earthquakes and volcanic activity at the time of the polar reversal? What formed the deep-sea canyons off of every large river around the world? Why are isotopes of heavy metals found in sediments associated with the ice ages and polar reversals? Finally, what caused the instant freezing of the mammoths in Siberia, northeastern Russia?

I will now briefly cover some of the competing theories about the ice ages. You will soon discover that they do not answer all of the phenomena associated with the polar reversals and ice ages. They only explain some of the observations but not all, using one coherent theoretical model.

In Chapter 3, I offerd an analogy of a 10,000-piece puzzle without the benefit of a finished picture that explains where the pieces fit together. A philosophy is like the finished picture. Without it you can spend a great deal of time trying to figure out where the individual pieces fit together. The standard philosophy that the field of geology has embraced is call the uniformitarian process, which calls for gradual processes over long periods of time. Only since the 1970s have they embraced the idea a meteor or comet hitting the Earth and changing the environment quickly.[3] This is a step in the right direction but still the underlining philosophy persists.

I have read most of the issues of the journals *Science, Nature,* and *Geology,* going back to 1958, and as I read them it occurred to me that those intelligent scientists were suffering from the lack of an accurate picture of how all these pieces/phenomena fit together. The *Theory of Multidimensional Reality* explains all of the mysteries associated directly with polar reversals and the ice ages. It even explains discoveries found on the Moon and Mars. It is my conclusion that unless you look at existence totally differently you cannot possibly figure out all the events that surround the polar reversal and ice age. Unfortunately, we as a society, are running out of time and we better agree on a philosophy very soon, that explains it or else this event will finish us off.

Competing Theories

The competing theories fit into four broad categories. They are: 1. The Earth received less energy output from the Sun, caused by a variety of reasons. 2. Continental drift causing land masses to move to colder latitudes. 3. The Earth's rotational poles changing, placing the north and south poles in different locations. 4. Changes in the ocean's circulation. I will cover some of the more popular theories and show where they fail to explain all of the observations known about ice ages and polar reversals.

The Earth received less energy output from the Sun

Some scientists theorize the Earth was hit by a comet or meteor 65 million years ago, and that finished off the dinosaurs by throwing vast quantities of dirt and dust into the atmosphere, thereby reducing the amount of sunlight reaching the earth. They cannot explain why these extinctions occurred cyclically throughout time and were always associated with polar reversals. Their theories cannot explain the creation of new species, based on previous ones, because only cosmic rays, gamma rays and ultraviolet light[4] can alter genes to create new species—and something hitting the Earth will not. It also does not explain the increase in volcanic activity, and why radioactive elements accumulated at the same time.

Another theory is referred to as the Astronomical Theory of Glaciation,[5,6] which states that the Earth and Sun passes through dense interstellar clouds that cause the Sun to increase its output, resulting in greater precipitation and ice accumulation.[7] The problems with this theory is that no spiral arms of dense dust have yet been detected in our Galaxy.[8] This theory also has the same problems of the previous theory, regarding a polar reversal occurring immediately before an ice age and the other associated facts.

Dr. Wallace Broecker advanced the theory that the colder climatic cycles are related to the variations in the Earth's tilt and precession, which occurs every 48,000 years.[9] Others have found strong peaks at ~100,000, 43,000 and 24,000 years.[10,11] By coincidence, this is a function of the 12,068-year cycle, which I discovered. Again his theory does not try to explain the sedimentary deposits on the Earth and the polar reversals. The cause-and-affect for the huge ice fields were avoided.

A theory similar to the previous one involves the Earth's poles wandering and an open Arctic Ocean.[12] This theory suffers from the same deficiencies as the previous ones.

The ocean current conveyor belt theory presents a model that states if the warm ocean current changed because too much fresh water entered the Atlantic, and shutdown the conveyor belt current, it would prevent warm tropical water from flowing to northern Europe. The same theory holds for the Indian and Pacific oceans but a different path for the conveyor belt in each ocean.[13] Again it cannot explain the polar reversal and creation of new species.

Causes of the Ice Ages and their dating

Approximately 12,000 years ago, there was an ice field 4,000 to 10,000 feet thick, from no less than 40-degrees latitude north,[14] and in the southern hemisphere from 40-degrees latitude south. The ice did not abruptly stop there. There are records of ice fields and thick snow all the way down to Mexico City![15] We know from studies of sediment, from all over the world, that the ice fields were deposited fast, within 300 years, ±200.[16] There is abundant sedimentary evidence from the Great Lakes areas of North America that shows the ice field there was deposited in less than 40 years. The mechanism I will describe will show that the snow and resulting ice field occurred within 11 years, and most likely covered most of the Earth from about 10 degrees north and south of the equator.

The sequence of events that created the Ice Age

I will divide the ice age and polar reversal into three time periods, the first being 50 years leading up to the polar reversal and the ice age. The second, the actual polar reversal, ice accumulation and other events that occurred within 11 years of the reversal and the last period being the aftermath. I will first present my theoretical model within this framework then I will present the physical evidence in support of the model.

50 Years before

The polar reversal is caused by the main clock cycle in the Diehold, crossing the x-axis (as depicted in Graph 3-17) occurring every 12,068 years. A complete cycle would represent two polar reversals, or 24,136 years.

The magnetic field of the Earth will have already started to decay within 50 years of the actual polar reversal, but the decay will start to exponentially decay as we get within 30 years of the Reversal. The magnetic field does not have to go to zero before it snaps to a reversed polarity. The matter part of the information, that makes up the Earth, acts like the secondary coil in a magneto, holding up the field until it snaps and create a big spike of energy, thereby putting energy back into our Universe. Paleomagnetic studies of sedimentary cores indicate that the magnetic field may go to about 15,000 gammas and then snap.[17] As the magnetic field decays, it will create increased potential in the core of the Earth, in the form of increased heat.[18] The additional heat will rise towards the surface and manifest itself in the form of increased volcanoes and earthquakes. The increase in earthquakes is a result of the continental plates sliding against each other more easily. The increased heat of the Earth's mantel, upon which the plates "float," become more lubricated, enabling them to move or fracture more easily.

The Earth's rotation will start slowing down in this period of time, resulting in the necessity to add seconds periodically to our clocks. Weeks, and maybe months, before the reversal the Earth's rotation will slow down markedly, maybe resulting in 28-hour days.

The Sun will also be affected by the collapsing magnetic field that makes it up. The Sun's output will start increasing over 140 years before the final reversal. The increased solar output will start heating up the surface of the Earth. Sea surface temperatures will rise during sunspot cycles, successively increasing, as we get closer to the reversal. The polar ice caps and glaciers will start melting, resulting in ocean levels rising. Before the polar reversal, most of the ice caps and glaciers may have melted. The increased sea surface temperature will create more violent and frequent storms worldwide. The increase in solar output will create additional levels of ultraviolet light hitting the Earth. This will deplete the ozone layers in the upper atmosphere during sunspot periods. It may also cause genetic mutations of amphibians, which lay clear eggs in water.

The Polar Reversal

The actual polar reversal occurs in one day. There is a complex series of events that will happen on that day. I will start with what happens to the Sun.

The collapsing magnetic field deep inside the Sun will create a large spike of energy that will cause the matter and dust shell on the surface, to expand very rapidly. I estimate a speed of 1,550 miles per second.[19] My best estimate for the energy output during the nova is no less than 2,000 times the normal energy output. It would not surprise me if it rose up to 50,000 times normal output. The north and south poles of the Sun will blow outward, as seen in some planetary nebula (Figure 9-1, Ant Nebula). The equatorial region of the Sun will blow outward along the planetary plane, hitting each planet as the dust shell expands rapidly. As the hot dust/matter shell hits each successive planet, it will push the planets a little further away from the Sun. After the reversal the planets will then receive a little less energy from the Sun. They will also loose some of their atmosphere and liquids on the surface. Eventually, the dust/matter shell will loose enough momentum and will stall out somewhere past Jupiter.

After the Sun's matter shell has been expelled, we will see the center and the real heat source of the Sun. A center modulation point where all the information that makes up the Sun is directed to a very bright point, giving off mostly ultraviolet light but not much radiant heat. The Sun will remain that way until the matter shell can form again, and that may take a number of sunspot cycles.

At the exact time of the polar reversal the Earth will stop its rotation and remain still for seven to eight hours. The forests and buildings on the Sun-side of the Earth will combust if there isn't flooding in the area. Not only will heat affect plants and animals, but the Sun will produce a massive dose of cosmic and gamma rays, which will reach the Earth within 10 to 15 minutes after the reversal. This cosmic particle pulse may last 10 to 30 seconds[20] with the potential to alter the genes of both plants and animals, including humans.[21,22]

The people on the Sun-side of the Earth will be able to see the Sun expanding, and the solar disk getting larger as it expands towards the Earth. I estimate the dust/matter shell will take between 17 and 18 hours to hit the Earth. The dust shell may not necessarily hit the side of the Earth that faced the Sun, at the moment of the polar reversal, because 18 hours may have passed and another side of the Earth would be facing the Sun. When the dust shell hits us, it will deposit vast quantities of dust and rock on one-half of our Earth. It will also evaporate at least 1,200 feet of ocean water, worldwide. Some of this water will be lost into space, carried away by the expanding dust shell. Some of it will appear to us as returning comets. Most of the water will remain in the atmosphere as superheated water for several days to a week. If you could see the Earth from space, you would see it with a thick tail, pointing away from the Sun. It will be made up of dust and ice crystals. Eventually the debris tail will disappear over several hundred years, as the Earth's gravity brings the material back into the upper atmosphere.

Just after the solar dust/particle shell passes the Earth, the Sun-side will have extremely low atmospheric pressure because the dust shell would have blown some of the atmosphere away, we just do not know how much. The backside of the Earth will temporarily have normal atmospheric pressure, but that condition will not last long. Very shortly after the dust shell passes by, the normal atmosphere on the backside, will expand very rapidly to fill up the front side (Sun-side) of the Earth. Two things will result from this process: The first will be extremely high-speed winds traveling around the Earth from all four corners of the globe, to fill up the extremely low pressure Sun-side of the Earth. The second consequence of this process is revealed in Boyle's law: *If a given weight of gas is considered and if its temperature is held constant, the pressure and volume of the gas will be inversely proportional.*[23] Applying this law to the conditions that will be present on the backside of the Earth, the atmosphere will expand very fast and, therefore, the temperature will drop to extreme levels, possibly below -170 degrees Fahrenheit below zero.[24] Any life forms caught out in the open, or even in the average building will be fast frozen almost instantly!

When the Earth stops its rotation, all the water in the oceans and lakes will continue to travel in the former easterly direction at the Earths former rotational velocity. It is like walking at 5 miles-per-hour (mph) with a pan of water and you stop abruptly. The water keeps on going at 5 mph. Just before the reversal, the Earth may only be rotating at 800 mph, instead of the normal 1,000 mph (at the equator), because the rotational rate will have slowed down considerably. I would expect the wave-speed to slow down considerably after the first hour on land. Many factors will determine the speed of the wave, such as land elevation, forest ground cover, gravity and the heat blast from the Sun. Depending on how long the Earth remains at rest will determine how far this immense Ocean wave will travel across the continents. When the Earth resumes rotation it will be in the reverse direction than previous. This action will make the tidal wave appear to once again speed up, but really its because the Earth would be now rotating in the opposite direction of the wave. I believe it is not unreasonable to assume the speed of this massive tidal wave would average about 350-miles-per-hour. Within a week, this massive ocean wave will have returned to the ocean basins. As it returns, it will naturally follow the river basins leading to the sea, cutting deeper into existing deep-sea canyons. Some deep-sea canyons currently cut down to over 11,000 feet below sea level.

The rotation of the Earth should resume within seven or eight hours after the reversal point, but in the opposite direction. Currently the Earth rotates from west to east. After the reversal it will rotate from east to west, so the Sun will rise in the west and set in the east. Not only are the oceans affected by the polar reversal, but also the crust of the Earth. Just as the Oceans "swim" on the crust of the Earth, the crust of the Earth "swims" on the heated-up mantle. The continental plates are also traveling at the Earth's previous rotational velocity at the time of the reversal. They will start banging up against each other thereby causing massive earthquakes lasting for weeks of constant shaking (over eight on the Richter scale) until the plates stabilize. Mountains will be pushed up and some may sink.

Volcanoes will erupt worldwide because of the increased heat in the Earth's core and mantle, and the pressure put upon the magma chambers in the crust. The dust from many erupting volcanoes will further add particles in the atmosphere, which will help seed the dense clouds already present as a result of the massive evaporation from the nova. The lava and mudflows from these eruptions will also further reshape the surface of the Earth.

Within a day of the nova and polar reversal, it will start raining scalding hot rain, which will turn to downpours of cooler and finally cold rain. Within 8 days it should start snowing worldwide because the Sun will not be giving off enough visible light and heat to warm the Earth. It will continue to snow until all the clouds and moisture are out of the atmosphere. This process may take 11 or more years. When it is done snowing the Earth will be mostly covered with snow and thick ice in the higher latitudes and or elevations. We will then in the grips of a full ice age!

After the Earth is hit by the dust shell, it will be pushed a little farther away from the Sun. Instead of being 93 million miles away from the Sun we may be 93.2 million miles away—or more.

The affects of the polar reversal on all forms of life, will be devastating. It is not difficult to understand why there is always a mass extinction of all types of species at the time of the polar reversal. Our job is to survive it, because the alternative is too depressing to think about. You now also understand the importance of what I discovered, and why God put the clues of this event within the Torah.

The Moon

The surface on the Moon will receive a dusting of particles, rocks and moltan glass from the Sun, in addition to the accumulation it has received from previous novas. There will be additional small craters from the larger rocks expelled from the Sun. Immediately after the nova, the Moon will appear red because the nova will turn the surface red-hot.

The Aftermath

Most, if not all of the Earth will get snowed on, some areas much less than others. The Earth will be in the grips of a full ice age. It should be colder than the last ice age because we will be a little farther away from the Sun. Dense clouds will cover most of the Earth for 20 or more years until all the moisture has been rung out of it. The Glaciers will block rivers and streams and create ice dams that will be present for hundreds of years. When the dams finally collapse they will cause huge floods on the river they dammed up. Eleven years after the polar reversal most of the snow and ice will have melted at the lower elevations near the equator (10-15° North and South). Animal life will start to come back in these areas. Small plants and trees will start to again appear. The oceans of the Earth will be over 700 feet below current sea level. Glaciers will be present in both Atlantic and Pacific oceans and extend down to 40° Latitude North and South. The occurrence of frequent major earthquakes should continue for 50 years after the polar reversal until the outer mantel of the Earth cools down and the plats stabilize. What will happen to Man? I do not know yet. It depends on whether anyone reads this book and believes me. Otherwise I don't know if anyone will make it. Not too many people made it last time and they had Noah to help them.

The rest of the Chapter will present most of the evidence that proves this model correct.

Problems with Carbon-14 dating and other forms of radioactive elements used for dating

The most popular method of dating artifacts and sediments is detecting the amount of radioactive carbon-14 (^{14}C) remaining in organic material. This isotope of carbon is believed to form in the upper atmosphere from the interaction of cosmic rays with nitrogen. Cosmic rays are nuclei from hydrogen and helium atoms, minus their electron. It is believed they come from stellar novas in our Galaxy and the Sun. The cosmic rays, moving at nearly the speed of light, hits a nitrogen atom and knocks a proton out, converting the nitrogen to carbon-14, which has a half-life of about 5,660 years. The older detection methods had an upper limit of 40,000 years, but newer methods extended it to 75,000 years.[25]

The main problem with this method of dating is that it makes the assumption that the creation and level of cosmic rays, from space and our Sun, have been uniform throughout time. The problem science currently has is that they do not recognize the possibility that our Sun novas periodically and gives off a dramatic increase of all types of highly energized particles, resulting in a wide variety of other radioactive elements such as Aluminum-26,[26,27] Berylium-10, Oxygen-18 & 16,[28] Magnesium-26,[29] Iron-60,[30] Thorium-230, etc. Some scientists have detected an increase of 50% to 80% of radioactive elements, 11,000 plus years ago on Kodiak Island.[31]

Two scientists from the University of Washington discovered a direct correlation between Carbon-14 production and the variability of our Sun during the Sunspot cycles. The ^{14}C levels were determined by examining tree rings in Douglas fir trees located in the Pacific Northwest.[32]

The Earth's magnetic field intensity is also an important factor in the amount of ^{14}C produced in the upper atmosphere. A decrease in the magnetic field will allow more cosmic rays to reach the atmosphere, therefore increasing the production of ^{14}C.[33,34,35] The inverse is true if the Earth's magnetic field increases. That means during the polar reversal, when the Earth's magnetic field goes to zero for about eight hours, there will be a large spike of cosmic rays from the Sun that will reach one side of the Earth. That means, in order to correctly calibrate a ^{14}C value, you must know which side of the Earth was facing the Sun during the last polar reversal and then assume a different side of the Earth was facing the Sun the previous reversal. I believe during the last reversal, India and China was facing the Sun when the Earth was hit by the solar cosmic rays at the exact time of the polar reversal. The ^{14}C values on that side of the Earth would appear younger than they

Depth (in m) and dates in yr B.P.	General lithology	Sediment description
		Cultivated horizon
2580 ± 60		Yellow silty clay with thin peat layer (2 - 3 cm).
10 900 ± 60 — 5		Brownish-yellow homogeneous and organic rich silty clay, with abundant Fe and Mn oxides and root traces. Silt lenses are present.
13 300 ± 60		
11 150 ± 45 6850 ± 80 11 460 ± 60		Gray homogeneous and organic rich silty clay. Some silt lenses observed. A few root traces and calcareous nodules are present.
— 10 10 200 ± 50		Light gray clayey silt and clay. Shell fragments (mostly gastropod) are rare. Root traces present.
7750 ± 50		Yellowish-gray clayey silt, with interbedded thin peat layers (1 - 3 cm). Root traces present.
7820 ± 35 7900 ± 35		Brown mud with intraclasts.
13 070 ± 60 — 15		Dark green stiff mud. Fe and Mn oxides and calcareous and phosphate nodules occur throughout. Ca-cemented tree roots (3 x 8 cm) are noted.

(Holocene / Pleistocene indicated along left margin)

Table 8-2: Carbon-14 dating of a core from south of Shanghai in eastern Yangtze delta, China.

really are and should be adjusted upwards in age. Caution should be taken when selecting sediments for ^{14}C analysis because the sediments laid down just before the reversal will have correct ^{14}C dating results, but any organic material laid down after, for an indeterminate number of years, will be affected by this spike of cosmic rays. For example Dr. Minze Stuiver of the University of Washington had tested peat bog from Whidbey Island. A sample from the top ^{14}C dated to 43,900 ± 1,000 years but a sample only 10 cm below it dated to 43,600 ±1,000 years, which is impossible. It is logical to assume that the lower sample was exposed to a higher concentration of ^{14}C. Another more recent example is displayed in Table 8-2 showing dated core samples from south of Shanghai in eastern Yangtze delta, China. Notice there is a 7,900 B.P. sample just above a 13,070 sample. But above these three younger samples are samples with dates of 10,200 B.P. to 13,300 B.P. It is impossible to have younger sediments below much older sediments, especially in a river delta. 12,068 years ago China would have received a high dose of ^{14}C at the time of the polar reversal so that is why these samples appear younger than they actually are.

The Evidence

The Ice Fields

I will begin by describing actual ice fields, which is what people think of when one discusses the ice age. The most recent ice age period is called the Pleistocene extending from 11,000 years before present (B.P.) to about 1,800,000 years B.P.

Most people who are asked what causes the ice age immediately think that there must first be cold. Not true—the cold comes later. First, you have to ask yourself: "How do you get an ice field 4,000 to 10,000 feet thick, covering approximately 30% of the Earth's surface? The answer—there must be precipitation in the form of snow, which means you must have clouds. Clouds form from condensed water vapor in the atmosphere. The water vapor comes from heat, applied to the oceans and lakes of the world, which is released into the atmosphere. So, in order to have an ice field 4,000 feet thick, you must first have a great deal of heat, applied quickly, to the surface of the ocean to cause fast and massive evaporation. The reason why there must be heat first is because; with a cold environment you get almost no evaporation. It is a simple formula, in order to have a given number of feet of ice, you must have an equal amount of heat to cause enough evaporation to change liquid water into water vapor that forms clouds which deposit snow on the Earth, which forms the ice sheets. The only heat source for the Earth's surface is the Sun. Heat is measured in units called calories.[36] If you analyze the math (Table 8-3), to calculate how much ocean water was evaporated to create an ice field that would cover 30% of the Earth's land surface, you would get about 400+ feet of ocean water. I came up with this depth very simply: The continental shelf off every continent goes down to a depth of 350 to 400 feet below the current sea level. Mature tree trunks have been found off the coast of the United States at these depths. The ^{14}C dating of these tree trunks was a little over 11,000 B.P. That means the ocean level was over 350 feet below the current level long enough, so the tree seeds were able to germinate and grow into a mature, full-grown tree, before it died from salt-water incursion. Where did the missing water go? It was held in the massive ice fields found all over the world. So it is safe to say that at least 350 feet of ocean water was evaporated to create the glaciers.

The amount of heat applied to every square foot of ocean water, to evaporate a column of water 350- cubic feet, is over 6.131 billion calories! Nothing but a stellar nova could cause such immense heat. After the Sun Nova's some of the evaporated water escapes the Earth and goes into space. I estimate that at least a third of the water comes back down on the Earth in the form of rain, before the temperature of the Earth cools down below freezing and it begins to snow. This process may take less than one week. It does not stop snowing until all the moisture is out of the atmosphere. It would not surprise me to discover that the initial Nova evaporated 1,200 feet of ocean water. I will examine sea levels a little later.

1 gal =	3,782.0 grams
1 cubic foot of water =	28.3 Liters
28.3 Liters =	28,300 grams
to evaporate 1 gm of water from 50°F/10°C degrees to boiling point =	619 calories
To evaporate 1 cubic foot of water =	17,517,700 calories
To evaporate 350-cubic feet of water =	6,131,195,000 calories

Table 8-3: Temperature it takes to evaporate 350 cubic feet of fresh water. It would take many more calories using sea water because of the sodium chloride content and other minerals in the water.

The next question is: "How much snow does it take to create an ice field 4,000 feet thick?" That number is found using another formula: Snow has a density of 50 to 300 kilograms per cubic meter, depending on how cold it is and the atmospheric humidity. Glacial ice is about 850 to 920 kilograms per cubic meter. If you estimate snow with an average weight of 175 kilograms, you end up with a 4.86 to one ratio of snow to ice. That means you have to have a snowfall of no less than 19,400 feet to end up with an ice field 4,000 feet thick. At this point you have to realize that this is not a normal snowfall. Something very special is happening here. You should also realize that the snow did not stop at 40-degrees latitude North and South. It must have snowed a lot farther south. There may not be much evidence of its presence after it melted. For instance, if a glacier is only 500 to 1,000 feet thick, and it forms on flat land, there is no reason for it to move. A glacier leaves very little evidence behind if it does not move. After melting it would leave only compressed clays, sand,[37] and soils.[38]

Dating Ice Fields

The Pleistocene cold period is divided into more than seven major ice age periods and they are divided into sub-ice ages. Dating the past ice fields is not that difficult. It just takes lots of digging and a sense of where to look. Geologists look for organic material under different types of sediment unique to ice fields such as glacial till, marine and sandy blue clays. The most common datable materials are woody peat, tree limbs, bones and seashells. Carbon-14 methods have indicated that ice fields have advanced and receded repeatedly during the Pleistocene period. Table 8-4 lists the radiocarbon dating (^{14}C) of organic materials found in sediments above and below the glacial till (sediment) or ice age clay sediment. The glacial till is usually a mixture of rocks and gravel, with little or no organic material. Generally, this method of dating is effective for periods of time less than 75,000 years. Short-term magnetic reversals are difficult to detect, because the magnetic field decays before the reversal, so the magnetic field is not very strong.[39] All magnetic polar reversals have preceded ice age sediments, and the reversals have been of nearly equal length.[40]

Peat bogs found five feet down on the Queen Charlotte Islands in British Columbia, Canada demonstrate the periodicity of the polar reversal. They have shown that the area was ice-free by 11,100 ± 90 years before present (B.P.) and 15,000 years B.P., with an ice field detected between the two periods. On nearby Graham Island, peat was dated to 12,400 years B.P.[41] In the Canadian high Arctic on Northeast Elsmere Island, just 13 miles from Greenland. Two shell layers under a glacial moraine were ^{14}C dated between 14,360 ±1,120 years B.P. and 23,300 ± 310 years B.P. That means the Robeson Channel, which separates the two countries, was ice-free between two known ice advances, over 12,000 and 24,000 years ago. Scientists estimated the ice sheet was 9,700 feet thick there during the last ice age.[42] In the State of Ohio ample evidence of multiple glacial advances are present, such as peat bogs found beneath two glacial tills. One was ^{14}C dated to 46,000 ± 2,000 years B.P. and the other between 30,000 to 39,000 years.[43] By my theory, there should have been polar reversals every 12,068 years—so the Ohio evidence shows that ice ages happened 48,272 and 36,204 years ago.

12,000 Years Before Present (B.P.)

Sample	Location	Years BP	Author	Journal	Vol.	Pg.	Date
Woody Peat Bed	Queen Charlotte Island	12,400	B. Warner et al.	Science	218	677	11/12/1982
Logs & Peat	East Otomona Valley, Irian	11,820±150	J. Peterson & G. Hope	Nature	240	37	11/3/1972
Shells	Kennedy Channel, Canada	14,300±1,200	J. England & R. Bradley	Science	200	265	4/21/1978
Peat	Lago Rupanco, Chile	12,200±400	J. Mercer el. al.	Science	182	1017	12/7/1973
Organic materal	Lapari Island, Italy	12,920	G. Bigazzi et. al.	Nature	242	322	3/30/1973
Carbonized plants	Santorini Is, Greece	12,950±756	H. Pichler	Nature	262	373	7/29/1976
Shell layer	W. Gezira, Sudan	11,300±400	M. Williams	Nature	211	270	7/16/1966
Mollusks	White Nile flood waters	11,950±550	D. Adamson	Nature	288	54	11/6/1980
Spruce Pollen	Rosebud South Dakota	12,580±160	P. Wells	Science	167	1574	3/20/1970
Organic matter	Ibyuk Pingo, Canada	12,000±300	J. Mackay	Science	176	1321	6/23/1972
Shells	Orcas Island, WA	12,350±400	D. Easterbrook	Science	152	766	5/6/1966

24,100 Years Before Present (BP)

Sample	Location	Years BP	Author	Journal	Vol.	Pg.	Date
Glaciation	Lake Weichselian, Norway	28,000	E. Pool	Nature	221	507	7/30/1966
Shells	Kennedy Channel, Canada	23,300±310	J. England & R. Bradley	Science	200	265	4/21/1978
Obsidian[1]	Lapari Island, Italy	21,000±4,000	G. Bigazzi et. al.	Nature	242	322	3/30/1973
Mammoth bone	Old Crow River, Yukon	25,750±1,500	W. Irving	Science	179	336	1/26/1973
Organic detritus	Stokes Pt., Arctic, Can.	22,400±240	J. Mackay	Science	176	1321	6/23/1972
Wood	Cape Cod, Maine	26,000±2,000	M. Bothner et. al.	Science	210	423	10/24/1980
Organic material	Hellenic Trench, Greece	24,000±1,000	D. Stanley et. al.	Nature	273	111	5/11/1978

[1]Fission Track Dating of obsidian checked with ^{14}C dating of material below.

36, 200 Years Before Present (B.P.)		Years BP	Author	Journal	Vol.	Pg.	Date
Glaciation	Lake Weichselian, Eng.	40,000	E. Pool	Nature	221	507	7/30/1966
Marine Mollusks	W. Pembrokeshire, Wales, UK	37,960±1,400	E. Pool	Nature	204	622	8/7/1965
Peat, wood	Strawberry Point, WA	35,600±300	M. Stuiver et. al.	Science	200	18	4/8/1978
Peat	Lago Rupanco, Chile	36,250±2,75	J. Mercer el. al.	Science	182	1017	12/7/1973
Peat	Marine Fauna	35,500±500	Y. Herman	Nature	238	394	1972
Carbonized trees	Santorini Is, Greece	36,700±950	H. Pichler	Nature	262	374	7/29/1976
Peat	Wolverhampton, Eng.	33,000±3,000	E. Poole	Nature	217	1138	3/23/1968
Bison bone	Old Crow River, Yukon	33,800±2,000	W. Irving	Science	179	336	1/26/1973
Shells	Garry Isl., Arctic, Canada	>35,000	J. Mackay	Science	176	1321	6/23/1972
Organic Materal	Roxana Silt, Illinois	36,000±1,000	R. Flint	Science	139	403	2/1/1963
Organic matter	SW British Columbia, Can.	35,400±400	J. Armstrong et. al.	Science	152	766	5/6/1966
Orcanic carbon	SE New England	35,300±1,100	M. Bothner et. al.	Science	210	424	10/24/1980
Marine molluse	S Wales, UK	*37,310±1,275	B. John	Nature	207	622	8/7/1965
Coral: Tridacna	Huon Peninsula, New Guinea	35,800±1,500	H. Veeh & J. Chappell	Science	167	862	2/8/1970
48,400 Years Before Present (B.P.)							
Wood	Pine River, MI	45,800±2,000	M. Stuiver et. al.	Science	200	18	4/7/1978
Peat	Strawberry Point, WA	43,900±1,000	M. Stuiver et. al.	Science	200	18	4/8/1978
Organic Materal	Gahanna, OH	46,000±2,000	R. Flint	Science	139	403	2/1/1963
Peat and silt	Yamal Peninsula, Russia	44,000±4,000	S. Forman, et. al.	Geology	27/9	807	Sept. 1999
Coral: Tridacna	Huon Peninsula, New Guinea	46,000±3,000	H. Veeh & J. Chappell	Science	167	862	2/8/1970

60, 300 Years Before Present (B.P.)		Years BP	Author	Journal	Vol.	Pg.	Date
Peat	Bogachiel River, WA	59,600±700	M. Stuiver et. al.	Science	200	18	4/7/1978
Peat	Odderada, Europe	60,500±600	M. Stuiver et. al.	Science	200	19	4/8/1978
Organic	Chilean glacier, Chile	62,600±1,00	M. Stuiver et. al.	Science	200	19	4/9/1978
Coral: Tridacna	Huon Peninsula, New Guinea	60,000±6,000	H. Veeh & J. Chappell	Science	167	862	2/8/1970
72,400 Years Before Present (B.P.)							
Wood	St. Pierre interstate, WI	74,700±2,000	M. Stuiver et. al.	Science	200	17	4/7/1978
Peat	Salmon Springs, WA	71,500±1,400	M. Stuiver et. al.	Science	200	18	4/8/1978
Organic material	South of Jamaica	75,000±8,000	D. Ericson et. al.	Science	162	1227	12/13/1968
Coral: Tridacna	Huon Peninsula, New Guinea	74,000±4,000	H. Veeh & J. Chappell	Science	167	862	2/8/1970

Table 8-4: ^{14}C dating of sediments found approximately 12,000 to 72,400 years ago.

Sediments from southern Illinois produced ^{14}C dates of 35,000 to 37,000 years B.P. An Illinois peat bog was dated to 11,200 years ago indicating that it took about 800 years for the ice to melt and once again permit plant life to grow.[44] Wood found above a blue silt layer (6″ to 15″ thick) near South Haven, Michigan was ^{14}C dated to 11,000 years B.P.[45] That means the ice field by Lake Michigan required about 1,000 years to melt before plant life was able to grow there again.

The work done by Stuiver, Heusser and Yang greatly helped extend the range of Carbon-14 dating to 75,000 years B.P. They also did a good job of dating peat bogs, under various glacial tills, in the state of Washington and the Great Lakes St. Lawrence region, covering five ice ages. They include 74,400 ± 2,000 years for deposits in the Great Lakes, and 71,500 ± 1,400 years B.P. for Salmon Springs near Tacoma, Washington. Near Forks, Washington, they dated peat at 59,600 ± 700 years old. Samples from the Chilean glacial advance was ^{14}C dated to 62,600 ± 1,300 years. From Pine Ridge, Michigan, a wood sample was dated to 45,800 ± 700 years and other sites, from Salmon Springs dated to 51,000 ± 400 years. Samples on Whidbey Island, Washington, was dated to 35,600 ± 300 years.[46] They have also dated separate ice age advances that occurred approximately 36,204, 48,272, 60,340 and 72,408 years, which correspond to my model. A little further southeast, the Caribbean deposits show a much colder period 20,000 and 60,000 years ago, using oxygen-18 (^{18}O) dating.[47]

In Wales, England near the Irish Sea, two ^{14}C samples were acquired from marine shells, dating to 37,960 ± 1,400 and 37,310 ± 1,275 years B.P.[48] Several other glaciations are known to have occurred in the Irish Sea, one estimated to have been between 57,000 and 42,000 years ago (averages to 49,500 years), but these are not ^{14}C dates.[49] Carbon-14 dating from Holland and France indicate a cooling period between 36,000 and 35,000 years ago.[50] These ^{14}C dates (36,000 to 35,000 years B.P.) are in accord with conditions of severe cooling indicated by marine cores from the eastern Mediterranean.[51] The main European Würm glaciation was estimated to have been most extensive between 25,000 and 17,000 years B.P.[52] Again, these last two dates are estimates, but you can easily see their proximity to my estimated polar reversal date of 24,126 years ago.

In New Zealand, the Franz Josef Glacier produced wood debris under a glacial moraine that ^{14}C dated to 11,150 ± 50 years B.P. The scientists theorized that the glacier acted like a bulldozer and pushed the dirt and rock over the trees as it advanced.[53] Another bog test site in New Zealand was found on North Island (38°40′ south latitude), at the 3,100-foot level. A dozen core samples indicated a dark brown peaty mud ^{14}C dated to 11,850 ± 60 years B.P. The sediment above showed a sharp decline in trees for 100-200 years, then an increase in grasses, herbs and shrubs, indicating a much colder period.[54]

Also in the southern hemisphere, peat deposits have been found under glacial moraine. In Chile by lake Largo Ranco (40° 15' S. Lat.), peat sediments in multiple locations and layers have ^{14}C dated to 12,200 ± 400 years, and 36,250 ± 2,750 years B.P. The first deposit was associated with white volcanic ash.[55]

I left the following two examples for last because I was very surprised to find evidence of massive glaciers so close to the equator. The first location is at the western side of the Island of Papua, New Guinea (West Irian) in the East Otomona Valley (3° S. Lat.). The Carstensz glaciers were present there, ^{14}C dated to 11,330 ± 150 years ago, and they left moraines and glacial till all the way down to the 5,500-foot elevation, less than 11,000 years ago. The ice field was believed to be at least 6,500 feet thick.[56] The highest peak in the mountain range is 15,584 feet. It is obvious that the snow and ice descended much farther down to the lower elevations, but the shocking thing is that it remained there for over 1,000 years. This gives us an idea how cold the Earth was after the last polar reversal.

The last location is in Africa between Lake Victoria (on the equator), to where the White Nile meets the Blue Nile, located in Sudan (16° N. Lat.). Two major glacier areas were present 12,000 to 11,300 ± 400 years ago in East Africa.[57] One glacier was near Lake Victoria, (Mounts Kilimanjaro, Kenya, and Ruwenzori),[58] the other was in the Ethiopian highlands. Geologists have found evidence of a lake 30 miles wide and about 20 feet above the prehistoric river levels, farther south where the Blue and White Niles meet. The prehistoric lakeshore extended from 16° to 13° Latitude North. A glacier from the Ethiopian highlands must have blocked the rivers for at least 700 years and created the lake. It would not surprise me if the depth of the lake was over 100 feet where the Blue Nile meets the White Nile. The fact that the lake existed was determined from freshwater snail shells discovered in clay sediments well above the existing high river mark. The shells were ^{14}C dated to 11,300 ± 350 years B.P. It is an established fact that the snowline in the Ethiopian highlands, some peaks as high as 15,100 feet, dropped to at least 3,400 feet.[59] Research also indicates that the White Nile had a major flood ^{14}C dated between 12,000 to 11,000 years ago.[60] When the ice dam finally melted and collapsed it flooded the Nile so enormously that it cut right down to the bedrock in many areas.[61]

The previous evidence irrefutably proves there is a 12,000±-year pattern to the ice ages. There is no argument that there were ice ages during the past 12,000, 24,000, 36,000, 48,000, 60,000 and 72,000 years ago. Others occurred in the long Pleistocene history of the Earth. No one can argue against the fact that ice fields are formed by excessive precipitation, in the form of snow, and that heat from the Sun causes evaporation of the oceans' surface. This is the only heat source available in such quantities. Scientists may argue about the calculated amount of heat necessary to cause the evaporation or the time required to cause it. But the model I have just described is irrefutable. Arguments to the contrary would be mere sophistry. It is like arguing the size of the iceberg, which sank the Titanic. Who cares? It still sank the Titanic! Each 12,068-year ice age also means that our Sun produced a large spike of energy that hit the Earth at the same time. This again proves that there is a clock cycle in the Universe affecting everything at exactly the same time everywhere.

The Oceans Reveal it All

The oceans cover two-thirds of the Earth's surface and supply the water for the glaciers. When the Earth stops its rotation at the time of the polar reversal, the oceans continue in their former direction and velocity. This, of course, causes a great flood, which is very destructive to all life on Earth. There is a flip side to this: The flood protects the land from the very harmful heat blast from the nova and rocks from the solar outburst when it reaches Earth. Without the flooding all plant life would be burned up on the Sun-side of the Earth. When the oceans pour out of their basins they empty on the opposite shoreline. For instance, during the last reversal the Mediterranean Sea went westward and emptied out through the Straits of Gibraltar (depth of 1070 feet). If you were on the West Coast of the United States at that time you would have seen the Pacific Ocean rushing away from you in a westerly direction. The next time it will be in the opposite direction.

After the Nova's dust shell hits the Earth, the oceans will loose at least 1,200 feet of water. I think at least 20 percent of it will fall back to Earth in the form of rain within a few days after the nova. Most of the rest falls in the form of snow, which creates the ice fields world wide. The evidence seems to show that it takes about 1,000 years for most of the glaciers and snow to melt, leaving us with the Earth as we know it today, with some glaciers, but not with most of the planet covered with snow and ice.

All of the continental shelves around the world were uncovered to about 400 feet below current sea level. The next section will reveal the evidence for this, as well as other geological discoveries.

Radioactive Aluminum-26 and Argon-36 and 38.

I stated earlier that the Sun gives off cosmic rays, which interact with particles in space and the upper atmosphere. This creates isotopes of various elements. One of them is ^{26}Al formed in micrometeorites in space.[62] Scientists had noticed an increase in the presence of ^{26}Al in sediment layers associated with past ice ages and polar reversals.[63,64] There are several theories why aluminium-26 (^{26}Al) shows up in these deposits, including major solar-flares[65] and a "nearby supernova explosion."[66,67] The other isotopes of Argon-36 and 38 have been found in sediments in Greenland and the Pacific Ocean indicating extreme solar activity.[68]

Sea-Level Changes

Some scientists, in the mid-1950s, had already noticed a correlation between sea-level changes and polar shifts. Their dating was off only because ^{14}C dating had not been perfected yet, so dating was done by estimates of sedimentation rates, which are inaccurate. But in spite of that, there was a clear relationship.[69] The sea levels have been falling gradually over millions of years. After every nova, we lose about two feet.[70,71] Table 8-5 shows sea-level depths from around the world derived from shells, coral reefs, peat and plants found at various depths on the continental shelves. Most of the dating comes from ^{14}C testing, but remember what I stated about the inaccuracies of any kind of radioactive isotope dating. In spite of that you can clearly see that the dating fits the global model I have laid out.

Graph 8-1 displays what I believe are Ocean levels over a full 12,068-year cycle. Notice that sea levels rise above existing levels just before the polar reversal because most of the existing glaciers and ice caps melt, due to solar warming before the Nova.

Geological evidence from South America also indicates the ocean level was at least 325 feet below current sea level. Extensive glaciers were present along the Andes Mountain range from the southern tip to Columbia in the north.[93] There is no question that this was a worldwide disaster affecting all plants and animals.

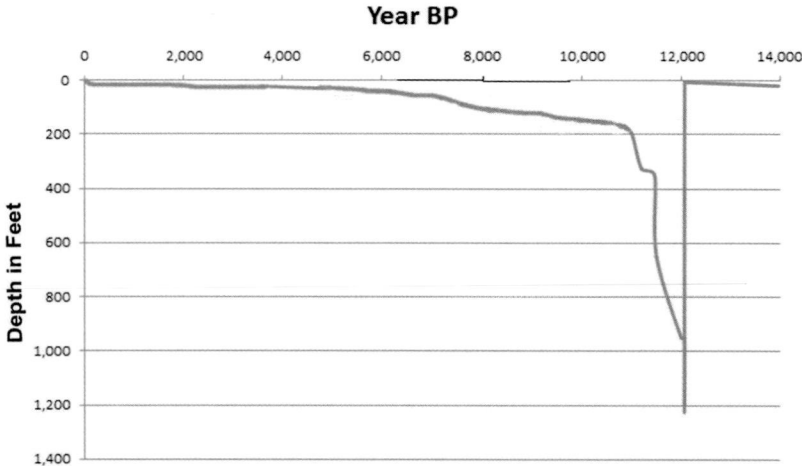

Graph 8-1: Graph showing the estimated sea level between two polar reversals.

Location	Sample dated	Est. Age B.P.	Depth (ft.)	Ref.
Atlantic Ocean				
Southeast coast of Florida	Coral reef	9,440 ± 85	-87	72
Coast of Maine	Barnacles	10,680 ± 160	-187	73
Coast of Connecticut (40°10' N)	Shells	10,850 ± 150	-282	74
New Jersey, Continental shelf (40°40' N)	Sand and shells	10,850 ± 500	-167	75
Coast of Connecticut (41°9' N)	Peat, Salt marsh	11,000 ± 350	-194	76
Argentine Shelf	Shells	11,100 (1)	-390	77
Connecticut Coast	Sedge peat	11,240 ± 160	-38	78
Coast of Maine	Shells	11,500	-197	79
Coast of Boston, Mass (40°9' N)	Shells	13,420 ± 210	-400	80
Argentine Shelf, Puerto Deseado	Shells	25,000 (1)	-492	81
Gulf of Mexico			0	
Texas shelf and coast	Organisms or plants	12,000 (1)	-184	82
Texas shelf and coast	Organisms or plants	13,000 (1)	-236	83
Orca Basin, Louisiana continental slope	Fine-grained silt	12,000 (1)	Not given	84
Mediterranean Sea				
Western Mediterranean	Marine Shells	23,800 (1)	-272	85
PacificOcean				
Santa Monica, California	Shells, V-17	10,165 ± 630	-67	86
Queen Charlotte Island, British Columbia, Canada	Coral	10,500 ± 350	-262	87
Queen Charlotte Island, British Columbia, Canada	Coral	10,700 ± 150	-361	88
Santa Monica, California	Shells, V-52	11,000 (1)	-166	89
Queen Charlotte Island, British Columbia, Canada	Coral	12,000 ± 200	-466	90
Queen Charlotte Island, British Columbia, Canada	Coral	12,400 ± 200	-502	91
S. California (33.7° N), USA (2 polar reversals ago)	Calcareous Algae	24,500 (1)	-392	92

Note 1: Non-Carbon-14 dating.

Table 8-5: Table of ocean level changes after the polar reversal.

Another method of dating and detecting ocean levels is finding glacial till and dating the organic material in the till. This is the case with glacial till found off the coast of New England. A sample of wood was found 195 feet down, in glacial outwash, which was ^{14}C dated to 26,000 ± 2,000 years B.P. Another sample was found at 159 feet and dated at 35,300 ± 1,100 years B.P.[94]

Sea Levels and Coral Reefs

In New Guinea, on the Huon Peninsula, a series of uplifted coral terraces are present that rise to over 2,275 feet. These terraces were formed by severe uplift of the land during multiple polar reversals, when the Earth stopped its rotation. A number of these coral terraces have been dated using both ^{14}C and [230]Th methods. The results revealed dates of 23,000 ± 2,000, 35,400 ± 1,300, 49,000 ± 3,000, 60,000 ± 6,000, 74,000 ± 4,000 years ago.[95] These dates correspond with five polar reversals in my model.

Our Shrinking Oceans

Numerous ancient terraced coastlines have been found—not entirely caused by uplifting but by the decrease in sea levels. This is the case with the following six examples of former lagoons and salt marshes found in Georgia, the first starting at 95 to 100 feet above sea level. The second site located is at 70 to 75 feet; the third, at 40 to 45 feet, the fourth, at 24 feet, the fifth, at 13 feet, and the sixth at 4.5 feet.[96] Each represents individual polar reversals and the resulting loss of some ocean water. The average incremental rise for these sites is 16 feet which I think is a little high and may indicate some uplifting in the areas during one or more of the past reversals.

Mastodon and Mammoth Teeth and what they mean

Teeth of both mastodons and mammoths were netted by fishermen in over 40 sites off the east coast of the United States. Shells found in the same silt were ^{14}C dated to 11,465 years B.P. The depth of where the teeth were found ranged from between 65 to 390 feet below sea level.[97] The significance of these discoveries is that these animals were tree eaters. Tree trunks and stumps have been found at these depths and the presence of mastodon and mammoth teeth confirms that the ocean level was below 400 feet deep. That means the ocean level worldwide, were that low long enough for seeds to germinate and grow to full-grown trees on the continental shelf. In corroboration of this, tree pollen from spruce, pine and fir trees were found in peat bogs on the northern side of Georges Bank found at a depth of 195 feet.[98] There ^{14}C dating proved to be 11,000 years B.P. It is a safe assumption that the same would be true for continental shelves worldwide. The Human survivors must have headed for the new coastlines to survive and for transportation. Fresh-water peat bogs were also found off the coast of Boston near the Georges Bank down to 192 feet, which were ^{14}C dated to 11,000 ± 350 years B.P.[99] Numerous other peat bogs have been found on the continental shelves worldwide that clearly show a much lower ocean level 12,000 to 11,000 years ago, which then gradually rose to our present sea level.[100]

Bay of Bengal and the Ice Age

Scientists have been able to determine if ocean water was predominantly cold or warm by counting the abundance of two types of small sea creatures called *Globigerian*. There are two types within this group: The *Globigerina rubescens* thrive in cold water and the *Globigerinoides tenellus* thrive in warm water. By counting both populations and their respective increases or decreases one can determine whether the ocean water was cold or warm, thereby indicating the presence of glacial ice. In the northern Indian Ocean by the Bay of Bengal deposits of these small animals have been counted from core samples. What they indicate is that as late as 8,775 ± 145 years ago (^{14}C dating), the colder species were predominant. Other earlier dates of 11,000 years have also been arrived at.[101] The same method has also indicated that there were cold periods 24,000 and 48,000-years ago, separated by warm periods.[102] The Ganges River flows into the Bay of Bengal and it drains northern India and the Tibetan mountains. This means that there were glaciers and snow in Tibet and northern India for 1,000 to 2,000 years after the polar reversal.

Lake Levels and the Rotation of the Earth

Currently our Earth rotates from west to east and the rotation controls the jet stream and weather patterns worldwide. Before the previous polar reversal, I theorize the Earth rotated in the opposite direction, so we would expect the weather patterns to be different. The Earth does not physically flip over. The reversal is a magnetic reversal which causes the rotation of the Earth to reverse. Areas that are now wet may have been dry 12,068 years ago. Such is the case between 15,000 and 13,000 years B.P., concerning the lakes in East Africa.[103] Lake Victoria and Lake Albert both exhibited extremely low water levels prior to 12,068 years ago. The White Nile "dwindled to a mere seasonal trickle."[104] That changed sometime after 12,000 years B.P. Even the Dead Sea in Israel had overflowed around 9,850 years B.P.[105] The same is true for lakes and plant-life in Northern Florida.[106] On the Artemisia steppe of Iran, it is known from pollen studies, that 13,000 years ago there was a cool climate, which changed to a warmer-wetter climate, more like a savanna, growing different types of trees.[107] These observations tend to support my theory that the Earth's rotation changes at the time of the polar reversal. But the best proof may come from small shell creatures called Planktonic Foraminifera (*Globorotalia menardii*). These creatures lived at the Pliocene-Pleistocene periods. The test sites where they were found were in the Caribbean and off the coast of Brazil. Core samples of the sea floor were taken and what was found was that plankton below the ice age polar reversal layer were coiled clockwise and the same species above the boundary layer coiled counter-clockwise.[108] This is like when you drain a sink, in the northern hemisphere, the water spirals down counter-clockwise and in the southern hemisphere it spirals clockwise. It appears these small creatures are telling us the Earth changed the direction of rotation after the polar reversal.

The Mediterranean Sea

During the last polar reversal, the Mediterranean Sea mostly emptied out when it flooded in a westerly direction. The Straits of Gibraltar currently have a depth of about 1,070 feet and that kept it from refilling with seawater until the ocean level went above that 1,070 foot level. It is possible that before the last reversal, the depth of the Straits was around 800 feet but the overflow event scraped out additional sediments to its current depth. We do not know how many years it took for the ocean level to rise above 1,070 feet but we do have proof that the Mediterranean had at least two infusions of freshwater for thousands of years due to the melting of the European ice caps. The first period was 80,000 to 84,000 years B.P. (seven reversals ago would equal 84,476 B.P.) and the most recent was estimated at 13,000 to 9,000 years B.P.[109] This discovery was made because of black organic-rich anaerobic sediment mud layers in the eastern Mediterranean. The sediments are rich in plankton that only grow in low salinity water. The same features were also discovered in the Gulf of Mexico (11,600 years B.P.).[110,111]

Another indication of much lower Mediterranean Sea levels were derived from the discovery of deep-sea brine, created by evaporated deposits rich in an isotope of sulfur, ^{34}S.[112,113] The same sharp increase in ^{34}S had been found in all the oceans and their peripheral basins.[114]

Submarine Canyons

All of the major rivers around the world cut deep-sea canyons (also known as submarine canyons) into the continental shelves. These sea canyons cut down to as deep as 11,000 feet below current sea level. Figure 8-1 illustrates a typical sea canyon. It was formed by the Delaware River off the coast of Delaware and Pennsylvania.[115] No one has understood how they are formed, because fresh water is lighter than salt water and it cannot sink down into ocean water to cut canyons. Rivers deposit silt and other debris to form an alluvial plain, such as in the Nile or Mississippi deltas, but they do not form canyons during "normal" times.

Submarine canyons are formed shortly after a polar reversal. This is how, as mentioned before: the oceans rush over the continents when the Earth stops rotating during the polar reversal. The oceans continue over the continents, moving a great deal of debris with the wave, cutting great canyons, such as the Grand Canyon—then finally returning back into the ocean basins. Depending on how wide the continent is, will determine whether the wave reaches the opposite

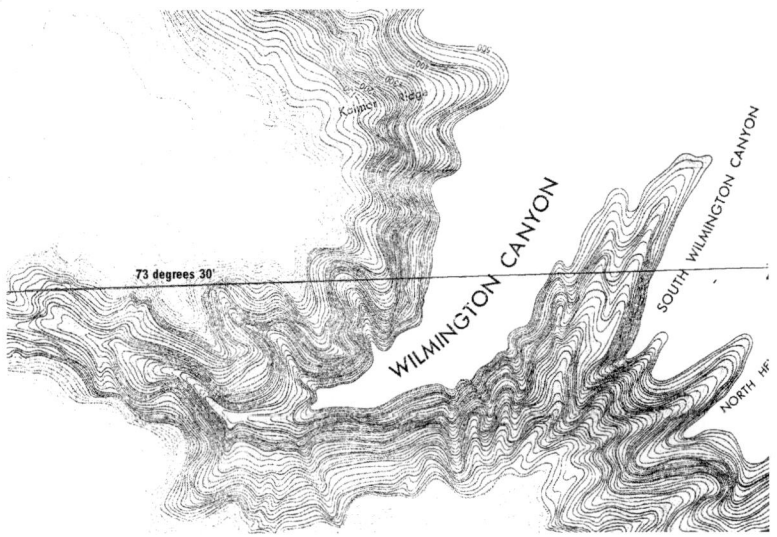

Figure 8-1: The Wilmington submarine canyon near the Delaware River descends to 12,000 feet below current sea level.

shore. I believe the actual reversal time will be a about seven to eight hours. During that period, the oceans will slow down their velocity because of gravitational pull, elevation and friction as the waters pass over the Earth. After the reversal period, the Earth will start rotating in the opposite direction, thereby temporarily increasing its relative forward velocity, to the ground, until gravity and friction take over and stop its advance. In China, during the last reversal, the water also drained into the Arctic Ocean.

The submarine canyons are formed when this massive ocean wave pours off the continent and again empties into the ocean basins. Remember that when the ocean water reaches the other shore, the ocean basin off the coast will be empty because that ocean has also rushed out of its basin, in the same direction. The canyon is further enhanced by massive runoff caused by torrential downpours, starting days after the nova and polar reversal.

There have been many dredged samples taken from ocean canyons. The research submarine *Alvin* retrieved samples from a small deep-sea canyon named *Oceanographer Canyon* on Georges Bank off the coast of Massachusetts. The *Alvin* dove to a depth of 4,750 feet. Researchers found no sediment in motion or relative motion between the sidewalls and the canyon walls, and no strong current was evident at the time. They discovered massive evidence of a downward-transport of all types of debris, from large stone blocks to various plant life. The age of those organic samples were from the Pleistocene era or younger—in other words, from previous ice ages.[116] The oil industry has also drilled in Lindenkohl Canyon and South Toms Canyon off the coast of New Jersey.

Pleistocene-age sediments, including plants, have been found at a depth of 975 feet, and extend another 1462 feet thick.[117]

In summary, the deep-sea canyons are formed when the ocean level has been reduced or emptied at the time of the reversal. As the ocean water returns into the ocean beds from the continents it cuts the canyons into the continental shelf. That is why the canyons have been found containing all types of debris. An example of this was found near Grand Banks, off the coast of Newfoundland. They found reef material 2.8 miles down[118] that must have lived near the sea surface before it was washed down when the oceans poured back into the basins.

The Earth's rotation has already started to slow down because of the Earth's decaying magnetic field.[119,120] At the time of the reversal, the Pacific Ocean will pass over the American Continent with an initial velocity of at least 400 miles per hour. The massive wave will be at least 5,000 feet high, because the Pacific Ocean is 12,100 to 20,700 feet deep, west of the American Continent. The wave will slow down because of gravity and the Rocky Mountains, but it will continue eastward for as many hours as the Earth remains at zero rotational velocity. How far this wave will reach is any one's guess, but if it manages to get past the Rocky Mountains there is nothing to stop it except the Appalachian Mountains. My educated guess is that it will reach the Atlantic Ocean because of the evidence already found in the deep-sea canyons off the continental shelf. The Atlantic basin should be mostly emptied because the Atlantic Ocean will have also surged in an easterly direction across the European and African Continents. Almost immediately after the nova, the radiant energy from the Sun will start evaporating the water on the Sun-side of the Earth. Within 18 hours after the actual polar reversal, the dust shell from the Sun will reach the Earth. The Ocean wave should still be passing over the continents when the extremely hot dust reaches the Earth and evaporates some of this water. When these hot gases hit us they will be hitting a very large surface area of ocean water no matter what side of the Earth is facing the Sun at the time.

In conclusion, what the oceans of the Earth do to the surface is to wipe out everything as far as the waves travel. But the ocean water also saves plant life from the nova.

The Sun and the Nova

The Sun affects so much on this planet, as shown in the previous section, that it is difficult to separate it from the other subjects in this Chapter. This section will focus on the evidence demonstrating that stars are not powered by nuclear fusion and on the evidence what stars leave behind when they nova. There is nothing in current matter theories of existence that explains how a star could periodically nova, or oscillate, as stars actually do. It is a problem for physicists.

Their matter philosophy prevents them from putting all of the observed pieces together and coming up with a coherent solution to the origin of the ice ages. My information theory of existence breaks through these obstacles and enables us to see a broader picture of what really happens, and come up with the correct solution.

What Powers the Sun?

The current theory for the heat source in the Sun is that it is derived from thermonuclear fusion generated from gravitational contraction. Fusion is the process of taking two, or more, lighter elements like hydrogen and, through the process of extreme heat caused by gravity, fuse them together to create a heavier element, such as helium. Fusion can theoretically occur for all elements lighter than iron. The theoretical byproduct of solar fusion is what is called neutrinos. This theoretical massless "particle" would travel at the speed of light and not become absorbed or stopped by most matter in the Sun. So, an experiment was setup by the Brookhaven National Laboratory in the early 1960s, to capture some of these neutrinos and count them—if they really existed. The purpose of the experiment was to see if their model of the Sun was really correct. The question arose as a result of the last Gleissberg Cycle that occurred 1958-1959. Solar activity was so great that scientists had a hard time understanding why it happened. Their answer was the construction of a 100,000-gallon tank of tetrachloroethylene (C_2Cl_4) and locate it in a mine 4,850 feet underground. They believed that when neutrinos hit clorene-37 atoms, that they would be transformed into radioactive Argon-37 isotopes, which could then be counted. The rate of collision of neutrinos with clorene-37 atoms was initially estimated at 35 in six days. Later estimates were gradually revised downwards until they reached 3 in six days and finally .3 solar neutrino units.[121] None had been detected.[122, 123] Some scientists suggested new tests[124] to find any neutrinos in order to save the "accepted" philosophies of solar physics. Currently they are drilling mile-deep test holes in the ice of the South Pole and inserting detectors as deep as they can, in order to again find any neutrinos. A journal article written by John Bahcall and Raymond Davis, Jr. in 1976, best summarizes this problem: "The conflict between observation and standard theory has led to many speculations about the solar interior that were advanced because their proponents believed that the subject is in a state of crisis."[125]

This experiment was the first clue to scientists that the center of the Sun was not powered by nuclear fusion. It was not the last. They have been trying to detect solar neutrinos for over 40 years and still had no luck. You think eventually someone would put a halt to this taxpayer raping and finally state the obvious—scientists do not have a clue what powers the Sun and start examining our philosophies of science.

Another problem astronomers have is with the chemical composition of the Sun. It appears, using some models of the Sun, that it has an abundance of iron and helium which cannot be explained using their models.[126] All this suggests that an information theory of existence will have a better chance of explaining the heat source of stars.

Solar Oscillations

There is *no* theoretical provision in solar physics which allows the Sun to oscillate, both short and long term periods. By oscillate I mean changing solar output during sunspot cycles—Gleissberg cycles (Table 3-3)—and the 12,068 year nova clock cycle. Some scientists have suggested that our Sun is like a pulsar because of very short term oscillations that have been detected (2 hours and 40 minutes).[127,128,129] Others have concluded the Sun is not powered by nuclear fusion because of these very short term pulsations.[130] It is an established fact that our Sun oscillates at about 20 different very short term frequencies. Size change also occurs ranging between .4 to 8 kilometers.[131] In Chapter 3, Figures 3-2 and 3-3, you can see one of the patterns these frequencies create.

The sunspot cycle of 11.092 years is short-term in galactic terms. In my Theory of Multidimensional Reality I conclude that this cycle is a resynchronizing frequency in the Diehold, just as the Gleissberg cycle is another longer term resynchronizing frequency. These frequencies cause an increase in the energy output of a star. In essence this puts energy back into the system (the Universe). I am not the only person who has made the analogy that the solar cycles seem to be controlled by some kind of chronometer inside the center of the Sun. Dr. R. Dicke of Princeton University's Physics Department, concluded that "No support is found for the conventional view of the sunspot cycle . . . Instead, both sunspots and the [D/H] solar/terrestrial weather indicator seem to be paced by an accurate clock inside the Sun."[132] The reason why astronomers have such a hard time understanding the output of our Sun and other stars is because they do not realize that the heat source is a center modulation point and not a nuclear reaction going on in the center. This is not to say that nuclear reactions are not happening in the Sun's corona, where it is hot enough for them to occur (approximately 10 million degrees Celsius).

Why a Star goes Nova

Most people have been taught that when a star novas, it is at the end of its solar life cycle. It has used up its hydrogen fuel and then blows up. We know that this is just *not true*. A few scientists have theorized "magnetic springs" periodically go off in the center of a star, and this mechanism would explain the tremendous dust shell velocities observed.[133] There are different levels and types of novas that have been observed. The most severe ones are called supernovas. This occurs when a star produces millions of times more energy in a catastrophic outburst, in only **one day**. These events are extremely bright and produce a great deal of ultraviolet light, X-rays[134] and gamma rays. A regular nova, produces only thousands of times more energy and is less severe. The star expels its outer shell of matter and dust in one day (Figure 8-2 and 8-3). Astronomers have one interpretation of these observations—and I have my own.

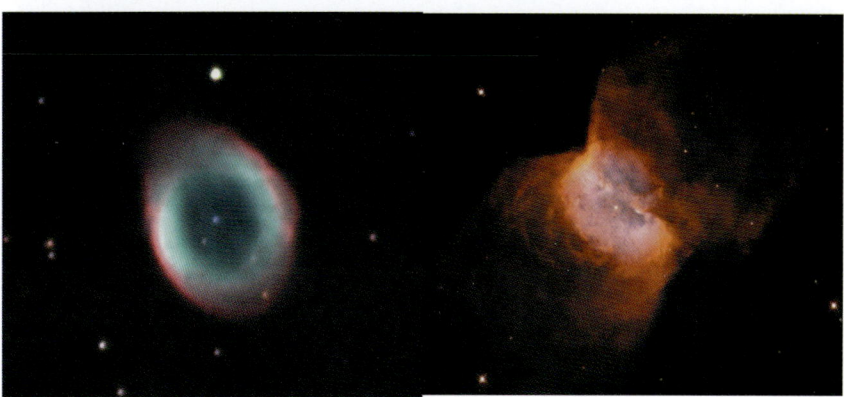

Figure 8-2: NGC 6720 Planetary Nebula in Lyra, the Ring Nebula, Hale Observatory.

Figure 8-3: NGC 7293, another Planetary Nebula. You are looking at the star from the perspective of its polar axis.

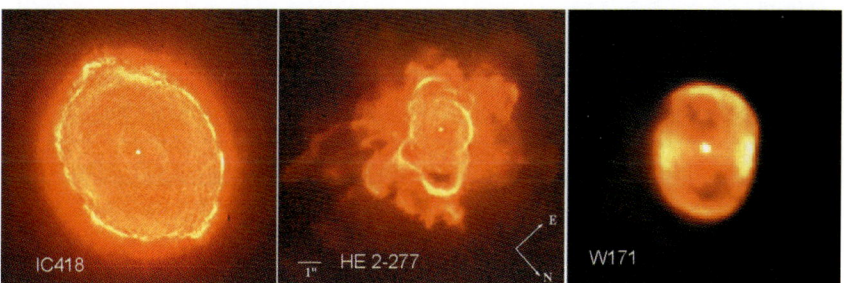

Figure 8-4: Three stars sometime after their novas. The stars are still visible in their shell.

The Theory of Multidimensional Reality calls for clock cycles that run through time and occur every 12,068 years. The clock cycles manifest themselves as a spike of energy, plus a reversal of polarity in stars, which we call a nova. This is similar to the graph in Chapter 3, Graph 3-4, which is a model of how matter is created in this dimension. What is true for the microcosmic is also true for the

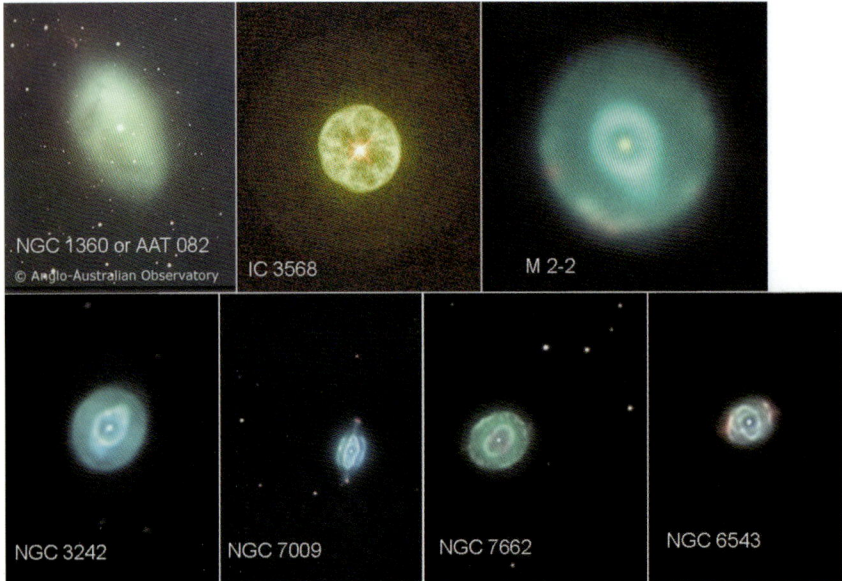

NGC 1360 or AAT 082
© Anglo-Australian Observatory

IC 3568

M 2-2

NGC 3242

NGC 7009

NGC 7662

NGC 6543

Figure 8-5: Picture of seven stars some time after their novas.

NGC 6826c

IC 4663

NGC 7027

Figure 8-6: Three planetary nebulae that have multiple dust shells, which may be the result of multiple novas over many hundreds-of-thousands of years.

macrocosmic, which shows you the reflexiveness of the Universe. As the information crosses the X-axis at the polar reversal it creates a very large spike of energy. This process puts energy back into the system. All stars novas at the exact same time including our Sun. The event happens in one day, but the effects last for hundreds of years. I will reveal the proof that our Sun novas. I have already shown in Chapter 3 that the clock cycle is found everywhere in our Universe. I have already described the effects of a nova on Earth's oceans, how it causes each recurring ice age, and creates isotopes of various elements in our atmosphere. The very presence of 4,000-foot ice fields proves that the Sun must

Figure 8-8: Nova photographs of star V838 from February, 2002 to December 2002, over an 11 month period.

have provided the massive heat necessary to create those ice fields. The regular cyclical appearance of the ice age proves that some kind of clock cycle is present in our Universe. Others have concluded the same: ". . . from core evidence, the climate seems cyclical—crudely sinusoidal, with perhaps systematically increasing amplitude and period for the later cycles. I suggest that some solar periodicity is the prime cause of the major climatic cycles, . . ."[135]

Figure 8-7: The dark spots are the Bok Globulars. They are found in all directions from our Sun.

I theorize that at the center of stars and planets there is a center modulation point, where all the information that makes up the star or planet is directed. The resulting spike of energy causes a star to nova. In Chapter 3, I presented how I discovered the six blank periods (Table 3-1 and Graph 3-1) in space, which proves all stars nova at the exact same time. The novas reveal the presence of the clock cycles in time. I also described the long term synchronizing frequencies called Gleissberg cycles of 88.735 years. Both of these observations cannot be explained when using the "accepted" physics theories.

The previous figures are examples of different types of novas photographed in our Galaxy (Figure 8-4). You will notice they are not all the same. They are all a little different, like people, who are all a little different.

If you were to see our Sun from space, a few hundred years after a nova, you would view it with a dust ring around it. In fact, that is why NASA had our Voyager-

2 satellite look for the Sun's remnant dust shell and photograph it in our solar system. Voyager-2 reached Saturn in August 1981 sometime after that and before it reached Uranus, it photographed the remnant dust shells.

The previous series of photographs show what I believe are stars shortly after a nova (Figure 8-4). These stars are for the most part larger than our Sun. Their novas do not conceal the exploding star.

The next series of photos also shows stars sometime after their novas (Figure 8-5 and 8-6). The time periods may be from thousands of years after a nova. We can see that stars expel their dust shells periodically. The stars shown are of all different sizes. Some are smaller than our Sun and others are much larger. The main point is that they all undergo this process, including our Sun, and that it is *not* a rare event in space.

It is also possible that some of these very large and elaborate nebulae may be the result of accumulation of many novas over millions of years.

Bok globules and Planetary Nebulae

In *Reality Revealed* I theorized that a clock cycle caused all stars to nova at the exact same moment in time. We observe them over time, because light travels at 186,000-miles per second, not instantaneously. At first, I looked for stars that novaed on one of these time lines, but back then I did not find any. In the mid-1980s I came across something called Bok globulars. They were stars of various sizes, appearing through a small dust cloud (Figure 8-7). The astronomer, Bok, had discovered over 450 of them, in all directions from our solar system. Their distances ranged between 13,000 to 14,500-light years from the Earth. In later years, some of these stars have been classified as planetary nebulae, pictured previously.

The traditional explanation for Bok Globulars is: They are young stars just forming from the intergalactic dust surrounding them. The problem with this theory is that it is contrary to the basic idea that Galaxies start as Quasars and then shoot off two jets of matter which eventually form the spiral arms of the galaxy. That means the oldest stars are located on the outside-edge of a Galaxy, the younger stars near the center. Our Sun is located about 30,000 light years from the center of the Milky Way. It is therefore fairly old. It does not make any sense to have very young stars in the center, and also further out with much older stars. Bok Globulars are found in all directions from our Sun. The discovery of star V838 in 2002 settled the issue because it clearly showed that we are looking at a star at various stages after it novaed.

Next I wondered: when a star novas, is it possible that it only produces 500 to 2,000 times more energy output in one day? With the discovery of Bok globulars I realized that when a star novas the matter shell around the star is expelled and obscures the light of the star for about 1,000 years. When the dust shell becomes thin enough the star becomes visible once again, as shown in Figures 8-5 and 8-6. So what I had to look for were blank places in space, where no stars were visible. My results were already shown in Chapter 3 (Table 3-2) and in Appendix A, which show the six blank places in space. These blank places in space cannot be explained using the "accepted" theories of Physics. Only an information theory of existence can explain them.

A star's behavior after the nova can be as follows: It would give off a sharp spike of energy in the form of visible and infrared light and perhaps look like Figure 8-4. After the nova, it will give off more ultraviolet light, then slowly lower to infrared light which is called a cool star.[136] Within a few months a thick dust shell conceals the star for about 1,000 years. Figure 8-7 shows what that looks like. After that, the dust shell will be visible for varying periods of time. Only after the Sun has regenerated the matter shell around it will it again give off "normal" light and heat.

Kuiper Belt

The dust clouds, which Voyager-2 photographed, were named the Kuiper Belt. It is also known as a nebula, and it expanded until it reached an equilibrium point between gravitational forces and momentum. This would be dependent on the composition and size of the star. Our own Sun has a dust shell around it and it is called the Kuiper Belt, which was discovered beyond Neptune's orbit.[137] The Voyager flights in the 1980s also detected dust rings beyond Jupiter.

An Observable Nova on a Clock Cycle

I found star V838 mentioned in the journals. It was observed on February 1, 2002. Figure 8-8 presents a series of photographs of V838 Monocerotis as it novaed. V838 is not like our Sun, which is categorized as a yellow dwarf star. V838 is a hot blue star.[138] We can clearly see the surrounding dust and matter-shell expanding rapidly into space. Over a period of eleven months the dust expanded past its own solar system. Any planets orbiting V838 Monocerotis would have been hit with hot dust and rocks from the dust shell.

The "accepted" explanation for what we are looking at is what some astronomers call a "light echo." The astronomers[139] believe that what we are seeing is the result of intense light passing through periodic shells of interstellar dust that might have been created by previous novas of the center star.[140] They all seem to want to avoid one obvious observable fact: We are looking at an expanding dust shell from the giant star in the center after a nova. As the dust shell expands outward the original cloud structure stretches, opening up holes in the dust cloud. If it was caused by light then why did the dust shell expand very little after

September 2? If it was light, by October the wave front would have been far beyond the shown visible field of view. Thus it is clear that the standard light hallow explanation of this nova is wrong.

Astronomers have had a very difficult time estimating the distance of star V838 from the Earth. Some calculated a minimum distance of 8,150 light years away.[141] Others gave a minimum distance of 15,648[142] to 19,560[143] light years away. Howard Bond's team gave it an upper limit of 26,080-light years.[144] Another distance mentioned was 35,860-light years.[145] Astronomer R. Tylenda summed it up quite well in his article, when trying to explain what we were seeing: "It cannot be accounted for by known thermonuclear events . . ." and "The outburst of V838 Mon[ocerotis] does not fit to any known class of out-bursting stars [novas], so determination of its luminosity is crucial for attempts to identify the mechanism of the event."[146] In other words, astronomers do not know what they are looking at.

I conclude that we are looking at a giant blue star which novaed about 24,136 years ago—and the light just reached us on February, 2002. The reason it may appear blue to us is because the star is in a stage where it is giving off mostly ultraviolet light. The reason why V838 was not totally obscured by dust clouds was because the size and power of the star is so much greater than a small star such as our Sun, so the dust shell was expelled far out into space, permitting us to see V838. I would expect that the dust cloud from our Sun would obscure the solar light for about 1,000 years, when viewed from outside of our solar system. I concluded this after finding the six blank periods shown in Chapter 3, Table 3-1. I would expect that it could take that long until the dust cloud expands sufficiently or until the outer planets' gravity attracts most of the dust so the Sun's light could be seen from outside of our solar system again.

There is a reason why astronomers do not know, or want to admit, what star V838 is clearly showing us. They have a problem with their explanation for Bok Globulars (Figure 8-7) and planetary nebulae. Their explanations for each phenomenon are different, and come from opposite ends of stellar evolution. They theorized that galaxies are formed from quasars and solar systems are formed from the primordial dust filling an area of space. As the dust came together, caused by mutual gravitational attraction, it formed stars and planets. The traditional explanation for Bok Globulars is that they are the remnant dust from the original galactic primordial dust that is just forming a new star. The explanation for planetary nebulae is that they are old dying stars which have expelled outer layers of their matter shell into space. In my view all three of these things are just different stages of a star after the polar reversal and nova.

Star V838 shows that stars do nova and expel their matter shell—and it is a common occurrence. V838 is just one instance of how scientists can filter out an obvious truth for an explanation that fits their "accepted" theories. The big problem in this case is their academic turf wars are going to cost us all our lives.

Remnants of Frequent Nearby Novas

I theorized that the Sun gives off a brief pulse of high-energy particles in the form of cosmic rays, gamma rays and x-rays. I concluded this because cosmic ray and x-ray bursts have been detected hitting the Earth from stars on a very frequent basis.[147] The bursts have been identified as compact stellar objects with the following characteristics: They have a short rise-time of less than a few seconds, and a decay time of several seconds to some minutes. These pulses may repeat themselves within two hours, to two days. Some bursts are regular and some are not. Over 30 were discovered by 1977.[148,149] The mechanism is not clearly understood, but the x-rays may be the result of streams of energy emitted by the center modulation point until such time as the matter shell again forms around the star and prevents the x-rays from reaching us.

The origin of cosmic rays are much easier to understand since it is an established fact they come from our Sun and other stars within our Galaxy.[150] Scientists have known that stellar outbursts, or nearby supernovas, could have caused such catastrophes, as those which caused extinctions ". . . in addition to close supernovas, solar flares and explosions at the galactic center should be borne in mind when considering possible catastrophes experienced by the Earth in its history."[151] Everyone seems to want to stay away from openly saying that our Sun does nova—repeatedly!

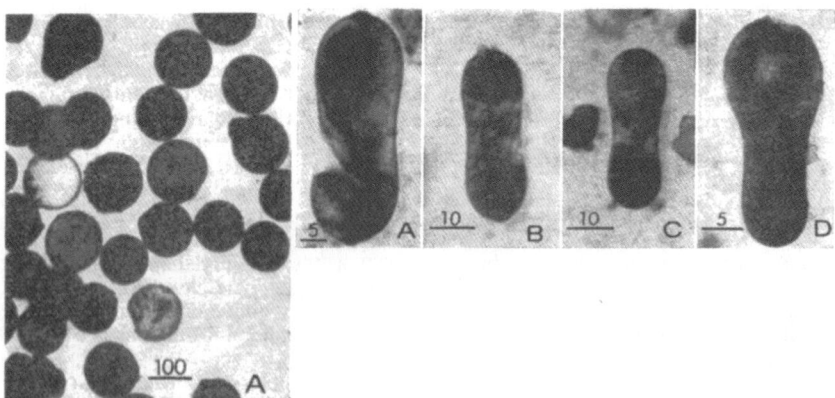

Figure 8-9: Glass tektites found on the Moon. Sizes range from .1 mm to 1 mm.

Figure 8-10: Glass tektites found on the Moon in the shape of a dumbbell. Size ranges 1 mm ± long. Credit: Apollo 11, George Muller, Univ. of Miami, Florida.

Another method of detecting supernovas and novas is by using a radio telescope and searching for the 408 and 5,000 MHz signal emanating from them. As of 1973, 27 such signals were found, of which there were nine with estimated distances falling into one of the 12,068-year time lines, or within 1,000 light years of it.[152]

The Sun's Fingerprint

The Sun's fingerprint is comprised of both the radioactive elements mentioned earlier, and small glass tektites with and without fission tracks. The expelled shell consists of glass beads, lumps of molten glass, dust and rocks[153] thrown out at very high temperatures. The nova also produces bursts of cosmic and gamma ray particles, x-rays and ultraviolet light. When the Sun novas it throws off its outer dust and matter shell into space at an estimated speed of 1,550 miles per second.[154,155] The dust shell contains small microtektites (glass beads .1 mm to 2.0 mm) with the Sun's "fingerprints" inside. The fingerprints are fission tracks within the glass beads put there by the cosmic ray burst the Sun gives off. A fission track is a small hole made in a mineral or other hard substance by a proton or atom. These atomic "bullets" are propelled at velocities approaching the speed of light and leave tracks in almost anything they hit. They are, in essence, the "fingerprint" of the Sun. The traditional academic explanation is that the hole was created by the decay of Uranium-238. The problem with the theory is that most of the examples I will present came from locations where there is no ^{238}U present. One of those locations is the Moon.

Glass Beads on the Moon

Samples of soil brought back from the Moon by the Apollo space flights had a large concentration of glass microtektites of various colors with fission tracks in them. NASA found that about 27% of the soil surface on the side of the Moon facing the Earth is made up of these glass beads. Figures 8-9 and 8-10 show microtektites found on the Moon. Scientists had noticed the similarity between Tektites found on the Earth and lunar glass beads.[156] The two basic shapes retrieved were round and elongated dumbbell-looking beads. The samples' sizes ranged from .01 mm to 1 mm in diameter. All of these microtektites were devoid of gas bubbles.[157] This is unlike volcanic-created obsidian beads, which have gas bubbles. The Moon beads are colored: clear, pale-green, red, orange, amber, dark-brown and black. Each color represents a slightly different chemical composition. The darker colors have more iron in them. They all had fission tracks[158] and a lack of bubbles.[159] Of the orange beads found, the chemical composition did not match the composition of the surrounding rocks. Drs. Roedder and Weiblen concluded that lunar beads could have been "ejected in an expanding cloud . . . where the individual particle leaves the central source of radiant heat at high velocity."[160] Other researchers have shown that some of the glass exhibits low-temperature shock cracks indicating collisions after the glass had solidified[161,162] such as in the flight to the Moon. The expanding dust cloud from a stellar nova would certainly fit these descriptions.

Samples brought back by Apollo 17 (December 1972) were almost entirely these tiny glass beads (0.1 to 0.2 mm, soil sample 74220.60) ranging in color from pale yellow to black, but most were orange. The orange beads indicate a high content of titanium, besides a very high content of zinc, copper and nickel, which excluded the possibility of volcanic origin for the glass beads.[163] To further reinforce this conclusion, analysis was done on the dumbbell-shaped beads (typically over 0.6 mm long). It was found that in order to form that shape the glass ball would have to rotate in excess of 500-1,000 revolutions per second, which eliminates the possibility of the dumbbells forming from volcanic activity or meteoric impacts.[164] The Apollo 16 Lunar samples indicated no evidence of volcanic rocks, but they did find rocks that were coated with dark-gray glass. They also showed signs of partial melting.[165]

Lunar Surface Melting

Ample evidence for surface melting was found on the Moon. Meteors and other materials found there had an abundance of Aluminum-26, which can only be produced from a nova or supernova. "It therefore seems that lunar melting provides additional evidence for a supernova preceding formation of the Solar System by no more than a few million years."[166] Thomas Gold, in 1969, was the first to suggest that a "flash from the Sun" could explain all the observations he saw of samples and pictures from the Moon, including rocks with a glaze mainly on top. He went as far as to write: "The phenomenon may not be in the nature of a flare, but in the nature of a very minor nova-like outburst of the Sun."[167] His job was to produce an explanation for small meteor impact craters found on the Moon. Their sizes ranged from 8 inches to 5 feet and exhibited glazing "apparently due to radiation heating; it suggests a giant solar outburst in geologically recent times."[168] He estimated that solar heating had occurred within the last 30,000 years because micrometeorites had not destroyed the sides and bottoms of the lunar craters. He also estimated the melting temperature was 1125°C and would be produced by a solar burst lasting 10 to 100 seconds. The Sun would have produced at least 100 times its luminosity output. He went as far as to theorize that stars the size of our Sun did this "every few tens of thousands of years."[169] Dr. Gold got it right. Over the next four years Dr. Gold was supported[170,171,172] and criticized[173] by many scientists. Many agreed with him that glazing on the lunar rocks was from the glass beads that landed on the Moon in a superheated state. The chemical composition of the glass beads had settled the issue because they did not match the surrounding rocks on the Moon.[174,175] Scientists from Owens-Corning Glass[176] and others[177] showed how the lunar glass beads were "formed by the atomization of a mineral melt in a high speed gas stream."[178] The small "meteorite" impact creators Dr. Gold studies, were created by super-heated globs of glass that splattered when they hit the moon. That is why no stone or metal meteorite was found in the bottom of the creator.

The surface of the Moon was repeatedly dusted by the hot dust cloud from the Sun when it novaed. Small rocks, possibly which may be in the form of liquid masses that are part of the nova's dust shell probably created the small craters. The initial nova explosion put the cosmic-ray particle tracks in the glass beads. I conclude that all the radioactive isotopes found in the glass beads were a result of thermal nuclear temperatures in the corona.

Tektites and Meteorites found on the Earth

Tektites are small, round and irregularly shaped glass objects that have fallen to the Earth. Sizes range from a few millimeters to as much as 10 centimeters. There are four general areas where tektites have been found, but tektites have been found all over the world. The youngest and largest field is called the Australasian tektite area. The field includes Australia, India, Indochina, China and the Philippines. The next field is in Africa by the Ivory Coast. The third field is located in the middle of Europe. The oldest field is in the United States, from Georgia to Texas and up the Mississippi River. Most tektite fields are found in late Pliocene to Pleistocene gravels and sometimes on the surface.[179] Obsidian artifacts found in Italy, with fission tracks, were associated with organic material ^{14}C dated to 11,400 ± 1,800 B.P.,[180] which places them at the time of the last polar reversal and ice age.

Most tektites are less than one millimeter in diameter, and these are called microtektites. They are mostly found in sediment layers associated with an ice age or polar reversal.[181,182] Figure 8-11 shows an example of microtektites found in Caribbean deep-sea cores.[183] These samples were found above magnetic reversals and at the same layer as the extinction of five radiolarian species with high concentrations of iridium. Figure 8-12 shows microtektites found in the Kharton Basin, south of Indonesia, and the Australian Basins. Microtektites have been discovered in core samples as far north as the Philippine Basin. The microtektites were located up to 50 cm above a magnetic reversal. The core sample found near Indonesia was also discovered near volcanic ash layers, indicating volcanic activity also occurred during the polar reversal, which I will describe later in detail.[184]

The sedimentary boundary layer between Cretaceous and Tertiary (65 million years ago, the K-T boundary) off the coast of New Jersey, has revealed similar

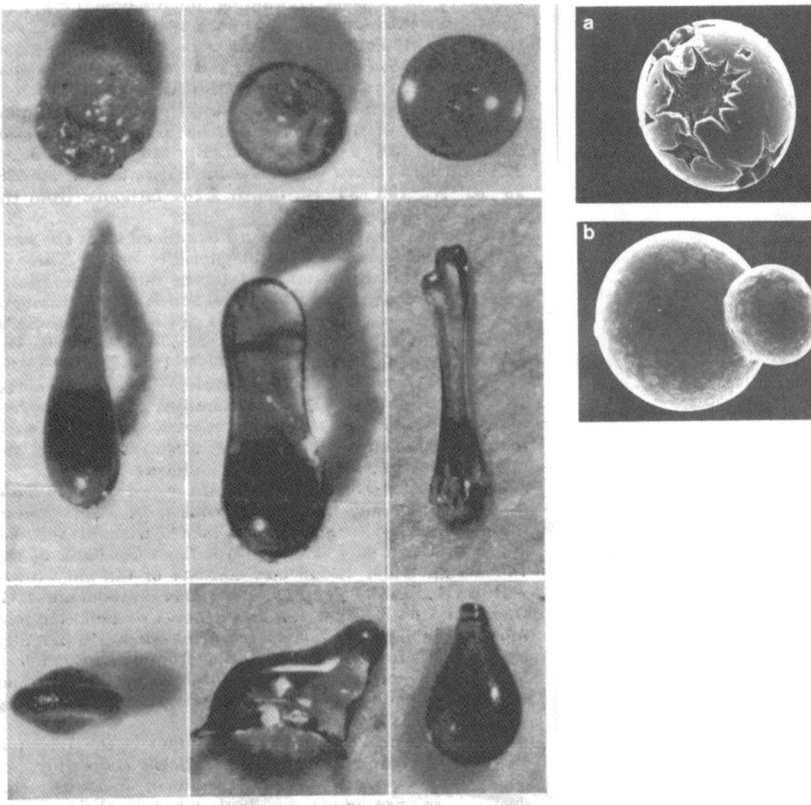

Figure 8-11: Right, Microtektites discovered in deep-sea drilling areas of the Caribbean (1 mm in size). Credit: Gerta Keller, US Geological Survey.

Figure 8-12: Left, Microtektites found in the Warton and Australian Basins. Their sizes range from .15 mm to .85 mm. Credit: B. Glass, Lamont Geological Observatory, New York.

glass beads of the same size and shape. Within six centimeters above these spherule beads are clay and other sediments that are a result of a large tsunami or other cataclysm.[185] These beads were found at different depths all the way along the Atlantic coast and along the Gulf of Mexico. Because of their age the beads had degenerated and do not have the smooth surface of glass beads found on the Moon or in other finds mentioned above.

Over 50,000 tektites were collected in the early 1960s, from South Vietnam. The samples included shapes that "indicated stretching: elongate drops, rods and straps that appeared to represent portions of the tails of drops that had been elongated to perhaps six inches or more, dumbbells, and discoidal shapes that showed whirlpool or swirling affects that had resulted from spinning."[186] Their conclusion was that the Tektites arrived in our atmosphere as individuals spinning very rapidly[187] like the glass beads found on the Moon. Tektites from Australia and the Ivory Coast were found associated with two magnetic reversals so it is assumed that the Tektites are a result of whatever mechanism happened at that time.[188] Another indication that the Sun novas at the time of the reversal. The Australian, Asian and Ivory Coast tektites differed in chemical composition but both were rich in rare-earth elements.[189] It does not surprise me to see chemical differences in these tektites, because there is no evidence that would exclude the possibility that the Nova's dust shell varies widely in its internal chemical composition. Tektites found in Thailand, the Philippines and Australia all had beryllium-10, one of the isotopes created by cosmic rays from space,[190] meaning the Sun. We do not know if there have been tests for aluminum-26. Other tektites have been found with both aluminum-26 and beryllium-10.[191,192]

Some tektites found in Luzon, Philippines, had small (.1mm to .5 mm diameter) iron-nickel spherules embedded inside.[193] The conclusion was that the spherules were extraterrestrial in origin, but not from the moon.[194] Iron and nickel are the same elements that make up most meteors, which I also believe originate from stars. The chondritic meteorites include metals and other elements that closely match the Sun.[195] The meteorites also exhibit cosmic-ray-induced particle tracks[196,197,198] that might come from the Sun.

I theorize they are materials thrown off by our Sun when it novas. The following is a description of what is known about them.

The chemical composition of tektites found in Australia and the Ivory Coast are chemically similar to other microtektites found on Earth and the Moon. They revealed an abundance of rare-Earth elements within a narrow range of atomic numbers, 57 to 71.[199] It has been suggested that these tektites are the product of volatile forces generating heat in excess of 2,800°C.[200]

The most telling evidence for the origin of microtektites is fission tracks found within them. A fission track is a very small hole produced as a by-product of nuclear reactions (they are also known as neutron particle tracks). Fission tracks were found in meteorites, tektites, mica and fossil bones. Giant fission holes have been found in some micas that "implies energies up to 14 million volts, much more energetic than those from any known natural α-emitting nuclei."[201] Figure 8-13 shows fission tracks in the meteorite Angra dos Reis.[202] Figure 8-14 shows a dinosaur rib found in the Gobi Desert in Mongolia. The same team of scientists also found the bones of turtles, small mammals, and rhinoceroses with fission tracks. Some of the sediment, the bones were found in, had little or no radioactive uranium ore.[203] In fact, most cosmic ray tracks found in meteorites are formed from iron and nickel atoms, not uranium.[204] Mica found in the eastern part of the United States, Québec, and Southern Rhodesia have fission tracks. Attempts have been made to date these samples, but the steady-state assumption of decaying ^{238}U has been inaccurate. I believe the source behind these neutron particles was cosmic ray particles produced by our Sun during a nova.

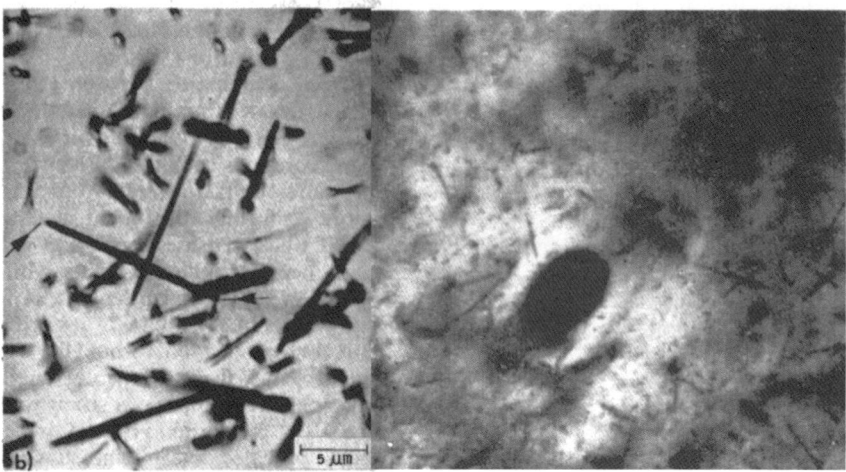

Figure 8-13: Fission tracks found in meteorite Angra dos Reis. Credit: N. Bhandari, Tata Institute of Fundamental Research, Bombay (Mumbai), India.

Figure 8-14: Fission tracks found in a dinosaur rib bone. Section size is 1 mm thick. Credit: Z. Kielan-Jaworoweka, Institute of Palaeozoology, Poland.

In summary, taking all of this physical evidence into account for these microtektites, it is my conclusion they are created by our Sun when it novas and that the fission tracks found in tektites on the Earth and the Moon were a result of cosmic ray particles thrown off by our Sun. It is, in essence, the "thumbprint" of our Sun.

Manganese Nodules

Manganese nodules have been found on the floor of the western Pacific Ocean in quantities great enough to be of economic value. They range in size from .1 mm to 20 cm. The average size is 5 to 10 cm. The two mysteries these nodules present are that they are not buried in the sediment but found on the surface. The second mystery is that they contain Aluminium-26, Berylium-10 and Thorium-230, three radioactive elements that are produce by cosmic rays from the Sun. The other point is these isotopes are only found on the first few surface millimeters.[205,206] This is just like the rocks found on the Moon that Thomas Gold analyzed. To try to solve this problem, scientists assume that the manganese nodules grew on the surface, but that would take tens of thousands of years. The incongruity of the problem brings them to a dead-end. The problem is solved if you realize these nodules are like the tektites and came from the nova. They are found in all the oceans and some lakes. As stated earlier, I believe rocks and bolder-sized debris are included in the ejected dust/matter shell of the nova. This would explain why the nodules are not buried and the presents of [26]Al and [10]Be that could only come from a star.

Evidence on Mars.

The model resulting from my theory calls for each of the planets in our solar system to be affected by recurring novas and its expanding dust shell. First of all the planets will loose some of their atmosphere when the dust shell hits. The planet will also be pushed a little further away from the Sun. Eventually, if the planet had water on it, most of the water will be lost and carried off into space in the form of comets. The surface of the planet would have numerous scattered rocks, which originated from multiple novas especially after the majority of the atmosphere had been removed over millions of years.

Man has learned at least three things from NASA's explorations of Mars:

1. Mars had oceans, rivers and evidence of water acting upon its surface—such as canyons and deltas.

2. The chemical composition of the soil tested was similar to the soil found on the Moon, and some layers on the Earth. The soils contained volatiles similar to those found on Earth. These volatiles were produced by cosmic rays interacting with the chemicals in the atmosphere.[207]

3. There were four-sided pyramids found on Mars that were over three kilometers high,[208,209] evidence indicating some highly advanced civilization once lived there.

Oceans and Rivers on Mars

Traditional theories have not been able to explain where all the water disappeared to that was formerly on Mars. The Viking I and II and Mariner flights have clearly shown that Mars once had large oceans and rivers, which also means that it once had an atmosphere on it.[210,211,212] The presence of rivers means that some time in the distant past it had clouds, rain and the normal weather cycle we have on Earth.

How can Mars have supported a civilization large enough to build pyramids, and other buildings, with a current atmosphere so thin and no water? Why are there rocks and dust spread over the surface of Mars? Figure 8-15 shows an ancient Martian ocean bed. Figure 8-16 illustrates a Martian riverbed, and Figure 8-17 shows some of the pyramids photographed on Mars.

Pyramid-shaped objects of considerable height were photographed on Mars (Figure 8-17). If true, this would mean that an advanced civilization had once

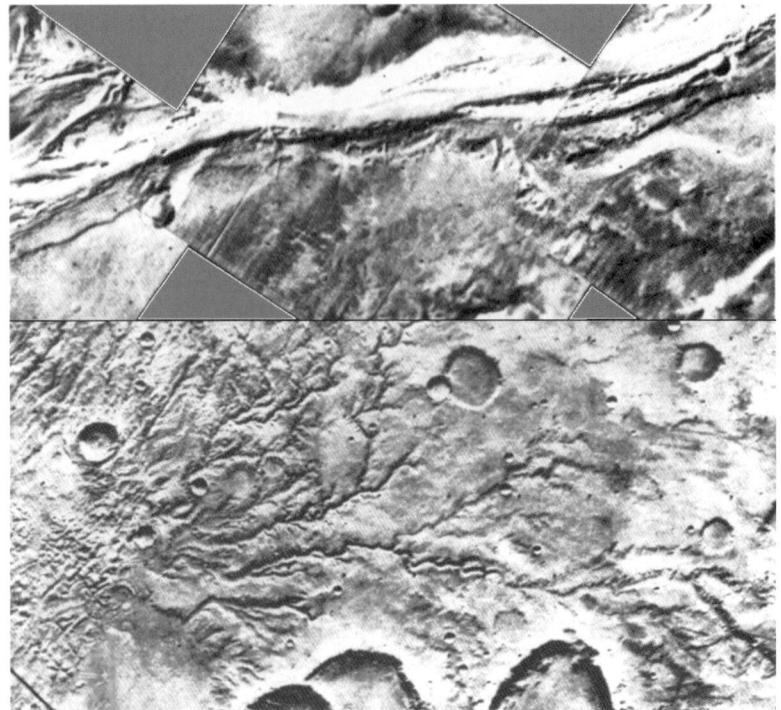

Figure 8-16: Top: An ancient Martian river basin showing stream terraces, bars, braided channels. Bottom: Viking frames 84A46-48 showing a classic meandering river bed. NASA.

Figure 8-17:
Pyramids and other ancient structures found on Mars. Frame 35A72; NASA.

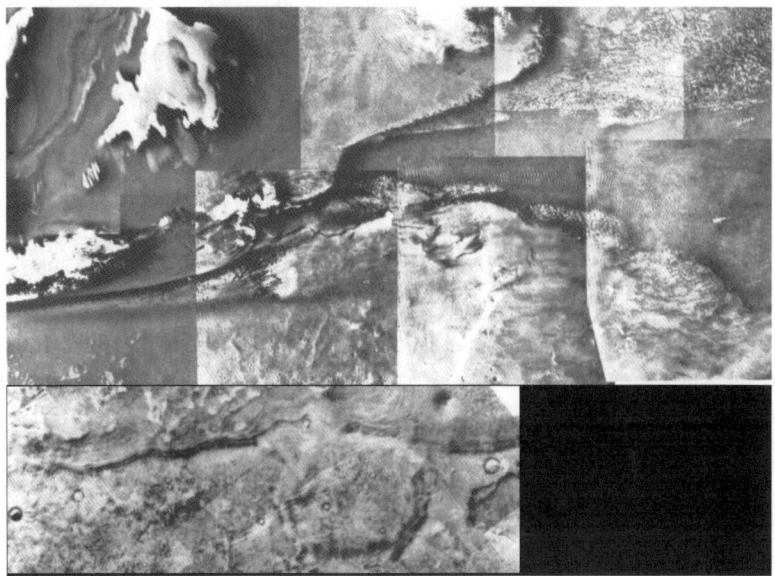

Figure 8-15: Top: Viking 2 frames 58B21 to 58B34 showing evidence of an ancient ocean channel. Bottom: Possible ancient seacoast. NASA.

lived on Mars. I believe such was the case, especially since my model concludes that Mars once occupied the same orbit the Earth currently has and had water on it. There is no reason to think it would not have developed life on it.

At one time, in the far-distant past, Mars was like Earth. Earth was closer to the Sun, at one time in the same orbit where Venus is today. During each nova, every 12,068 years, the exploding dust shell pushed the planets a little farther away from the Sun. The nova's deposited rocks, dust, and glass beads over the surfaces of the planets. Eventually, over hundreds-of-millions of years, the planets lost most of their water. They are eventually pushed too far away from the Sun to support life. Intelligent life-forms on the planet must have emigrated off of their dying planet and moved to the next one, closer to the Sun. This is very likely what happened to the civilization on Mars, and we Earthlings will have to do the same in the distant future. This is really a story of evolution. You must evolve to how the Universe really works.

The Atmosphere of Mars

The atmosphere of Mars has many elements common to those on Earth. There is a similar abundance of noble-gases[213] and volatiles (H_2O-water, CO_2-carbon-dioxide, Cl-chlorine, N_2-nitrogen, S-sulfur, Ar, F, H, B) on Mars and Earth.[214,215] A few of these gases can be explained by the verifiable volcanic activity on Mars, but they do not explain all of them.[216] The only source for the presence of these volatiles and noble gases is from the Sun, or cosmic rays from deep space. The rocks strewn on the surface of Mars seem to indicate they came from the Sun.[217] The rocks are present because, as the atmosphere got thinner, more rocks were able to penetrate the thin atmosphere and reach the surface.

Comets

Astronomers today still do not really know what creates short- and long-period comets.[218,219] I previously theorized that after each nova, the Earth looses some of its water. Guess where it goes? A comet is a big dirty snowball. It is mostly made up of water, dust and rocks. Every time a comet travels close to the Earth, NASA seems to do everything to avoid sending a test probe through the tail of one, to bring back a sample. I think they avoid it because they know what they will find!

After a nova the expanding hot dust from the Sun passes by the Earth evaporating great amounts of ocean water. The water is evaporated into the upper atmosphere, where it eventually changes into thick clouds, after the Earth cools down. But some of the water is captured by the dust shell and goes out into space. That is why comets have radically elliptical orbits, which eventually orbit the Sun. This is the obvious and simple explanation of what happened to Mars, and what will eventually happen to our Earth.

Effects on the Earth

Global Warming and the Sun

Today the public debate is whether there is global warming and, if there is, has man been the cause behind it by putting excessive amounts of carbon dioxide into the atmosphere? Carbon dioxide is supposed to keep solar radiation inside the lower atmosphere, instead of reflecting it out into space. Personally, I do not understand how their theory can explain the amount of increase, claimed. Within my model, the Sun's output will start increasing at least 100 years before the polar reversal, so I would expect the Earth to be heating up now. I do accept the fact the Earth is heating up. The question: Is man responsible for the warming, and what percent of it? This section will show how we measure solar output, its affects on the Earth, and answer this important question.

The main heat source for the surface of the Earth is our Sun—no debate there. We measure solar output by measuring five indicators: 1.) sunspot numbers; 2.) solar winds (including cosmic rays); 3.) solar output in the form of electromagnetic waves; 4. solar luminosity (Irradiance); 5. Sea Surface Temperatures (SST).

Sunspot Numbers

The Sun has two well known cycles, which I have mentioned previously in Chapter 3: The 11.09-year sunspot cycle, and the Gleissberg cycle[220] (eight sunspot cycles, Table 3-2). As shown in Table 3-2, the sunspot number has been increasing since 1894. The last Gleissberg cycle occurred in 1958 when the Sun erupted/flared-up so much that the science community wondered if their model of the Sun was correct. The next sunspot cycle after the Gleissberg cycle was predicted to drop down to a level that corresponded to the first sunspot cycle (1884) in the previous Gleissberg cycle, but it did not. The sunspot cycle in 1969 was 180% greater, and the third in the cycle (1991) was 254% higher than the 1905 cycle. This was a dramatic increase in solar activity, and most likely an omen of cycles to come.

Solar Winds and Cosmic Rays

The Sun throws off some of its matter (10^{-14} solar mass per year), in the form of solar winds, which correlate with a sunspot cycle.[221,222] The solar winds include electrons and cosmic rays which in turn create ^{14}C production[223,224] in the upper atmosphere. Dr. T. Wigley, stated: "By re-examining the possible link between major atmospheric ^{14}C anomalies and climate, I concluded that this link was real. This implies that significant irradiance variations occur in parallel with the solar fluctuations responsible for ^{14}C anomalies."[225] Over the past 100 to 150 years, the solar winds have increased along with sunspot numbers. He also stated that: "The agreement between times of major ^{14}C anomaly and times of globally-advanced glaciers (i.e. cool summers) is shown to be statistically significant."[226] It is becoming obvious that our Sun is a variable star increasing during sunspot maximum, and decreasing after it. But overall, it has increased more than it has decreased. There have also been discussions pertaining to whether the increase in solar irradiance is due to an increase in the solar diameter (.3 to .4%) over the past 300 years.[227]

It has been known that the magnetosphere reflects solar wind activity, and that in turn, means we can extrapolate what the solar wind was like back in 1900.[228] There have been "considerable changes in the solar wind over the past century. This implies some change in the state of the solar corona itself."[229] If you are wondering how the size of the corona affects you—this is what happens. I have wondered why the Moon was the exact correct distance between the Earth and Sun, so that it would perfectly conceal the photosphere of the Sun during a solar eclipse. Then all we see is the corona. I believe that God has built in a warning system into our solar system to warn us of an impending nova. At some point, the photosphere and corona will expand so much that at the time of a lunar eclipse we will see a large glowing doughnut around the Moon. At that point, no matter how stupid a civilization may be it will know that something terrible is about to happen to them.

Solar Output, Electromagnetic Waves

The radio noise (10.7 cm wave length and others) the Sun produces during solar storms linked to sunspots, manifests itself by ionizing the various layers of Earth's upper atmosphere. The solar radio noise has mirrored the steady increase in solar outputs of all types including the electromagnetic noise propagated from the Sun.

Solar Luminosity

The Sun's output changes over relatively short periods of time. Scientists have been monitoring solar luminosity using three satellites since 1978. Over just 20 years they had observed the Sun's luminosity increase.[230] Two rocket flights equipped with sensing equipment, measured for over 30 months, detected a .4% increase in solar luminosity.[231] Other scientists have made the connection between solar irradiance and sunspot cycles,[232] and are also connected to the Gleissberg cycle, with an increase in sea surface temperatures.[233]

Another method of measuring changes in solar luminosity is to measure the reflective brightness off of the planets Uranus and Neptune, and the moon of Titan orbiting around Saturn. From 1950 to 1975, measurements had shown an increase during sunspot maximum or decrease following the sunspot maximum, but overall there has been an increase since 1950.[234] This increase will continue until the polar reversal and the coming nova, which will be the ultimate in solar brightness.

Sea Surface Temperatures

The surfaces of the oceans are directly heated by the Sun. So by measuring the sea surface temperatures you can indirectly find out if the solar output has changed. The Sun heats up the surface of the oceans, which in turn evaporates more water vapor into the atmosphere, which creates clouds and weather fronts that circle the Earth. Therefore the Sun has a direct effect on the oceans and our weather. There are over 1,000 scientific articles that show a cause and affect relationship between Sun, weather and climate.[235] One such study revealed the correlation, covering 120 years, between sea surface temperature and the 88-year Gleissberg cycle. An article by George Reid proposed that the increase was related to a small increase in solar irradiance accompanying the solar cycle.[236] He wrote, "A positive correlation between solar luminosity and the intensity of solar activity was invoked by Eddy[237] to explain the apparent coincidence between the Maunder minimum in solar activity [very few if any sunspots] and the peak of the Little ice age in Europe and North America, as well as similar coincidences in early times between various climate indicators . . ."[238] As early as 1973 the correlation between solar activity, weather, and the growing season, was established. There was a clear relationship between the length of the growing season and the 11.09-year sunspot cycle. During the Gleissberg Cycle of 1958-59 the growing cycle was extended by 30 days compared to the sunspot minimum period.[239] A perfect correlation between the monthly-mean sunspot number and the mean-seasonal temperature variation clearly shows that solar activity is the majority cause behind global warming or cooling, depending upon solar activity.[240] The author of that article also felt that the velocity of the solar wind, or solar flux, rather than the number of sunspots should measure solar output.

I have done an analysis comparing sunspot numbers and the frequency of tornadoes from 1916 to 1994 and found that tornados increased at the same time as sunspot maximum (Graph 8-2). The reason for this could be that there is an increase in air temperature because of sea surface temperature increases, which creates more storms. Storms create tornadoes—and that might be the connection.

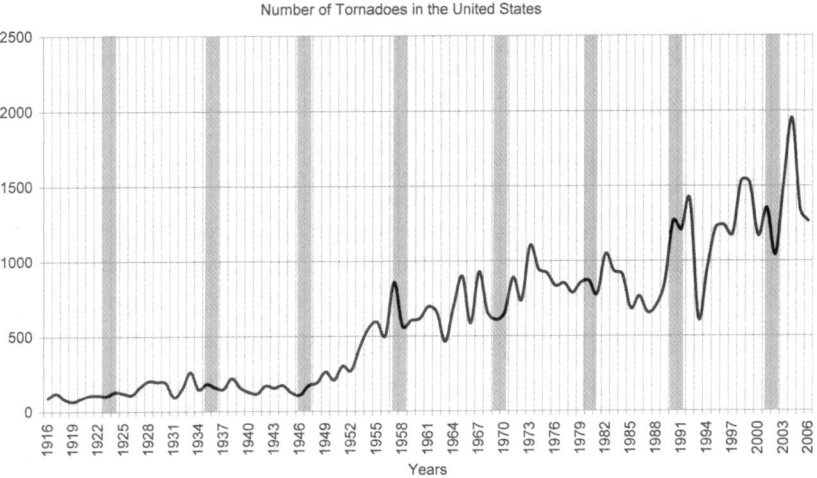

Graph 8-2: The vertical bars are the two years of greatest solar activity during sunspot maximum. It always precedes or accompanies an increase in tornado activity.

George Reid and Kenneth Gage in 1988 worked for the National Oceanic and Atmospheric Administration (NOAA). They performed one of the most complete studies on the connection between solar activity and sea surface temperature.[241] Their study was designed to compare the effects on global warming from the Sun and from CO_2. It clearly illustrated a direct correlation between increased solar activity (sunspots) and an increase in Sea Surface Temperature

(SST) (Graph 8-3). Their study also showed that when the Sun decreased its output after sunspot maximum the global average temperature also decreased (Graph 8-4). They concluded that "a modulation of the Sun's luminosity with a period of about 80 years and an amplitude of about 0.5% is consistent with the globally averaged direct SST measurements . . ." The conclusion of their paper stated that CO_2, if it does have an effect on air temperature, cannot explain the great increase in global warming over the past 100 years, and not on the sea surface temperatures.[242] Sea surface temperature is a greater determinant of future weather than CO_2 levels.

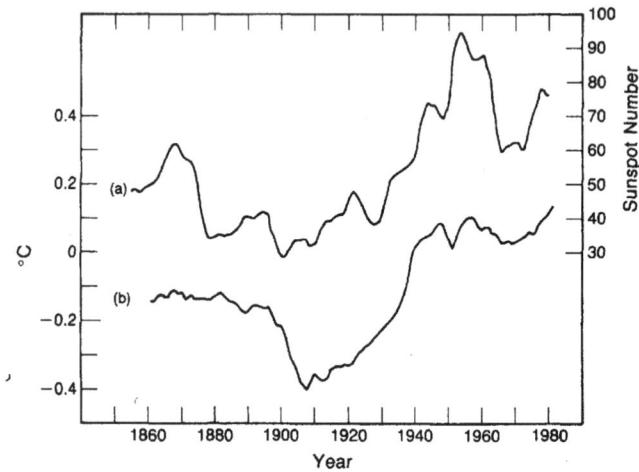

Graph 8-3: 11-year running statistical means of (a) sunspot number, and (b) global average SST anomalies. Credit: George Reid & Kenneth Gage, NOAA.

There is a conflict in the science community between two factions; on one side are those who say CO_2 is the major cause behind our current temperature increase. The other side believes that the evidence shows the current increase is due to natural cyclical increases in solar output that have occurred before. The importance of what Graphs 8-3 and 8-4 show is that CO_2 is not the dominant factor in global warming. If the pro-CO_2 side were right, world temperatures would not have gone down after sunspot maximum in 1958, but should have remained stable or continued to rise. As I remember, we did not stop driving our big gas-guzzling cars in the 1960's and coal-fired electric plants did not stop belching out CO_2.

One of the scientists in the forefront of the CO_2 global warming argument was Dr. Wallace Broecker from the Lamont-Doherty Geological Observatory, Columbia University. In his early journal article of August, 1975 he rang the alarm bells to the effect that CO_2 emissions, which industrial countries are putting into the atmosphere, were causing global warming.[243] I have several problems with some assumptions he made, as well as his conclusions. First of all, he made the

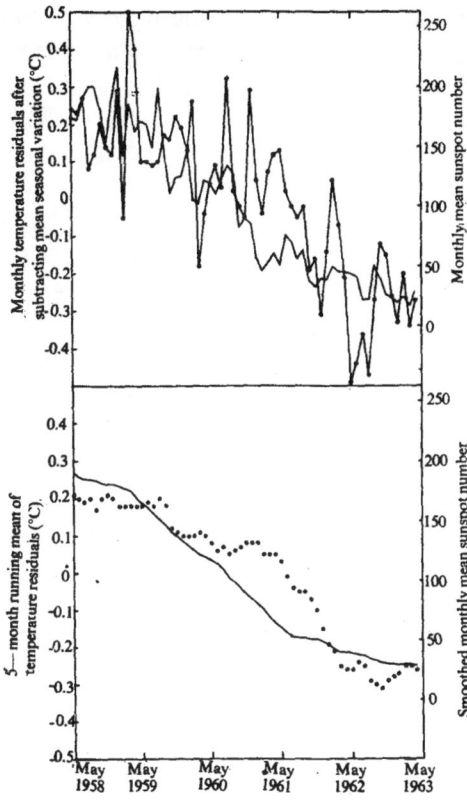

Graph 8-4: The top graph compares the average monthly temperature with the monthly mean sunspot number. The smooth line represents the sunspot number and the dotted line is the temperature line. The lower graph shows the same information, but the data points were smoothed to better show the trend. January 1958 was sunspot maximum. Credit: J. King Radio & Space Research Station, England. Also V. Starr and A. Oort.

assumption that 50% of all CO_2 emissions from the past and future will remain in the atmosphere—but there is no supporting documentation backing up this assumption. Nobody knows how much CO_2 remains in the atmosphere. He stated that the oceans and the plant life on the Earth absorb most of this CO_2. He cited another scientist[244] who estimated that the oceans remove 35 ± 10 percent of the CO_2 emissions. He does not give a percentage for the Earth's biomass (plant life) contributing to the additional removal. The point being, these are all assumptions and gross estimates. Nobody knows for sure what the CO_2 percentage is. Broecker gave estimates for temperature increases over the next 35 years (from 1975), showing an increase up to 1.1°C by 2010. We have had temporary increases during sunspot maximum, and several years after, before the temperature lowered again. When the actual weather history from 1900 to 1975 did not quite fit his model, he complained that the only accurate weather gauge is ice core samples from Greenland.[245] The ice cores contain the increase and decrease of the Oxygen-18 isotope. The graph he showed displayed the increase and decrease of this isotope compared to regular Oxygen-16. He then stated that the curves fit an 80- and 180-year cycle but did not mention that this is the Gleissberg cycle of 88.73 years. The ice cores are really showing the solar activity during the Gleissberg cycles because oxygen-18 is produced in the upper atmosphere by cosmic rays from our Sun. Which brings me to the last point—and the most surprising thing about his paper—Broecker did not mention the words "Sun" or "solar" anywhere. Keep in mind I previously mentioned that over 1,000 journal articles have been written about the Sun-weather connection, but I guess he chose to ignore them all. It is my opinion that Dr. Broecker's paper resembles a political document rather than a true science paper, testing both sides to arrive at the truth.[246]

The Ozone Layer

Ozone is three atoms of oxygen and it is expressed as O_3. The ozone layer is located in the lower part of the stratosphere (10 miles to 21 miles) above the Earth. Ozone is measured in Dobson Units (2.69×10^{16} ozone molecules per square centimeter). Ozone serves the purpose of preventing harmful amounts of ultraviolet light from reaching the surface of the Earth. Over the past 27 years there has been great concern that our ozone layer was disappearing at the two poles because of man-made chemicals for refrigeration. NOAA has had research stations at the South Pole monitoring the ozone layer.[247,248] Graph 8-5 shows the ozone layer decreased the most during two sunspot maximums. The sunspot cycle of 1991 started early, and that may explain the decrease in ozone starting in 1987. This observation tends to show that there is a solar cause-effect relationship explaining the decrease in the ozone layer at the poles. Other researchers have also seen this relationship, including solar proton events.[249,250,251]

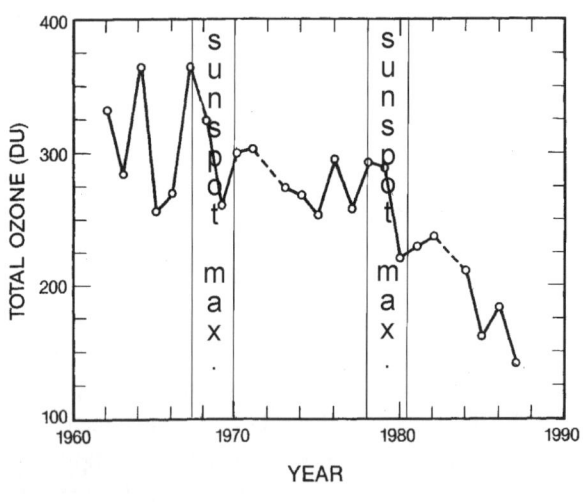

Graph 8-5: The decrease in ozone over the South Pole from 1962 to 1987. Credit: W. D. Komhyr, R. D. Grass, P. J. Reitelbach, S. E. Kuester,, *et al.*, NOAA.

Section Conclusion

In summary, the Earth's weather and SST is directly affected by the solar output. All of the above listed solar indicators of output, show an increase over time, and the increases have accelerated since 1958. I would expect that as we get closer to the polar reversal the solar output will dramatically increase, causing the oceans to warm up (maybe over 6° C). This will, in turn, melt the ice caps and warm the atmosphere. The warming will continue and ocean levels will exceed the present until the polar reversal. The important point to remember is that burning fossil fuels in our cars or coal-fired electrical plants does not cause the major portion of global warming. The vast majority of warming is caused by the variations in solar output that mirror each sunspot cycle. Current global warming would be happening now if *nobody* lived on the Earth, because it is independent from man's activities. In my opinion, it appears the CO_2 argument for global warming is really a political argument used by those who wish to control more of people's behavior so that power and money flow towards them.

The Magnetic Field of the Earth

The dominant theory for the magnetic field of the Earth is that there is a geodynamo generated by the liquid iron magma in the outer core of the Earth. The theory also holds that the center core is a cooler solid iron core (a stable core). Many have criticized this theory because of the melting point of iron and the temperature at which iron can no longer be magnetized (the Curie temperature).[252,253] Their theory for this geodynamo is dependent on convection currents being driven by seismic earthquakes, but that is pure conjecture. Earthquakes could not generate a fast-enough convection current to maintain a stable magnetic field. There are a lot of theoretical problems with the toroidal geodynamo model for the magnetic field of the Earth. The biggest problem their theory has is that "toroidal vector fields cannot act as dynamos"[254] because when magnetic fields cross themselves, they cancel themselves out and produce no voltage. The other problem is the heat source for the outer core. You can solve some of the theoretical problems by allowing for a heat source from below, but that means you have to have a very hot center core. It would have to be above the Curie temperature, which in turn means no magnetic field. Needless to say, their model has problems. Two observations prove the geodynamo model wrong. They are the radio noise given off by the Earth and the historic path of the geomagnetic dipole axis. Both will be covered later.

The Heat Source and the Magnetic Field of the Earth

I have already stated that my theory calls for a center modulation point in the Earth. All of the information for the planet is directed to that point in time and space. I believe the potential in the center is so great that the area would best be described as a plasma under pressure. For this reason I theorize the matter is not formed until it reaches the area we call the lower level of the outer core, some 700 miles from the center. In essence, it works on the same principle as the heat source in the center of a star. Later I will cover the correlation between sunspot maximum and an increase in volcanic activity (Graph 8-8). The center modulation point should also give off radio noise perpendicular (out along the equator) to the direction the information is being directed into this dimension (north and south poles), which should be identifiable. The modulation point does not necessarily have to remain fixed in its axis. It could vary a small amount, but in a recognizable pattern (Graph 8-6).

Earthquakes and the Core of the Earth

I should first explain what P-waves and S-waves are: When there is an earthquake it generates three types of earth movements. The P-waves are compression waves because they transmit through the Earth by pushing against the medium, and therefore travel fastest. The S-waves are shear waves because their waveform is from side to side, and therefore slower. The slowest are surface waves. Geophysicists would disagree with my assertion that the center core is a high-pressure gas plasma, because they would say that the P-waves created by earthquakes travel faster than the S-waves, and the P-waves are assumed to travel through the core. Scientists would not have considered that a compression wave could travel as fast through a plasma as it can through a solid medium, because their philosophy does not allow for it, therefore it does not come to mind. But I believe P-waves do travel just as fast through a gas plasma under high pressure as they can through a solid.

The Magnetic Field

I mentioned in Chapter 3 that magnetism is still a phenomenon in physics. Nobody knows why iron can display something we call magnetism. Only a few other elements display paramagnetic properties, and the rest do not. Now you may begin to wonder how geophysicists can theorize about the Earth's magnetic field without first knowing what magnetism is. Now you understand *why* they have not discovered what causes the Earth's magnetic field,[255] not to mention why it periodically reverses itself every 12,068 years.

Our Earth's magnetic field intensity has evidently not changed much since the Carboniferous period (280 to 345 million years ago) and the Permian period (220 to 280 million years ago).[256]

There are three observations that help prove my theory: radio noise detected around the Earth, the dipole wobble of the field, and the repeated polar reversals that occur cyclically. All three of these observations the geodynamo theory cannot answer.

Radio noise around the Earth

Spacecrafts orbiting the Earth have detected a multitude of radio frequencies in the ionosphere and magnetosphere and near the magnetic equator.[257] One type of wave is a ring of radio noise detected 5° north and south of the equator, something like the rings of Saturn. Another type of emission is categorized in three broad groups of natural origin. The first is called ULF (Ultra low frequencies, 10^{-3} to 1 Hz),[258] ELF (extremely low frequencies, 5 to 3000Hz) and VLF (very low frequencies, 3 to 30 kHz). These radio noises have been described as a hiss, a chorus, noise, and a whistling sound.[259] ULF waves and others are detected on the Earth's surface. ULF and VLF frequencies show up in the D-layer (30 to 55 miles altitude), E-layer (55 to 73 miles), and F1 and F2 layers (73 to 244 miles) in the ionosphere. There are several theories for the cause of these radio frequencies, but none have worked with existing models of the core and geodynamo theory of the Earth. What we do know is, "these waves are linearly polarized along the magnetic field, which in turn implies that they are propagating perpendicular to the magnetic field. . . ."[260] That means these radio waves are coming from the direction of the Earth's core.

My explanation for the radio noise propagating along the equatorial disk is that if you make the assumption that the direction for the Earth's information comes from the direction of the poles towards the core of the Earth then the information would create an electrostatic field (radio noise) perpendicular to the axis of the Earth, and would propagate outward from the core like a disk or the rings of Saturn.

The other radio noise found in different layers of the ionosphere is like group-waves forming around a center frequency source. The multiple frequencies are not difficult to understand, since the amount and variety of information that make up the Earth match the wide spectrum of frequencies emitted from it. I know that much of this discussion is technical, but the point you should get out of it is that there is a simpler, more logical explanation for all of the observations of what we know about the Earth's magnetic field and its heat source.

The Dipole Wobble

The magnetic field of the Earth does not stay in one place—it moves. Graph 8-6 shows the estimated path the magnetic North Pole takes over an estimated 1,500 years.[261] This movement of the magnetic poles is called a wobble. Notice

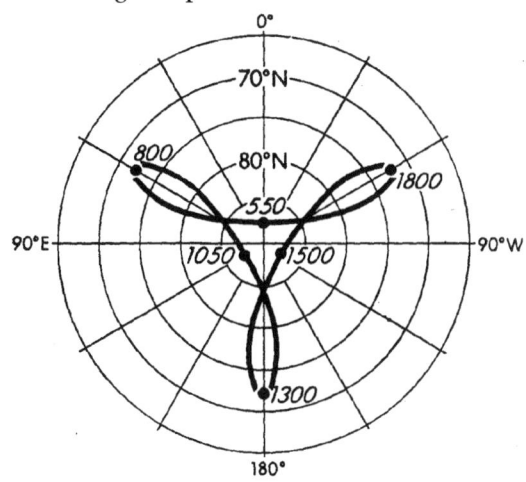

Graph 8-6: A model of the path of the magnetic North Pole over the past 1,500 years. Dates were from data derived by N. Kawai and K. Hirooka, 1967. Credit: J. Jacobs, University of Alberta, 1971.

the path is not random but indicates a very organized pattern. What is also very interesting is that the estimated 1,500 years to make a complete cycle is very close to 1,508.5 years, which would be one-eighth of a 12,068-year clock cycle.

Polar Reversals

There is no debate that the Earth's magnetic field reverses itself periodically. The only debate is how often it reverses and how to date the sedimentary layers between reversals. As I stated earlier, the polar reversal occurs within one day. "A decrease in the [magnetic] intensity of the field of about an order of magnitude occurs immediately before the reversal, while its orientation remains substantially unchanged. The onset of the reversal is marked by abrupt swinging of the virtual geomagnetic pole along an arc of a great circle."[262] Current theory states that the last polar reversal was about 700,000 years ago,[263] but a great deal of field evidence has shown that reversals have happened much more recently. The short-term reversals are called "magnetic excursions," presumably to avoid the philosophical problem of proving that the "accepted theory" is obviously wrong. The generally-accepted practice is that, if the magnetic reversal time is shorter than 100,000 years it is called an *excursion*. If longer than that it is classified as a polar reversal.[264,265] "Should a completely antipodal field direction be attained [polar reversal], then in order to be classified as an excursion, and not a reversal of the field, it should not persist for longer than about 10^4–10^5 [10,000 to 100,000] years. This is considered to be the usual time for a complete oscillation in the intensity of the global field."[266] As you can plainly see the length of times given are assumptions with no hard evidence for the dates. I assume this is done to ignore short-term reversals that conflict with "accepted" theory. Geophysicists have come to the realization that the field does reduce before the reversal,[267] but they are not sure if a polar reversal accompanies the magnetic field reduction. Now we can begin to understand why geophysicists have not figured out why there are polar reversals every 12,000+ years. What is written in textbooks is something contrary to what is found in the field and the journals. So students learn that the last polar reversal happened 700,000 years ago, which is just not true.

Another dating problem arises because scientists estimate an ocean sedimentation rate of only 2 ± 2 millimeters per 1,000 years. This rate is far less than actual, once you realize the mechanism that occurs during the reversal. Some scientists have been more direct and did define the technical problems they face: ". . . the Earth's field operates like a remarkably symmetrical oscillator or, more precisely, like a bistable flip-flop circuit." ". . . short polarity intervals occur, with durations of less than 10^5 [100,000] years. The latter have proved to be the most difficult to measure."[268] "Events shorter than 3×10^3 [30,000] years are below the level of experimental noise . . . there is a loss of information about short polarity intervals because the magnetizing processes in sediments . . ."[269] Dating of sediments for periods shorter than 30,000 years is difficult because magnetic artifacts are below the level of experimental noise, so shorter periods have not been conclusively resolved using magnetic profiles.[270] As Allan Cox flat-out stated ". . . early researchers were justifiably conservative in not identifying short polarity intervals unless they rose clearly above this noise level."[271] Translation: when scientists discovered a magnetic field reversal and it was weak, they ignored it, instead of trying to find a stronger sample.

Variations in the rate of sedimentation, noise due to stratigraphic gaps, chemical changes in the ferromagnetic minerals, and the effects of organisms have also made dating short events difficult, though not impossible.

This gives you a good picture as to why geologists have problems measuring and recognizing short-term polar reversals. It is because of technical and philosophical reasons, not because the evidence is not there. Currently accepted theories do not allow for cyclic magnetic reversals occurring evenly throughout time. Once you accept the reality of this phenomenon, you inevitably come to the realization that this clock mechanism must be coming from outside of our reality. This is also why geophysicists have not figured out what causes the heat source and magnetic field of the Earth.

As I have stated repeatedly before, the polar reversal is initiated by the clock cycle in the Diehold, which occurs everywhere in the Universe, every 12,068 years. The effects of the polar reversal on the Earth are immediate, resulting in magnetic field reversal and the Earth's axial rotation stopping and then reversing direction. The Earth does not flip over as some might think. The actual reversal time may last seven to eight hours, during which the Earth would not rotate at all. I have already mentioned the other events that occur at this time, so I will not repeat them. The next section will present evidence that polar reversals occur every 12,068 years.

Short-term Magnetic Reversals

Scientists determined the duration of magnetic polarity by examining magnetic residue left in magnetized rocks and sediments. Sometimes they are helped by finding organic material they can date using ^{14}C or ^{37}Ar methods. These methods are much more accurate than making assumptions about rates of sediment accumulation. Table 8-5, in the previous section on ice fields, listed a large amount of evidence which clearly shows ice ages occurring roughly every 12,000 years. Table 8-6 shows magnetic reversals using sedimentary evidence, but no glacial moraine samples as was covered in Table 8-5.

Our own magnetic field is decaying at an increasing rate. One geologist had calculated, using a linear decay rate, that the Earth will have a polar reversal about 2,230 C.E.[272] However, magnetic fields decay exponentially so it will happen

12,000 Years Before Present (B.P.)

Sample	Location	Years BP	Author	Journal	Vol.	Pg.	Date
Clay, lavas, lake sediments	Mt. Kullen, Scania, Sweden	12,400-12,100	R. Thompson	Nature	263	490	10/7/1976
Marine clay & silt	Gothenburg, Sweden	>12,230	Morner, Lanser, Hospers	Nature	234	173	11/26/1971
Various	California	13,300	M. Barbett, M. McElhinny	Nature	239	327	10/6/1972
Marine clay & silt	Viby, Sweden	12,400	M. Noel, D. Tarling	Nature	253	705	2/27/1975
Lake Muds	Lake Erie, USA	12,500	M. Noel, D. Tarling	Nature	253	705	2/28/1975
Lake Muds	Mono Lake, USA	<13,300	M. Noel, D. Tarling	Nature	253	705	3/1/1975
Lake Muds	Lake Windermere, UK	13,400 ± 400	M. Noel, D. Tarling	Nature	253	705	3/2/1975
Lake Muds	Lake Chalco, Mexico	14,450	M. Noel, D. Tarling	Nature	253	705	3/3/1975
Marine muds	Gulf of Mexico, USA	12,500	M. Noel, D. Tarling	Nature	253	705	3/4/1975
Varved clay	Blekinge, Sweden	12,077-12,103	M. Noel, D. Tarling	Nature	253	705	3/5/1975

24,100 Years Before Present (B.P.)

Sample	Location	Years BP	Author	Journal	Vol.	Pg.	Date
Various	California	30,400	M. Barbett, M. McElhinny	Nature	239	327	10/6/1972
Human cremations	Lake Mungo, Australia	26,270 ± 470	M. Barbett, M. McElhinny	Nature	239	327	10/7/1972

48,400 Years Before Present (B.P.)

Sample	Location	Years BP	Author	Journal	Vol.	Pg.	Date
Baked trees, rock	Laschamp, France	45,400 ± 2,500	C. Hall, D. York	Nature	274	462	8/3/1978

Table 8-6: Dating non-glacial evidence of magnetic reversals.

a lot sooner than that. Sediments listed in the tables were found near polar reversal or ice age boundary layers.

The Jurassic Period

The Jurassic period is famous for its dinosaurs and how the period ended. Some think it came to an abrupt end when a comet or meteor hit just south of the Gulf of Mexico and killed most life on the planet. At least that is what the science community theorizes. I don't think so. The close of the Jurassic period saw the creation of flying reptiles, birds and, perhaps mammals. Something hitting the Earth will not cause a magnetic reversal, or create new species, or the alteration of existing species. Only cosmic and gamma rays from a solar nova will alter genes. Data from Colorado "indicate that the Late Jurassic was a time of frequent reversals of the magnetic field with reversed polarity predominating. Thirteen intervals of polarity can be defined."[273]

Plate movement and Sea Floor Spreading

The Earth is made up of a number of continental plates. Where these large plates meet, lava and magma seep up to the surface and push the plates apart. This process is called sea floor spreading.

The South Atlantic Ocean ridge has been studied for a distance of 1,500 kilometers, perpendicular to the ocean ridge. Slight changes in the magnetic field are detected by towing a magnetometer near the bottom of the Ocean. The one problem is that there is a lot of "geological noise" which prevents detection of magnetic reversals shorter than 20,000 years.[274] Results from lava flows clearly indicate that polar reversals have occurred in equal time intervals. It has been found that this linear relationship exists on ocean floors all over the world.[275] Graph 8-7 shows two charts of magnetic changes on the sea floor, spreading over the mid-Labrador sea ridge. Notice the even durations of magnetic polarity.[276]

Estimates for the rate of sea floor spreading have been made for all the oceans. The North Atlantic sea floor is spreading apart at the slow rate of 1 to 2 cm per year. That translates to 396 feet to 792 feet per 12,068-year polar reversal

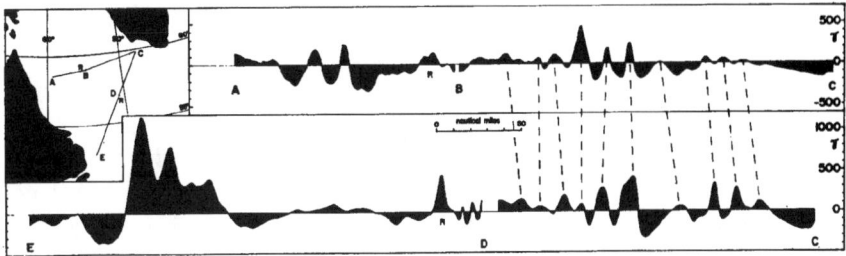

Graph 8-7: Sea floor spreading and magnetic reversal remnants found in lava flows. Credit: P. J. Hoon & E. Godby, Geological Survey of Canada.

cycle. The South Atlantic and Indian oceans are spreading apart at 2 to 3 cm/yr (1,207 feet per reversal cycle), and the Pacific Ocean is at 4 cm/yr (1,590 feet per reversal cycle).[277] These measurements do not preclude the distinct possibility that just before, during and shortly after a polar reversal, it spreads at a much greater rate—we just do not know how much. But taking the existing estimates, you would expect to detect a different polarity every 792 feet to 1580 feet, depending on the location. That means, if there is too much sediment over one segment, the detector may not pick it up and one would miss a whole magnetic reversal. This is one of the problems scientists face detecting these remnant magnetizations. Regardless, the evidence is clear that magnetic reversals occur equally through time, which shows a definite clock cycle at work here.

A Warmer Earth in our Distant Past

Scientists know the Earth was warmer (7 to 14 degrees centigrade)[278] and wetter during the Cretaceous period 100-million years ago,[279,280] which suggests that our Earth was a lot closer to the Sun. The Cretaceous period was dominated by reptiles, which are cold-blooded animals that need higher levels of ultraviolet light to help digestion. These attributes imply that the Earth was a lot warmer and closer to the Sun. The repeated novas are the reason why the Earth is farther away from the Sun today than it was in the distant past. "On a geological time scale, the Earth's climate has been getting rapidly cooler."[281] What happened? The expanding dust shell from the nova hit the Earth and pushed it a little farther away from the Sun. The nova also evaporated a great deal of water. Some of this water was carried away from the Earth into outer space.

Our Earth is currently experiencing another type of global warming. This time it is due to solar output increasing since the 1890s. This is caused by the approaching clock cycle as magnetic fields (information) on the Sun and Earth start collapsing.

Global Increase in Volcanic Activity

At the time of the coming polar reversal the Earth will stop its rotation for about eight hours and then rotate in the opposite direction.[282] This reversal action will cause tremendous Earth movements, because the continents, in essence, "float" on a sea of magma. At the time of the reversal, the continents will have the same forward velocity of the previous rotational direction and will bang into each other until they assume the direction of the new, reversed rotation. The plate movements will cause increased volcanic activity because the movement squeezes the magma chambers in the crust of the Earth. Years before the polar reversal there will be an increase in volcanic activity because the mantel will have heated up and, I am sure, for years after.[283] I did an analysis comparing sunspot cycles from 1890 to 1992, with volcanic eruptions (Graph 8-8), and found that every single sunspot maximum was accompanied by an increase in volcanic activity.

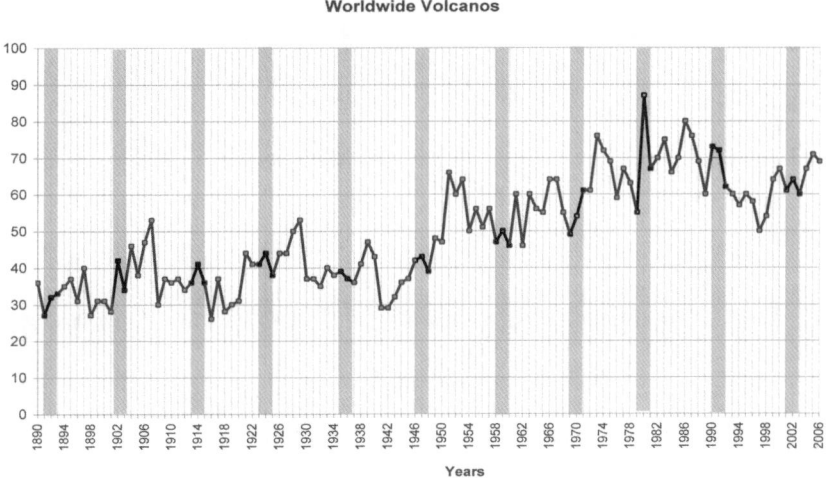

Graph 8-8: Sunspot cycles compared to volcanic activity worldwide. Every sunspot maximum accompanies or precedes and increase in volcanic activity. The gray bars are the 2 years of maximum sunspot activity.

Geologists have recorded that over the past two million years there has been increased volcanic activity at the same time as the ice ages[284,285,286,287] and polar reversals.[288,289,290] "The possibility that large Pleistocene ice sheets were initially triggered by massive volcanic eruptions is not contradicted by the geological record. There is an apparent correlation between major phases of global ice advance and ash eruptions in New Zealand, Japan and southern South America over the past 12,000 years."[291] They found volcanic ash just above the reversal sediment, and below the glacial till. I theorize that the heat source at the center of the Earth is a center modulation point. At this point many frequencies are directed. When the main clocking frequency collapses it will create a tremendous spike of energy. The energy potential created will in turn create additional heating of the outer core and mantel of the Earth. As the Earth gets closer to the polar reversal there will be an increase in earthquakes and volcanic activity. The increased heating of the Earth's outer core and mantel will cause increased volcanic activity.

Table 8-7 illustrates magnetic reversal dating, using volcanic sediments and flows. A combination of dating methods was used, but those listed are all occurring on, or near, the estimated date for a past polar reversal.

Volcanic activity at the time of a polar reversal 12068 B.P.

Objects Dated	Location	C14 Age/B.P.	Author Ref.	Journ.	Vol.	Pg.	Date
Organic material	Gothenburg, Sweden	13,750–12,350 B.P.	R. Farbridge	Nature	265	430	1977
Burnt trees	Laschamp volcano area, France	12000 B.P.	M. Noel, A. Cox	Nature	253	705	1975
Marine clay	Gothenburg, Sweden	12,400 B.P.	N. Morner, et. al.	NPsci	234	173	1971
Lake sediments	Lake Windermere, UK	<13,000 B.P.	K. Creer, et. al.	EPSL	14	115	1972
Various	Gothenburg Excursion	12,350 B.P.	N. Morner, et. al.	NPsci	251	408	1974
Lake sediments	Lake Eire, USA	12,500 ± 1,500	K. Creer	EPSL	31	37	1976
Sediments	Blekinge, SE Sweden	12,103 ± 150	M. Noel	GFForh.	97	357	1975
?	Japan	12,700 – 13,100	J. R. Bray	Nature	252	679	12/20/1974
Carbonised plants	Santorini Volcano	12,950 ±756	H. Pichler, W. Friedrich	Nature	262	373	1976
?	Southern South America	13,500 – 14,000	J. R. Bray	Nature	252	679	12/20/1974
?	New Zealand	13,800 – 14,700	J. R. Bray	Nature	252	679	12/20/1974
?	New Zealand	12,500 – 13,400	J. R. Bray	Nature	252	679	12/20/1974

24,136 B.P.

Objects Dated	Location	C14 Age/B.P.	Author Ref.	Journ.	Vol.	Pg.	Date
?	Japan	24,100 – 24,900	J. R. Bray	Nature	252	679	12/20/1974
?	New Zealand	26,300 – 27,900	J. R. Bray	Nature	252	679	12/20/1974
^{14}C	Aira-Tn & Ito, Japan	21,000 – 22,000	M. Rampino, S. Self., et.al.	Science	206	826	11/16/1979
^{14}C	Ata, Japan	24,500 ± 900	M. Rampino, S. Self., et.al.	Science	206	826	11/16/1979

36,204 B.P.

Objects Dated	Location	C14 Age/B.P.	Author Ref.	Journ.	Vol.	Pg.	Date
Fireplace sediment	Lake Mungo, Australia	30,780 ± 520	M. Barbetti	Nature	239	329	10/6/1972
Organic	Czechoslovakia	32,000 ± 4,000	V. Bucha	Geomag	22	253	1970
Carbonized tree	Santorini Volcano	36,700 ± 950	H. Pichler, W. Friedrich	Nature	262	373	1976
Baked clay	Massif Central, France	33,000 ± 4,000	J. Huxtable, M. Aitken	Nature	275	207	9/21/1978
^{14}C	Campanian Tuff-Ischia, Mediterranean	30,000 – 35,000	M. Rampino, S. Self., et.al.	Science	206	826	11/16/1979

48,272 B.P.	Location	C14 Age/B.P.	Author Ref.	Journ.	Vol.	Pg.	Date
Volcanic material	Laschamp volcano, France	46,700 ± 2,100	C. Hall & D. York	Nature	274	462	8/3/1978
Organic	Lake Biwa, Japan	49,000	K. Yaskawa	JGGeog	25	447	1973
60,340 B.P.							
Volcanic material	Laschamp volcano #443	53,500 ±12,800	C. Hall & D. York	Nature	274	462	8/3/1978
EPSL: Earth Planet. Sci. Letters	NPSci:Nature Phys. Science		GF Forh.: Geol. Foren. Firh.				

Table 8-7: Dating of volcanic sediments occurring about the same time as a polar reversal.

The Mass Extinctions and Creation of New Species

This has to be one of the most baffling secrets, if you try to use traditional theories of what causes ice ages. What is known is that a fossil of one species is found below the polar reversal layer. Then, one can see clay and/or glacial till material above that.[292] New species related to the former species are then found above that. After a polar reversal many species are reported to have changed to varying degrees. After major reversal periods, scientists have seen plants, fish, mammals and reptiles change, some more than others. Extinctions have been correlated to low sea levels,[293] which is what happens after the nova. Some species of Antarctic Radiolaria disappeared at the time of magnetic reversals during the past five million years.[294,295] "Radiolaria are holoplanktonic protozoa widely distributed in the oceans. They occur throughout the water column from near surface to hundreds of meters depth."[296] The mechanism for the extinctions is simple to figure out, once you understand what happens during the reversal. It is amazing anything survived this devastating event.

Two respected scientists of note are John Sepkoski and David Raup (University of Chicago). They have one of the most complete data sets of marine fossil records. They discovered a repeating pattern of extinctions. They found that "a 26-million-year cycle through the past 240 million years, is some kind of statistical artifact, . . . So far the signal has refused to be statistically massaged out of the data. 'Although it causes me some considerable philosophical anguish,' says Sepkoski, 'the periodic signal does begin to look real.'"[297] It is very possible that there is a clock cycle in the Diehold which causes a very large nova that does considerable damage to our planet. There is enough evidence, I have found, to conclude that there are cycles within cycles, so it is entirely possible that one of these clock cycles could be devastating.

The last polar reversal and ice age in North America saw the extinction of 57 species of large mammals and 21 small species.[298] They included the American ground sloth,[299] horses, mastodons, mammoths, saber-tooth tigers, and others. Man has been ruled out as the major cause of this because some species of birds became extinct at the same time, and it is not believed that birds were on cave men's dinner menu. Besides, the current theory assumes man did not repopulate North America until several thousand years after the polar reversal. Man, in substantial numbers, was not present until 2,000 years after the reversal and that was about the end of the ice age.

Man's physical evolution was also due to polar reversals and the amount of exposure humans endured, and what side of the Earth they were on at the time of the polar reversal and nova. Neanderthal bones were found below sediments of the last ice age, and modern man was found above. But there is no missing link. The last polar reversal appears to mark the end of the Neanderthals.[300] There has been an academic battle, since Darwin, between the Creationists and the Evolutionists. I hope to settle this battle here and now.

The reason there is no missing link—both sides are partially right and partially wrong—is because evolutionary changes occurred during the polar reversal, and "thought" has something to do with how we evolved. There are only three ways, in nature, to alter the DNA pattern that makes up all living things: cosmic rays, gamma rays, and ultraviolet light. All evolution has been upward. No de-evolution has ever been found. When the Sun novas it produces lots of cosmic and gamma rays. That is what put particle holes into tektites and fossils. This radiation was certainly random when it hit. In order to have upward evolution, you must have selected impacts to alter the correct chromosomes to make these changes. At this point it becomes obvious that "thought" or "consciousness" takes part in the process! It is not important for us, at this stage of our evolution, to know exactly how this process happens. It is essential, however, to recognize that this is an important link.

Extraterrestrial Causes

I am not the only researcher who has come to the conclusion that a nova or supernova could have caused mass extinctions.[301,302,303,304] The problem is, the others do not want to say that our Sun is the culprit except a few, and they voice it very gingerly. The article by Dale Russell and Wallace Tucker states:

> These affects are not necessarily unique to a supernova explosion. An intense source of high-energy radiation is all that is required. Another possibility is an anomalously large solar outburst. . . . the extinctions were of unusual magnitude and geologically brief duration, and may have been accompanied by a thermal drop. . . The supernova theory does not conflict with the record of extinctions at the Cretaceous-Tertiary boundary as at present understood. It has merit in that its predicted affects can be compared to the record to a greater extent than some other theories of mass extinction.[305]

It would not surprise me to find that these two authors, as well as Thomas Gold, knew that the Sun novaed but they could not figure out the mechanism so they only hinted at the possibility.

Fast-Frozen Mammals

On the northern slopes of Siberia, thousands of frozen mammoths and other mammals were found. During the 19th century, the Russians collected ivory tusks from these frozen animals and sold them all over Europe. The flesh from these animals was still fresh enough to feed their sled dogs. Undigested grasses were found in the mammoths' stomachs. Scientists had concluded these immense animals were fast-frozen at the time of the last ice age, but no one has ever figured out the mechanism that caused it—until now.

During the last polar reversal, the Sun's dust and particle shell hit over the Atlantic Ocean.[306] It removed much of the upper atmosphere on the Sun-side of the Earth. After the dust shell passed by, the Earth was left with one side having extremely low atmospheric pressure and the other side with "normal" pressure. What happened next was the atmosphere on the backside of the Earth expanded rapidly outward to fill up the low-pressure side of the Earth. Boyle's law states that temperature goes down when a gas expands. The coldest temperatures recorded on the Earth are in Vostok, Antarctica, where it has gone down to −129°F. In Northern Alaska temperatures go down to −70°F in the winter but during this cataclysmic event the temperature on the backside of the Earth must have dropped very rapidly freezing within minutes everything that was exposed and in the open. It would not surprise me if within one hour after the dust shell passed the Earth it got below −200°F on the back side of the Earth. If people are not in a deep cave, or a well-insulated waterproof structure, they will surely freeze to death within minutes, depending on how well the shelter is protected.

When is the Next Polar Reversal?

This is the all-important question. To be forewarned is to be forearmed is my best advice. I know this book will scare the living daylights out of most intelligent people, because the proof is irrefutable. We have seen how these different ice age events fit together into one coherent philosophical model. Nevertheless, ask yourself: "What would happen if I had never written this book?" I believe that none of you would know that there is going to be a polar reversal in your immediate future. In all probability, academia will fail you, and governments (world-wide) will keep quiet and try to save themselves, if they find out at all! There is nothing like a shelter filled with bureaucrats. How would you like to start a new civilization with only those people? If I had not written this book, most likely, none of you, or your children, would survive the coming polar reversal!

You now have a very complete picture of what will happen during the next polar reversal and what happened in the past. I have even given you a date-range, September to December 2046, when the next Gleissberg cycle maximum will occur. I also stated in Chapter 3, that the Torah gives the exact date for the next polar reversal. One of the principles I have mentioned about the Torah is that whenever it mentions something for the first time it is of paramount importance. The first time it mentions a month and day is when the sky opened up during the time of Noah's flood, found in Genesis 7:11-12.

> [11] In the six hundredth year of Noah's life, in the second month, on the seventeenth day of the month, on the same day were all the fountains of the great deep broken up, and the windows of heaven were opened. [12] And the rain was upon the Earth forty days and forty nights.

First, ask yourself: "Do we really care what the month and day it was for the polar reversal at the time of Noah, 12,000-plus years ago? And the answer is, No! We do care what year it was, but the month and day has little significance for us today. Another clue is given in Noah's age of 600 years, which converts into the exact number of years between the polar reversals $(600 \times 24.136 = 14,481.6 \div 12 = 1,206.8$, or $12,068)$. So, based on this assumption, I believe God was revealing to us exactly when the next polar reversal will occur. I went to the NASA website and went to their 5,000-year lunar calendar, to see when the new Moon would occur in September, 2046. I then calculated when the first month in the Jewish calendar would occur. It turned out that the end of the first month happens on September 29, 2046, so the first day of the second Hebrew calendar month starts on September 30. The 17th day of the second month in the Hebrew calendar will be October 16, 2046.

Gen.	Name	years to first birth
1	Adam	130
2	Seth	105
3	Enosh	90
4	Kenan	70
5	Mahalalel	65
6	Jared	162
7	Enoch	65
8	Methuselah	187
9	Lamech	182
10	Noah	500
11	Shem	100
12	Arpachshad	35
13	Shelah	30
14	Eber	34
15	Peleg	30
16	Reu	32
17	Serug	30
18	Nahor	29
19	Terah	70
20	Abraham	100
	Total years	**2046**

My next challenge was to see if the Torah contained the actual year for the coming polar reversal. In Chapter 2, I showed the multiple occurrences of the number 12,068 (in code) found in the Hebrew Scriptures. Two of the occurrences were in the generations of Adam and Shem (Chapter 2, Table 2-1). It took me several years to discover how the actual year for the polar reversal was encoded but I found it in 2001. The date was found by totaling the years when each generation had his first-born son from Adam to Abraham. Adam is the exception because the Torah does not tell us how old Adam was when he had Cain and Abel. So we count from his third son, Seth. We arrive at the correct number: 2046. Table 8-8 lists those generations:

Table 8-8: Table shows the age at which each patriarch fathered his eldest son. The total years gives the exact year for the next polar reversal This again shows the Torah is entirely a code book and was Divinely inspired.

What is so profound about this table is that the number 2046 is a date using the Christian calendar, which was created by the Catholic Church after 563 C.E.

Moses could not have known what this number meant, nor would any of the other priests following him. The message is obviously intended for us, in our time. The number is there to warn us when the polar reversal will occur.

I believe the correct date for the next polar reversal—or as the Hebrew Scriptures calls it, "God's Day of Judgment"—will be October 16, 2046. Now you may ask, "Are there any ways of checking the importance of this date?" There is one, but I am going to describe that in Chapter 11, on "Time Lines and Patterns."

Who Else Knows about this Great Secret?

To discover any great secret, you must first find the event or action that caused someone, group, or government to react. The event that began the space program, and all of the research on the ice age and solar studies, was solar cycle number 19, the last Gleissberg Cycle. The Sun produced more flares, sunspots and output than it had ever produced before. So questions were raised: Is our model of the Sun correct? Is it powered by nuclear fusion? What was the cause of the last ice age? Could the Sun have played a part in it?

Over the past 35 plus years, I had studied all the issues of the science journals, *Science, Nature,* from 1958 to 1993, and *Geology* from 1965 to 2002. Over the years, I have also read hundreds of journal articles from astrophysics, geophysics and space journals. I have come across articles which suggested the Sun had novaed. Sometimes they did not directly say "nova;" instead, they wrote things such as: "giant solar outburst in geologically recent times," or "very minor nova-like outburst."[307] A group of scientists from NOAA, and the National Center for Atmospheric Research, which I referenced before, jointly wrote an article entitled, "Influence of ancient solar-proton events on the evolution of life."[308] In this article they presented proof that evolution was influenced by solar activity. The foreword to the article stated:

> "There is mounting evidence that past extinctions of faunal species have oc-
> curred in near coincidence with reversals in polarity of the geomagnetic field.
> Could the link lie in catastrophic depletion of stratospheric ozone caused by
> solar-proton irradiation over a reduced geomagnetic field?"

They are saying—a Nova! I had a good laugh when I found one scientist who wrote the following to avoid saying the word "nova": "magnetic springs that periodically go off in the center of the Sun."[309] As you may suspect, this subject must be a very hot-potato for scientists If they write that our Sun novas, they will receive a lot of criticism from other scientists, because they are going to have to explain the mechanism. How can the Sun nova periodically? It cannot be explained using the matter-oriented theory of existence.

Other scientists have avoided the problem by guessing that the mass extinctions of dinosaurs were caused by a nearby supernova.[310] Another scientist wrote, a "nearby supernova explosion" could explain the radioisotopes of [10]Be and [26]Al found in deep-sea sediments.[311] All of these scientists hit it right on the head—but did not follow through, because their philosophies, professional and institutional pressures, restricted them.

NASA Knows Too!

After researching the subject for over 35 years, I am absolutely convinced that NASA knows that our Sun novas and causes all the ice ages. There is no indication they know what the mechanism is which causes the Sun to nova, but they know that it does. I deduced this by recording the types of research the federal government funded and the number of journal articles published on those subjects. The resulting data gave me an idea of what kind of questions they wanted answered. The event that clinched it for me was something NASA did on the Voyager-2 flight as it past Neptune. They instructed the Voyager-2 camera to be turned perpendicularly to the radius of the Sun and, at certain times, to photograph the remnant dust shells which our Sun had left from past novas. They would not have done this unless they knew *exactly* what they were looking for!

Many have wondered why our manned space program ended after we landed on the Moon and sent Viking to Mars. I believe the primary purpose behind the U.S. space program was to see if the Sun had novaed and caused the ice ages. Once NASA landed on the Moon and Mars they got their answer. It was obvious what had happened—our Sun novaed. After that it was no longer necessary to spend billions on the space program. Remember, governments only spend billions to buy votes and get answers to important questions. Once the answer was confirmed, the spending was no longer necessary.

My Conclusion

I am sure many of you are troubled, and perhaps scared, by what you have just learned so far in this book. You now realize that I was not just affecting braggadocio in my Introduction. The publication of this book occasions the first time anyone has actually identified and described what events occur during *God's Day of Judgment*. You now know that there are consequences for mankind not evolving to what we are supposed to—in the eyes of God. Chapter 10 will show that the Hebrew priests were told about the next cataclysm, and accurately wrote about it. Chapter 10 will also reveal the greatest secret the Torah has concealed within it—the secret of redemption, or how man can save himself from this terrible and terrifying event.

Chapter 9:
Philosophies of the Universe

Other Civilizations — same Philosophy

When I researched and wrote *Reality Revealed* in the 1970s, I studied the mythologies and religious books of almost all major cultures around the world. All of them agreed that a great cataclysm had occurred sometime in the past. Some knew roughly when it had happened. Some knew only a few of the events. The accuracy of their history was dependent on how advanced their civilization was 12,100 years ago.

The summary of events listed in the previous chapter will be helpful in understanding the philosophies in this chapter.

In the following sections, I will cover four civilizations whose mythological accounts and/or philosophic orientations were most proximate to what had really occurred. The four cultures are: the Hebrews, Greeks, the Hindus, and China. The one culture, which figured it all out—including how many years there were between the reversal cycles—were the Hebrews. I firmly believe this because of what Abraham discovered in the cave.

The Hebrews Knew the Whole Thing

Many ancient Jewish legends and prophecies describe the same events. While reading the Jewish legends, ask yourself: "How did they know about these events, unless their history was accurate?" While reading the Jewish prophecies, ask yourself: "How could people, living over 2,600 years ago, know about some of these events without an advanced civilization, or God, telling them?" Also, remember that the nova did not cause the polar reversal on the planet. The polar reversal was caused by the clock cycle in the Diehold, and that also caused the nova and the Great Flood on the Earth.

I have divided the Hebrew[1] sources into two types: the oral tradition encompassed in the legends and the Hebrew Scriptures. I have already presented in Chapter Two, the secret coded data found in the Hebrew Scriptures. It proves that the priest class, up until 587 B.C.E., knew about the number of years between the cataclysms. They did not label it that, but rather, *God's Day of Judgment* — which it truly is.

Legends of the Jews

The Jewish legends mirror the Biblical surface stories in the Torah but with some important additions. Let us start with the story of creation as it describes the first day of creation. It is written that God created seven heavens, each having a different purpose, and God created light and darkness. (My comments are in brackets.)

> "Though the heavens and the Earth consist of entirely different elements, they were yet created as a unit [one synchronistic system]. The heavens were fashioned from the light of God's garment."[2]

I interpret this to mean that the "seven heavens" represent the seven dimensions of existence meaning the second to the eighth dimension. The number seven may also be a reference to Mount Sinai. When the legend speaks of the heavens and the Earth as made from different elements, but not yet brought into physical being, we can interpret this to mean that in the beginning of time the information for all existence was present in the Diehold, but not yet transmitted or played out and made physical. The Universe was still only information in the Diehold. The idea that the Universe was formed from the "*light of God,*" would be similar to saying that "the Universe is a thought-form of God" and, as you now know, light is the representation of this information.

The creation legend continues by stating that on the second day the information took another form.

> "The firmament is not the same as the heavens of the first day. It is the crystal stretched forth over the heads of the Hayyot, from which the heavens derive their light, as the Earth derives its light from the Sun. This firmament saves the Earth from being engulfed by the waters [random potentials] of the heavens; it forms the partition between the waters above [the information] and the waters below. It [information] made to crystallize into the solid [matter] it is by the heavenly fire, which broke its bounds, and condensed the surface of the firmament [became matter]. Thus fire made a division between the celestial and the terrestrial at the time of creation."[3]

This legend is attempting to explain the difference between the first dimension (the storage dimension); the second dimension (transmission dimension); and, of course, the third dimension, which is the first level at which matter is modulated into existence. I have not found this concept in any other legends or mythologies in the world. The information handed down by Abraham is scientifically correct, according to my information theory of existence, even though it is stated in terms that pre-scientific, primitive people can still comprehend.

What the Torah Says

The beginnings of Genesis has two curious verses, 1:6 to 7, which appears to be troublesome for translators. The verses are currently translated as follows. The word firmament is the alternative translation:

> [Genesis 1:6] And God said, be there the expanse [firmament] in the midst, of the waters, and let it cause a division between the waters and the waters. [7] And God made the expanse [firmament], and caused a division between the waters, which were under the expanse, and the waters, which were above the expanse: and it was so.

The two words they are having problems translating are "expanse/firmament" and "waters." The Hebrew word they are translating to mean "expanse" or firmament is לרקיע. This is actually two words. The ל ל before a word means *to, for, belongs to* or *in regards to*. The dictionary translation of the word רקיע is *vault of Heaven, firmament,* or *sky.* I showed in Chapter 4 that this word and several letters after it form a full four-sided pyramid. That shape is equivalent to the carrier wave of the Diehold in this dimension and hence equivalent to the word and concept of the Diehold.

The second word waters (מים) is a little more difficult to figure out. It is currently translated as water, but water does not make any sense within the context of the two sentences. There are no two types of water, one on the ground and one somewhere in space. Water is water and why does it say that water is above or below something? So it occurred to me that Moses did not really imply water at all but something that is analogous to water. After all, we cannot forget we are dealing with late Bronze Age people so, abstract ideas and terms must be presented using analogies they can understand. So next I analyzed the qualities of water and compared them to whatever had similar qualities. Water is necessary and present in all life forms. It has no shape itself but it takes the form of whatever container it is poured into. This last characteristic rang a bell. A very close analogy to water is information. Information is common to all manifest matter in the Universe. Information creates the matter world and therefore takes any kind of programmed form but has no form itself. For these reasons I believe Moses was trying to describe information and not literally water. With these two interpretations for "expanse/firmament" and "waters" I believe the correct translation for verses 1:6-7 should be:

> [1:6]And God said, let there be the *Diehold* in the center, of the *information*, and let it cause a division between *information* and *another information*. [7]And God made the *Diehold*, and caused a division between the *information*, which were *outside* the *Diehold*, and the *information*, which were *inside* the *Diehold*: and it was so.

This interpretation makes much more sense than the accepted translation because it scientifically describes how the Universe came into being and where we come from. With this interpretation we can conclude the Hebrew's were the first peoples told and knew how the Universe really worked.

Previous Civilizations

The idea of previous worlds and civilizations on the Earth was well known to the ancient Hebrews. They believed that previous civilizations had all been destroyed by cataclysms brought down upon them by God. They believed that some people perished by the deluge, and others were consumed by fire. It was believed that God summoned "the Angel of the Face [the Sun] . . . to destroy the world. The angel opened his eyes wide, and scorching fires and thick clouds rolled forth from them."[4]

As the legend states, when Noah was born, his father, Lamech, and grandfather, Methuselah, noticed something very unusual about him. Noah's father told Methuselah the following:

"He is not like a human being, but resembles the children of the angels of heaven, and his nature is different, and he is not like us, and his eyes are as the rays of the Sun, and his countenance is glorious."[5]

After Methuselah heard this he, in turn, went to Enoch, a wise man. Enoch told Methuselah that Noah's birth was a sign of things to come.

"The Lord will do a new thing in the Earth. There will come a great destruction on the Earth, and a deluge for one year. This son who is born unto thee will be left on the Earth, and his three children will be saved with him, when all mankind that are on the Earth shall die. And there will be a great punishment on the Earth, and the Earth will be cleansed from all impurity."

The Torah informs us that Noah prophesied to the people of his time for 100 years,[6] before the deluge, to try to make them turn from their evil ways. This legend incorrectly states that it was 120-years. It was said that during the 120 years, the Sun rose in the west and set in the east.[7] One week after Methuselah died, the deluge struck the Earth. The obvious conclusion is that when the Earth stopped rotating during the polar reversal period, the oceans continued moving in a westerly direction, resulting in the inundation of all the Earth's continents. The legend continues by telling us what happened the day Noah entered the ark:

"The Sun was darkened, and the foundations of the Earth trembled, and lightning flashed, and the thunder boomed, as never before. And yet the sinners remained impenitent. In naught did they change their wicked doings during those last seven days." Hot rains came down from the heavens, scalding the flesh of the sinners. The entire time the deluge lasted: "the Sun and the moon shed no light." . . . "The flood was produced by a union of the male waters, which are above the firmament [the rain] and the female waters issuing from the Earth [the oceans]. The upper waters rushed through the space left when God removed two stars out of the constellation Pleiades."[8]

Jewish legends have all the main factors that occur during a polar reversal. The Sun disappeared or produced very little visible light, the earthquakes, and of course, the great flood. Notice that the number seven is mentioned multiple times. The hot rain is of course from the ocean water instantly evaporated by the solar Nova. As the water vapor started to cool it first rained down scalding hot.

Hebrew Scriptural Proofs of the Cataclysm

The Hebrew Scriptures refer to the cataclysm as "the End of Days," "the Lord's day," or "God's day of judgment." These terms refer to the polar reversal and the nova. To prove my point, I will quote some verses from the Hebrew Scripture, which exemplify some aspects of the polar reversal.

Earthquakes

ISAIAH 24:18-20:

For the windows on high are opened, and the foundations of the Earth do shake;

The Earth is broken, broken down, the Earth is crumbled in pieces, the Earth trembleth and tottereth,

The Earth reeleth to and fro like a drunken man, and swayeth to and fro as a lodge;

JEREMIAH 4:24:

I beheld the mountains, and lo, they trembled, and all the hills moved to and fro,

The End of a Cycle

ISAIAH 13:6:

Howl ye; for the day of the Lord is at hand; as destruction from the Almighty shall it come.

JEREMIAH 23:20:

. . . in the end of days ye shall consider it perfectly.

JEREMIAH 25:33:

And the slain of the Lord shall be at that day from one end of the Earth even unto the other end of the Earth; they shall not be lamented, neither gathered, nor buried; they shall be dung upon the face of the ground.

JEREMIAH 30:7:

Alas! for that day is great, so that none is like it; and it is a time of trouble unto

Jacob, but out of it shall he be saved.

Joel 1:15:

Alas for the day! For the day of the Lord is at hand, and as a destruction from the Almighty shall it come.

Joel 2:1-2:

. . . let all the inhabitants of the land tremble; for the day of the Lord commeth, for it is at hand; A day of darkness and gloominess, as blackness spread upon the mountains;

Zephaniah 1:14-15:

The great day of the Lord is near, it is near and hasteth greatly, even the voice of the day of the Lord, wherein the mighty man crieth bitterly.

That day is a day of wrath, a day of trouble and distress, a day of wasteness and dissolution, a day of darkness and gloominess, a day of clouds and thick darkness,

The Sun and the Nova

Genesis 7:11:

In the six hundredth year of Noah's life, . . . on the same day were all the fountains of the great deep broken up, and the windows of heaven were opened.

Deuteronomy 28:22-24:

The Lord will smite thee with consumption, and with fever, and with inflammation, and with fiery heat, . . .

And thy heaven that is over thy head shall be brass, and the Earth that is under thee shall be iron.

The Lord will make the rain of thy land powder and dust; from heaven shall it come down upon thee, until thou be destroyed.

Isaiah 29:5-6:

But the multitude of thy foes shall be like small dust, and the multitude of the terrible ones as chaff that passeth away; Yea, it shall be at an instant suddenly—

There shall be a visitation from the Lord of Hosts with thunder, and with earthquake, and great noise, with whirlwind and tempest, and the flame of a devouring fire.

Isaiah 30:26-27:

Moreover the light of the moon shall be as the light of the Sun, and the light of the Sun shall be sevenfold, as the light of the seven days . . .

Behold, the name of the Lord cometh from far, with His anger burning, and in thick uplifting of smoke; His lips are full of indignation, and His tongue is as a devouring fire;

Isaiah 34:8-10:

For the Lord hath a day of vengeance, . . . and the streams thereof shall be turned into pitch, and the dust thereof into brimstone, and the land thereof shall become burning pitch.

It shall not be quenched night nor day, the smoke thereof shall go up for ever;

The Flood
Isaiah 30:25:

And there shall be upon every lofty mountain, and upon every high hill, streams and watercourses, in the day of the great slaughter when the towers fall.

Amos 5:8:

Him that maketh the Pleiades and Orion, . . . that calleth for the waters of the sea, and poureth them out upon the face of the Earth; the Lord is His name;

Nahum 1:7-8:

The Lord is good, a stronghold in the day of trouble; . . .

But with an overrunning flood he will make a full end of the place thereof, and darkness shall pursue His enemies.

Volcanoes
Deuteronomy 32:22:

For a fire is kindled in My nostril, and burneth unto the depths of the nether-world, and devoureth the Earth with her produce, and setteth ablaze the foundations of the mountains.

The Sun Going Out and No Stars Visible Because the Earth is Inside the Stellar Nebula
Isaiah 13:10:

For the stars of heaven and the constellation thereof shall not give their light; the Sun shall be darkened in his going forth, and the moon shall not cause her light to shine.

Jeremiah 4:23:

I beheld the Earth, and lo, it was waste and void; and the heavens, and they had no light.

Jeremiah 4:27-28:

For thus saith the Lord: the whole land shall be desolate; yet will I not make a full end.

For this shall the Earth mourn, and the heavens above be black; because I have spoken it, I have purposed it, and I have not repented, neither will I turn back from it.

JOEL 2:10-11:

Before them the Earth quaketh, the heavens tremble; the Sun and the moon are become black, and the stars withdraw their shining.

JOEL 3:2-4:

In those days will I pour out My spirit.

And I will show wonders in the heavens and in the Earth, blood, and fire, and pillars of smoke.

The Sun shall be turned into darkness, and the moon into blood, before the great and terrible day of the Lord come.

There is much more, but I do not want to beat you over the head with it. I just want to present enough proof so the average person realizes that the Hebrews knew about what happened during the last cataclysm, what will happen in the next, and that it happens in one day.

The most revealing observation listed above is that the Sun gave off no light and the stars disappeared! This is preposterous to modern man that our Sun would actually give off no light and the stars became invisible. But that is exactly what happens after the nova.

There are only three ways the Hebrews could have known this: If they had an accurate history of the past; if a more advanced civilization told them; or if God, the Lord of Hosts, told them. From what I know now, I believe it was God through the Ark!

The Greeks

Greek mythology is most commonly known as classical mythology. The three great Greek philosophers, who wrote or spoke about ancient cultures and past world cycles, were Hesiod (around 700 B.C.), Solon (630-560 B.C.E.), and Plato (428-347 B.C.E.). Notice that Solon lived and wrote at the time of the destruction of the first Temple, and during the rise of Babylon. I will elaborate on that later. I also found many similarities between Plato's philosophy and the philosophy of the Hebrews. I did not think it was a coincidence, but the connection was not easy to find. I eventually discovered it, and will reveal it later. Solon, Socrates and Plato each expressed theories on how the Universe was created, and its relationship to God. I will focus only on Plato because only his writings exist today.

In Plato's book, Timaeus, he informs us that the Universe has a soul, and that this soul is God. Plato looks at spirit as good and matter as evil. In *The Dialogues of Plato,* Benjamin Jowett analyzes Plato's conception of the Universe:

"The astronomy of Plato is based on the two principles of the same and the other, which God combined in the creation of the world. The soul, which is compounded of the same, the other, and the essence, is diffused from the center to the circumference of the heavens. We speak of a soul of the Universe; but more truly regarded, the Universe of the Timaeus is a soul, governed by mind, and holding in solution a residuum of matter or evil, which the author of the world is unable to expel, and of which Plato cannot tell us the origin. The creation, in Plato's sense, is really the creation of order."[9]

Plato's writings are sometimes not the easiest to understand because he is trying to describe very abstract ideas with analogies that man studied 2,360 years ago. I intend to show that Plato's philosophy of the Universe was derived from a philosophy similar to my information theory of existence, described in Chapter 3 and also has its roots in Hebrew philosophy. I will now quote some sections from Plato's *Timaeus,* and will insert my interpretations (and clarifications) in brackets. The first selection deals with the creation of the Universe. Please note the close similarity between Plato's conception of the Universe and the Hebrews' philosophy. I believe that it is not a coincidence.

". . . we must first ask concerning it [the Universe] that primary question which has to be investigated at the outset in every case, namely, whether it has existed always, having no beginning of generation, or whether it has come into existence, having begun from some beginning. It has come into existence; for it is visible and tangible and possessed of a body; and all such things are sensible, and things sensible, being apprehensible by opinion with the aid of sensation, come into existence, as we saw, and are generated [created into existence]. And that which has come into existence must necessarily, as we say, have come into existence by reason of some cause. Now to discover the maker and father [singular God] of this Universe were a task indeed; and having discovered Him, to declare Him unto all men were a thing impossible [this is an allusion to the Jews]."[10]

He says the Universe is a copy of something that God had modeled it after. The Greeks definitely attribute the creation of the Universe to a God: "God created Becoming and the All." The Greek god Pluto sat in front of the gates of *The Land of the Blessed,* [heaven]. His other name was "The Host of Many." One of the Hebrew names for God is "Lord of Hosts." This is too close to be a coincidence.

The idea that man is a reflection of the Universe was also found in the *Timaeus.*

"But we shall affirm that the Cosmos, more than aught else, resembles most closely that Living Creature of which all other living creatures, severally and generically, are portions. For that Living Creature embraces and contains within itself all the intelligible Living Creatures, just as the Universe contains us and all the other visible living creatures that have been fashioned. For since God desired to make it resemble most closely that intelligible Creature [the Diehold] which is fairest of all and in always most perfect, He constructed it as a Living Creature, one and visible, containing within itself all the living creatures which are by nature akin to itself."[11]

Plato described a singular-perfect Universe, creating time and objects. He also hinted at programming permeating the whole Universe.

". . . the Maker made neither two Universes nor an infinite number, but there is and will continue to be this one Heaven [Universe], unique of its kind. . . The soul of the Universe [God the Operating System] is woven through the entire Universe; the soul is invisible [this might be a way in which Plato is describing the operating system] but in as much as the nature of the Living Creature [God or the Diehold] was eternal, this quality was impossible to attach in its entirety to what is generated wherefore He [God] planned to make a movable image of Eternity [the Diehold], and as He set in order the Heaven, of that Eternity which abides in unity He made an eternal image, moving according to number, even that which we have named Time. For simultaneously with the construction of the Heaven He contrived the production of days and nights and months and years, which existed not before the Heaven came into being. And there are all portions of Time: even as "Was" and "Shall be" are generated forms of time.[12]

The past and future are both results of the "motions of time":

"Time, then, came into existence along with the Heaven, to the end that having been generated together they might also still be dissolved together, . . ."

Plato described something similar to the concept of information. He states that there was something that formed everything, and it takes all shapes. It is obvious he did not understand it fully himself, but he attempted to explain it anyway.

"And of the substance which receives all bodies the same account must be given. It must be called always by the same name; for from its own proper quality it never departs at all; for while it is always receiving all things, nowhere and in no wise does it assume any shape similar to any of the things that enter into it. For it is laid down by nature as a molding-stuff for everything [information], being moved and marked by the entering figures, and because of them it appears different at different times [the three different dimensions]. And the figures that enter and depart are copies of those that are always existent, being stamped from them in a fashion marvelous and hard to describe, which we shall investigate hereafter.

"For the present, then, we must conceive of three kinds,—the Becoming, that "Wherein" it becomes, and the source "Wherefrom" the Becoming is copied and produced. Moreover, it is proper to liken the Recipient to the Mother, the Source to the Father, and what is engendered between these two to the Offspring; and also to perceive that, if the stamped copy is to assume diverse appearances of all sorts, that substance wherein it is set and stamped could not possibly be suited to its purpose unless it were itself devoid of all those forms which it is about to receive from any quarter."

". . . Wherefore, let us not speak of her that is the Mother and Receptacle of this generated world, which is perceptible by sight and all the senses, by the name of Earth or air or fire or water, or any aggregates or constituents thereof (the matter world): rather, if we describe her as a kind invisible and un-shaped, all-receptive, and in sore most perplexing and most baffling way par-taking of the intelligible, we shall describe her truly."[13]

I believe he is trying to describe a concept like the Diehold, and a primary form of information of all *things*. Since he did not have the analogy of a computer there was no way Plato could have fully understood the concept he was trying to describe above. The philosophers of Taoism also tried to describe the concept of information, but all they accomplished was to describe its attributes.

Finally, Plato tried to describe the idea of a Diehold, a computer that contains all the information for the entire Universe.

". . . One Kind is the self-identical Form [the Diehold], ungenerated and indestructible, neither receiving into itself any other form any quarter nor itself passing any whither into another, invisible and in all ways imperceptible by sense, it being the object which it is the province of Reason to contem-plate; . . . and a third Kind is ever-existing Place [the Diehold], which admits not of destruction, and provides room for all things that have birth, itself being apprehensible by a kind of bastard reasoning by the aid of non-sensa-tion, barely an object of belief; for when we regard this we dimly dream and affirm that it is somehow necessary that all that exists should exist *in* some spot and occupying some *place*, and that which is neither on Earth nor any-where in the Heaven is nothing. . . ."[14]

You can see Plato was having a difficult time understanding and describing the concept of a Diehold, but he knew it was not in heaven nor in our dimension. It was just somewhere else to him, comprehensible by mind only as an idea.

In Chapter 3 (section: *Proof of the Carrier Wave of Existence*), I describe pyramids as the three-dimensional representation of the carrier wave which the information is modulated onto. I believe I found a sentence in the *Timaeus* which states the same thing: "that solid which has taken the form of a pyramid shall be element and seed of fire…"[15]

Previous Cycles in Time

Most of the Greek philosophers wrote of four previous ages on the Earth. The first was known as the golden age of mortal man. In *The Statesman*, Plato theorized that the cataclysm was the result of the Earth reversing its motion during a special point in time. He is describing the polar reversal!

> "There was a time when God directed the revolutions of the world, but at the completion of a certain cycle he lets go; and the world, by a necessity of its nature, turned back, and went round the other way. For divine things alone are unchangeable, but the Earth and heavens, although endowed with many glories, have a body, and are therefore liable to perturbation. In the case of the world, the perturbation is very slight, and amounts only to a reversal of motion. . . . But the truth is, that there are two cycles of the world, and in one of them it is governed by an immediate Providence, and receives life and immortality, and in the other is let go again, and has a reverse action during infinite ages. This new action is spontaneous, and is *due to exquisite perfection of balance,* to the vast size of the Universe, and to *the smallness of the pivot upon which it turns* [the polar reversal happens in one day]. All changes in the heaven affect the animal world, and this being the greatest of them, is most destructive to men and animals. At the beginning of the cycle before our own very few of them had survived; and on these a mighty change passed."[16] [Emphasis added.]

Plato is stating the same thing I am, except that I know the scientific reason why it happens, whereas he merely repeated what he was taught. Plato referred to the Sun rising in the west in the Atreus myth. The Greek philosopher Censorinus called these periods of cataclysms, the "Supreme Year." The Supreme Year marks the close of one age and the beginning of a new world age. The Greek word *'kataklysmos'* was associated with the Supreme Year, and it meant a great winter, or what we would call the ice age. Another Greek word, *ekpyrosis*, was also associated with the Supreme Year, which meant "combustion of the world."[17] Not exactly a subtle clue.

Solon's Story of a Previous Civilization and the Cataclysm

Plato recounted in the *Timaeus* the story of Solon's meeting in Egypt with a group of Egyptian priests. The story includes many descriptions of the last polar reversal.

> ". . . Thereupon one of the priests, who was of a very great age, said: O Solon, Solon, you Hellenes are never anything but children, and there is not an old man among you. Solon in return asked him what he meant. I mean to say, he replied, that in mind you are all young; there is no old opinion handed down among you by ancient tradition, nor any science which is hoary with age. And I will tell you why. There have been, and will be again, many destructions of mankind arising out of many causes; the greatest have been brought about by the *agencies of fire and water*, and other lesser ones by innumerable other

causes. There is a story, which even you [the Greeks] have preserved, that once upon a time Paethon, the son of Helios [the Sun], having yoked the steeds in his father's chariot, because he was not able to drive them in the path of *his father [the sun], burnt up all that was upon the Earth,* and was himself destroyed by a thunderbolt. Now this has the form of a myth, but really signifies a declination of the bodies moving in the heavens around the Earth, and a *great conflagration of things upon the Earth, which recurs after long intervals;* at such times those who live upon the mountains and in dry and lofty places are more liable to destruction than those who dwell by rivers or on the seashore. And from this calamity the Nile, who is our never-failing savior, delivers and preserves us. When, on the other hand, the gods purge the Earth with a deluge of water, the survivors in your country are herdsmen and shepherds who dwell on the mountains, but those who, like you, live in cities are carried by the rivers into the sea. Whereas in this land, neither then nor at any other time, does the water come down from above on the fields, having always a tendency to come up from below; for which reason the traditions preserved here are the most ancient.

The fact is, that wherever the extremity of winter frost or of summer does not prevent, mankind exists, sometimes in greater, sometimes in lesser numbers. And whatever happened either in your country or in ours, or in any other region of which we are informed—if there were any actions noble or great or in any other way remarkable, they have all been written down by us of old, and are preserved in our temples. Whereas just when you and other nations are beginning to be provided with letters and the other requisites of civilized life, after the usual interval, the *stream from heaven, like a pestilence, comes pouring down,* and leaves only those of you who are destitute of letters and education; and so you have to begin all over again like children, and know nothing of what happened in ancient times, either among us or among yourselves. As for those genealogies of yours, which you just now recounted to us, Solon, they are no better than the tales of children. In the first place *you remember a single deluge only, but there were many previous ones*; in the next place, you do not know that there formerly dwelt in your land the fairest and noblest race of men which ever lived, and that you and your whole city are descended from a small seed or remnant of them which survived. And this was unknown to you, because, for many generations, the survivors of that destruction died, leaving no written word. For there was a time, Solon, before the greatest deluge of all . . ." [Emphasis added.]

Where did this Philosophy come from?

A philosophy has a "fingerprint." What do I mean? When you read it it has certain components that are unique to the philosophy. So when another person or writer expounds upon the idea, you can recognize its real origin. I believe that you would agree with me that the similarities between Hebrew philosophies of the Universe and God, and the philosophies Plato expounded are just about the same. So I investigated this to determine if there actually was a connection.

Where did Plato get his knowledge?

What we know about Plato is that he lived from 428 to 347 B.C.E. He wrote extensively about Socrates (469 to 399 B.C.E.) because he was Plato's teacher. We know Socrates was forced to commit suicide because he was "corrupting his students." But that seems to be a poor explanation of what really happened. Plato also quoted Solon, who lived from 630 B.C.E. to 560 B.C.E. Solon died 26 years after the fall of Judah and Jerusalem to the Babylonians. Solon lived in the Nile Delta, in Egypt, just before the Babylonians invaded.

Plato wrote in *Timaeus* that Solon was at the summer home of the Pharaohs in *Tanis*. Jeremiah, Chapter 43:7, says that some of the Jews, including Jeremiah and Baruch, escaped to the town of "Tahpanhes" in Egypt. *Tahpanhes* and *Tanus* are the same word and place.[18] Thus we have some indication that Solon and Jeremiah were both in Tanis at the same time.

The next job was to see if I could trace from Solon to Plato. If I could do this, it would tie Plato's philosophy ultimately to Solon and his contact with the Jews. We can assume Solon wrote his thoughts and philosophies down because Plato and others mention his writings. Solon's student was most likely Anaximander (610 to 546 B.C.E.) because his writings discuss similar cosmic questions as Solon and Plato. Anaximander's students could have been Anaximenes (588? to 528 B.C.E.) and Xenophanes (570 to 505 B.C.E.) and after them could have been Parmenides (b.510?). All three of the previous men deal with cosmological subjects which border on a similar theory but we have so little of their writings, directly from them, that we just do not know.

The next philosopher who definitely has a philosophy similar to Solon's and Anaximander's is Anaxagoras (500 to 428 B.C.E.). Plato's teacher was Socrates. Socrates' teacher appears to have been Anaxagoras. His writings also state that thought was the cause of all things in the Universe.[19] The obvious solution to the problem is that Solon's work kicked around Greece for about 100 years after Anaximander, until Anaxagoras understood them and adopted them for his school. During the missing 100-plus years, some philosophers may have accepted parts of Solon's writings, and some just did not understand them until Socrates and Plato arrived.

There are enough philosophical similarities between Plato and his predecessors, and the Hebrew philosophy of the Universe, to indicate Solon received his knowledge of creation from the Hebrews during his stay in Tanis. There is another more interesting possibility— which I will explore later.

Plato wrote something very interesting about why Solon was in Egypt:

> Yes, Amynander, if Solon had only, like other poets, made poetry the business of his life, and had completed the tale which he brought with him from Egypt, and had not been compelled, by reason of *the factions and troubles which he found stirring in **his own country when he came home***, to attend to other matters, in my opinion he would have been as famous as Homer or Hesiod, or any poet.[20] [Emphasis added.]

Plato implied that Solon was not Greek by birth and that Egypt also was not his native land but rather a country with major troubles, such as a war or occupation by an invading army. The trouble Plato wrote about had to be the Babylonian invasion of Judea and Philistine cities along the coast. Later, I will cover the question: "Who really was Solon."

Historians know the Greek aristocracy enjoyed the benefits of the hallowed institution of slavery, as did most other kingdoms of the time. Hebrew religious philosophy did not permit permanent slavery because of its concepts about the Universe and God. The Greek religion was also against the Jewish religion because they did not believe God should interfere with man's freewill. The Jewish Religion was God centered and did believe God had a much greater role in man's existence. The Greek leadership certainly felt it had a reason to want an end to Jewish religious ideas which threatened their economic power base and were diametrically opposed to Greek pagan religious beliefs.[21]

I found a curious statement in Plato's *Seventh Letter,* where he writes, between the lines, that he was forced to conceal some philosophies because they were politically unacceptable, and not permitted:

> I found myself obliged to say, in praise of the kind of philosophy, that this alone can give us insight into public and private justice; and that consequently the human kind in every land will have no cessation from evil until either the kind of men who rightly and truly pursue philosophy shall acquire authority in the state, or the ruling statesmen shall by some divine dispensation be real philosophers."

Realizing that Plato was not able to write openly, I then knew he had to write between the lines and get out as much of the truth as he could. With this in mind, I discovered in Book 6 of the *Laws,* a number of lines, out of the mouth of Socrates, saying to the effect that the Greeks came from twelve proud tribes. The various Greek peoples were divided into two main groups, the Dorian's and the Ionian's. The Dorian's were divided into three tribes and the Ionians into four. In the late 6th century B.C.E. they were renumbered to ten tribes.[22] But at no time were the Greeks ever 12 tribes. Only the Hebrews were known to have come from 12 tribes.

The last significant clue was written by Socrates in the book *Meno:*

> "Seeing then that the soul is immortal and has been born many times, and has beheld all things both in this world and in the *nether realms*, . . ."[23] [Emphasis added.]

The "nether realm" means the underworld. The Hebrews also referred to a place called "the nether-world." This was a place where souls stayed in limbo. It was not in heaven nor in the realm of the living. These two terms are too close to be a coincidence.

Who Was Solon?

I was really curious about Solon after finding out he was not a Greek and was in Tanis at the same time as Jeremiah and Baruch. So I looked into his name to see if it was Greek and found out that it was not. So I spelled Solon in Hebrew, as well as I could, and came up with שׁוּלוּן. The name is made up of two names put together, a technique used before by the priests for their prophet's or priestly names. The first word is spelled שׁוּע, which means rich, or *Noble.* The second word is לוּן, which means to *mutter* or *murmur.* Together the name means *a rich or noble person that has to keep quiet.*

So I looked further to figure out who it could be and the only book that has the clues is Jeremiah, starting from Chapter 40. The story is highly coded, including name changes but it comes down to this: some of the Judean leadership, including priests, escaped to Tanis after Gedaliah was murdered.

Jeremiah had three sons: Gemeriah, Jaazaniah (Jezaniah) and Ahikam. Gemeriah, under the name Seraiah, was the High Priest at the fall of Jerusalem. He and his son, Nethaniah, were killed by the Babylonians after the fall of the Temple. Jeremiah's third son was Ahikam whose son was Gedaliah, who was appointed Governor of Judea after the Babylonians occupied it. Gedaliah was murdered in turn by Ishmael, the grandson of Gemeriah to take revenge for the murder of his father and grandfather. Needless to say some of these members of the priest family did not want to go back to Judea for several very good reasons. I can trace the destination of Ishmael because Jeremiah tells us that he went to the land of the Ammonites. That leaves only Jaazaniah and his family. The Hebrew Scriptures have very little information about him, which leads me to believe that he disappeared sometime after Tanis, Egypt. His age would match Solon's and he was of the royal line. He was also a priest and would have the same education and philosophy as the rest of the priest class. I think he took what money he had from the Temple treasury and left for Greece when the Babylonians invaded Egypt a few years later. I think when he arrived in Greece he used the only intellectual asset he had to make a livings—his priestly education. I believe he started a school in Greece and taught some aspects of Jewish philosophy that he packaged in a way that was not a threat to the Greek ruling class. I also believe he handed down his school to his descendants, as most family business are.

If I am correct then that would explain the strange things Plato writes about 12 tribes and the similarities between Hebrew philosophy and Plato's own.

Conclusion

It is my conclusion that Plato's philosophies of the Universe and creation originated in part from Hebrew philosophies from the priest-class. I believe Socrates was killed because he was teaching these forbidden ideas to the children of the Athenian aristocracy, whose fathers did not like it one bit. They had supported the war against the Kingdom of Judea, and Socrates was teaching their children Hebrew philosophies of the Universe, and who knew what else? After all, Plato wrote that he could not write the whole truth, so we do not know what Socrates actually taught.

Philosophies and Mythologies of India

The Hindu's of India have a rich selection of ancient philosophies that mirror my philosophy of the Universe. They also have very good descriptions of the polar reversal incorporated in their mythologies. The Hindu writings state, "At first the Universe was not anything. There was neither sky, nor earth, nor air. Being non-existent, it resolved, 'Let me be.' It became fervent. From that fervor, smoke was produced. It again became fervent. From the fervor, fire was produced. Afterwards the fire became 'rays' and the 'rays' condensed into a cloud, producing the sea."[24] They also express the belief that the Universe was created by mind which is comparable to how I describe the Operating Systems creates everything in the Universe. They say the force that comes from the "mind" is called "Brahma" and it permeates and creates everything in the Universe. The Brahma is comparable to my theory that the matter world is created from the information transmitted from the Diehold. The term "Brahma" is an attempt to explain information that makes up all existence. There are three ways the Hindu sages could have come to this concept of the Universe. One, they could have evolved to it; they found records left by a previous advanced civilization; or they came in contact with another civilization that already had the information.

The other philosophical point I do not believe a non-technologically advanced people could evolve to is the concept of what the smallest piece of matter is, the atom. The Greeks have been credited with the development of the concept of the "atom." But I am not so sure of that, because of what is written in the *Vedanta-Sutras* and *Adhyaya*. In this quotation they are philosophizing about the parts that make up matter:

As we observe four elementary substances consisting of parts, viz. earth, water, fire, and air (wind), we have to assume four different *kinds of atoms.* *These atoms marking the limit of subdivision into minuet parts cannot* *be divided themselves; hence when the elements are destroyed they can* *be divided down to atoms only*; this state of atomic division of the elements constitutes the pralaya (the periodical destruction of the world). After that when the time for creation comes, motion [karman] springs up in the aerial atoms. This motion which is due to the unseen principle [soul] joins the atom in which it resides to another atom; thus binary compounds are produced, and finally the element of air. In a like manner are produced fire, water, earth, the body with its organs. Thus the whole world originates from atoms. From the qualities inhering in the atoms the qualities belonging to the binary compounds are produced, just as the qualities of the cloth result from the qualities of the threads. Such, in short, is the teaching of the followers of Kanâda.[25] [Emphasis added.]

The Theory of Multidimensional Reality agrees with their concept of the atom, because the atom cannot be divided, as explained in Chapter 3. It is obvious from this quotation that they had the principle that when a soul or "conscious energy" is superimposed with atoms, it produces motion, which is life, caused by time. Their reference to binary compounds may be an attempt to explain what we call the electrostatic field and the magnetic field that are a function of the information. The Hindus' explanation of the atom, presents a problem for scholars explaining how the ancient Hindus came to this body of knowledge. The only ways a civilization could obtain this state of knowledge was list previously. Their views are very similar to the Hebrews and there are several ways they could have come in contact with them between 951 B.C.E. to 500 B.C.E. One could have been Solomon's trade routs to the Orient. They could have gotten it from the Greeks, during the time of Alexander the Great, but we just do not know. If not from the Hebrews then the question is "Were the Hindus 12,000 years ago, an advanced civilization or was there another previous civilization?"

The Hindus also have a detailed description of three previous ages before ours. The Hindus mark blocks of time by calling them katurgas, which are equal to 12,000 years of the "gods." Even though they associate it with their gods, I believe this is a misinterpretation of the original information and should be interpreted as 12,000 human years. With this in mind, the following mythology is attributed to the Indian sage, Markandeya, believed to be one of the survivors of the last cataclysm. He gives a great description of the concept of long periods of time truncated by cataclysms.

In the beginning there existed a supreme being: great, incomprehensible, wonderful, and immaculate, without beginning and without end.... He is the Creator of All, but is Himself Increate, and is the cause of all power.

After the Universe is dissolved, all Creation is renewed, and the cycle of the four Ages begins again with Krita Yuga. 'A cycle of the Yugas comprises twelve thousand divine years. A full thousand of such cycles constitutes a Day of Brahma.' At the end of each Day of Brahma comes 'Universal Destruction.'

Markandeya goes on to say that the world grows extremely sinful at the close of the last Kali Yuga of the Day of Brahma. Brahmans abstain from prayer and meditation, and Sudras take their place . . .; all men degenerate and beasts of prey increase. The earth is ravaged by fire, cows give little milk, fruit trees no longer blossom, Indra sends no rain; the world of men becomes filled with sin and immorality. Then the earth is swept by fire, and heavy rains fall until the forests and mountains are covered over by the rising flood. All the winds pass away; they are absorbed by the Lotus floating on the breast of the waters, in which the Creator sleeps; the whole Universe is a dark expanse of water."[26]

Markandeya goes into greater detail of what happens during the destruction:

"After a drought lasting for many years, seven blazing suns will appear in the firmament; (the sun novas) they will drink up all the waters. Then wind-driven fire will sweep over the earth, consuming all things; penetrating to the nether world it will destroy what is there in a moment; it will burn up the Universe. Afterwards many-coloured and brilliant clouds will collect in the sky, looking like herds of elephants decked with wreaths of lightning. Suddenly they will burst asunder, and rain will fall incessantly for twelve years until the whole world with its mountains and forests is covered with water. The clouds will vanish. Then the Self-created Lord, the First Cause of everything, will absorb the winds and go to sleep. The Universe will become one dread expanse of water."[27]

Markandeya's description of the destruction, covers many of the same events I calculated occurred during the polar reversal, covered in Chapter 8. The same subject is also elaborated on in the Hindu Book, *Bhagavata Purana.* It tells us there were four previous ages, which ended by a conflagration, flood, and hurricane. The Indian books, *Ezour Vedam* and *Bhaga Vedam*, also write of four previously expired ages. In the book, *Visuddhi Magga* (World Cycles), it says "There are three destructions: the destruction by water, the destruction by fire, the destruction by wind."[28] It also tells us of seven past ages, all of which ended in a world cataclysm.

In Brahmanism it is thought that after each cycle, "the Universe becomes water as in the beginning."[29] This expression is, of course, referring to the period after the great flood when the earth appears to have been entirely covered with water by the oceans rushing over the continents.

The Hindu book, *Vedanta-Sutras,* also states that the world is periodically devastated and changes both name and form. The surface of the earth dissolves and is later produced anew. All contradictions to truth pass away.[30] In the book, *Satapatha-Brahmana* there is an excellent description of what the sun does at the end of a yuga (world age, polar reversal). This also indicates that at the moment of the nova/polar reversal, the sun was somewhere over India.

> "That one [the sun] bakes everything here, by means of the days and nights, the half-moons, the months, the seasons, and the year; and this [Agni, the fire] [the sun] bakes what is baked by that one: 'A baker of the baked [he is],' said Bhâradvâga of Agni; 'for he bakes what has been baked by that [sun].'
>
> In the year these amounted to ten thousand and eight hundred: he stopped at the ten thousand and eight hundred."[31]

The Brahmans are telling us that the Sun destroys one-half of the earth at the end of a cycle. They believe that this event occurs every 10,800 years. This is not very far off from our theory that it occurs every 12,068 years. Also notice they are 1,200 years shout of 12,000 years.

In the third section called the *Brahmana* in the same Hindu book, they give us another description of the Sun when it nova's:

> "Now yonder burning [sun] doubtless is no other than Death; and because he is Death, therefore the creatures that are on this side of him die. But those that are on the other side of him are the gods, and they are therefore immortal.
>
> And the breath of whomsoever he [the Sun] wishes he takes and rises, and that one dies. And whosoever goes to yonder world not having escaped that Death, him he causes to die again and again in yonder world, even as, in this world, one regards not him that is fettered, but puts him to death whenever one wishes."[32]

Figure 9-1: The Ant Nebula with the added illustration showing what an equatorial plane of stellar dust and "rocks" would look like as it expands outward along the planetary plane.

You will notice there is no disagreement as to what causes the destruction of the earth.

The principle of 12,000 years between world cycles shows up in Brahman literature in the *Rik* Verses numbering 12,000, each verse representing a year.

In the Hindu book, *Prasna-Upanishad*, we are told that in the past the Sun rose in the west and set in the east, "Now Aditya, the Sun, when he rises, *goes toward the East*, and thus receives the Eastern spirits into his rays."[33] This condition would have existed before the last polar reversal. The Hindus are not the only ones that knew the Sun rose in the west and set in the east before the last cataclysm. I fount it in nine other civilizations' mythologies.

From the Hindu book, *Avatar*, we are given a good description of what the Sun looked like when it novaed and how many hours it took for it to expand. What is important about this legend is that you can get a rough estimate for how long it takes for the hot dust shell from the Sun, to reach the earth. This was explained in Chapter 8.

> "By the power of God there issued from the essence of Brahma a being shaped like a boar, white and exceeding small [the Sun]; this being, in the space of an hour, grew to the size of an elephant of the largest size, and remained in the air.

> Brahma [the name of the Hindu king] was astonished on beholding this figure, and discovered, by the force of internal penetration, that it could be nothing but the power of the Omnipotent which had assumed a body and become visible. He now felt that God is all in all, and all is from him, and all in him; . . .

> That vara [the Sun] figure, . . . again made a loud noise, and became a dreadful spectacle. Shaking the *full flowing mane which hung down his neck on both sides*, and erecting the humid hairs of his body, he proudly displayed his *two most exceedingly white tusks*; then, rolling about his wine-colored [red] eyes, and erecting his tail, he descended from the region of the air, and plunged headforemost into the water. The whole body of water was convulsed by the motion, and began to rise in waves, while the guardian spirit of the sea, being terrified, began to tremble for his domain and cry for mercy.[34] [Emphasis added.]

This mythology is very important because it is describing what the Sun looked like at the time of the nova. On the surface you would think the author is taking great liberties with literary artistic style but he is describing something that only someone who saw a stellar nova up close, would know. Figure 9-1 shows the Ant Nebula[35] (Menzel 3). This is a star after its nova, which has blown stellar material from its north and south poles in a very beautiful pattern. I have added two white lines coming out from the star's equatorial region to show what the Brahmans meant by two "white tusks." I added this because I have seen many photos of planetary nebula (a stellar nova) that have dis-played this characteristic such as seen in the Egg Nebula (catalog #WFPC2) and Eta Carinae Nebula.

The excerpt also tells us that a great deal of noise was associated with the Sun's nova. When it refers to the Sun having a "full flowing mane" and tusks coming from two sides, this is probably an attempt to describe the dust shells of the Sun as they expanded. Toward the end of the legend, we see clear reference to the oceans of the earth also convulsing at the time of the polar reversal. It must have been a truly frightening sight to the people on that side of the earth.

Philosophies and Mythologies of China

The most ancient records of Chinese mythologies were destroyed under the Emperor Tsin-Khin-Hong (246-209 B.C.E.). He had ordered all of the ancient history and astronomy books, as well as the ancient classics, to be burned. A few of the original manuscripts survived in the possession of the Chinese scholar, Confucius. Other scholars rewrote original ancient manuscripts from memory, but much had been lost. Even with these handicaps, the rewritten legends and mythologies, which have survived to the present day, clearly chronicle a past destruction that occurred on the Earth.

During the last polar reversal, China was inundated by the Pacific Ocean and was on the back-side of the Earth when the dust shell hit the Earth. The few survivors probably took refuge on the highest mountains, or in the deepest caves or the large pyramid mounds found in parts of China.[36,37] Chinese mythology describes Fû-hsi as one of the survivors of the last destruction. He is considered the Chinese "Adam." Sometimes described as some sort of deity, he was essentially a highly enlightened individual and perhaps a scholar who survived the last cataclysm. Some Chinese scholars placed Fû-hsi at about 3,322 B.C.E, but, in actuality, no one knows how long ago Fû-hsi lived. We know from the writings of Kwang-Tze (6th century B.C.E.) that Fû-hsi lived during the age of "perfect virtue," which was his description of the last civilization.[38]

The Taoist scholar, Lâo-Tze (born 604 B.C.E.) gave us a description of scientific philosophy at the end of the last age. He wrote that in the paradisiacal state, the Tâo ruled man. The Tâo, or Taoism, is the oldest of Chinese religions. Lâo-Tze's ancient classic, *Tâo Teh King,* tells us even more about the philosophies of this antediluvian world:

> The skillful masters (of the Tao) in old times, with a subtle and exquisite penetration, comprehended its mysteries, and were deep (also) so as to elude man's knowledge. As they were thus beyond man's knowledge.[39]

Kwang-Tze says a bit more about Fû-hsi: "Fû-hsi received the Tao (the teachings of) and by it penetrated to the mystery of the maternity (creation) of the primary matter."

The scholar Lieh-Tze wrote about a land of bliss, which preceded this civilization. This was thought to be five islands supported underneath by five large turtles. These islands were known as the "Islands of the Blest." They were:

> . . . inhabited by white souls of saintly sages who have won immortality by rendering their bodies transparent . . . [perhaps sixth dimensional beings].

> At one time the islands drifted about on the tides of [the] ocean, but the Lord of All who controls the Universe, having been appealed to by the Taoist sages who dwelt on the isles, caused three great Atlas-turtles to support each island with their heads so that they might remain steadfast.[40]

It was believed that every 3,000 years, the tortoises rose to the surface and turned over to see the Sun. There are four cycles in the Chinese ages, so a complete cycle is 12,000 years, which is close to what I calculate. The legend continues by saying:

> Once upon a time, . . . the Atlas-turtles that support the Islands of the Blest suffered from a raid by a wandering giant [the Sun]. This giant, it is said, went fishing for these turtles. The giant accomplished hooking two of the turtles, which left those two islands to drift toward the north, where they were stranded among the ice fields. The white beings that inhabited these islands were thus separated from their fellow saints on the other three islands, . . .

> In the North, these saints suffered from the evils of the North (frost and darkness). This legend was an attempt to recall that after this antediluvian period had ended the skies were black and the Earth was cold, which, of course, described the ice age.

I will now cover how the Chinese mythologies described the actual destruction of the Earth. First, I will give you a quote from the Chinese Encyclopedia on the Emperor Kang-Hsi, 1662 B.C.E.

> In traveling from the shores of the Eastern Sea [Pacific Ocean] toward Che-Iu, neither brooks nor ponds are met with in the country, although it is intersected by mountains and valleys. Nevertheless, there are found in the sand, *very far away from the sea, oyster-shells and the shields of crabs.* The tradition of the Mongols who inhabit the country is, that it has been said from time immemorial that in a remote antiquity the waters of the deluge flooded the district, and when they retired the places where they had been made their appearance were covered with sand. . . . This is why these deserts are called

the "Sandy Sea," which indicates that they were not always covered with sand and gravel.[41] [Emphasis added.]

Another Chinese encyclopedia called the end of a world cycle a "great year." "The cosmic mechanism winds itself up and 'in a general convulsion of nature, the sea is carried out of its bed, mountains spring out of the ground, rivers change their course, human beings and everything are ruined, and the ancient traces effaced.' "[42]

In the Chinese deluge legend, we read of the mythical Empress Nu-Kwa (royal lady of the West—the Sun). In her kingdom it was alleged:

Three rebels had conspired with the demons or gods of water and fire to destroy the world, and a great flood came on. Nu-Kwa caused the waters to retreat by making use of charred reeds [the burning forests]. Then she re-erected one of the four pillars of the sky against which one of the rebels, a huge giant, had bumped his head, causing it to topple over [the sky fell down].[43]

The Emperor, Yahou, was believed to have lived shortly after the last great cataclysm. Many times the cataclysm was called the time of Yahou. It is most probable that he was one of the other survivors of the cataclysm or a later descendant of Fû-Hsi. It is obvious from the following annals of Yahou that he was alive when evidence of the destruction was still present on the Earth:

At that time the miracle is said to have happened *that the Sun during a span of ten days did not set,* the forests were ignited, and a multitude of abominable vermin were brought forth. In the lifetime of Yao [Yahou] the Sun did not set for ten full days and the entire land was flooded.

An immense wave "that reached the sky" fell down on the land of China. The water was well up on the high mountains, and the foothills could not be seen at all . . .

"Destructive in their overflow are the waters of the inundation," said the emperor. "In their vast extent they embrace the hills and overtop the great heights, threatening the heavens with their floods."[44] [Emphasis added.]

The legend clearly connects three major events of the cataclysm, the Sun novaing, the oceans flooding the contents, and the rotation of the Earth stopping, all at the same time. The flood was not caused by rain, but rather by the Pacific Ocean rapidly moving westward over China. This implies that the Earth rotated in the opposite direction than today. Notice that the floodwaters are described as "a mountain of water." They were not kidding.

The philosopher Wen-Tze gave us the following additional description of what happened during the last cataclysm:

> When the sky, hostile to living beings, wishes to destroy them, it burns them; the Sun and the Moon lose their form and are eclipsed [give off no light]; the five planets leave their paths; the four seasons encroach one upon another; daylight is obscured; glowing mountains collapse; rivers are dried up; it thunders then in winter, hoarfrost falls in summer; the atmosphere is thick [dust in the air] and human beings are choked [with dust]; the state perishes; the aspect and the order of the sky are altered; the customs of the age are disturbed [thrown into disorder] all living beings harass one another.[45]

The Chinese legend, "Reign of the Chaos," states that in the beginning the world was in darkness. Smoke and clouds covered the Earth. After the Age of Chaos, heaven and Earth separated. "Records had not yet been established or inscriptions [writing] invented. At first even the rulers dwelt in caves and desert places, eating raw flesh and drinking blood."[46]

A great deal of significance is attached to dragons in China. It is thought that when dragons fight fireballs fall to the ground and a strong wind prevails in the heavens. When dragons' eggs hatch lightning flashes, thunder bellows, and darkness comes to the Earth.[47]

Chinese dragon lore is quite extensive—too extensive for a long discussion here. I believe these dragon stories were the result of what was seen in the sky during the nova. China was on the backside-edge of the Earth when the Sun's dust shell reached the Earth. The expanding solar dust shell passed around the Earth, and some of the dust shell formed what looked like a large smoke ring. The fiery smoke and dust ring looked like a "dancing" fiery snake in the sky. The Voguls, in Russia, were also on the backside of the Earth, and they also write that the Earth was "burning at both corners of the sky."[48]

It is a shame that the Emperor Tsin-Khin-Hong destroyed the original early accounts of the last cataclysm. I am sure if these records were available to us today, they would give us a much more accurate and detailed description of the sequence of events during the cataclysm, but what I have described previously should be sufficient for anyone to see the similarities between what the Chinese, Hindus, Greeks and Hebrews say happened. The stories described in our mythologies are in essence our ancestors screaming out at us, telling us what happened a very long time ago. We should not ignore them just because our current science cannot conceive of the mechanism, which caused such destruction.

These mythologies are very descriptive and demonstrate that our ancestors were screaming from their graves to tell their future generations what had happened to them a very long time ago. We must heed their warnings, no matter which race we come from because we are all in the same situation and there is no place on Earth you can escape it unless you prepare of it.

Taoism

The religious teachings of Taoism preserved what the survivors of the cataclysm thought about existence, and how the original creation came about. Pure Taoism was first taught by Fû-hsi (the Chinese Adam). This implies that Taoism was actually the teachings or philosophy of the previous civilization. As mentioned earlier, Lao-Tze and Kwang-Tze both wrote essentially the same thing. At this point a description of Taoism is necessary:

The translation of the word Tao is "the way of nature, her processes, her methods, and her laws."[49] When translated as a force or power, it means the power that works in all created things, producing, preserving and giving life. The Tao is invisible and is, therefore, not of this dimension. The following is a collection of descriptions written by Lao-Tze and Kwang-Tze. You will notice the only concept that can possibly fit their description of the Tao is what my theory calls "the information that makes up all things in the Universe." Taoist philosophers sometimes mistakenly associated the Tao with both the first and second dimensions. In my Information Theory there is a clear differentiation between information in a storage dimension and in a transmission dimension.

- The Tao permeates the smallest of things, as well as the largest of things.
- The Tao does not have a positive existence nor a negative one. It is rather a mode of being.
- The Tao is the maker and transformer.
- The Tao is hidden and has no name, but it is the Tao which is skillful at importing to all things what they need and making them complete.[50]
- The Tao is the spontaneously operating cause of all movement in the phenomena of the Universe.
- The Tao does not decay.
- The material world springs from the immaterial (the Tao).
- It does nothing and has no bodily form. It may be handed down (by the teacher) but may not be received (by his scholars). It may be apprehended (by the mind), but it cannot be seen. It has its roots and ground (of existence) in itself.[51]
- The Tao is older than the highest antiquity, and yet could not be considered old (the Tao is independent of time and space).
- No one knows its beginnings, no one knows its ends.
- Creation passes through one portal. The root of all things emerges through one gate [the Diehold].[52]

- The supreme Tao begets all creation and causes all phenomena.

Neither Lâo-Tze nor Kwang-Tze associated the Tâo with anything similar to the western concept of God. I believe this was their major mistake. You cannot have information without a creator and processor of that information.

Taoism does not have a creation mythology, such as the one described in Genesis. Instead, they describe the creation of the Universe as a result of the Tâo coming into existence. As you will notice from Lâo-Tze's and Kwang-Tze's explanations of the creation of the Universe, they are very similar to my concept of how the Universe came into being.

> In the beginning of all things there was nothing in all the vicinity of space, there was nothing that could be named. From this came the first existence but it was without bodily shape. This was the formation of the Universe under the guidance of the Tao. It was evolution, not a creation. The Tao, they do not say came into existence but rather into operation."[53]

Lâo-Tze tells us that existence and non-existence give birth to one another, the one to the idea of the other. He also stated that man is composed of body and spirit, each not necessarily dependent on the other. The idea of a spirit or soul superimposed on three-dimensional matter was echoed by Kwang-Tze. He tells us:

> As things were completed (came into being) there were produced the distinguishing lines of each (the shape of matter), which we call the bodily shape. That shape was the body perceiving in it the spirit.[54]

The similarities between early Taoism and the *Theory of Multidimensional Reality* are not merely coincidental! There is one philosophical point stated by Lao-Tze, which definitely proves that the origin of Taoism came from a civilization with a highly advanced science. I say this because the only way to know this principle of the Universe is to have sophisticated astronomical equipment to make the observations. Lao-Tze said, "(So heaven) diminishes where there is superabundance, and supplements where there is deficiency. It is the way of heaven to diminish superabundance and to supplement deficiency."[55] He is enunciating the same theory, which I had formulated about the Universe.[56] The Diehold will never permit too much or too little information being directed toward a given area in time and space. It will correct any such disproportion over time. For me to arrive at this conclusion, I had to have knowledge about our expanding Universe, Quasars and, most importantly, to know what gravity truly is. The founders of Taoist philosophy must have been familiar with the same scientific observations as us today. Otherwise, I do not see how they could have come up with this principle.

An important point to be drawn from the principles stated in Taoism is that the best legacy we have from previous civilizations (the one before the last cataclysm) is not the pyramids or caves left by them, but rather the philosophical ideas of what existence is.

This observation would fit one of our basic premises, which is that ideas always transcend material things. In conclusion, Taoism had attempted to explain the abstract idea that the Universe is the product of information, and that the Tao is their name for information.

Chapter Conclusion

In this Chapter, I have attempted to show that previous philosophies of the Universe were very similar to the *Theory of Multidimensional Reality*. It is also obvious that these examples from the Hebrews, Chinese, Hinhus, and Greeks show that they knew about the cataclysm in some detail, and this knowledge was wrapped into their theologies. My first book, *Reality Revealed,* contained a very long chapter with extensive references to mythologies from around the world, which stated just about the same thing. Almost all ancient cultures knew there were repeated cataclysms that hit the Earth, occurring after long cycles of time. We are now—today—dealing with a real tragedy for all of mankind. The survivors of past cataclysms may or may not have been literate or educated. If they were, we received an accurate description of what happened to them. If they were not, we got a very sketchy picture of the sequence of events.

Today, man is in worse shape than ever before. The science philosophies taught in schools and colleges are wrong. Most of the great religions are man-made creations, tools to capture power over the little people by forcing them to believe in theologies their leadership knows are human creations. We are in the *End of Days,* within 39 years of the next polar reversal, and nova—and no one knows it is going to happen. Needless to say, this situation is sad. You may ask yourself: Whom should we blame for our predicament? A Roman family richly deserves the blame, but that explanation will be for another book.

Finally, how do the Jewish people fit into this whole mix? That will be covered in the next chapter.

Chapter 10:
Prophecies of the End of Days

Defining a Prophesy

Let us start with a description of a real prophecy. There are so many misconceptions of what a prophecy is and how someone has one. The place to start is the Torah, because it defines the three ways God communicates with man:

> [6] And He [God] said: "Hear now My words: if there be a prophet among you, I the Lord do make Myself known unto him in a *vision*, I do speak with him in a *dream*. [7] My servant Moses is not so; he is trusted in all My house; [8] with him do I *speak mouth to mouth*, even manifestly, and not in dark speeches; and similitude of the Lord doth he behold . . ."[1] [Emphasis added.]

God states that he gives some people prophetic visions or dreams but the main system He provided was communicating through the Ark of the Covenant. I explained how it worked in Chapter 7. The reason God instructed the Hebrews to construct the Ark and perform the sacrifices with the smoky fire, was so He could communicate with Moses and the High Priests through the Ark. The main job of the High Priest was to make sure there was sufficient smoke inside the Tabernacle, in case God "spoke" the priest could write it down. In other words, the High Priest was taking dictation. After he wrote down the message(s), he added connecting sentences and some history, so the message was comprehensible to all, and in the correct context.

The Jewish priests followed this practice through the time of Baruch. After the destruction of the Temple Baruch put the Ark back into the cave. He was the last Prophet because no one else has had access to the Ark since then. All those who have come after him, who claim to be prophets, do so presumptuously and out of self-interest.

There is a statement in Jeremiah about the lack of prophets towards the End of Days.

> [Jeremiah 23:20] The anger of the Lord shall not return, until He have executed, and till He have performed the purposes of His heart; In the *End of Days* ye shall consider it perfectly. [21] I have not spoken to them, yet they prophesied. [Emphasis added.]

Prophetic Dreams and Finding Mount Sinai

I do not discount dreams altogether. I classify dreams into two categories: The majority seem to be personal interpretations of past, present or possible future events. The second category is rarer and usually very vivid. The dream may be of some future event or of a past-life experience. I believe some dreams are important because it was a series of similar dreams, which my friend and I had, when we were both in our teens, which—coupled with a number— enabled me to find the real Mount Sinai. We did not know each other then and we lived 3,000 miles apart, but we had the same kind of dreams, about the same place, and doing the same thing. It was this dream, in conjunction with a number, which has followed me throughout my life that enabled me to find the real Mount Sinai. Therefore, I know first hand that some dreams mean something. I used my dreams as clues to find the real Mount Sinai.

The other "evidence" which leads me to accept that some dreams are important: I have monitored an Internet news group called alt.dreams.prophetic, and a dream web site. A few of the dreams described some aspect of the cataclysm, such as a great wave coming from the west and covering the land, or the sky turning red and the ground shaking for days on end. One person described the Tabernacle constructed on top of a hill and named the color of the roof perfectly. In that dream, the dreamer described the surrounding land and appearance of Mount Sinai perfectly, and he knew his dream had something to do with God. He had no idea what the Tabernacle looked like, or that it was the Tabernacle. I can say that some people are having accurate dreams about the future cataclysm and they do not realize what they are "seeing." I would expect, as we get closer to 2046, people will increasingly have these kinds of dreams. At some point, everyone will understand why.

The one thing these people all have in common is that they had no idea what they were looking at. They just reported what they saw during an emotional slice of time. Over the years, I have tried to analyze why most of us have had a "prophetic dream," at least once, in our lives. The only explanation I can offer is that we are the best recipients of our own thoughts from the future. As I stated in Chapter 3, we exist in the Diehold as a unique program, which is, in essence, our soul. When an emotional event happens to us, it is possible our thoughts transmit and ripple through time. If we are in deep sleep at the time, our subconscious minds might be receptive to such signals from our future. Your subconscious may or may not filter out some of this information. The result is, you have a *special vivid dream* that you know is different.

Biblical Texts and Chapters Excluded from Prophecies

Trying to predict the future is difficult enough, but if, unbeknownst to you, the "holy books" you are relying on for reference are written by some Romans or Arab sheiks, your conclusions will not be worth the paper their books are printed on! The following Old Testament books and chapters have been excluded from my analysis of Jewish prophecies because they are *not* Jewish books[2]: Isaiah (Chapters 9, 14, 19, 24, 27, 40 - 66); Daniel, Jonah, Habakkuk, Zechariah, Malachi, Job, Ruth, Ecclesiastes, First Chronicles and a number of chapters in Psalms. The entire New Testament is likewise excluded, for similar reasons.

I also exclude Jewish books written after the fall of the Temple, including Ezra's, and Nehemiah, except Baruch's, which I will explain next.

The Books of Baruch

The most important Hebrew prophets who wrote about events set to occur in the End of Days are: Moses in Deuteronomy, Isaiah, Hosea, Jeremiah and Baruch, aka Ezekiel, aka Joel, aka Obadiah.

Baruch deserves special mention because he was the priest who put the Ark back into the Cave and sealed it up. God must have told him that one day He would reveal the location of the Holy Mountain and Cave, so that man would eventually know the truth, and its treasures, which will be available to the descendants of Jacob (Israel) during their time of need. I cited numerous legends in Chapter 7 that describe the same subject but one such verse sums it up: ". . .it was the wish of God that the place of hiding should remain a secret until the redemption, and God Himself will make the hidden things visible."

God told Baruch to take the Ark out of the Temple in Jerusalem and take it to the family cave down in the Sinai desert. The coded story is found in Ezekiel, Chapter 12:3, where Baruch writes, "prepare thee stuff for exile, and remove as though for exile by day in their sight . . ." The "stuff" is the Ark of the Covenant and other items inside the Temple. I covered this in Chapter 7 on the Ark, but I did not cover what Baruch did when he arrived at the cave. I estimate he stayed there about a month and set up the Tabernacle, which I think was already stored there.[3] He then sat in front of it and "took dictation" from God, until He was finished. After that, Baruch had the cave entrance sealed to conceal its location. He then proceeded to Tanis,[4] Egypt, to meet up with his grandfather, Jeremiah. After the Babylonians invaded Egypt he traveled to Babylon, where he wrote the books Ezekiel, Joel, Obadiah, some chapters of what is now called Psalms, finished the last parts of Jeremiah, Second Kings and Second Chronicles. He also rewrote the Torah, inserting word breaks, final letter forms, and used more accurate letter designs.

Baruch was our last true prophet, one who we know took dictation from the voice of God. After him—nobody can claim that!

The Prophecies Covered

The Hebrew Scriptures cover 7 different prophecies, but one of them is so obscure and difficult to figure out, you must first discover what the Hebrew alphabet and the Torah actually is. Once you understand that, there is a slight chance you can figure out the last prophecy—which is the secret of our redemption!

The seven prophecies are:

1. The coming cataclysm, described as *God's Day of Judgment*. I described many of these prophecies in the previous Chapter. The purpose of these prophecies is to warn people what will befall them in the future, if man does not learn what he is supposed to in time.
2. The return of the Jews to their ancestral homeland, Israel, in the End of Days so Mount Zion can be built.
3. Battles between Israel and the surrounding Arab states, and a much larger battle involving nations outside of the Middle East. The last battle will take place in northern Israel, at a site known as Megiddo.
4. A new philosophy of religion and science will be discovered and taught. The old philosophies will be discredited and discarded.
5. The rediscovery of Mount Sinai.
6. A New Jewish Temple to be built in Israel, as described in Chapters 40-48 of Ezekiel.
7. How to save one's self from *God's Day of Judgment*.

I will elaborate on each of these seven prophecies in the following sections. Finally, I will explain how God knows these events are going to happen, and the mechanism necessary, in the Diehold, to make these prophecies happen.

Description of the Events Included in God's Day of Judgment

I have already included, in Chapter 9, most of the prophecies describing individual events which will be part of the Cataclysm. Additionally, there are general prophecies describing human psychological conditions at the time of the polar reversal. One of the descriptions is found in Chapter 2 of Isaiah (Isaiah's priestly name was Hilkiah and his given name was Jeiel).[5] The chapter places the time of these future events in the block of time called: "And it shall come to pass in the End of Days . . ." By verse 12 one gets the feeling that everything man thought and falsely valued is about to change, quickly, by force:

> [Isaiah 2:12] For the Lord of hosts hath a day upon all that is proud and lofty, and upon all that is lifted up, and it shall be brought low: . . .

> [16] And the loftiness of man shall be bowed down, [17] and the haughtiness of man shall be brought low; and the Lord alone shall be exalted in that day. [18] And the idols shall utterly pass away. [19] And man shall go into the caves of the rocks, and into the holes of the earth, from before the terror of

the Lord, and from the glory of *His majesty*, When He raiseth to shake mightily the earth. [20] In that day a man shall cast away his idols of silver, and his idols of gold, which they made for themselves to worship, to the moles and to the bats; [21] To go into the clefts of the rocks, and into the crevices of the crags, from before the terror of the Lord, and from the glory of *His majesty*, when He shake mightily the earth. [Emphasis added.]

Similar verses are found in Chapter 31:7. Importantly, God reveals that before the polar reversal, man will still be devoted to man-made gods and to materialism. But what is materialism—gold, money, expensive cars, big houses—when the all-destructive cataclysm is about to happen? It means nothing! Your survival, your family and close friends mean everything! Isaiah's words might have meant nothing to you before I defined what God's Day of Judgment is. But now you know that Isaiah is describing a real and terrifying event. Most people could not imagine an event so destructive and terrible that people would not care about their wealth and escape into caves. Now you know why God gave this text to Isaiah. Chapter 31 is not quite over. Verse 22 gives us a prophecy about how many people will survive the cataclysm: "Cease ye from man, in whose nostrils is a breath; for how little is he to be counted!" This verse was so important that I checked the translation and then changed "accounted" to "counted," because that is what the Hebrew actually states. What is shocking about the verse is that it says, in an awkward way, that there will be very few living, breathing human beings left after the next cataclysm, not even worth counting. The problem with many translations of the Hebrew Scriptures is that the translators cannot imagine such a terrible event so they translate it away and substitute what they think it should read. This is obviously a mistake.

The next quote from Isaiah Chapter 13, starts off sounding like it was a prophecy about Babylon, but it is not, because when we get to verse 10, it describes the polar reversal and the sun's light as diminished. This can only happen after a nova. The metaphor God used here treats the forces of nature as His mighty army gathered to do battle against the earth.

[13:5] They come from a far country, from the end of heaven, even the Lord, and the weapons of His indignation, to destroy the whole earth. [6] Howl ye; for the day of the Lord is at hand; as destruction from the Almighty shall it come. [7] Therefore shall all hands be slack, and every heart of man shall melt. [8] And they shall be affrighted; Pangs and throes shall take hold of them; They shall be in pain as a woman in travail; They shall look aghast one at another; Their faces shall be faces of flame. [9] Behold the day of the Lord cometh, cruel, and full of wrath and fierce anger; To make the earth a desolation, and to destroy the sinners thereof out of it.

The previous description is certainly frightening, but probably most college-educated people would regard this as nothing more than primitive man describing some battle between two empires, rather than what it really is—the coming cataclysm.

I described Verses 10 through 12 in the previous Chapter, so I will start with verse 13, which further describes the astronomical events which we will see: "Therefore I will make the heavens to tremble, and the earth shall be shaken out of her place, for the wrath of the Lord of Hosts, and the day of His fierce anger."

The next prophecy comes from the prophet Zephaniah. His priestly name was Hilkiah (the second Hilkiah) and his given name was Azaliah. He was the father of Jeremiah and the high priest. The kingdom of Israel had already been defeated and carried off into the land of the Assyrian Empire. All was not well in Judea because at the time there was still the worship of Baal. God warned the people what he was about to do to them, unless they stopped. But his prophecies concerning the kingdom of Judea were prefaced by two verses which described what could only happen during the polar reversal. God first warned man: "This is what I am going to do to you in the far future and this is what I am going to do to you in the near future, if you do not shape up."

> [Zephaniah 1:2] I will utterly consume all things from off the face of the earth, saith the Lord. [2] I will consume man and beast, I will consume the fowls of the heaven, and the fishes of the sea, and the stumbling blocks with the wicked; and I will cut off man from off the face of the earth, saith the Lord.

The next prophecy was written by Jeremiah, whose given names were Shaphan and Azzur. His priestly name was Mahseiah. His prophecy seems to be the result of images He saw above the Ark, images of our world after the nova and the polar reversal:

> [Jeremiah 4:23] I beheld the earth, and, lo, it was waste and void; and the heavens, and they had no light. [24] I Beheld the mountains, and, lo, they trembled, and all the hills moved to and fro. [25] I beheld, and, lo, there was no man, and all the birds of the heavens were fled. [26] I beheld, and, lo, the fruitful field was a wilderness, and all the cities thereof were broken down at the presence of the Lord, and before His fierce anger. [27] For thus saith the Lord: The whole land shall be desolate; yet will I not make a full end. [28] For this shall the earth mourn, and the *heavens above be black*; because I have spoken it, I have purposed it, and I have not repented, neither will I turn back from it. [Emphasis added.]

Well, it is nice to know that God is not going to make a full end of us, but it is still a very bleak picture of the future.

Baruch, writing as Joel, wrote the last prophecy covered in this section. The prophecy tells us at what time of year the polar reversal will occur, namely in the fall after the harvest, when the corn is dried. The prophecy in Chapter 4:15 accurately states that after the sun novas and the dust shell passes the earth we will not be able to see the stars because the dust will be too thick.

[Joel 1:15] Alas for the day! for the day of the Lord is at hand, and as a destruction from the Almighty shall it come.

[1:17] The grains shrivel under their hoes; the garners are laid desolate, the barns are broken down; for the corn is withered.

[4:15] The sun and the moon are become black, and the stars withdraw their shining.

All of these quotations accurately describe one or more of the events which occur only during the cataclysm. Before you read my book, these prophecies were very likely impossible to understand. How could they happen? What sequence of events could cause such total destruction? Now you know.

The Prophecy of the Return of the Jews to Israel

The return of the Jewish people from their forced exile around the world began before 1939 and World War II. But the two big returns occurred after the war, and then after the creation of the State of Israel. The return from the Arab countries was almost total, since Arab countries expelled their Jews with nothing more then the clothing on their backs. From the 1930s to 1945, the European Christians tried to murder as many Jews as they possibly could, and steal their property and money. At the close of World War II, after 6,000,000 Jews had been murdered, the Christian countries of Europe and of North and South America felt guilty about what they had done. They permitted the creation of the state of Israel on November 29, 1947 (the date of the United Nations vote). The destruction and the Jews' return were prophesied in the Hebrew Scriptures as far back as Moses. Here are two prophecies in Deuteronomy:

[4:27] And the Lord shall scatter you among the peoples, and ye shall be left few in number among the nations, whither the Lord shall lead you away. [28] And there ye shall serve gods, the *work of men's hands*, wood and stone, which neither see, nor hear, nor eat, nor smell. [29] But from thence ye will seek the Lord thy God; and thou shalt find Him, if thou search after Him with all thy heart and with all thy soul. [30] In thy distress, when all these things are come upon thee, in the *end of days*, thou wilt return to the Lord thy God, and hearken unto His voice; [31] for the Lord thy God is a merciful God; He will not fail thee, neither destroy thee, nor forget the covenant of thy fathers which He swore unto them. [Emphasis added.]

Both Christians and Muslims have subjected Jews to forced conversions or death over the past 1900 and 1400 years respectively. Not just Jews have been seduced into worshiping materialism—all of mankind is diluted by "things." Man's drive to posses more is insatiable if he does not have a clear picture of what existence truly is. The underlying cause is man's insatiable ego.

Moses' next prophecy seems to echo the previous one with the addition of a vague reference to the Holocaust:

> [31:29] For I know that after my death ye will in any wise deal corruptly, and turn aside from the way which I have commanded you; and *evil will befall* you in the *end of days*; because ye will do that which is evil in the sight of the Lord, to provoke Him through the work of your hands. [Emphasis added.]

The next prophet was Hosea. His priestly name was Azariah and his given name was Amoz. He was the father of Isaiah and he wrote until about 725 B.C.E. His prophecy echoes the others with a theme of Jews having no country of their own, no Ark, no high priest, but eventually returning to their own country with a strong desire to return to believing in God. Non-Jews seldom know that after World War II and the Holocaust, many Jews did not believe in God, their reasoning being: "How could God do this to us and let such evil succeed?" Of course, their logic is right but there is a reason why it happened and I will explain that at the end of this section and Chapter Eleven.

> [Hosea 3:4] For the children of Israel shall sit solitary many days without king, and without prince, and without sacrifice, and without pillar, and without ephod or teraphim; [4] afterward shall the children of Israel return, and seek the Lord their God, and David their king; and shall come trembling unto the Lord and to His goodness in the *end of days*. [Emphasis added.]

Azariah's son, Isaiah, wrote the next prophecy. His prophecy clearly predicts that many Jews would be killed, which can only be interpreted as the Holocaust. But by verse 23 the extermination is referring to mankind worldwide during the End of Days:

> [Isaiah 10:20] And it shall come to pass in that day, that the remnant of Israel, and they that are escaped of the house of Jacob, shall no more again stay upon him that smote them; but shall stay upon the Lord, the Holy One of Israel, in truth. [21] A remnant shall return, even the remnant of Jacob, unto God the Mighty. [22] For though thy people, O Israel, be as the sand of the sea, only a remnant of them shall return; an *extermination* is determined, overflowing with righteousness. [23] For an *extermination* wholly determined shall the Lord, the God of hosts, make in the midst of all the earth.

His next prophecy foretells the creation of the State of Israel. This is indicated by the references to near Eastern states where Jews had lived until they were expelled after World War II. The first return occurred when the Jews returned from Babylon with Ezra. The "ensign" must be the flag of the State of Israel.

[11:11] And it shall come to pass in that day, that the Lord will set His hand again *the second time* to recover the remnant of His people, that shall remain from Assyria, and from Egypt, and from Pathros, and from Cush, and from Elam, and from Shinar, and from Hamath, and from the islands of the sea. [12] And He will set up an *ensign* for the nations, and will assemble the dispersed of Israel, and gather together the scattered of Judah from the four corners of the earth. [Emphasis added.]

Jeremiah is the next Prophet. He presided at the end of the kingdom of Judea and received many prophecies from God. The first prophecy includes statements that the Jews would not have the Ark and they would never think about it. It also prophesies that the House of Judea would come together with the House of Israel and return to the land of Israel. One must remember that the kingdoms of Judea and Israel fought each other, from the time of the first King of Israel, until Assyria captured Israel and exiled its people.

[Jeremiah 3:14] Return, O backsliding children, saith the Lord; for I am a lord unto you, and I will take you one of a city, and two of a family, and I will bring you to Zion; [15] and I will give you shepherds according to My heart, who shall feed you with *knowledge and understanding*. [16] And it shall come to pass, when ye are multiplied and increased in the land, in those days, saith the Lord, they shall *say no more*: The ark of the covenant of the Lord; neither shall it *come to mind*; neither shall they make mention of it; neither shall they miss it; neither shall it be made any more. [17] At that time *they shall call* Jerusalem the throne of the Lord; and all the nations shall be gathered unto it, to the name of the Lord, to Jerusalem; neither shall they walk any more after the stubbornness of their evil heart. [18] In those days the house of Judah shall walk with the house of Israel, and they shall come together out of the land of the north to the land that I have given for an inheritance unto your fathers. [Emphasis added.]

What is important about this prophecy is that it implies that the knowledge they had before will be replaced by a new body of knowledge which they will learn in the End of Days. I will expand on this subject later. The other important point appears in verse 17, which states: "They [man] shall call Jerusalem the throne of the Lord." It does not say, "God said Jerusalem is the throne of the Lord," implying that the throne will be somewhere else.

Jeremiah's next prophecy describes a much larger return to Israel than has ever been seen before:

[16:14] Therefore, behold, the days come, saith the Lord, that it shall no more be said: As the Lord liveth, that brought up the children of Israel out of the land of Egypt, [15] but: As the Lord liveth, that brought up the children of Israel from the land of the north, and from all the countries whither He had driven them; and I will bring them back into their land that I gave unto their fathers.

Isaiah repeated the return prophecy in Chapter 30:1-3. But by verse 7 to 11, God includes the cataclysm theme along with the return of the Jewish people, and the punishment of the nations which had maltreated them. There is no question the prophecy is about the End of Days. I have italicized the phrase "for I am with thee," for a good reason, since I believe it is a clue to being saved (redeemed) from the next cataclysm. Again, mass destruction is implied. The name of King David is used here as a clue to the date of the next cataclysm. David died around 971-970 B.C.E. If we add 3,017 years to his date of death, we wind up with the year 2046! This must therefore be the year of the next cataclysm. I think this is why God used David's name in conjunction with the End of Days and God's Day of Judgment.

> [30:7] Alas! for that day is great, so that none is like it; and it is a time of trouble unto Jacob, but out of it shall he be saved. [8] And it shall come to pass in that day, saith the Lord of hosts, that I will break his yoke from off thy neck, and will burst thy bands; and strangers shall no more make him their bondman; [9] But they shall serve the Lord their God, and *David their king*, whom I will raise up unto them. [10] Therefore fear thou not, O Jacob My servant, saith the Lord; neither be dismayed, O Israel; for, lo, I will save thee from afar, and thy seed from the land of their captivity; and Jacob shall again be quiet and at ease, and none shall make him afraid. [11] *For I am with thee*, saith the Lord, to save thee; for I will make a *full end of all the nations* whither I have scattered thee, but I will not make a full end of thee; for I will correct thee in measure, and will not utterly destroy thee. [Emphasis added.]

The final Prophet whom I will include is, of course, Ezekiel, whose given name was Baruch and whose priestly name was Seraiah. The first prophecy of his, which covers the return to Israel, seems to be talking about a third return of the remaining Jews in the world. The previous verse speaks about God again becoming King of the Jews. In verse 35, He speaks of communicating "face to face," which also implies the new (third) "Temple" would already be built, and that has not yet happened.

> [Ezekiel 20:34] ...and I will bring you out from the peoples, and will gather you out of the countries wherein ye are scattered, with a mighty hand, and with an outstretched arm, and with fury poured out; [35] and I will bring you into the wilderness of the peoples, and there will I plead with you face to face. [36] Like as I pleaded with your fathers in the wilderness of the land of Egypt, so will I plead with you, saith the Lord God. [37] And I will cause you

to pass under the *rod*, and I will bring you into the *bond of the covenant*; [38] and I will purge out from among you the rebels, and them that transgress against Me; I will bring them forth out of the land where they sojourn, but they shall not enter into the land of Israel; and ye shall know that I am the Lord. [Emphasis added.]

The next prophecy gives us the reason why God returned the Jews to Israel. There is an additional mention of materialism as the modern-day "god." The word *profane* is used a number of times in these prophecies, so it is appropriate to define the word here so you know what is meant in this context. This is the dictionary definition: "forth from the temple, hence, not sacred, common, . . . 1. Not concerned with religion or religious matters; secular. 2. Irreverent towards God or holy things; speaking, spoken, acting, or done in contempt of sacred things."[6] After reading this definition, most of us could fit into this definition, but why were the Jews singled out to be held to such a high standard, which few of us could live up to? How were the Jewish people supposed to support themselves in a hostile world, where they were not permitted to own land?[7] It is obvious that God chose the Hebrews for a specific assignment and punished then when they deviated from it. So the question is: What is the function of the Jews in a world designed by God? I found the answer in Exodus Chapter 19. Moses had just arrived at the base of Mount Sinai with the whole congregation. He went up into the mount and God "spoke" to him:

[Exodus 19:5] "Now therefore, if ye will hearken unto My voice indeed, and keep My covenant, then ye shall be Mine own treasure from among all peoples; for all the earth is Mine; [20] and ye shall be unto Me *a kingdom of priests*, and *a holy nation*. These are the words which thou shalt speak unto the children of Israel." [Emphasis added.]

This was God's entire plan for the Hebrews, to become a nation of priests! Ask yourself: "Can there be a nation full of priests—all chiefs and no commoners?" Who is going to be the congregation? Also, God did not delineate between the Levite class and the other tribes. God seemed to be saying, "They will all be priests but the only tribe that would directly access the Ark and communicate with God would be the priests descended from Aaron." The other tribal members would be priests to the rest of humanity. As the centuries rolled on, people began to forget about their religion, not believing their history nor in the miracles God had performed for them. The materialistic world blinded them, and many Jews took the wrong path. As a result, God had to change the course of history so that it would still end up with the results He desired. The next question is: What result did He want? The picture that comes to mind, after studying all of the prophecies, is that God is preparing us for His grand entrance. I will elaborate on this idea in greater detail later, under the heading of The New Temple.

[Ezekiel 36:20] And when they came unto the nations, whither they came, they profaned My holy name; in that men said of them: These are the people of the Lord, and are gone forth out of His land. [21] But I had *pity for My holy name*, which the house of Israel had profaned among the nations, whither they came. [22] Therefore say unto the house of Israel: Thus saith the Lord God: I do *not* this for your sake, O house of Israel, but for *My holy name*, which ye have *profaned* among the nations, whither ye came. [23] And I will sanctify My great name, which hath been profaned among the nations, which ye have profaned in the midst of them; and the nations shall know that I am the Lord, saith the Lord God, when I shall be sanctified in you before their eyes. [24] For I will take you from among the nations, and gather you out of all the countries, and will bring you into your own land. [25] And I will sprinkle clean water upon you, and ye shall be clean; from all your uncleanness, and from *all your idols*, will I cleanse you. [Emphasis added.]

[32] Not for your sake do I this, saith the Lord God, be it known unto you; be ashamed and confounded for your ways, O house of Israel.

In summary, the previous prophets accurately described future events which have already happened in the past 70 years, and described likely events that could happen in the near future. What is interesting is that these prophecies are not flattering towards the Jewish people at the time of Ezekiel, which leads me to believe they are what Baruch took down from God's dictation and Baruch was obliged not to change any of it. In fact, God even states that He does these acts for "His good name," not because of the Jews at all. I do not know of any other peoples' "holy books" which contain derogatory statements about their own people.

Battles Over Israel

The nation of Israel has been involved in four wars (1948, 1956, 1967, and 1973) with her neighbors since her inception. All of these wars occurred before the End of Days, which started in 1996. The wars that are described in the Hebrew prophecies occur only in the End of Days period.

We live in a time when Arabs from Egypt, Jordan, Syria, and Lebanon—who immigrated to the land of Israel over the past 100 years—are now calling themselves "Palestinians" and claiming title to the land of Israel. Before the 1967 war the Jews were called "Palestinians" by the British, and the others were just called "Arabs," because everyone knew they did not originally come from Israel. The point now is that everyone is fighting over who owns the land.

I will go back to the recent past, during and just after the First World War. The British were given the territory by the League of Nations. The Balfour Declaration of 1917 stated that there was to be a Jewish state there. The British were assigned by the League of Nations to bring this about as their "mandate." The Jewish state was supposed to be "in" all of "Palestine," including all the land east of the Jordan River. But in 1921-22 the British administrator, Winston Churchill, gave 80% of "Palestine," specifically all the land east of the Jordan River, to an obscure Arabian Prince as "payment" for supporting the British in World War I. The prince or king was appointed by England, which chose someone from the Saudi royal family. This part of "Palestine" is now called "Jordan." The Jewish State was supposed to have been created after the close of World War I, but after the discovery of oil in the Saudi Kingdom in the 1920s, the British decided that it was not such a good idea to create a Jewish State, especially if they intended to develop and market Saudi oil. So by the late 1930s, the British actively kept the Jews out of Palestine—so they could be slaughtered by the German Nazis, thus bringing an end to the "Jewish problem."

I am focusing on the British because after reading the history and the documents of the period, you will also come to realize that the British are just as guilty as the Germans for what happened to the Jews in Europe between 1932 to 1948.

I am not going to start debating people on the subject of who has better title to the Land of Israel, other than the Jews. I really do not have to. Israel is the only land that God specifically gave to any people! He repeats the boundaries at least three times in the Hebrew Scriptures. There is a basic legal principle that states that whoever possesses stolen property (real or personal) does not have good title to the stolen property no matter how many years after the theft. After the third Jewish revolt against Rome (about 132-135 c.e.), Emperor Hadrian had the remaining Jews deported to other parts of the Roman Empire. He renamed Jerusalem to Alelia Capitolina, and Judea was renamed Syria Palaestina. Rome's objective was to murder all the Jews and erase their existence from the region.

The land of Israel, formerly known as Judea, was stolen from the Jews by the Romans (who then became the newly created Catholic church), because the Jews refused to accept Arius Calpurnius Piso's fictional story. Islamic armies, under the command of Saladin, captured Palestine in 1187 c.e., stealing it from the eastern Christians. The British stole it away from the Muslim Turks after 1918, and finally, the Jews fought for it and won it back in 1948.

I mention this history now because I believe there are two great battles over Israel that were prophesied. The first war will be more regional, and will settle the issue about who owns the land. The war will directly involve Egypt, Lebanon, Syria, Iran, Saudi Arabia and Jordan.

The second war will be a much larger battle, involving world powers that are more distant. These will seek something far more valuable than money or the land of Israel and couldn't care less about the Jews or the truth.

The reason I think there will be two wars is because the prophetic writings concerning the larger second war state that Israel will be attacked from countries in the north. These prophecies do not mention Egypt, Syria, Jordan and Lebanon at all! That means that those countries will be knocked out long before the larger second battle. You will see what I mean from the following prophecies.

How to Start a War

The first prophecy describes the first battle, involving Arab states now surrounding and threatening Israel. There are numerous reasons why this war could start. It could start by a "Palestinian" state-sponsored terrorist attack, which could kill hundreds or thousands of Israelis. Israeli would then attack the new Palestinian state, which would bring the Syrians, and Iranians, into direct conflict with Israel. It could also cause the Egyptians to remilitarize the Sinai desert. Israel could also be hit by rocket attacks from Hezbollah in southern Lebanon. This would result in an Israeli reprisal attack on Lebanon and Syria. After which, all hell would break loose.

The second scenario could happen if the US and or Israel attacked Iran's nuclear facilities to prevent them from building nuclear bombs. Iran has already threatened Israel, stating that Iranian-backed Hezbollah terrorists in southern Lebanon would launch their short-range and midrange rockets towards Israel. Israel would in turn attack Lebanon and Syria. The "Palestinian" Arabs would most likely step up fifth-column attacks on Israel. The Egyptians would remilitarize the Sinai desert. The Jordanians and Saudis would most likely join the war, if they thought Israel could be beaten. Israel cannot fight a three-front war, or even a two-front war. At this point, Israel would have to make a decision. Do they knock out one attacker and do a holding action against the other side, until they destroy the first. This is what I believe is going to happen, and the side they will hit first will be the Egyptians in the Sinai desert because they pose the greatest threat. The following prophecies will describe what I mean.

The First War in the End of Days

Only three prophets—Isaiah, Jeremiah and Baruch writing as Joel—describe the countries involved in the first battle during the 50-year period called the End of Days (1996 to 2046). The first was Isaiah, writing in Chapter 11, verse 14-15. The verse refers to the Arab Palestinians as "Philistines," maybe because Arafat, their late terrorist leader, tried to convince the outside world, that they were "descended" somehow from the Philistine who lived in Gaza. The Gaza Strip will be the first part of any new Palestinian terrorist state. Isaiah's previous verses spoke about the mass return of Jews from the surrounding Arab states. That took place after the 1948 War of Independence.

[Isaiah 11:14] And they shall fly down upon the shoulder of the Philistines on the west; together shall they spoil the children of the east [Jordan and the West Bank]; they shall put forth their hand upon Edom and Moab [Jordan]; and the children of Ammon shall obey them. [15] And the Lord will utterly destroy the *tongue of the Egyptian Sea*; and with His scorching wind will He *shake* His hand over the *River*, and will smite it into *seven streams*, and cause men to march over *dry-shod*. [Emphasis added.]

The Nile River was referred to as the Egyptian Sea, "the tongue of the Egyptian Sea." The Nile Delta looks like a tongue sticking out into the Mediterranean Sea. The word *"tongue"* also implies eating food, and the Nile Delta has been the primary food supply for Egypt since the beginning of Egypt. The Nile has had as many as eight branches, but today it is divided into seven.[8] "River" is also a direct reference to the great Nile River. The prophecy states that the great River will be reduced to seven small streams. The explanation for "dry-shod" will be given after the next quote because it is very important, but I want to include Baruch's prophecy next. It is short but right to the point.

[Joel 4:19] Egypt shall be a *desolation*, and Edom [Jordan] shall be a desolate wilderness, for the violence against the children of Judah, because they have shed innocent blood in their land. [20] But Judah shall be inhabited forever, and Jerusalem from generation to generation.

I believe that if Israel is faced with an attack from three sides, it will have to deliver a decisive blow to one or more sides, in order to win the war, survive, and not be exterminated. The Egyptians are Israel's most dangerous threat because they have been supplied with modern American tanks, aircraft, helicopters, ground-to-ground missiles, as well as ground-to-air missiles.

One of two things could happen. The first: a massive earthquake might destroy the Aswan High dam, allowing Lake Nasser (Figure 10-1) to rush down the Nile to the Mediterranean Sea. The phrase "He shakes His hand over the River" may predict such an earthquake. Lake Nasser is over 310 miles long, up to 16 miles wide, and up to 460 feet deep at its greatest depth. The average depth is about 90 feet. Over 90% of Egypt's population lives within 12 miles of the Nile, and a large portion of it lives in the Nile delta.

The second: Israel might blow up the dam during the next war and cause the disaster. There is no question that if the Aswan High Dam were destroyed it would mean the end of Egypt. Egypt would lose most of its population within two days, and would never be a threat to Israel again.

The previous prophecies seem to predict a major disaster will befall Egypt. The prophecy says that the disaster will "cause men to march over dry-shod" because before the dam was built the Nile would sometimes dry up during the

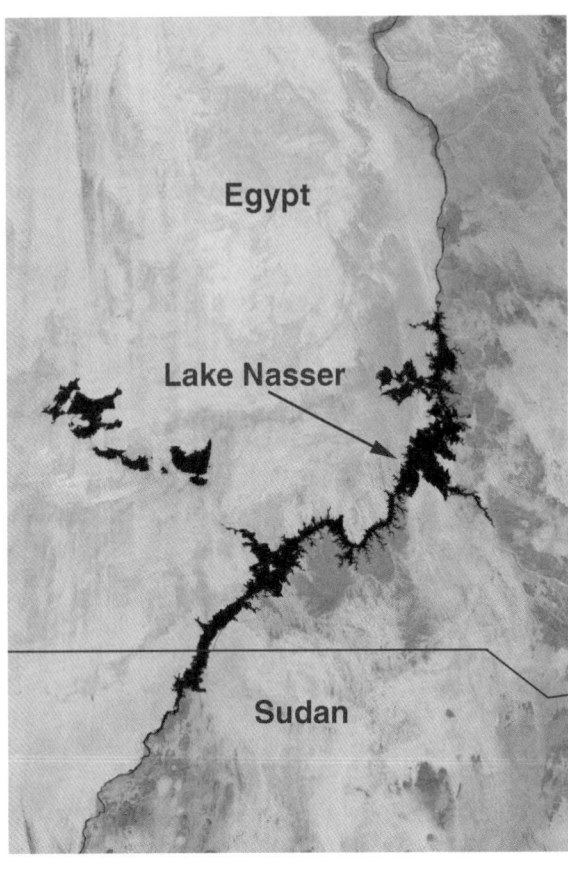

Figure 10-1: Satellite photo of Lake Nasser, the world's largest man-made lake, located in Egypt and Sudan. Courtesy of NASA.

dry season. The flooding of the Nile used to take place on the summer solstice. The prophecy may be telling us at what time of year the next war is going to happen.

Isaiah and Jeremiah mentioned that Syria would also be destroyed, but it is not listed among the nations attacking Israel in the larger second battle. Therefore it will probably be destroyed in the first war. Both prophecies take place in the End of Days.

> [Isaiah 17:1] The burden of Damascus [the capitol of Syria]. Behold, Damascus is taken away from being a city, and it shall be a ruinous heap. [2] The cities of Aroer are forsaken; they shall be for flocks, which shall lie down, and none shall make them afraid.

> [Jeremiah 49:24] Damascus is waxed feeble, she turneth herself to flee, and trembling hath seized on her; anguish and pangs have taken hold of her, as of a woman in travail. [25] "How is the city of praise left unrepaired, the city of my joy?"

BARUCH'S PROPHECIES

Baruch wrote under different prophets' names because each book he wrote described a different aspect of the cataclysm. The book of Joel describes physical aspects of the polar reversal, but says nothing of events leading up to the destructions of the Temple and of Babylon.

The verses in Joel which deal with the first war are found in Chapter 4, verses 4 to 8. Tyre is now part of Lebanon, and "Philistia" is today's Gaza strip.

> [Joel 4:4] And also what are ye to Me, O Tyre, and Zidon, and all the regions of Philistia? Will ye render retribution on My behalf? and if ye render retribution on My behalf, swiftly, speedily will I return your retribution upon your own head. [5] Forasmuch as ye have taken My silver and My gold, and have carried into your temples My goodly treasures; [6] the children also of Judah and the children of Jerusalem have ye sold unto the sons of Jevanim [Greeks], that ye might remove them far from their border; [7] behold, I will stir them up out of the place whither ye have sold them, and will return your retribution upon your own head; [8] and I will sell your sons and your daughters into the hand of the children of Judah, and they shall sell them to the men of Sheba [Yemen], to a nation far off; for the Lord hath spoken.

The previous verses seem to indicate that the Arabs are going to be expelled from Israel. Some may be exiled to Yemen, but I am sure Syria, Lebanon and Jordan will receive their share of their returning Arab brothers. Baruch, writing as Ezekiel echoed the same prophecy in Chapter 34:23, where he described the future when God will again be King of Israel. In verse 25, he refers to "evil beasts" being cast out of Israel. The phrase is very interesting because it is used only two times in the entire Hebrew Scriptures (Leviticus 26.6 and Ezekiel 34:25), and in both occurrences it refers to men as the "beasts." It is also easy to discover to whom God referred, because there are no wild beasts in Israel today, and a wild beast cannot be considered evil. An animal eats to live. It does not have hate in its heart towards its food. Only man can do evil and the Arabs today—in Israel—are encouraged to sacrifice their own children for their religious beliefs, just as the Canaanites did 3,300 years ago. Arab parents encourage their teenage children to strap on vest bombs, sneak into Israel, and blow themselves up in crowds of Jews, so as to murder as many as they can. It does not get any sicker than that. I would definitely classify that as *evil*.

> [Ezekiel 34:25] And I will make with them [the Jews] a covenant of peace, and will cause *evil beasts* to cease out of the land; and they shall dwell safely in the wilderness, and sleep in the woods. [Emphasis added.]

The previous prophecies indicate there will be a regional war which will render Egypt a desolate country, never to rise again. The prophecies also predict that Lebanon, Jordan and Syria will likewise be defeated in this war.

The Final Battle, the War of Decision

According to prophecy there is going to be very large battle between Israel and a large number of other countries, led by a powerful country in the north but not necessarily directly north of Israel. The battle is described as taking place in the Valley of Jehoshaphat, which name means *"The Lord judgeth."* There is no valley named "Jehoshaphat," but you can discover where it is located. There was a king of Judah named Jehoshaphat, who reigned from 873 to 849 B.C.E. He lived in the territory of Issachar, located southwest of the Sea of Galilee. There is a large valley in that territory called Emeq Yizre'el. On a hill north of the valley is the town of Nazareth. On the west side of the Valley is the site of the historic battle of Megiddo. Because Nazareth was supposedly the birthplace of the central figure of Piso's religion. I believe the battle will be in the valley of Emeq Yizre'el. It will be there because it has always been the natural invasion route into Israel and because Baruch has called it "the valley of decision." Man must decide which philosophy of God is the correct one.

I will list most of the prophecies which describe this last big battle. After studying all of them I believe there are two reasons why Israel will be attacked. The prophecies make it sound like many countries from different parts of the world will take part in this battle, but when you plot the countries on a round globe, you realize it involves Muslim countries from Pakistan to North Africa, and some countries within the European Union.

So the questions I have are: What would cause so many countries to want to war against little Israel, with such a small population, and no natural resources to speak of? Why would an attacking force need to be so large against such a small country? Amassing an army costs many billions of dollars. There is no financial reason for such a large group of countries to want to attack little Israel so why do they want to attack? I believe I have discovered two reasons which will be explained in the conclusion of this section.

These prophets mention a number of countries cryptically, by their tribal origins. The following is a list of names alongside today's names for the region or country.

1. ***Cush***: Sudan and Ethiopia.
2. ***Dedan***: Saudi Arabia.
3. ***Gog***: Southern Russia, Kazakhstan, Uzbekistan, Tadzhikistan, Kirghistan, Turkmenistan, the Caucasus Mountains and westward along the shores of the Black Sea to the Danube.
4. ***Gomer***: Ukraine or Crimea
5. ***Jevanim***: Greece
6. ***Meshech***: Eastern Turkey and Georgia east of the Black Sea.
7. ***Persia***: Iran.
8. ***Put***: Northern Africa, including Libya, but not Egypt.
9. ***Sheba***: Yemen and Southern Saudi Arabia.
10. ***Tarshish***: Spain and/or maritime Western Europe.[9]
11. ***Togarmah***: Armenia

12. *Tubal*: Turkey

Three of the clues God gave us include the name of the town north of the valley, the repeated use of the phrase "a controversy with the nations," and the mention of Mount Zion. I will begin with Isaiah and proceed forward in time to Baruch. I am sure the European Union will be involved because Spain and Greece are members and Turkey has applied to become a member. In addition, we do not know which countries are included in "maritime western Europe." They could be France and Italy.

ISAIAH:

[29:5] But the multitude of thy foes shall be like small dust, and the multitude of the terrible ones as chaff that passeth away; yea, it shall be at an instant suddenly—[6] There shall be a visitation from the Lord of Hosts with thunder, and with earthquake, and great noise, with whirlwind and tempest, and the *flame of a devouring fire*. [7] And the multitude of all the nations that war against Ariel [Israel], even all that war against her, and the bulwarks about her, and they that distress her, shall be as a dream, a vision of the night. [8] And it shall be as when a hungry man dreameth, and, behold, he eateth, but he awaketh, and his soul is empty; or as when a thirsty man dreameth, and, behold, he drinketh, but he awaketh, and, behold, he is faint, and his soul hath appetite—so shall the multitude of all the nations be, that *fight against Mount Zion*.

[34:1] *Come near, ye nations*, to hear, and attend, ye peoples; let the earth hear, and the fullness thereof, the world, and all things that come forth of it. [2] For the Lord hath indignation against all the nations, and fury against all their host; He hath *utterly destroyed them, He hath delivered them to the slaughter*. [3] Their slain also shall be cast out, and the stench of their carcasses shall come up, and the mountains shall be melted with their blood. [4] And all the host of heaven shall moulder away, and the heavens shall be *rolled together as a scroll*; and *all their host shall fall down*, as the leaf falleth off from the vine, and as a falling fig from the fig-tree. [Emphasis added.]

MICAH:

[4:1] But in the *end of days* it shall come to pass, that the *mountain of the Lord's house* shall be established as the top of the mountains, and it shall be exalted above the hills; and peoples shall flow unto it.

[4:11] 'And now many nations are assembled against thee, that say: 'Let her be defiled, and let our eye gaze upon *Zion*.' [12] But they know not the thoughts of the Lord, neither understand they His counsel; for *He hath gathered them as the sheaves to the threshing-floor*. [13] Arise and thresh, O daughter of *Zion*; for I will make thy horn iron, and I will make thy hoofs brass; and thou shalt beat in pieces many peoples; and thou shalt devote their gain unto the Lord, and their substance unto the Lord of the whole earth. [Emphasis added.]

JEREMIAH:

[25:30] Therefore prophesy thou against them all these words, and say unto them: The Lord doth roar from on high, and utter His voice from His holy habitation; He doth mightily roar because of His fold; He giveth a shout, as they that tread the grapes, against all the inhabitants of the earth. [31] A noise is come even to the end of the earth; for *the Lord hath a controversy with the nations*, He doth plead with all flesh; as for the wicked, He hath given them to the sword, saith the Lord. [32] Thus saith the Lord of hosts: Behold, evil shall go forth from nation to nation, and a great storm shall be raised up from the *uttermost parts of the earth*. [33] And the slain of the Lord shall be at that day from *one end of the earth even unto the other end* of the earth; they shall not be lamented, neither gathered, nor buried; they shall be dung upon the face of the ground. [34] Wail, ye shepherds, and cry; and wallow yourselves in the dust, ye leaders of the flock; for the days of your slaughter are fully come, and I will break you in pieces, and ye shall fall like a precious vessel. [Emphasis added.]

BARUCH AKA EZEKIEL, JOEL AND OBADIAH:

I bring your attention to Ezekiel, Chapters 38 and 39. Baruch loads the chapters up with ancient biblical names of the invading countries along with an excellent description of God's punishment on the invading army and their homelands. Baruch named all 12 of the countries previously listed. The first prediction in verse 2 mentions the land of Magog. This translates as *"the land of Gog,"* meaning the lands in which the descendants of Gog had settled.

[Ezekiel 38:2] Son of man, set thy face toward Gog, of the land of Magog, the chief prince of Meshech and Tubal, and prophesy against him, [3] and say: Thus saith the Lord God: Behold, I am against thee, O Gog, chief prince of Meshech and Tubal; [4] and I will turn thee about, and *put hooks into thy jaws*, and I will bring thee forth, and all thine army, horses and horsemen, all of them clothed most gorgeously, a great company with buckler and shield, all of them handling swords: [5] Persia, Cush, and Put with them, all of them with shield and helmet; [6] Gomer, and all his bands; the house of Togarmah in the uttermost parts of the north, and all his bands; even many peoples with thee. [7] Be thou prepared, and prepare for thyself, thou, and all thy company that are assembled unto thee, and be thou guarded of them. [8] After many days thou shalt be mustered for service, in the *latter years* thou shalt come against the land that is brought back from the sword, that is gathered out of many peoples, against the mountains of Israel, which have been a continual waste; but it is brought forth out of the peoples, and they dwell safely all of them. [Emphasis added.]

The phrase, *"put hooks into thy jaws"*, gives me the impression that God will be baiting them to attack Israel! By verse 13 God gives us a hint of the bait He uses: "Comest thou to take the spoil? hast thou assembled thy company to take the prey? To carry away silver and gold, to take away cattle and goods, to take a great spoil?"

The gold and great spoil He refers to is the gold, silver, and other portable wealth which Joseph embezzled when he was Prime Minister of Egypt.[10] The location of this wealth is also tied to the location of Mount Sinai. However, there is something more valuable than that, something so valuable that any industrial country would want it at any cost.

The next excerpt specifies the time period.

[Ezekiel 38:14] Therefore, son of man, prophesy, and say unto Gog: Thus saith the Lord God: In that day when My people Israel dwelleth safely, shalt thou not know it? [15] And thou shalt come from thy place out of the uttermost parts of the north, thou, and many peoples with thee, all of them riding upon horses, a great company and a mighty army; [16] and thou shalt come up against My people Israel, as a cloud to cover the land; it shall be in the *end of days*, and *I will bring thee against My land*, that the nations may know Me, when I shall be sanctified through thee, O Gog, before their eyes. [Emphasis added.]

The prophecy clearly sets the time period, *i.e.*, in The End of Days. It also repeats the theme that God is baiting these people to invade Israel. He repeats this theme again in verse 38:17 just to ensure the reader understands who is pulling the strings. The next part of the prophecy goes into the actual event which will happen during God's battle with these people. I call it "God's battle" because He states, in so many words, that He will do this to redeem His holy name.

[Ezekiel 38:18] . . . My fury shall arise up in My nostrils. [19] For in My jealousy and in the fire of My wrath have I spoken: Surely in that day there shall be a *great shaking* in the land of Israel; [20] so that the fishes of the sea, and the fowls of the heaven, and the beasts of the field and all creeping things that creep upon the ground, and all the men that are upon the face of the earth, *shall shake* at My presence, and the mountains shall be thrown down, and the steep places shall fall, and every wall shall fall to the ground. [21] And I will call for a sword against him throughout all my mountains, saith the Lord God; every man's sword shall be against his brother. [22] And I will plead against him with pestilence and with blood; and I will cause to rain upon him, and upon his bands, and upon the many peoples that are with him, an *overflowing shower, and great hailstones, fire, and brimstone.*

[23] Thus will I *magnify Myself*, and sanctify Myself, and I will make Myself known in the eyes of many nations; and they shall know that *I am the Lord.* [Emphasis added.]

These prophecies lead me to the unambiguous conclusion that God is very vengeful, and what will occur is definitely not a nuclear war. Its causes are more astronomical than terrestrial—and it is obvious that God wants man to know that He is the One who performed these actions, not any person or country.

Ezekiel's next chapter almost repeats the same prophecies, but describes the reasons why.

[Ezekiel 39:1] . . . I am against thee, O Gog, chief prince of Meshech and Tubal; [2] *and I will turn thee about and lead thee on,* and will cause thee to come up from the uttermost parts of the north; and I will bring thee upon the mountains of Israel;

[4] Thou *shalt fall upon the mountains of Israel*, thou, and all thy bands, and the peoples that are with thee; I will give thee unto the ravenous birds of every sort and to the beasts of the field, to be devoured. [5] *Thou shalt fall upon the open field*; for I have spoken it, saith the Lord God. [6] And I will send a *fire on Magog*, and on them that dwell safely in the *isles* [Greece]; and they shall know that I am the Lord.

[9] And they that dwell in the cities of Israel shall go forth, and shall make fires of the weapons and use them as fuel, both the shields and the bucklers, the bows and the arrows, and the hand-staves, and the spears, and they shall make fires of them seven years;

[12] And *seven months shall the house of Israel be burying them*, that they may cleanse the land. [13] Yea, *all the people of the land shall bury them*, and it shall be to them a renown; in the day that *I shall be glorified*, saith the Lord God. [14] And they shall set apart men of continual employment, that shall pass through the land to bury with them that pass through those that remain upon the face of the land, to cleanse it; after the end of *seven months* shall they search. [Emphasis added.]

The previous verses tell us that a "fire" of some sort will hit a very wide expanse of territory. I plotted the countries previously listed and they stretch from Afghanistan to Spain. What is very interesting is that if you take a string and stretch it from Pakistan to Spain, it will pass over every country listed, except the countries in Africa. It also includes other countries that have tormented and murdered the Jewish people such as Rome (Italy), and southern France.[11] I will later expound on what I believe will cause the destruction.

Verses 4 through 14 describe what will happen to the armies of the invading forces. According to the prophecies the slaughter will be so immense that it will take the entire population of Israel a full seven months to bury most of the dead of the armies of Gog and their allies. I also get the impression these soldiers will just drop dead in a single moment. The legends state the same idea. Our technology today could not accomplish this, but an advanced technology of 10 million years in the future could!

The next quote gives us some insight why God would do this kind of destruction:

> [Joel 4:1] For, behold, in those days, and in that time, when I shall bring back the captivity of Judah and Jerusalem, [2] I will gather all nations, and will bring them down into the valley of Jehoshaphat; and I will enter into judgment with them there for My people and for My heritage Israel, whom they have scattered among the nations, and divided My land. [3] And they have cast lots for My people; and have given a boy for an harlot, and sold a girl for wine, and have drunk

> [Ezekiel 39:21] And I will set My glory among the nations, and all the nations shall see My judgment that I have executed, and My hand that I have laid upon them. [22] So the house of Israel shall know that I am the Lord their God, from that day and forward. [23] And the nations shall know that the house of Israel went into captivity for their iniquity, because they broke faith with Me, and I hid My face from them; so I gave them into the hand of their adversaries, and they fell all of them by the sword.

> [Ezekiel 28] 'And they shall know that I am the Lord their God, in that I caused them to go into captivity among the nations, and have gathered them unto their own land; and I will leave none of them any more there; [29] neither will I hide My face any more from them; for I have poured out My spirit upon the house of Israel, saith the Lord God.'

Baruch's book of Joel contains only four chapters, but it is loaded with prophecies of events at the End of Days. The prophecies highlight events about the coming cataclysm and the big battle with the armies of Gog, and also mention the Greeks as the sons of Jevanim. Baruch again gives us the distinct impression that the nations will be baited into invading Israel.

> [Joel 4:9] Proclaim ye this among the nations, prepare war; stir up the mighty men; let all the men of war draw near, let them come up. [10] Beat your plowshares into swords, and your pruning-hooks into spears; let the weak say: "I am strong." [11] Haste ye, and come, all ye nations round about, and gather yourselves together; thither cause Thy mighty ones to come down, O Lord! [12] Let the nations be stirred up, and come up to the valley of Jehoshaphat; for there will I sit to judge all the nations round about. [13] Put

ye in the sickle, for the harvest is ripe; come, tread ye, for the winepress is full, the vats overflow; for their wickedness is great. [14] Multitudes, multitudes in the valley of decision! For the day of the Lord is near in the valley of decision.

This is similar to what he inserts into Ezekiel.

God's Reasons for the Second War of Decision

There are two reasons behind this last War. The first is obvious, once one looks at the list of Middle Eastern Muslim countries which will be involved. The question is: What will provoke them to attack Israel, and why is the United States— supposedly Israel's "supporter"—not mentioned as coming to the aid of Israel? The answer is hinted at in the previous and later prophecies. Imagine, if you will, that Mount Sinai's existence is revealed to the world, with all the altars Moses built there. At the same time, the cave of Machpelah is discovered with all the extremely important artifacts and graves in it. The following objects should be there: The burial tombs of Abraham, Isaac, and Jacob, 11 of Jacob's 12 sons and their wives and perhaps other descendants; also the Tabernacle, the Ark of the Covenant, the candle stand and the incense table—all ready to work. That is what should be found just in the first cavern.

This type of limestone deposit forms a series of huge caverns which go on for miles. Over a period of millions of years a series of connecting tunnels are formed by rushing water as the land rises. The second cavern room should contain the valuables Joseph hid in the family burial cave system (gold, silver, precious stones and anything else that was of great value and was portable).[12] I estimate that there should be at least two tons of valuables in the Cave. One more cavern will be discovered, but I will cover that after I complete this section.

After all these religious objects are discovered and made known, I am certain they will amaze and shock the world! The effect these discoveries will have on the Jewish people worldwide will be overpowering and emotionally exhilarating. For the first time in 2,600 years, Israel will know for sure that the Exodus, Abraham, Isaac, Jacob, Moses, Mount Sinai, the Ark, the Torah and, most importantly, God—are all real and true and that academia was wrong! These discoveries, coupled with the Jews in Israel finding out what the Calpurnius Piso's of Rome wrote, will "put them over the top!" The Jews will know for sure that the world has been ruled by fraud, fiction and fabrication, and I do not want to leave out vanity and emptiness. What I believe will happen is the Jews from around the world will start the last return to Israel.

Now let us assume that someone gets the Ark to work and God speaks from it. There is a very good chance He will have a lot to say. One of those statements just might be to order the Jews to cleanse the land, just as He told Joshua to cleanse the land of Canaanites. I believe that will include destroying all remnants of other religions including churches and mosques in Israel, including the Dome of the Rock. They might just throw out the entire Moslem population plus the Christian leadership out of Israel.[13] That action would then be the major reason why the rest of the Islamic world would attack Israel. I am sure the United Nations would sanction Israel and, with European Union (EU) backing, there would be a vote to attack Israel. The United States would either vote against Israel or just abstain. Additionally, I am sure there would be a complete trade embargo against Israel, which I believe the U.S. State Department would gleefully support.[14] For example, after President Truman recognized the newly-formed State of Israel, the State Department succeeded in implementing an arms embargo on the entire region knowing very well that America's British ally would continue supplying arms to the surrounding Arab states which had already attacked Israel. Some unites of the Jordanian army were lead by British generals. The only conclusion one can come to is that the diplomatic leadership in Washington decided that the few surviving Jews, who had escaped the European Holocaust should now be finished off by America's real friends, the now oil-rich Arabs. So the Jews went to the one superpower who had no friends in the area, The Union of Soviet Socialist Republics (U.S.S.R.). Joseph Stalin sold them arms and airplanes leftover from World War II.

Joseph and how God Moves in Mysterious Ways

I think it is important, at this time, to explain my statement about Joseph's gold and other valuables embezzled from Egypt when he was Prime Minister of Egypt, because this directly ties into what I just predicted about a trade embargo against Israel. The lesson to learn from what I am about to write is that you cannot fool or trick God because He has the ability to manipulate events in time so they come out the way He wants. My discovery started in 1999 when I began to research the Egyptian side of the Exodus story and then decoded the Genesis story. I am not going to describe the entire Biblical narrative of Joseph in Egypt in this book because it is too long and is not on the subject of this book, but I will tell you the following.

Joseph's Egyptian name in Egyptian history was Senmut. He was the lover and Prime Minister of the only female Egyptian Pharaoh, Hatshepsut. Hatshepsut had a dream (in 1488 B.C.E.) about seven years of plenty and seven years of drought. Joseph interpreted the dream correctly. Hatshepsut made Joseph Prime Minister of Egypt and changed his name to Senmut. The whole story was coded into the Torah and in Egyptian history. It was the reason why Moses was raised in the Pharaoh's house. The whole story will be told in my next book.

This was an act of God. God created Joseph as an egocentric individual, and made him very good looking. He was sold into slavery by his brothers and wound up being purchased by the wife of the then Pharaoh, Tuthmoses II. She purchased Joseph because he was so handsome and she used him as her personal "sex toy," as the Torah implies. The Pharaoh died about 1492 b.c.e., and Hatshepsut gained all the power, eventually naming herself Pharaoh. She had her famous dream, which Joseph interpreted correctly, and so she appointed Joseph Prime Minister to carry out his plan to grow as much food as possible in Egypt so they could sell it back to a starving world, including the starving Egyptians. Twice Genesis informs us that Joseph personally received the money from the sale of all that grain,[15] implying that Joseph may have appeared as the landowner of the grain and not the Pharaoh. The legends state that he took two-thirds of it and hid it away. It went into the family burial cave. Small periodic withdrawals were made by his descendants (through the line of Ephraim) up to and including Aaron and Moses.[16] After Moses and the congregation left Mount Sinai, Moses and Aaron sealed up the cave tunnel, leading to the deeper caverns. There was a reason why it had to be sealed up and forgotten, but that explanation will be for another book. I will hint that Korah's rebellion had something to do with it.[17] The following legend sums it up:

> ". . . during the seven years of plenty, he [Joseph] had divided it [the gold] into three parts. The first part he surrendered to Pharaoh. The second part he concealed in the wilderness, where it was found by Korah, though it disappeared again, not to come to view until the Messianic time, and *then it will be for the benefit of the pious*. The third part Joseph hid in the sanctuary of Baal-zephon, whence the Hebrews carried it off as booty,"[18] [Emphasis added.]

I found Baal-Zephon during my third expedition to Egypt in 2002. It was a depleted gold mine. Joseph temporarily hid the gold and silver there before removing it again and transporting it to its final location, the family burial cave.

Now I will explain what has just been revealed to you: God knew that the present-day Jews would have a difficult time in the End of Days because of the economic embargo which will be placed on Israel for the reasons earlier theorized. Therefore, God created the perfect set of conditions so that hard currency would be available when Israel needed it the most. The lesson here is that God can reach back in time and create and ensure the right set of future conditions. You cannot fight an Entity that can reach back in time to change events to come out the way He wants.

The Europeans

Now that I have given a logical hypothetical reason for the Muslims to attack Israel, what would be the motivation for other countries to attack Israel? It could be any number of industrial countries and I do not leave out any of the European countries or China, India, Russia or any other country. Their motivation would be for a very practical reason, and that involves what is hidden much deeper in the cave system. That is the real prize for an industrial power that wishes to become a world power.

If the clues which Moses gave us are correct, a cavern full of highly advanced technology—perhaps over 10 million years more advanced than ours—will be found. This is the ultimate prize for any country or alliance of countries, to acquire. The country which possesses it and understands how to use it would be "top dog" for thousands of years. The advanced technology is the bait which God will use to gather the harvest into the sickle. It would not surprise me if China, Russia and the United States went after the cave individually and even fought each other to steal it. Little Israel could end up in possession of it all. Little will any of these "Super Powers" realize that they are not really in control, and I will prove to you what I mean in the next Chapter on Timelines. It may be the reason why China and the United States are not mentioned directly in the war prophecies. They could be fighting each other in another part of the world, each trying to prevent the other from stealing the advanced technology.

In summary, it is obvious the Arab Muslim countries will enter into an alliance with other countries to defeat Israel, destroy the remaining Jews and take the spoil. The Arabs and Europeans have the motivation, plus a long history of anti-Semitism. Once they learn about this treasure trove of high technology, getting their hands on it will be their number one goal.

How the Invading Army and their Countries will be Destroyed.

According to Biblical prophecies, the invading armies of Gog and its allies seem to be killed all at once. This could be accomplished by nuclear weapons but I do not think this will be the case because God keeps repeating in the verses that everyone will know that He is the one who did it. The prophecies seem to indicate that the new Temple will already have been built when the attack occurs. *The Legends of the Jews* says:

> Eldad now began to make prophecies, saying: . . . "At the *end of days* there will come up out of the land of Magog a king to whom all nations will do homage. Crowned kings, princes, and warriors with shields will gather to make war upon those returned from exile in the land of Israel. But God, the Lord, will stand by Israel in their need and will slay all their enemies by *hurling a flame from under His glorious Throne*. This will consume the souls in the hosts of the king of Magog, so that *their bodies will drop lifeless upon the mountains of the land of Israel*, and will become a prey to the beasts of the field and the fowls of the air."[19] [Emphasis added.]

The king, the legend refers to, may be whoever becomes the president of the EU or Russia or it could be a popular Muslim Middle East Leader. The reference to "hurling flame" echoes the phrase "overflowing shower, and great hailstones, fire, and brimstone," which appears in various forms throughout the prophecies. This facet of the destruction I will describe next, but I do not think it will hit the invading army in Israel, because that would render the land somewhat unusable, and I do not think that is what will happen in Israel. Therefore, we are left to next examine what could kill an entire army, spread over a very wide expanse, instantly. You cannot think in terms of our technology, or even one 20 or 40 years in the future. You have to think in terms of technologies millions of years into the future, a technology built upon a totally different philosophy of science than what is taught today. With that in mind, think of a technology that deals with all of us as information in the Diehold.

I will explain what I think is described as "overflowing shower, and great hailstones, fire, and brimstone." I do not believe that they involve nuclear bombs because those do not cause an overflowing shower or hailstones raining down, and "brimstone" is burning sulfur. Considering the length and path of possible destruction could be over 3,500 miles, there could be only two possibilities, which fit the prophetic descriptions given, and they are a solar flare or a comet hitting the Earth.

A solar flare looks like Figures 10-2 through 10-4. If the Sun threw off something denser than these, but was in a tubular shape, as in Figure 10-4, it could do the damage described in the prophecies. These events are not that rare. The largest known flare occurred on October 26-29, 2003, (Figure 10-3). The other two occurred within the past 10 years. This event is called a solar ejecta, containing charged particles, dust and other molten matter. By the time a Solar flare hit the Earth, some of the molten material would have cooled to become

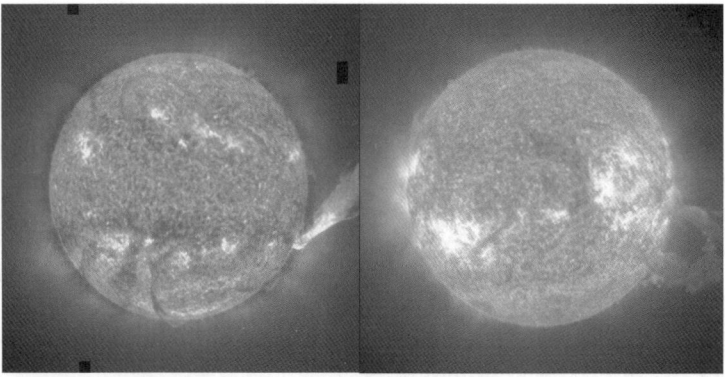

Figures: 10-2 and 10-3: Examples of solar flares.

rocks, but burning hot dust and gases would rain down on a portion of the Earth, leaving it a burning mess. Interstellar dust does contain some amounts of sulfur.[20]

If a Comet hit the Earth, it might cause the same effect but add a lot of water vapor to the mix. If one of these events ever happen, there will be no debate that it was an act of God. What will be left of these invading countries, after such an event, is anyone's guess, but I do not think they would be willing to try another attack against Israel!

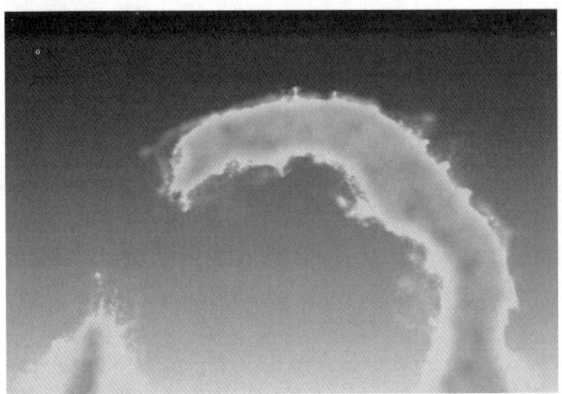

Figure: **10-4:** Example of a solar flare shaped like a tube.

A New Body of Knowledge Will Be Taught

The prophets Isaiah, Micah, Jeremiah, and Baruch all hint at a new body of knowledge that will be taught, and it appears to be far different from the body of knowledge taught before. Many of the following prophecies mention the "mountain of the Lord," or Zion. Both are references to the new "Temple" that will be built. This structure will exemplify the new scientific philosophy, which I will describe in the next section. Isaiah 29: 9-14 is important because it flat-out tells us that there will be a book that changes everything, and it will disturb a lot of people. Verse 14 describes the current Jewish leadership, and unambiguously tells us that their body of knowledge is wrong.

> [Isaiah 2:2] And it shall come to pass in the *end of days*, that the mountain of the Lord's house shall be established as the top of the mountains, and shall be exalted above the hills; and all nations shall flow unto it. [3] And many peoples shall go and say: "Come ye, and let us go up to the mountain of the Lord, to the house of the God of Jacob; and *He will teach us of His ways*, and we will walk in His paths." For out of Zion shall go forth the law and the word of the Lord from Jerusalem.

> [Isaiah 29:9] Stupefy yourselves, and be stupid! Blind yourselves, and be blind! ye that are drunken, but not with wine, that stagger, but not with strong

drink. [10] For the Lord hath poured out upon you the spirit of deep sleep, and hath closed your eyes; the prophets, and your heads, the seers, hath He covered. [11] And the vision of all this is become unto you as the words of a writing that is sealed, which men deliver to one that is learned, saying: "Read this, I pray thee" and he saith: "I cannot, for it is sealed"; [12] and the writing is delivered to him that is not learned, saying: 'Read this, I pray thee'; and he saith: "I am not learned." [13] And the Lord said: Forasmuch as this people [the Jews] draw near, and with their mouth and with their lips do honor Me, but have removed their heart far from Me, and *their fear of Me is a commandment of men learned by rote*; [14] Therefore, behold, I will again do a marvelous work among this people, even a marvelous work and a wonder; and *the wisdom of their wise men shall perish, and the prudence of their prudent men shall be hid.* [Emphasis added.]

The prophet Micah, son of Isaiah, in Chapter 4:2-3 says almost exactly the same things as Isaiah 2:2-3.

The current state of the Jewish religion is best summarized in the line, *"their fear of Me is a commandment of men learned by rote."* Every year the Torah is read completely through, to be read again the next year. The prayers and rituals are repeated again and again, but very little understanding is extracted from the words the Rabbis recite. They read the surface stories and only perceive the surface meaning of the books, including the prophets. They cite the opinions of learned Rabbis of 300, 1,000 and 1,400 years ago, even though those Rabbis knew nothing about Egyptian History, never went to seek the real Mount Sinai, or tried to solve the puzzle of the Hebrews in Egypt by going to Egypt. The Jewish religion should never have degenerated into a religion primarily of ritual, tradition, dietary laws and rabbinical rulings from the Talmud—not to mention all the damage the Piso's did to the religion by adding so many Roman writings to the Hebrew Scriptures in the First and Second Centuries C.E. The Hebrew religion, as given to us by God and Moses, should have been much more. The Jewish religion is the hope of the world because it provides the clues to save us from the cataclysm. It is also the only religion that proves the existence of God and that is what I think a religion is supposed to do.

The next prophet, Jeremiah, wrote about a completely new covenant with Israel, a far more advanced covenant than the first one described in the Torah, as you will see.

[Jeremiah 31:31] Behold, the days come, saith the Lord, that I will make a *new covenant* with the house of Israel, and with the house of Judah; [32] not according to the covenant that I made with their fathers in the day that *I took them by the hand* to bring them out of the land of Egypt; forasmuch as they broke My covenant, although I was a *Lord over them*, saith the Lord. [33] But *this is the covenant that I will make* with the house of Israel after those days, saith the Lord, I will put My law in their inward parts, and in their heart will I write it; and I will be their God, and they shall be My people; [34] and they shall teach no more every man his neighbor, and every man his brother, saying: "Know the Lord?"; for they shall all know Me, from the least of them unto the greatest of them, saith the Lord; for I will forgive their iniquity, and their sin will I remember no more. [Emphasis added.]

These verses are very significant because they hint at the nature of the New Covenant, or new knowledge. Ask yourself, "Whose hand do you have to hold?" You hold the hands of children who need guidance, to lead them to the right path. God is telling us that the surface stories of the Torah were intended to be understood by people with limited scientific knowledge. The next covenant/philosophy must teach man what God actually *is*, and what man's place is in the universe.

Verse 34 is a bit cryptic, but it is predicting that the Jews will not have to ask people, "Do you know what God is?" because the new philosophy will have spread all over the world, so nearly everyone will be familiar with the new philosophy, which will explain what God actually is.

The last prophecy in this section came from Baruch in Psalm 119. He wrote the following:

[Psalm 119:18] Open Thou mine eyes, that I may behold wondrous things out of *Thy Law*. [19] I am a sojourner in the earth; hide not Thy commandments from me. [Emphasis added.]

When Baruch writes, "The Law" it usually means the Torah, but in this case I believe it refers to God's New Covenant.

Mount Sinai Found

The Torah, and many books of the Hebrew Scriptures, give clues as to where Mount Sinai is located. It is only reasonable to assume that God would eventually make sure Mount Sinai would be rediscovered in the End of Days. Isaiah gives us one prophecy which I believe says the real Mount Sinai will be found. The area it describes is in a desert, and the term "in the wilderness," as well as Zin (which later became Zion), had been used by Moses to describe the location of Mount Sinai. The discovery of the real Mount Sinai, with all the altars visibly there, would certainly bring joy to the Jewish people as a reconfirmation of their belief in the Torah and the Lord of Hosts.

[Isaiah 35:6] . . . for in *the wilderness* shall waters break out, and streams in the desert. [7] And the parched land shall become a pool, and the thirsty ground springs of water; in the habitation of jackals herds shall lie down, it shall be an enclosure for reeds and rushes. [8] And a highway shall be there, and a way, and it shall be called *The way of holiness*; the unclean shall not pass over it; but it shall be for those; the wayfaring men, yea fools, shall not err therein. [9] No lion shall be there, nor shall any ravenous beast go up thereon, they shall not be found there; but *the redeemed shall walk there*; [10] And the ransomed of the Lord shall return, and come with *singing unto Zion*, and everlasting joy shall be upon their heads; they shall obtain gladness and joy, and sorrow and sighing shall flee away. [Emphasis added.]

The New "Temple" — Zion.

The most important prophecies stated in the Hebrew Scriptures are the construction of the New "Temple."[21] Most people think it will be built in Jerusalem, but I do not believe so. I believe the Biblical clues indicate it will be built elsewhere. You will also learn what its purpose actually is. The following prophecies include a wide variety of subjects involving the new "Temple:" its location, what will be inside, what it will look like and the function of the priestly class.

I have divided this section into eight parts. In the first part I will discuss some background history, what God's relationship was with the Hebrews. Then I explain the rest.

1. Background History, from Genesis to Ezekiel

The Genesis Garden of Eden story tells us that man once had the ability to communicate directly with God. In the center of this "garden" was the "tree of knowledge, good and evil." The "tree" was a terminal to the Diehold, as I wrote in Chapter 4, where all information in the Universe was accessible. A fruit tree was chosen in the story because the cross-section of an apple looks somewhat like the cross-section of a toroid that produces the Hebrew alphabet (Figure 10-5). The seeds in the center are analogous to the waveform when it crosses the X-axis and creates the matter world we live in—in essence the seeds of continuous creation.

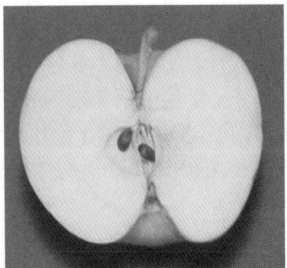 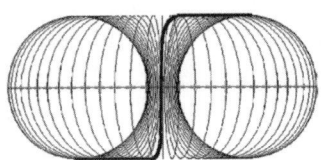

Figure 10-5: A Cross-section of an apple that resembles the cross-section of a toroid.

In the Genesis story, man did something counterproductive to God's plan. Adam and Eve (man) changed from one state into a lower state of existence. I do not use the word "evil" because I do not sense that is what God told them when He informed them of the numerous consequences of their actions. God just stated what life would now be like after their self-imposed de-evolution. This action resulted in their departure from the Garden of Eden. At some point in history, perhaps after a previous polar reversal, man lost the ability to communicate directly with God. In the book Ezekiel (Chapters 40 to 48) Baruch informs us that someone in the future will build a new "Temple," in which God will once again communicate directly with man, and bring peace and a single religious truth to mankind. Hence, the Biblical story comes full circle.

2. A Time when God was King of the Hebrews

The system set up by Moses was to have God become Israel's King. The High Priest was God's administrator who in turn gave instructions to an appointed judge who would carry out God's instructions. There was very little centralized governmental structure. About 1021 B.C.E. the Hebrews had problems with the Philistines, so they went to the prophet Samuel to ask him to appoint a human king, like all the other countries. The conversation went as follows:

[First Samuel 8:6] But the thing displeased Samuel, when they said: "Give us a king to judge us." And Samuel prayed unto the Lord. [7] And the Lord said unto Samuel: "Hearken unto the voice of the people in all that they say unto thee; for they have not rejected thee, but *they have rejected Me, that I should not be King over them*. [8] According to all the works which they have done since the day that I brought them up out of Egypt even unto this day, in that *they have forsaken Me*, and served other gods, so do they also unto thee. [9] Now therefore hearken unto their voice; howbeit thou shalt earnestly forewarn them, and shalt declare unto them the manner of the king that shall reign over them." [Emphasis added.]

This event in Jewish history marks the worst decision the Hebrews ever made! This date in history was very important to God because it was exactly 3,017 years (1/4 of 12,068 years) from 1996, which was the first year of the beginning of the End of Days! The previous quote shows that God considered himself to be King over the Hebrews and that He was "hurt" when they chose a human king to lead them. The next verse reveals just how mad He really was.

[Hosea 13:10] . . . now, thy king, that he may save thee in all thy cities! And thy judges, of whom thou saidst: "Give me a king and princes!" [11] I give thee a king in Mine anger, and take him away in My wrath.

Another significant event tied to 1996 is that a red heifer was born in April of 1996—the tenth red heifer born in Israel since the time of Moses, and it was predicted to be the harbinger of the Messiah and the End of Days:[22,23]

The end of the Jewish Kingdom occurred 434 years later, with the fall of Jerusalem. Baruch escaped with the Ark and placed it back into the family cave, where he spent about 1 month[24] taking dictation from God while in front of the Ark. His prophecies in Ezekiel describe a future time when God will again be King over His people. Those prophecies will be covered later.

3. Exact Location of the New "Temple"

The general consensus, over the past 2,950 years, has been that the New "Temple," as described in Ezekiel, Chapters 40-48, will be built in Jerusalem on the same site as the previous two Temples. We will now examine the evidence for this prophecy and settle the issue of where the new "Temple" will be built, and if there will be only one Temple.

First of all, the New "Temple," described in Ezekiel by Baruch, will be 1005 feet[25] across at the base. In addition, there will be a courtyard around it, measuring 100 feet[26] on all sides. The building and courtyard would be 1,205 feet square. The Temple Mount in Jerusalem is not large enough to accommodate such a massive structure. With this major consideration in mind, it is no wonder that many Bible scholars, and I, have concluded that Ezekiel's new "Temple" will not be built in, or around, Jerusalem. There are other major factors that exclude Jerusalem, and I will describe them later in the section on the description of the new "Temple". I write the word "Temple" in quotes because the structure Baruch describes seems to have a very different purpose. One class of priests appears to have the one function of servicing and maintaining the "Temple," but the priests descended from the priest Zadok will have access to the inner chambers. The ceremonies will be different, with no mention of the Ark, the laver, the bread table, the altar of incense, and changes to the central altar. For these and other reasons, Ezekiel's "Temple," appears to have a very different purpose from the First Temple. The second Temple did not have the Ark so it did not have the same function as the first temple.

The conclusion I have come to is that there will be two temples built, a smaller-sized Temple built in Jerusalem, which may resemble Solomon's Temple design, and a much larger New "Temple" described by Baruch in Chapters 40 to 48.

4. The Word Zion

Much confusion has been caused by the word *"Zion"* or *"Mount Zion."* Over a period of 420 years the word evolved into three meanings. No scholar seemed to know where the word originated. It first appeared in 1004 B.C.E. in

Second Samuel 5:7, and II Chronicles 5:2 when King David captured a citadel south of the present day Temple Mount in Jerusalem. In the following two cases, the word *Zion* was obviously used for the name of the hill south of Jerusalem.

> [II Samuel 5:6] And the king and his men went to Jerusalem against the Jebusites, the inhabitants of the land, who spoke unto David, saying: "Except thou take away the blind and the lame, thou shalt not come in hither"; thinking: "David cannot come in hither." [7] Nevertheless David took the *stronghold of Zion*; the same is the *city of David*. [Emphasis added.]

The word "*Zion*" next shows up in First Kings 8:1 (written about 958 B.C.E.):

> Then Solomon assembled the elders of Israel, and all the heads of the tribes, the princes of the fathers' houses of the children of Israel, unto king Solomon in Jerusalem, to bring up the ark of the covenant of the Lord out of the *city of David, which is Zion*. [Emphasis added.]

The next time the word appeared, about 780 B.C.E., it was used by Isaiah[27] 27 separate times. He referred to *Zion* 17 times to describe the Hebrew people and/or country, six times he meant the hill south of Jerusalem. Finally, in four lines, he linked the word with the End of Days, as follows: The first time he used it, he wrote that the Law comes out of Zion and the word of the Lord comes out of Jerusalem.[28] The next time Isaiah used the word it referred to a structure called "Mount Zion" assigned to the time period called the End of Days.[29] The description of this "Temple" is unlike any structure ever built before:

> [Isaiah 4:5] And the Lord will create over the whole habitation of Mount Zion, and over her assemblies, a cloud and smoke by day, and the shining of a flaming fire by night; for over all the glory shall be a canopy. [6] And there shall be a pavilion for a shadow in the daytime from the heat, and for a refuge and for a covert from storm and from rain.

Isaiah also wrote a verse describing the "controversy of Zion,"[30] meaning Israel's controversy with the other nations and cultures about the definition of what, and who, God is. In Chapter 28:16, Zion was linked to something that was built:

> [Isaiah 28:16] Therefore thus saith the Lord God: Behold, I lay *in Zion* for a foundation a stone, a tried stone, a costly corner-stone of sure foundation; he that believeth shall not make haste. [Emphasis added.]

Thus, the word "*Zion*" changed definitions over a period of 420 years. Isaiah's son Micah wrote the word "Zion" eight times, and linked six of those to the End of Days. Three of the six were associated with a physical rather than social reality, such as Mount Zion or "look upon Zion." The other three occurrences were associated to the country Israel or the Hebrew people.

The prophets Zephaniah and Jeremiah used the word "Zion," but only Zephaniah used it in conjunction with the End of Days, using the word to describe a nation or the Hebrew people.

Finally, Baruch used "Zion" only in his writings of Obadiah and Joel. You would think that he would also have used it in Ezekiel, but he does not. In Obadiah and Joel, he uses the word "Zion" to describe the new "Temple" and prophesized that it would be linked with salvation and God's presence in Zion.

> [Obadiah 1:17] But *in Mount Zion* there shall be those that escape, and it shall be holy.

> [Joel 3:3] And I will show wonders in the heavens and in the earth, blood, and fire, and pillars of smoke. [4] The sun shall be turned into darkness, and the moon into blood, before the *great and terrible day of the Lord come*. [5] And it shall come to pass, that whosoever shall call on the name of the Lord shall be delivered; for *in Mount Zion* and in Jerusalem there shall be those that escape . . ." [Emphasis added.]

In conclusion, the name *"Zion"* had three separate meanings: One for the name of a hill just south of the Temple mount, another as the generic name for the country or people of Israel, and, finally, as the name of the new "Temple" structure where God's presence will be in the future.

THE ORIGIN OF THE WORD ZION

"Zion" was derived from the name "Zin," as in, "the wilderness of Zin." Moses used seven interchangeable names to identify Mount Sinai, or the area around it. One of the names was the *wilderness of Zin*. The next step was to take the word apart. *"Zin"* is spelled two ways: צן or ציּן. *Zayon* (ציּוּן) means *dryness* or *desert*. Another variant of the word is *ziyot* (ציּוּת) which means *"dry desert"* or *"parched earth."* The connection between the desert and Mount Sinai is obvious. It appears that the priest Zadok, aka Nathan, created the word as an allusion to honor Mount Sinai and gave the word a double meaning for other priests to use in the future. Zadok wrote that just as Mount Sinai was where man communicated with God, so will the hill just south of Jerusalem, because that was where David had brought the Ark, to his house located on the hill renamed "Zion." This is the real derivation of the word.

5. The Location

The Ezekiel story was not the only place in the Hebrew Scriptures which foretold that a "Temple" or structure was to be built. Genesis Chapter 28 tells the story of Jacob's escape from the wrath of his brother. The former left the family homestead and traveled to Haran in southern Turkey. The first night away from

Beer sheva (Beersheba), he camped in an area called Luz, and had a dream. The story relates only a small portion of what he saw in his dream:

> [Genesis 28:12] And he dreamed, and beheld a ladder set up on the earth; and the top of it reached to heaven; and behold the angels of God ascending and descending on it.

> [16] And Jacob awaked out of his sleep, and he said: "Surely the Lord is in this place; and I knew it not." [17] And he was afraid, and said: "How full of awe is this place! This is none other than the *house of God*, and this is the *gate of heaven*." [Emphasis added.]

The dream was a vision of something that was predicted to happen in his future, but no clue was given as to when. Notice the expressions "house of God" and "gate of haven." These are additional allusions to a structure to be built and a link to the Diehold and God the Operating System. The dream took place one long day's journey from Beer sheva, which means it could not possibly be located in Jerusalem. With a donkey, Jacob could have covered about 22 miles a day, which puts him in a valley just north of Tel Arad, in the middle of Israel. This is important to remember, because of what you will read in the prophecies listed later. He could not possibly research Jerusalem in one day from Beer sheva.

Found in Second Samuel is a story about king David. He wanted to build a Temple for the Ark of the Covenant so that God (the Ark) could dwell in, but God told king David, through the priest Nathan, that he will not build the "Temple," but someone else descended from him in the far future.[31] Nathan wrote:

> [II Samuel 7:11] . . . Moreover the Lord telleth thee that the Lord will make thee a house. [12] "When thy days are fulfilled, and thou shalt sleep with they fathers, I will set up thy seed after thee, that shall proceed out of thy body, and I will establish his kingdom. [13] *He shall build a house for My name*, and I will establish the throne of his kingdom for ever." [14] "I will be to him for a father, and he shall be to Me for a son; if he commit iniquity, I will chasten him with the rod of men, and with the stripes of the children of men;"

> [16] "And thy house and thy kingdom shall be made sure for ever before thee; thy throne shall be established *for ever*." [17] According to all these words, and according to all this vision, so did Nathan speak unto David. [18] Then David the king went in, and sat before the Lord; and he said: "Who am I, O Lord God, and what is my house, that Thou hast brought me thus far? [19] And this was yet a small thing in Thine eyes, O Lord God; but Thou hast spoken also of Thy servant's house for *a great while to come*; and this too after the manner of great men, O Lord God." [Emphasis added.]

Again nothing was revealed as to when the structure would be built. Of course, Solomon, with his immense ego, thought he was the one who would build it. But by verse 19, it clearly states that a far-distant descendant of David would be selected to build it. His kingdom was supposed to last forever. But Solomon's kingdom ended after his death, when it split into two kingdoms—the Kingdoms of Judea and Israel. That proved Solomon was not the descendant God desired to build the "Temple." Solomon's Temple was destroyed in 587 B.C.E., which again proves that his Temple was not the "Temple" prophesized.

The following prophecies clearly show that the "Temple," which God spoke of, would be built in the "middle" of Israel. The traditional rabbinical interpretation for these verses are that the "Temple" would be in the "middle" of the people's hearts. They do not interpret the prophecies literally but rather metaphorically. I have learned that we should first look for a literal interpretation. That is usually what the Biblical writers intended:

> [Ezekiel 37:28] "And the nations shall know that I am the Lord that sanctify Israel, when My sanctuary shall be *in the midst of them* for ever."

> [Ezekiel 39:7] And My holy name will I make known *in the midst of My people Israel*; neither will I suffer My holy name to be profaned any more; and the nations shall know that I am the Lord, the Holy One in Israel.

> [Joel 2:26] And ye shall eat in plenty and be satisfied, and shall praise the name of the Lord your God, that hath dealt wondrously with you; and My people shall never be ashamed. [27] And ye shall know that *I am in the midst of Israel*, and that I am the Lord your God, and there is none else; and My people shall never be ashamed. [Emphasis added.]

There are more prophecies which relate to the same message, but I have listed them in other categories because of other contents of the verses.

The next clue was found in Ezekiel 47: 5-12. There we learn that the new sanctuary will produce a great deal of water flowing from the four sides of its base. The clue: The water will flow to the east into the Dead Sea—Baruch writing as Ezekiel mentions the town of En Gedi—and also to the west into the Great Sea, which is the Mediterranean. There is a five-mile stretch that would fit these predictions. The next step for me was to go to Israel and see the area first hand.

The next clue in Ezekiel was found in Chapter 40:2: "In the visions of God brought He me into the land of Israel, and set me down upon a *very high mountain*, whereon was as it were the *frame of a city on the south*." The verse describes a high hill just north of a city which, in turn, is near the "Temple" Baruch described. The city is the City of God, described later by Baruch.

Figure 10-6: The valley north of Arad, Israel where I think Ezekiel's "Temple" will be built.

My family and I traveled to Israel in April 1996 to see relatives, travel all over the country, and see the sights. It was the 3,000-year anniversary of King David choosing Jerusalem as his capital, so it seemed like a good time to go. One of the places I wanted to see was the place where I thought Ezekiel's "Temple" will be built. Figure 10-6 shows a picture of the area. The valley in the picture has an elevation of 613 meters.

This number perked my interest because there are 613 Commandments in the Torah. So I converted 613 meters (1 meter = 39.37007874 inches) into inches, and got 24,133 inches. That was really close to 24,136 inches, and I knew that God was perfect, so I researched more to see what I would find.

First, I decided to research how the "meter" was derived. I had to research French post-revolutionary history. The French wanted to invent a measurement system that would be different from the British system, so they sent survey teams out and endeavored to measure the Earth, from the North Pole to the Equator, using a meridian running through the *MOST* important place on the Earth, at least from the French point of view. Yes, the meridian ran through Paris! So I decided to use the same philosophy, but asked myself: "What would be God's most important place on the Earth?" I deduced it would be the site where Jacob had his dream locating the new "Temple." I call this "Luz," because that name appeared in the Jacob story. So I headed off to the map library at the University of Washington and collected elevation measurements at every 30 minutes of longitude along the $6,200 \pm$ mile meridian. The average elevation turned out to be approximately 1,350 feet. I added that number to the distance from the North Pole, to the equator at sea level and then divided by 10,000,000, and came up with a new value of 39.3735 inches. That resulted in the 613-meter elevation, totaling 24,136 inches, not 24,133 inches. That converts into 1,000 sacred cubits! Could God have made the valley's elevation measurable in inches so that it would prove God's existence? It certainly looks that way.

THE LEGENDS OF THE JEWS

The legends of the Jews provided a very good clue about the future location of the new "Temple," but first I have to lead you into it. The descendants of Moses' father-in-law, Jethro, were the Kenites. The following verse details for us where they moved to in the Judges period, (before 1021 B.C.E.):

> [Judges 1:16] And the children of the Kenite, Moses' father-in-law, went up out of the city of palm-trees with the children of Judah into the wilderness of Judah, which is in the *south of Arad*; and they went and dwelt with the people. [Emphasis added.]

The following legend incorporates a number of clues not mentioned in the Hebrew Scriptures, *e.g.*, it mentions a Messiah who is supposed to build Ezekiel's "Temple." The idea of a Messiah is *not* a Jewish concept, but was adopted by the tenth century Jewish leadership, from Christian ideas. Therefore when the legend mentions the "Messiah" being in the Arad area, it is really saying that the new "Temple" will be built in the same area—and I would rather you focus on that than a person.

> Balaam furthermore announced the events that would come to pass at the time of David's sovereignty; and also what will happen at the *end of days*, in the time of the Messiah, when Rome and all other nations will be destroyed by Israel, excepting only the descendants of Jethro, who will participate in Israel's joys and sorrows. Yea, the Kenites are to be the ones to announce to Israel the arrival of the Messiah, and the sons of the Kenite Jonadab are to be the first at the time of the Messiah to bring offerings at the Temple and to announce to Jerusalem its deliverance?" This was Balaam's last prophecy.[32] [Emphasis added.]

In conclusion, I believe there will be two Temples built in the End of Days. There will be a "regular" Temple, built on top of the Temple Mount in Jerusalem, and that Temple will house the Ark of the Covenant and the other sacred objects recovered from the Cave after it has been uncovered.

The second New "Temple" will be built in the middle of Israel in an uninhabited valley near Arad.

6. God will be King Again

God was King of the Hebrews, up until 1021 B.C.E., when the people asked Samuel to appoint a man as king so he could fight the Philistines. There are numerous prophecies stating that in the End of Days, after Ezekiel's "Temple" is built, God will again be King of the Jewish people and will directly communicate with man. Hence, the Bible story comes full-circle, from the Garden of Eden story to Ezekiel. What the Hebrews lost in 1021 B.C.E., they will regain sometime in the End of Days.

The next quote from Solomon refers to when he is sitting in front of the Ark, asking God whether He will some day again dwell on the Earth, as God did during the days of Adam and Eve:

> [First Kings 8:27] But will God in very truth dwell on the earth? Behold, heaven and the heaven of heavens cannot contain Thee; how much less this house that I have built!

The following verses reinforce the prophecy that God will someday have a presence on the Earth again:

> [Isaiah 11:9] They shall not hurt nor destroy in all *My holy mountain*; for the earth shall be full of the knowledge of the Lord, as the waters cover the sea.

> [Micah 4:2] And many nations shall go and say: 'Come ye, and let us go up to the *mountain of the Lord*, and to *the house of the God of Jacob*; and He will teach us of His ways, and we will walk in His paths'; for *out of Zion shall go forth the law*, and the word of the Lord from Jerusalem. [3] And He shall judge between many peoples, and shall decide concerning mighty nations afar off; and they shall beat their swords into plowshares, and their spears into pruning hooks; nation shall not lift up sword against nation, neither shall they learn war any more.

> [Ezekiel 20:33] As I live, saith the Lord God, surely with a mighty hand, and with an outstretched arm, and with fury poured out, will *I be King over you*; [34] and I will bring you out from the peoples, and will gather you out of the countries wherein ye are scattered, with a mighty hand, and with an outstretched arm, and with fury poured out;

> [Ezekiel 20:40] For in My *holy mountain*, in the mountain of the height of Israel, saith the Lord God, there shall all the house of Israel, all of them, *serve Me* in the land; there will I accept them, and there will I require your have-offerings, and the first of your gifts, with all your holy things.

> [Ezekiel 34:23] And I will set up *one shepherd* over them, and he shall feed them, even My servant David; he shall feed them, and he shall be their shepherd. [24] And I the Lord will be their God, and My servant David prince among them; I the Lord have spoken. [Emphasis added.]

These prophecies can only come about if the new "Temple" is built in the End of Days. In Micah, 4:2 he implied that there will be a new body of knowledge which will be taught sometime in the last 50 years: "He will teach us of His ways." In Ezekiel, 34:23 when it refers to "one shepherd," that can be interpreted to be one High Priest who will communicate with God, and then to everyone else, what God's wishes are. In turn, the governmental leader, is supposed to obey whatever He says.

Regarding the word "*prince*" used here, the Hebrew word for "prince" נשיא can also mean "*an exalted one,*" "*prince,*" "*king*" or "*chief.*" In this case, I think in both cases it should say "*an exalted one.*" I also want to stay away from the word "prince" because it implies that the man is the "son" of God the King, and since my definition of God states He is the Operating System of the Universe, God is in essence the Father of ALL of our souls, not just one persons.

7. Description of Mount Zion

The book of Ezekiel, by far, has the best description of the new "Temple." Earlier prophets, such as Isaiah, Micah and Jeremiah, only describe the "Temple" as resembling a hill or mount. It is my impression that God did not give a detailed description of the new "Temple," until after Solomon's Temple was destroyed. Baruch seems to have received a complete description of the new "Temple," but he felt it necessary to split up all the information from God into four separate prophetic writings, Ezekiel, Joel, Obadiah and in parts of Psalms. He may have done this for his own safety and ultimately for the safety of the Ark and everything else stored in The Cave. If he had written all of this under one name the Babylonians could have figured out that he knew the location of the Ark.

The following prophets gave us a description of the new "Temple," and what its purpose will be.

ISAIAH

The second reference to the new "Temple," but the first description, appears in the writings of the High Priest Hilkiah, writing as Isaiah about 752 B.C.E:

> [2:2] And it shall come to pass in the *end of days*, that the *mountain of the Lord's house* shall be established as the top of the mountains, and shall be exalted above the hills; and many nations shall flow unto it. [2:3] And many people shall go and say, "Come ye, and let us go up to the *mountain of the Lord*, to the *house of the God of Jacob*; and he will teach us of his ways, and we will walk in his paths." For out of *Zion* shall go forth the law, and the word of the Lord from Jerusalem. [2:4] And he shall judge among the nations, and shall decide for many peoples; and they shall beat their swords into plow-shares, and their spears into pruning hooks; nation shall not lift up sword against nation, neither shall they learn war any more. [2:5] *O house of Jacob,* come ye, and let us walk in the light of the Lord.

[Isaiah 4:3] And it shall come to pass, that he that is *left in Zion*, and he that remaineth in Jerusalem, shall be called holy, even every one that is written unto life in Jerusalem; [4] when the Lord shall have washed away the filth of the daughters of Zion, and shall have purged the blood of Jerusalem from the midst thereof, by the spirit of judgment, and by the spirit of destruction. [5] And the Lord will create *over the whole habitation of Mount Zion*, and over her assemblies, *a cloud and smoke by day, and the shining of a flaming fire by night*; for over all the glory shall be a canopy. [6] And there shall be a pavilion for a shadow in the day-time from the heat, and for a refuge and for a covert from storm and from rain. [Emphasis added.]

This is the first time we are given a time-frame, namely "the End of Days." I believe Hilkiah used the phrase "house of Jacob" as an allusion to Jacob's ladder dream which describes God's structure as mentioned earlier. We also discover our first connection to the new "Lord's house" will resemble a mountain but with some very unusual powers. I describe this connection a bit later, but first we will continue with additional references to the new "Temple," and see what else is associated with it. The following are additional references, also found in Isaiah:

[Isaiah 11:9] They shall not hurt nor destroy *in all my holy mountain*; for the earth shall be *full of the knowledge of the Lord*, as the waters cover the sea.

[25:6] And *in this mountain shall the Lord of hosts* make unto all people a feast of fat things, a feast of wines . . .

[25:10] For *in this mountain* shall the hand of the Lord rest, . . .

[30:29] Ye shall have a song, as in the night when a holy solemnity is kept; and gladness of heart, as when one goeth with a pipe to come *into the mountain of the Lord*, to the Rock of Israel. [30] And the *Lord shall cause His glorious voice to be heard*, and will show the lighting down of His arm, . . . [Emphasis added.]

These four excerpts further link the new structure with the appearance of a mountain, but we know that it will not be a mountain because the prophets repeatedly state, "into the mountain," and in verse 30:29 it states that God's "voice" can be heard. A similar story is found in Ezekiel, Chapter 43:2, which states that His voice sounded like "many waters."

MICAH

Hilkiah's son, Oded—aka Eliakim—wrote as Micah the prophet, and he repeated what his father wrote in some prophecies, word for word. It was as if

God gave two versions of the same story to two different priests nearly at the same time.

[4:1] But in the *end of days* it shall come to pass, that *the mountain of the Lord's house* shall be established as the top of the mountains, and it shall be *exalted above the hills*; and peoples shall flow unto it.

BARUCH

Baruch's book of Ezekiel, Chapters 40 to 48, presents us with an incredibly detailed description of at least five new structures. The dimensions are given in sacred cubits (1 sc = 24.136") and are used as measurements to describe everything from large plats of land to the building dimensions. Four structures which measure 100 sc square are on each side of the main central "Temple" structure.[33] The central "Temple" building will be of massive size, 500 sc square (1,005 feet). It also gives a detailed description of a "City of God," located north of the "Temple" complex, measuring 4,500 sc square.[34]

Chapter 44 (of Ezekiel) details the new rules and procedures for this new "Temple." The tract of land to be occupied by the "Temple" complex will be located inside a larger tract of land which will belong to the "Prince" (an exalted one). The tract measures 10,000 reeds (1 reed = 6 sc) by 25,000 reeds (60,000 sc by 150,000 sc). If Ezekiel's New "Temple" were built in Jerusalem all of Jerusalem would have to be evacuated to make room, which does not make any sense. This is just one more reason why it is obvious that this "Temple" will not be built in—or even near—Jerusalem. Supposedly, there are going to be six tracts of land (all 10,000 reeds by 25,000 reeds), north of the common area (500 reeds by 25,000 reeds), and six south of the "Temple" complex, which is included within the "princes" tract of land. That translates into 137-plus miles south of the "princes'" land, and 150 miles north. The total length of all 14 tracks of land would be 305.5 miles by 57.1 miles. These dimensions are larger than the current size of Israel—which leads me to believe that the land divisions are intended for after the next polar reversal, when all evidence of houses, cities and people will have all been swept away by the Mediterranean Sea and the Atlantic ocean.

When writing Psalms, Baruch inserted two references to the *"City of God,"* and tied one of them to the final battle with Gog, which means the new "Temple" will be built before the larger second battle:

[Psalms 46:5] *There is a river*, the streams whereof make glad the *City of God*, the holiest dwelling-place of the Most High.

[Psalms 48:2] Great is the Lord, and highly to be praised, in the *city of our God*, *His holy mountain*, [3] Fair in situation, the *joy of the whole earth*; even *Mount Zion*, the uttermost parts of the north, the city of the great King.

[4] God in his palaces hath made Himself known for a stronghold. [5] For, lo, *the kings assembled themselves*, they came onward together. [6] They saw, straightway *they were amazed*; they were affrighted, they hasted away. [7] Trembling took hold of them there, pangs, as of a woman in travail. [8] With the east wind Thou breakest the ships of *Tarshish*. [9] As we have heard, so have we seen in the *city of the Lord of Hosts, in the city of our God—* God establish it for ever. [Emphasis added.]

The Ark is not mentioned in Ezekiel's "Temple" at all, which implies that another technical device will be employed to enable God's presences to appear. I found in Ezekiel 43:1 the following description of the workings of the "Temple." I get the impression from verse 5 that in the center of the structure there is a tremendously bright light generated when the Lord of Hosts appears. This is reminiscent of the description in Exodus which reveals that man could not look upon God because the light (or information) was too great:

[Ezekiel 43:1] Afterward he brought me to the gate, even the gate that looketh toward the east: [2] and, behold, the *glory of the God of Israel* came from the way of the east; His voice was like a *noise of many waters*; and the *earth shined* with His glory . . .. [5] And a spirit took me up, and brought me into the inner court; and, behold, the *glory of the Lord filled the house.* [Emphasis added.]

Baruch gives a detailed description of a very large altar in the center of the new "Temple"[35] (Figure 10-7 shows my interpretation of his dimensions). It will be 52 feet across at the base and 15 feet high with steps and a slope on its east side. This altar is entirely unlike any other Hebrew altar built in the past. The technology inside the new "Temple" building is supposed to cause the light of God to appear above this altar.

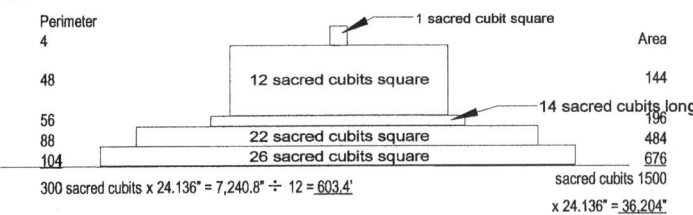

Figure 10-7: Drawing of the center altar in the new "Temple."

Baruch's book of Joel includes a similar story about God's voice coming from this holy structure and, for the first time, the structure is linked to the word "Zion." Baruch makes sure the reader does not become confused by thinking it is a structure in Jerusalem. This is one of the reasons why Bible scholars believe that two temples will be built. A traditional temple in Jerusalem will contain the

Ark of the Covenant, and a much larger "Temple"—Mount Zion—will be located in the middle of Israel. The result? God will have a real presence on the Earth in Mount Zion, and He will also be heard from a Temple in Jerusalem. From this point on, in my book I will refer to the new "Temple" as Mount Zion, exactly as the prophets have done.

> [Joel 4:16] And the Lord shall roar *from Zion*, and utter *His voice from Jerusalem*, and the heavens and the earth shall shake; but *the Lord will be a refuge* unto His people, and a stronghold to the children of Israel. [17] So shall ye know that I am the Lord your God, *dwelling in Zion My holy mountain*; then shall Jerusalem be holy, and there shall no strangers pass through her any more. [18] And it shall come to pass in that day, that the mountains shall drop down sweet wine, and the hills shall flow with milk, and all the brooks of Judah shall flow with waters; and *a fountain shall come forth of the house of the Lord*, and shall water the valley of Shittim. [Emphasis added.]

We know that Baruch wrote the book of Joel because for one thing he placed parallel stories in both Ezekiel and Joel. In Joel 4:18 he wrote about water streaming from Mount Zion. In Ezekiel 47, Baruch described a river of water issuing from the four sides of Mount Zion. He also repeated the same prediction that God will dwell in Mount Zion and his voice will be heard from Jerusalem, implying two different locations.

Chapter 44 of Ezekiel goes into a detailed description of the priests and the new ceremonies to be performed at Mount Zion. In Chapter 45:10-25 Baruch describes the offerings to be given by the people, and in Chapter 46 he describes the offerings. These offerings are given to the "prince" (an exalted one), who enters via the eastern entrance, to offer them in front of God. We are told that the "prince" will be only man who is allowed to enter and exit the eastern entrance, and that he will eat with God: "As for the prince, being a prince, he shall sit therein *to eat bread* before the Lord."[36] This is an amazing idea to conceptualize, that a human being could eat in the presence of God, but that is what is stated.

The strangest description of Mount Zion is how the people enter and leave the structure.

> [Ezekiel 46:8] And when the *prince* shall enter, he shall go in by the way of the porch of the gate, and he shall go forth by the way thereof. [9] But when the people of the land shall come before the Lord in the appointed seasons, he that entereth by the way of the north gate to worship shall go forth by the way of the south gate; and he that entereth by the way of the south gate shall go forth by the way of the north gate; he shall not return by the way of the gate whereby he came in, but shall go forth straight before him. [10] And *the prince*, when they go in, *shall go in, in the midst of them*; and when they go forth, *they shall go forth together*. [Emphasis added.]

These verses describe procedures so different from those in the past. The new practices have never been seen in any Jewish Temple at any time. I believe there is a reason why people will be instructed to go in from one door and later leave from the opposite door. I will cover the reason later.

What Will be Built

All of the previous prophetic quotes can be summed up as follows: In the future, during the block of years called the *End of Days*, a new "Temple" called Mount Zion will be built to enable man to communicate again with the Operating System of the Universe—God. The "Temple" will also be used to save mankind during the next polar reversal. The structural dimensions given for Mount Zion are huge, and the descriptions are unique in history. Baruch wrote that the base of the "holy place" is 500 sacred cubits square[37] (12,068 inches)! Repeatedly, the verses likened the appearance of Mount Zion to a mountain. But what sort of building resembles a mountain or hill? I believe Figure 10-8 provides the answer—a pyramid! The Great Pyramid of Giza is 755-feet across at the base, and 454.5 feet high. The new Mount Zion would be a 660-foot high pyramid, which would qualify it as a hill.[38] The reader will recall that two of the shapes which are formed by 12 of the Hebrew letters are a pyramid and a full octahedron, both having the exact same slope-angles as the Great Pyramid of Giza, in Egypt— 52.66 degrees.

The close proximity between Mount Sinai and Giza brings one to the reasonable conclusion that whoever built the Great Pyramid also placed the Tables of the Law and other advanced technology in the cave that Abraham would later purchase for 400 pieces of silver. The Great Pyramid of Giza and the Hebrew alphabet both reflect a common philosophy of the Universe derived from an ancient civilization which existed sometime before the last polar reversal. This

Figure 10:8: Southern side of the Great Pyramid of Giza, Egypt.

378 | God's Day of Judgment

civilization may date back two or more polar reversals, which the Garden of Eden story implies (*i.e.,* over 24,136 years ago). Whatever the case, we know that we are dealing with a totally different philosophy of the Universe compared with what is taught by academia today.

What I have covered still does not explain how this building "works," or what its real function will be. We only know that someone will build a very large square-based building, which will most likely be a pyramid, built in the middle of current-day Israel, and it will be referred to as Mount Zion.

8. Future Role of the priestly class.

The new priest class and their future functions will not be too much different from what they were from the time of Joshua through the first fall of Jerusalem. Baruch gives us descriptions of the tasks the priests will perform. However, those descriptions seem to be centered on the functioning of Mount Zion. It does not inform us who will be ministering to the rest of the world. The problem today is that most Jews do not know what tribe they are descended from—including the vast majority of rabbis. The descendants of Zadok, the High Priest, are today called "Cohanim." They were the priestly class who were permitted to sit in front of the Ark.

The new priest class will be divided into three basic groups. Descendants of Aaron through Zadok will be the High Priests who are allowed to administer directly to God, in front of the Ark. The majority of them will most likely be in charge of the Temple in Jerusalem. It appears the "prince" is the only person who will have direct contact with God, in the center of Mount Zion. The rest of the priests from the line of Aaron, will administer and perform the housekeeping of the Temple in Jerusalem and Mount Zion. The third group may be comprised of the rest of the Jewish men who are to be appointed priests to the rest of the world's population. If you think this is unlikely, I cite what Moses wrote in Exodus, 19:5, which I quoted before. He had just arrived back at the base of mount Sinai with the entire congregation. He went up and entered the Cave, where he spoke to God. God told him aboout His long term plans for the Hebrew people:

> [Exodus 19:5] "Now therefore, if ye will hearken unto My voice indeed, and keep My covenant, then ye shall be Mine own treasure from among all peoples; for all the earth is Mine; [6] and ye shall be unto Me a *kingdom of priests*, and *a holy nation*. These are the words which thou shalt speak unto the children of Israel."

> [Deuteronomy 7:6] For thou art a *holy people unto the Lord thy God*: the Lord thy God hath chosen thee to be His own treasure, out of all peoples that are upon the face of the earth. [7] The Lord did not set His love upon you, nor choose you, because ye were more in number than any people—*for ye*

were the fewest of all peoples—[8] but because the Lord loved you, and because He would keep the oath which He swore unto your fathers, hath the Lord brought you out with a mighty hand, and redeemed you out of the house of bondage, from the hand of Pharaoh king of Egypt. [Emphasis added.]

As you can see, it appears that God has had a plan for the Hebrews for over 3,300 years, but I believe it goes much farther back. The plan, which is called the Abrahamic Covenant, goes back to 1620 B.C.E. It began when God lured Abraham out into the Sinai desert, under the pretense of making him sacrifice his only begotten son, Isaac. The trip to the Land of Moriah had two purposes: One was to test Abraham and the other was to help him discover the Sacred Cave. Read the following prophecy by God Himself, after Isaac was spared:

[Genesis 21:17] ". . . that in blessing I will bless thee, and in multiplying I will multiply thy seed as the stars of the heaven, and as the sand which is upon the seashore; and thy seed shall possess the gate of his enemies; [18] and in thy seed shall *all the nations of the Earth be blessed*; because thou hast hearkened to My voice." [Emphasis added.]

The reason I believe God's plan started then is because of what is written in verse 18. Today in 2007, the Jews are not considered the blessing of the world. Uniformly, in the Arab/Muslim world, they are taught that the Jews are the "sons of Satan."[39] It is the same in some Christian circles. You must ask yourself: "What set of circumstances would so dramatically change the current viewpoint and state of affairs?" I believe the set of conditions that will cause the changes are the following:

1.   To be forewarned is to be forearmed. When the rest of the world finds out that the human race is headed for a cataclysmic polar reversal. That our science did not discover it and the popular religions did not have a solution to save us, and that it was the Jews and the Hebrew Scriptures that warned mankind, this will help change their attitude towards the Jewish religion and the Hebrew Scriptures.

2.   If the Ark is found, Mount Zion is built, and God is again able to communicate with humanity through a High Priest, all people will be humbled and grateful to the Jews for keeping the Torah in spite of 1900 years of Christian and Muslim persecution and murder.

I quoted a prophecy earlier that said: "for the earth shall be full of the knowledge of the Lord, as the waters cover the sea."[40] This means that once all this happens, all existing religions will pass away like a bad dream in the night. Once you define what God is you have also defined what God is not. What God is saying to Moses in the book of Genesis and Exodus is that all of the tribes of Israel will be priests for the rest of mankind.

The Secret of Redemption

I should define what is meant by the word *"redemption."* The dictionary definition is: *"deliverance; rescue, deliverance from sin, salvation."* So, what will we be redeemed from? The answer is: God's Day of Judgment, which includes the polar reversal, the Sun's nova, the flood, the ice age, the earthquakes, and all the rest. In short, how do we save ourselves from these terrible events? Make no mistake about it: God wants man to survive the polar reversal. That is why He directly intervened and lead Abraham to the Cave, and why He gave Moses the two tablets with the symbols on them, which became the first alphabet, which directly helped spread and advance knowledge worldwide. Many of the prophecies listed before say God gives refuge to people who will be saved, but the question is: How? What mechanism will be necessary to enable this?

Mount Zion and the secret of redemption are indelibly tied together. They are the same because of what is contained within Mount Zion. Eight clues follow:

First Clue:

In the second lunar year on the fiftieth day, Moses and the entire congregation pulled up tent stakes, left Mount Sinai, and traveled south three days.

> [Exodus 23:20] Behold, *I send an angel* before thee, to keep thee by the way, *and to bring thee into the place, which I have prepared.* [21] Take heed of him [the angel], and hearken unto his voice; be not rebellious against him; for he will not pardon your transgression; for *My name is in him.*

> [Exodus 32:34] "And now go, lead the people *unto the place of which I have spoken unto thee; behold, Mine angel shall go before thee*; nevertheless in the day when I visit, I will visit their sin upon them." [Emphasis added.]

Moses had to leave Egypt after he killed an Egyptian. I estimate that he was about 20-years old at the time.[41] He traveled to Midian, located partially on the west side of the Gulf of Aqaba, south of Elate (Eilat). He got married and lived as a shepherd, grazing sheep throughout the Sinai desert for 60 years. He had no problem traveling from Media to Mount Sinai in the north, and later into Egypt. Moses did not need anyone to show him the way from the coast of the Bay of Suez, through the pass and into the interior of Sinai, and then to Mount Sinai. He must have known the Sinai desert better than he knew the back of his hand. He knew where every water hole was and where to find food for thousands. So why did God provide him an "Angel" to guide him through the Sinai? This makes no sense, unless this "Angel" had to lead them to a very specific spot that no one else knew. To reinforce that idea it clearly states in both quotes, Moses was told of a specific place where God wanted him to bring the people. In fact, God says it is a place "I have prepared." That does not sound like random wandering throughout the Sinai desert. God wanted them in a very specific place.

Why does God say "My name is in him?" Is God saying that the name of the "Angel" was the same as part of the whole unspoken name of God?"

Second Clue:

The following verses elucidate the second clue:

> [Numbers 14:26] And the Lord spoke unto Moses and unto Aaron, saying: [27] "How long shall I bear with this evil congregation, that keep murmuring against Me? I have heard the murmurings of the children of Israel, which they keep murmuring against Me. [28] Say unto them: As I live, saith the Lord, surely as ye have spoken in Mine ears, so will I do to you: [29] your *carcasses* shall fall in this wilderness, and all that were numbered of you, according to your whole number, from twenty years old and upward, ye that have murmured against Me; ... [31] But your little ones, that ye said would be a prey, them will I bring in, and they shall know the land, which ye have rejected. [32] But as for you, your *carcasses* shall fall in this wilderness. [33] And your children shall be wanderers in the wilderness *forty years*, and shall bear your strayings, until your *carcasses* be *consumed* in the wilderness.

> [Numbers 32:13] And the Lord's anger was kindled against Israel, and He made them wander to and fro in the wilderness forty years, until all the generation, that had done evil in the sight of the Lord, was *consumed*. [Emphasis added.]

Here, the Torah leads us to believe that all of the generations, *i.e.*, everyone over 20 years of age, that left Egypt and rebelled against God were "consumed" by the desert over the next 40 years. Several details stand out within these two Torah quotes. Why did Moses use the Hebrew word that is translated "*consume*" בעיני found in verse 32:13, instead of the word "*to die,*" מות or מתה, or "*death,*" מת or מתיח, which seems to fit the intent of the writer, and which had been used previously? Using (or translating the word as) "*consume*" brings to mind, "to change something from one state into another." That is what happens when you consume food, or when a fire consumes fuel. It changes its state. That is different from dying and death. What becomes significant is when you take the word בעיני apart, the first part of the word בע means "*prayer,*" and the second half, נ means "*wailing*" or "*lament.*" Taken together, the word seems to indicate that the people were '*praying and wailing*" for having done evil in the eyes of God. That is far different from being "*consumed.*" After examining the different places where the words "*wasted*" and "*consumed*" appear, it is my conclusion that the Biblical scholars just do not know how to translate that word or phrase.

The next detail was the word "*carcasses,*" פגריכם. Moses was the person who created the original words in the Hebrew language, which we have today. He is the one who created the technique of using smaller words and combining them to make different, larger words, but with similar meanings. In other words, he had total control over which sequence of letters became the new Hebrew words, and their meanings. In Appendix C, I describe the code systems Moses created in the Torah. One of those systems was that of adjacent letter swapping. He allowed the reader to use a letter before or after the displayed letter, to derive a new word. The new word was the real meaning Moses wanted to convey. An example of this is with the word "*carcasses.*" The first part of the word is derived from פגי, which means "*body,*" "*corps,*" or "*dead body.*" That word fits the meaning of "*carcass.*" The second half of the word, יכם, is not a word, but if you swap the מ (m) for an ל (l), you get the word יכל, which means "*to prevail*" or "*overcome.*" The new meaning of the whole word is "*to overcome death.*" The clue is that Moses made the two component words mean something totally opposite to that of the longer word, *carcass*?

The previous verses imply that the whole congregation that left Egypt, over the age of 20, were supposed to be dead by the end of the 40 years. The book of Deuteronomy starts by saying that it was written at the end of the 40 years in the deserts of Sinai. Moses recaps to the people what they had gone through over 40 years. In the following verses, Moses was supposed to be speaking directly to the whole congregation. You will begin to see the pattern Moses sets up.

> [Deuteronomy 1:30] The Lord your God who goeth before you, He shall fight for you, according to all that He did for you in *Egypt before your eyes*; [31] and in the wilderness, where *thou hast seen how* that the Lord thy God bore thee, as a man doth bear his son, in all the way that ye went, until ye came unto this place.

> [4:33] Did ever a people hear the voice of God speaking out of the midst of the fire, as *thou hast heard, and live*? [34] Or hath God assayed to go and take Him a nation from the midst of another nation, by trials, by signs, and by wonders, and by war, and by a mighty hand, and by an outstretched arm, and by great terrors, according to all that the Lord your God did for *you in Egypt before thine eyes*?

> [5:4] The Lord spoke with *you face to face* in the mount out of the midst of the fire [5] I stood between the Lord and you at that time, to declare unto you the word of the Lord; for ye were afraid because of the fire, and went not up into the mount . . .

> [6:22] And the Lord showed signs and wonders, great and sore, upon Egypt, upon Pharaoh, and upon all his house, *before our eyes*.

[7:15] And the Lord will take away from thee all sickness; and He will put none of the evil diseases of Egypt, *which thou knows*, upon thee, but will lay them upon all them that hate thee.

[7:18] thou shalt not be afraid of them [the other nations]; thou shalt well remember what the Lord thy God did unto Pharaoh, and unto all Egypt: [19] the great trials which *thine eyes saw*, and the signs, and the wonders, and the mighty hand, and the outstretched arm, whereby the Lord thy God brought thee out; so shall the Lord thy God do unto all the peoples of whom thou art afraid.

[9:17] And I took hold of the two tables, and cast them out of my two hands, and broke them *before your eyes*.

[10:21] He is thy glory, and He is thy God, that hath done for thee these great and tremendous things [in Egypt], which *thine eyes have seen*.

[11:2] And know ye this day; for *I speak not with your children that have not known, and that have not seen the chastisement of the Lord your God*, His greatness, His mighty hand, and His outstretched arm, [3] and His signs, and His works, which He did in the midst of Egypt unto Pharaoh the king of Egypt, and unto all his land; . . . [7] but *your eyes have seen all the great work of the Lord* which He did.

[29:1] And Moses called unto all Israel, and said unto them: *Ye have seen all that the Lord did before your eyes in the land of Egypt unto Pharaoh*, and unto all his servants, and unto all his land; [2] the great trials which *thine eyes saw*, the signs and those great wonders; . . . [9] Ye are standing this day *all of you before the Lord your God*: your heads, your tribes, your elders, and your officers, even *all the men of Israel*. [Emphasis added.]

I believe you can see from these 10 quotes what the pattern is. I included so many because of the importance of what it reveals. These verses were written at the end of the 40 years. All of the Exodus generation over 20 years old are supposed to be dead but as you can plainly read in these ten citations, they are ALL alive and well at the end of the 40 years. Verse 29:9 plainly states "all of you" and the rest of the verses say "*with your own eyes or ears.*" Verse 11:2 states Moses is NOT talking to their children but to the adults that left Egypt. Moses makes sure the reader knows he is not talking about the children.

The clue: Why did he insert these verses in the Torah, and why are they contradictory to the whole idea that the entire Exodus generation was to be "consumed" by the desert? They were obviously still alive at the end of 40 years!

Statistically speaking, if you have a population of even a 100,000 people, you will have at least 30-percent of them still alive after 40 years.

Archaeologically speaking, there has never been any evidence found of a large population living in the Sinai Desert, especially near Kadesh Bernea, where the whole congregation was supposed to have stayed for 38 years—but it gets only two lines of mention in the Torah.[42] However, we found huge worked-flint fields around Mount Sinai. Our Egyptian guide told us that he had found large flint fields south of Elate in the area of what was western Midian. We found no ancient graves, no garbage dumps, no evidence of long-term settlements, of any type.

A code method used by Moses, in the Torah, is to grossly understate or overstate a place or person, so the reader will notice and focus on it. He did this with the name *Kadesh Bernea*. If they had in fact lived in *Kadesh Bernea* for 38 years, Moses would have written much more about the place then a very few lines.

We know they were in the Sinai because of what we found during the three expeditions. We found evidence they were there from the remains of altars and the numerous flint fields around Mount Sinai. So where did the Hebrews go?

Clue Three

The Hebrews left Mount Sinai on the 50th day of the second year. They traveled for three days into the wilderness of Paran,[43] which is towards the south.[44] The Hebrew word for Paran is פָּארָן. The first part of the word פֿא can be pronounced פֶּה or as פִֿיָה, which means "*mouth*" or "*opening*" (keep in mind the word "*consume*" mentioned earlier). רָן means "*rejoicing*" or "*song of joy.*" Together they could be interpreted to mean "*joyful opening.*" That is a very strange name for an empty wilderness.

Clue Four

The Hebrews left Mount Sinai and traveled south three days. They camped, and suddenly strange things started happening that cannot be logically explained in our normal reality. The strange events are as follows:

1. A fire started in the uppermost part of the camp.[45] There is nothing strange about that, but the fire went out when Moses prayed. That is impossible in our reality. He then called the place Taberah, תבערה.[46] The word is made up of two words: The first word is תב. If we swap the letter ב for the preceding letter א, it becomes תא, which means "*room*" or "*chamber.*" The second word, ערה, means "*to uncover.*" The word Taberah thus means they "*uncovered a room*" or "*chamber.*" Now you know why an "angel" had to lead them through the desert to a special place prepared by God.

2. The people began to complain that they were sick of eating just manna, a food that fell like dew in the early morning. They wanted meat, so God intended to teach them a lesson. God told Moses that they will be forced to eat meat for 30 days. The next day, a wind brought a huge number of quails inland. They supposedly dropped dead a day's journey (4.5 miles) on both sides of the encampment. The pile of dead birds was said to be 2 cubits (4.2 feet) high, which is an incredible number of birds in an area that has very few birds today. That would be impossible in our reality.

3. God was not through with them yet. As the meat was still in their mouths, the Scriptures state that the people started dying from a very fast plague.[47] No one dies that fast from food poisoning, therefore this is also impossible in our reality.

4. Aaron and his sister Miriam decided that God also talked to them. God became angry at them for thinking that and for challenging Moses.[48] The end result was that Miriam came down with an instant case of leprosy. No one catches leprosy that fast. This is impossible in our reality.

5. Korah and 250 ($250 \times 24.136 = 6,034$) other princes of the other tribes assembled themselves against Moses and Aaron. God again became angry and the Earth opened and swallowed them up, including all of their tents, belongings and families.[49] The earth does not open up and an earthquake happens instantly on cue. This is impossible in our reality.

6. God still was not done administrating retribution on the congregation. A fire came down from the sky and killed another 250 men who offered incense in front of the tent of meeting.[50] The ground instantly opening up and fire falling down from the sky just do not happen on command, in our reality.

7. The next day the congregation again came to the entrance of the tent of meeting and challenged Moses because 250 of their leaders had been killed the previous day. God told Moses and Aaron to get away from them, so He could kill them all. Moses sent out Aaron with an incense tray to make atonement for the people. A plague had already started by the time Aaron went among the people. Scripture reports that 14,700 people (the total number of days in 40 years plus 90 days) had died that fast.[51] This is impossible in our reality.

There are more strange events, but I believe I have made my point. What happened to the people, three days after they left Mount Sinai, cannot be explained in our normal reality. What I found in my research about Mount Sinai and Egypt, is that Moses did not lie to us. He did encode the story, to ensure that future scholars would gets the true meanings out of the surface story. He did not lie about what they built around Mount Sinai because I discovered all of it, and the altars were exactly where he said they would be, and they matched the exact dimensions and directional orientations. I know Moses did not lie. Our job is to decipher what he is really telling us.

Clue Five

The following two quotes come from Deuteronomy. Moses was once again speaking to the whole congregation at the end of 40 years supposidly wandering in the desert:

> [8:4] Thy raiment [clothing] waxed *not old upon thee*, neither did thy foot swell, these forty years.

> [29:4] And I have led you forty years in the wilderness; your clothes are *not waxen old* upon you, and thy *shoe is not waxen old* upon thy foot. [Emphasis added.]

I have traveled and explored the Sinai desert three times, for a cumulative total of five weeks. I can tell you honestly and with certainty that your shoes or boots would definitely wear out within one year of walking on the sharp limestone chips, flint stones and other rocks. There is no way any pair of shoes or sandals could last 40 years of walking on that hard stony ground. Your clothing would fall apart after just a few years, exposed to the elements and ultraviolet light from the harsh sun of the Sinai. There is no way your clothing would last 40 years, or even five years! Why did Moses write something that is impossible, unless he was forcing us to think about these lines so we would discover the great secret? We are supposed to come to the conclusion that something very important was revealed to us about the "living" conditions the Hebrews experienced for 40 years after they left Mount Sinai.

Clue Six

The next quote relates to the angel who led Moses and the congregation through the Sinai Desert for 40 years.

> [Deuteronomy 29:1] And Moses called unto all Israel, and said unto them: Ye have seen all that the Lord did before *your eyes* in the land of Egypt unto Pharaoh, and unto all his servants, and unto all his land; [2] the great trials which thine eyes saw, the signs and those great wonders; [3] but the *Lord hath not given you a heart to know, and eyes to see, and ears to hear, unto this day.* . . . [5] Ye have not eaten bread, neither have ye drunk wine or strong drink; that ye might know that *I am the Lord your God.* . . . [9] Ye are standing this day *all of you before the Lord your God*: your heads, your tribes, your elders, and your officers, even all the men of Israel, . . . [28] *The*

secret things belong unto the Lord our God; but the things that are revealed belong unto us and to our children for ever, that we may do all the words of this law. [Emphasis added.]

This very complex chapter in Deuteronomy is so strange and surprising. I have already quoted other parts of Chapter 29, but these verses are the most significant to me. First of all, the Chapter begins with Moses speaking to the entire congregation which left Egypt. In verse three, Moses told the congregation that they were not given an explanation for what they saw and heard over the past 40 years after they left Mount Sinai! This was an incredible admission on Moses' part. He basically told the congregation they had no idea where they were, what they saw, heard and did in those 40 years. By verse five, we realize the speaker is now God himself speaking to the congregation.

The second most important aspect of this clue is given in verse 28. He admits there are "secret things" which belong only to God. Moses flat-out tells us that there were secret events, which occurred here that the Hebrews had no knowledge of and could not possibly understand, because they had no advanced scientific knowledge as a prerequisite for comprehending them.

Clue Seven

The next clue is very strange and is usually passed over by most rabbis as mere symbolism. Some ancient priests used it as their basis for the idea that the dead will be brought back to life in the End of Days, after Mount Zion is built. The quote is in Deuteronomy 30:40. It is not an accident that it immediately follows the strange material covered in the previous chapter. The chapter starts off by saying that all the blessings and curses will have passed, implying the time period is the End of Days when God will again bring all the Jews back to the land and they will believe in God the way He wants them to.

[30:3] That then the Lord thy God will turn thy captivity, and have compassion upon thee, and will return and gather thee from all the peoples, whither the Lord thy God hath scattered thee. [4] If any of thine that are dispersed *be in the uttermost parts of heaven*, from thence will the *Lord thy God gather thee, and from thence will He fetch thee*. [5] And the Lord thy God will *bring thee into the land* which thy fathers possessed, and thou shalt possess it; and He will do thee good, and multiply thee above thy fathers.

On the surface, it sounds as though the dead will come back, and live with the living, which is impossible. This world is made for the living, not for the dead, and only the living can have children and multiply. So it has to mean something else very important. The Hebrew word used for heaven is שָׁמַיִם. The word is also used in Genesis Chapter 1:8, but appears as אַרְקִיעַ שָׁמַיִם (Rakiyeh Shemayim), which means "vault of heaven." A direct translation of the meaning would parallel my concept of the Diehold and the carrier-wave, the shape of a pyramid. I believe Moses, in Chapter 30, does not mean the "dead," but people somewhere else who will be retrieved from the Diehold. I believe the people the verse is referring to are the missing generation Moses was talking to. I will explain this later.

Clue Eight

My last clue is found in Chapter 37 from Baruch's book of Ezekiel. It is a very strange story. It describes the strangest vision in his entire book. The prophet wrote that God carried him to a desert valley full of dry bones. The very dry valley is a very accurate description of the Sinai desert. God asked him: "Son of man, can these bones live?" The answer seems obvious to Baruch, so he answered "No." Then God states:

> [37:4] "Prophesy over these bones, and say unto them: O ye dry bones, hear the word of the Lord: [5] Thus saith the Lord God unto these bones: Behold, I will cause breath to enter into you, and ye shall live. [6] And I will lay sinews upon you, and will bring up flesh upon you, and cover you with skin, and put breath in you, and ye shall live; and ye shall know that I am the Lord." [Emphasis added.]

So Baruch did what God had instructed him in this vision, and the bones came to life with breath in them:

> [37:10] So I prophesied as He commanded me, and the breath came into them, and they lived, and stood up upon their feet, an exceeding great host. [11] Then He said unto me: "Son of man, *these bones are the whole house of Israel*; behold, they say: Our bones are dried up, and our hope is lost; *we are clean cut off*. [12] Therefore prophesy, and say unto them: Thus saith the Lord God: Behold, I will open your graves, and cause you to come up out of your graves, O My people; and I will bring you into the land of Israel. [13] And ye shall know that I am the Lord, when I have opened your graves, and caused you to come up out of your graves, O My people. [14] And I will put My spirit in you, and *ye shall live*, and I will place you *in your own land*; and ye shall know that I the Lord have spoken, and performed it, saith the Lord." [Emphasis added.]

The next clue: "We are clean cut off." That is what happened to the lost Exodus generation, which could not enter the Promised Land and was supposed to have been "consumed" by the desert. The next clue is an obvious one. "These bones are the whole house of Israel." The Jews of today do not know what tribes they came from. The only ones who definitely knew were the lost Exodus tribes. This brings us to the twelve large tracts of land listed in Ezekiel, each dedicated to a different tribe of Israel. How would any of the current-day Jews know what tribal lands to settle in? The answer is, they wouldn't. These verses in Chapter 37 seem to correspond to people in "heaven" (another state of existence, but not dead) coming back, mentioned previously in Deuteronomy, 30:4.

The Answer to the Riddles and the Secret of Redemption

I have spent enough time presenting most of the clues I found. I am sure there are a lot more, but it would take a lifetime to find them all, and it probably would not matter. Now it is time for the answer to the clues, and what kind of structure the new Mount Zion will be, and how I believe it will work. Remember, sometimes the answer is derived when all other possibilities are eliminated. Also, you must think in terms of an information theory of existence. You may love the "matter world" you live in, but remember that that is not how you are going to solve this puzzle and survive the cataclysm.

There are five components needed to save most of mankind, but I will list only four of them: First, a computer program that will communicate with the Diehold; Second, a light based crystal computer that has the program embedded into it; third, a pyramid that has the same slope angles as the carrier wave of the Diehold; and the fourth component, an interface device.

Mount Zion

Ask yourself: "What would be the greatest gift God could give mankind?" The clues are found in the Garden of Eden story and in Chapters 40 to 48 in Ezekiel. The greatest and most important gift would be the ability to communicate with God—the Operating System of our Universe. This would mean building a terminal to the Diehold.

I wrote in Chapter 4, on the formation of the Hebrew Alphabet, that I would explain the secret of redemption, and now is the time. The Torah is the programming used to program a light-based computer, placed inside a large pyramid, which enables it to become a "terminal" to the Diehold. The slope-angle of the pyramid must be 52.66-degrees because it has to produce the carrier wave signal of the Diehold. When the computer inside is turned on, it will enable man to "see" God's presence inside, and will be able to communicate with Him directly. What is attributed to Mount Zion leads me to believe that it will become a new terminal to the Diehold. Such a terminal does so much more then just communicate with the Lord of Hosts, and you will soon find that out.

The Polar Reversal and Survival

There are two ways to survive the polar reversal and the resulting cataclysm. The hard way is building a very strong waterproof shelter, either above ground or in a reinforced cave in a mountain. Next, you will have to store enough supplies and food for 10-to-20 years. The shelter should be built within 10-degrees north or south of the equator because of the snow and ice fields. There are more things you have to prepare for, and I suggest you have a second look at Chapter 8 and let your mind wonder what you have to prepare for and where you should build your shelter. As you can see, this will not be an easy task—but not impossible.

The second way to save mankind is hinted at in the Hebrew Scriptures, as in these two verses:

> [Isaiah 25:9] And it shall be said in that day: "Lo, this is our God, for whom we waited, that He might save us; this is the Lord, for whom we waited, we will be glad and rejoice in His *salvation*."

> [Baruch aka Joel 3:5] And it shall come to pass, that *whosoever shall call on the name of the Lord shall be delivered*; for *in Mount Zion* and in Jerusalem there *shall be those that escape*, as the Lord hath said, and among the remnant those whom the Lord shall call. [Emphasis added.]

What man is being saved from is the polar reversal. That is why Moses referred to "those that escape." Those will escape from the effects of the polar reversal and the ice age. That is also why the number 12,068 in encoded so many times into the Torah—because that is the number of years between polar reversals.

The phrase "whosoever shall call on the name of the Lord shall be delivered" has had many interpretations by other religions, so I feel it is necessary to define what Baruch meant by this. He is referring to the Jewish definition of God as expressed in Numbers and Deuteronomy.

> [Numbers 23:19] God is not a man that He should lie; neither the son of man that He should repent: when He hath said, will He not do it? Or when He hath spoken, will He not make it good?

> [Deuteronomy 4:15] Take ye therefore good heed unto yourselves—for ye saw no manner of form on the day that the Lord spoke unto you in Horeb out of the midst of the fire—[16] lest ye deal corruptly, and *make you a graven image, even the form of any figure, the likeness of male or female*, [17] the likeness of any beast that is on the earth, the likeness of any winged fowl that flieth in the heaven, [18] the likeness of any thing that creepeth on the ground, the likeness of any fish that is in the water under the earth; [19] and lest thou lift up thine eyes unto heaven, and when thou seest the sun and the moon and the stars, even all the host of heaven, thou be drawn away and worship them, and serve them, which the Lord thy God hath allotted unto all the peoples under the whole heaven. [Emphasis added.]

In other words, God is not anything you see in this dimension other then the radiated light from His presents inside the future Mount Zion.

How man is going to be saved

I will first present you with a common computer analogy using the Windows operating system. Let us say you are in a text-editing program and you want to save a block of text. You would first highlight the desired text with your mouse device, and then hit the Control C keys if you want to copy the text or Control X keys if you wanted to remove it from the document, but wished to place it into another program. What happens when you hit Control X or C is the information for the text goes into a buffer, which is a block of memory in the computer. You can now go into another program and hit Control V and the text will reappear in the other document.

A terminal to the Diehold, Mount Zion in this case, should theoretically be able to allow you to move yourself out of this reality framework and into another created reality. I would not expect it to be a complete recreation of our reality, but it would be very close to it. It would be like a buffer in a computer. Time would probably "play out" much more slowly than in normal reality. The idea of a buffer came to me because of some statements in *Legends of the Jews*. It states that after Aaron "died" the Congregation wanted Moses to take them to where Aaron was buried and, when they walked to the Cave where he was buried, that was the first time they saw a clear sky with no clouds above them. The legend says they never saw a clear sky and stars above them for the 40 years they were in the desert, which of course is impossible in the Sinai in normal time and space! There may be limitations to this other state of existence because it is intended like a buffer only as a temporary "waiting area." I would also expect that "time" would play out at a much slower rate. This temporary area would not regenerate the sky because just as we cannot see the stars in daytime, they are still there and that would be too much information for the system to regenerate. In other words, the 40 years of wandering after they left mount Sinai, were really spent in a buffer which created a near perfect reality except for the heavens,

References to Another State of Existence

A previous quote from Deuteronomy referred to people in the "uttermost parts of heaven" coming back. Ezekiel 37 concerns a valley full of dry bones that come to life. Both of these verses are hinting at another "state of existence" other than death.

This other state of existence had been translated as the "netherworld," תחתית שאול, used by Moses,[52] and sometimes as "*Sheol*," "netherworld," or "hell," שאול, in other parts of the Hebrew Scriptures. Moses was the first and only writer to use the Hebrew phrase שאול תחתית found in Deuteronomy Chapter 32:22. The first word, שאול, has been incorrectly translated in the context Moses intended. In Hebrew, each letter has its own meaning. ש means "Tooth," as in using your teeth to *consume* food. That is why the word "consume" was used in connection with the Hebrew generation that was "consumed" in the Sinai Desert. The definition of the remaining letters אול is "belly," "mighty," or "powerful person," as in God. The resulting phrase could literally mean: "*consumed by God.*" This would be a likely description of people going into a special place for a period of time. The Hebrew word, תחתית, means "the lower" or "lowest part." The phrase is shortly followed by the Hebrew word for "earth" ארץ. The final meaning for the verse should be: "Consumed to the lower earth by God," which means they went into some kind of cave or opening and disappeared!

The word *Sheol* (שאול) was later used in Second Samuel, Isaiah, Amos, and Ezekiel, but the meaning of the word changed. It is obvious, from reading Samuel that the word was part of their vocabulary, with a mythological meaning, but the people no longer really knew what it meant. There are some books written after Ezekiel, which also use the word, but since they are not prophetic books, or their authorship is questionable, I will not cover them. Unfortunately, translations and interpretations after Baruch/Ezekiel have taken on a different meaning than what I believe Moses had in mind. The only writer, after Moses, to use the word, תחתית, was Baruch in Ezekiel, where he used it five times in Chapters 31 and 32. He does not use the same phrase as Moses but uses the words בתוך, which means "*middle*" or "*between*," and בארץ, which means "*in the Earth,*" in different combinations. In 31:14 Baruch uses these words, in conjunction with others, in a phrase that literally means "*towards Earth, the lower part between in the valley of Adam towards into the opening.*" What is he saying? That people are in a cave someplace in the "*valley of Adam,*" which has to be the Garden of Eden. In 31:16 he repeats a similar meaning but includes the word Eden, just to make sure the reader gets the clues. In these two chapters Baruch uses some of these verses to hint at where he hid the Ark by stating that he put something in a cave in the vicinity if Eden.

The Promise of Redemption

The eight clues presented can be understood if one realizes the missing Exodus generation was placed within a portal that put them into another state of existence, inside a terminal to the Diehold. This terminal must still exist, south of Mount Sinai. It is most likely the pyramid (tree) described in the center of the Garden of Eden story. I could only speculate about why God may have placed these people there. Maybe it was to retrieve them in the future, to prove to us that this is the way to save ourselves from the impending cataclysm. We are not going to know for sure until my theory is tested and a portal to another time and space is found.

A buried pyramid dating back over 24,000 years would explain why an "angel" had to lead the Hebrews to a specific spot, which was probably a portal into another state of existence. All the strange occurrences which happened after they arrived three days south of Sinai can be explained if the congregation had entered a portal, which put them into another "created reality." In that case, anything can happen because everything they saw and did was in a different "created reality." That is why their clothing did not wear out because to them they were there for only 31 days inside the buffer.

In Ezekiel 37 and Deuteronomy 30:4 it is implied that someone will discover the other terminal to the Diehold and bring the Hebrews out of it. That is the only conclusion I can reach after putting all the clues together. Otherwise we have to ask ourselves: "Why is Moses or God giving us all these clues?"

The clues for the other terminal are also presented in the Jewish legends. They repeatedly state that Mount Sinai was in the vicinity of Paradise. The word "Paradise" is synonymous with the Garden of Eden and as I have stated before, the "tree of knowledge of good and evil" is a terminal to the Diehold built by a highly advanced previous civilization.

This raises the key question: How can we save ourselves from the next polar reversal? The easy way to save us would be to enter the new terminal to the Diehold, Mount Zion, and go into this other temporary reality for a short period of time until the worst effects of the polar reversal are over. At the appropriate time, we would walk out of Mount Zion, back into this time-space reality.

Time

This brings us to the question of time in the other state of existence. I am sure time "plays out" much more slowly in this buffer (other existence) than ours. The technical reason is that the Diehold does not have to generate a "complete" reality for the information in the "holding" buffer. The Diehold only has to propagate existence at a much slower rate, therefore time (which is related to the propagation rate of information) should be much slower. This method in essence would save computing resources in the Diehold.

Such a conclusion was hinted at in the book Numbers 14:34 when God said to the congregation that because they did not heed Joshua's and Caleb's report of the Promised Land God was going to have them "wander" the wilderness for 40 years. Joshua and company had spied out the land for 40 days. God said the congregation's punishment would be "After the number of days which you spied out the land, forty days, for every day a year, shall you bear your iniquities even *forty years* . . ." [Emphasis added.] But we are told earlier, in 11:18 - 20, the congregation was there for a total of 31 days.[53] I believe the 31 days refer to time inside the portal or buffer and the 40 years refer to real time outside the portal/

buffer. If you convert 40 years into days you get 14,610, and then divide by 31 days, you get a time ratio of 471 to one. That means that every minute inside this buffer is the outside equivalent, in normal time, of 7 hours and 51 minutes! Their 31 days inside the portal was the equivalent to 40 years outside. This is perhaps why Moses wrote that their clothing and shoes did not wear out. They would not have in only 31 days.

Remember that the congregation in Sinai never read the Torah so they had no idea what Moses wrote in there. No one then would have noticed that Moses has two stories going on here. The congregation thought they were there for only 31 days before their children left the portal/buffer to travel to the Promised Land, but they were actually there for 40 years in "real" time.

If I am correct and the lost generation is in such a device, and if they were led out of it in 2006, they would think only seven years had passed since their children had left Mount Sinai!

I know this is beyond even most science-fiction literature themes, but that is where Moses' clues lead us. We know the Hebrew alphabet was created by a highly advanced previous civilization, which used an information theory of existence, so we cannot discount the possibility. This portal could save every person and animal on Earth. If it was decided to emerge from Mount Zion 200 years after the cataclysm, people would have to stay inside for only 155 days.

How does Prophecies Happen?

This entire chapter has focused on prophecies, but have you ever asked yourself; "How does God know that a specific event will occur the way He wants in the future?" After all, so many variables could happen over 100 or 3,000 years. For instance, a man may not meet a woman, and they, in turn, would not create a son who does something marvelous or terrible. We have always assumed that we, as humans, have free will. When you add that into the equation of predicting the future, you realize how difficult it is to predict with certainty an event such as the return of a people to their ancient homeland. In the conclusion of this chapter, I am going to tackle what God does in order to make sure that what He says will happen, will actually happen. I am first going to present two phenomena that will help answer this question.

Reverse Speech

Reverse speech is the discernment of recognizable phrases when a person's recorded speech is played backwards. David John Oates of Australia discovered this phenomenon in the 1980s. The subject matter of the reverse speech usually matches that of the regular forward speech. Oates's definitive work is found in his book, *Reverse Speech: Voices From The Unconscious*,[54] which is an excellent reference for the study of reverse speech. He has placed a large number of sound files as examples, on his website: http://www.reversespeech.com. One famous example is of Neal Armstrong's famous first walk on the Moon. Neil Armstrong said: "That's one small step for a man, one giant leap for mankind." Reverse speech revealed him saying: "Man will space walk."

What is amazing and significant about reverse speech is that the person always tells the truth—they cannot lie. I believe John Oates has proven that humans have recognizable speech running in an opposite direction, counter to the way we perceive time. Reverse speech phrases may be the reason we "feel suspicious" about someone after we meet and talk to them. We may be subconsciously picking up the reverse speech message, and if it does not match the regular speech, we "get a funny feeling about that person."

Prophetic Dreams

It is not uncommon to have a prophetic dream at least once in your life. There are so many variables which could change and prevent it from unfolding, but sometimes it just happens. There is no logical necessity for a dream eventually to come true. The question is: "What mechanism is necessary for something like this to happen?"

Tenses and the Torah

There is an interesting observation about the Torah—not known to many—concerning tenses. Moses was writing the biblical surface story of the Torah between 1306 to 1267 B.C.E. Some of the tenses utilized in the accounts of events in Moses' past were written in future tense and future events are written using past tense, and nobody knows way.[55] Even the unspoken holy name of God, Yahova (יהוה) is sometimes interpreted as being past present *and future.* I interpret His name as the combination of yud, י, being a shortened name for God, and Hova, הוה, meaning "Builder." So the two words together mean "God the Builder." And what does he build? The Universe!

There is a reason why the tenses are reversed, and I believe Moses was telling us something very important about how God works and how the Universe really works.

The Only Answer

The first two phenomena are related, and they are also related to the tenses in the Torah and how God knows that His prophecies will come true. We perceive time in only one direction. Events that have already happened we call the past, and events that have yet to happen, we call the future. It is impossible to consciously compose a sentence in reverse of time at the same time you are talking forward in time. The same thing is true for those pesky prophetic dreams which we have all had. Why do our past dreams mirror our future?

This is the only logical answer. The Torah is telling us something very important. God has programmed our existence from the end of time to the beginning of time. That is how He knows it will come out the way He wants. If our free will alters something in the past which changes God's ultimate goal, He just reprograms events from that point in time forward, to make sure it comes out the way He wants. The next chapter may be one of the more disturbing chapters in the book because it deals with the issue of free will and proves that God is dynamically programming our existence whether we like it or not!

My objective

I am sure readers will wonder what my long-term goals are for the information I have discovered. I have three main goals. The first: to explain to people that there is a God and who He is in relation to the Universe He controls. The second being, to try to warn as many people as possible so they will "encourage" government to try to save as many people as possible before October 2046. The third goal is to try to preserve the Republican form of government the founding fathers of the United States created for us. I believe our form of government is the reason why this country has been so successful over the past 220 plus years. It has guaranteed the greatest degree of freedom (both religious and political) of any other country, which has stimulated progress, invention, and scientific discovery. Over the years, I have analyzed many forms of government and I have concluded that a parliamentary form of government is an illusion of democracy because the people are given only the choice of voting for a political party and not for an individual.

In order for our form of government to survive the Polar Reversal, the people must know why this form of government is worth saving and it must be instilled in their hearts. I would not want mankind to sink back to tribalism, then feudalism, and finally dictatorships. Mankind spent over 11,800 years under those forms of rule before a small collection of men got together in Philadelphia and came up with the Constitution of the United States. One of the things worth saving from this civilization is our Constitution. For all our government's faults, it is still the best one out there—short of having God as your King.

Chapter 11:
Timelines and Patterns

The Issue of Free Will

The one right all humans assume we have is free will. We do not want to think that our lives are predestined. The concept of free will may have originated with the religion of the ancient Geeks. They wanted to separate man from God, so God would not interfere with man.[1] The Greeks felt that people had a right to make their own mistakes without God's "interference." They were against the Jewish religion because the Jews claimed descendency from Noah and the Jewish religion was God-centered, not man-centered. The Greeks claimed their descendency from Cane, Adams first son, and the Hebrews claimed theirs from Seth, Adams third son.

This chapter is going to present something that has never been discovered before in history. How God controls events and to the extent he does it. My one major purpose in this book is to prove the existence of God and reveal who He is in relationship to the Universe. This Chapter will definitively demonstrate that we are in a computer—the Diehold—and there is a God who controls it.

Cycles

I presented a videotape analogy in Chapter 3 to help describe our reality. I then posed the question: "How do you know you are a created *being* and that your existence comes from somewhere else?" The answer: You look for cycles in existence which logically should not be present if everything is supposed to be random. If you discover patterns in time, this is proof that some entity is programming events to occur at specific points in time. The entity is, of course, God—so this chapter will complete my proof that our Universe is the product of information, which exists in the Diehold, and that God is the Operating System of our Diehold.

Two Parts to My Discovery

There are two parts to the discovery. First, there are 12,068-day (33.04 years) cycles in history discernible by counting backwards from the cataclysm date of October 16, 2046. An important event occurs every 12,068 days. More often than not, the event has been or will be directly or indirectly related to the Jewish people (Table 11-1).

There are, of course, problems with determining the exact dates for many ancient events, because we just do not know the exact day an event occurred. We may know the month or the season while sometimes the year may be up for debate. The dates I provide, which are older than 970 b.c.e., derive from my calculations, from my research into Jewish history, and from the Oxford University chronology of Jewish history. Another important point to remember is that sometimes the date was when an agreement was made, or a decision on the event was made.

The second part of the discovery was to find the "holy numbers" 3,017, 6,034, 12,068, 24,136, and 36,204 showing up as the total number of days between two important dates, and which were also incorporated in measurements, such as for distances. For an example, let us say 50 years separate two important events, so $50 \times 24.136 = 1,206.8$, which is the main coded message of the Torah.

How I Discovered It

I discovered the number 12,068 embedded in the Torah in 1994. In 1989 I had already figured out the date for the next Gleissberg cycle (between September to December 2046). The Genesis story of Noah gave me the month and day when the next polar reversal will occur. Naturally, I wanted to see how old I would be when this event will happen. It turned out I will be 99.12 years old. For some reason, I decided to convert 99.12 years into days and found that it equals 36,204 days ($3 \times 12,068$)! At that point, it got personal and, I have to admit, it did startle me for a while. I had to think about what I had discovered for a couple of weeks. It did not stop my research but changed it. I started to look at the cycles in a different way. I thought that maybe other events occurred on one of these 12,068-day cycles, counting back from my then tentative polar reversal date. Table 11-1 lists the dates and events that I have found so far. My only problem was finding research books with important dates in history specifying the exact month and day. Another problem was that we do not know *where* to look for an important event, which God made happen on that date. Normally, you would search for the date of a battle, or signing of an important document, or a meeting, etc., or the birth date of someone who has done something important. I also discovered that some very significant events occurred on a half-cycle and quarter-cycle of 12,068.

Cycles Through Time

The following table illustrates important events which occurred exactly on the 12,068[th] day or on a factor of it starting from October 16, 2046 and counting back in time. Some might say it is a coincidence, and that would be true if it was only two or three dates, but I have found over 40 important dates. I am sure there are many more, but research books with dates older then 1600 c.e. are hard to come by. I have chosen to list all the whole-cycle dates. If you find other significant events which occurred on whole-cycle or half- or quarter-cycle dates please write to the publisher with your findings.

Most of these dates are very important in history. One of the most important dates in the twentieth century was August 19, 1914. If President Wilson had declared the United States was not going to be neutral and would join the war in Europe, I believe the Germans would have sought a way to end the war early, and it would not have dragged on to November 1918. The peace treaty eventually resulted in the economic collapse of the German Republic. The Depression brought about the election of Adolf Hitler and therefore the Holocaust. The Europeans and the American Christian leadership were ashamed of what had happened to the Jews, so they voted for the creation of the State of Israel in the United Nations on November 29, 1947.

I have a difficult time believing that these events are random and mealy a coincidence. They each fall on one of the cycle dates and it is not by accident. It is by design.

Related Dates

Another discovery I made was finding the number of years between two important events in history, revealed one of the sacred numbers. The following events are arranged in chronological order, from the oldest date to the present day.

1606 B.C.E.: FROM ISAAC TO THE EXODUS

I estimate that Isaac was born in 1606 B.C.E. and he was a little over 13 years old when his father was instructed by God to sacrifice him. The Exodus occurred in September 1306 B.C.E. The difference is 300 years (300 x 24.136 = 7,240.8 ÷ 12 = 603.4).

MARCH 1306 B.C.E.: FROM MOSES TO MY EXPEDITION

In earlier Chapters I referred to the Exodus as occurring in 1306 B.C.E. I had calculated that the exact date of the Exodus was September 29, 1306 B.C.E. I had also figured out the date that Moses and Aaron cursed the waters of the Nile,[2] which occurred on the summer solstice, which occurs in the third week of June. Three months before this, Moses was at mount Sinai grazing his flock of sheep. He went there because the area was a savanna, at the time, with grasslands and water in the wadis.

On my first expedition to Mount Sinai we arrived at the base of Mount Sinai on November 29, 1997. If my calculations are correct, Moses had to be at Mount Sinai on March 25, 1306 B.C.E., exactly 1,206,800 days before we arrived there. The reason I am sure about this is because Moses was there long enough for the

Table 11-1:

Cycle	Days from Cataclysm	Day/Month	Year	The 12,068-day cycles and the events occurring on those dates.
0	0	16-Oct	2046.79	
1	12068	1-Oct	2013.75	
2	24136	17-Sep	1980.71	The Iran-Iraq War started. Libya attacked a U.S. EC-135 reconnaissance plane over the Mediterranean resulting in the U.S. attacking Libya.
3	36204	2-Sep	1947.67	I was born in New York City, USA. First person to develop an information theory of existence. First person in 2,600 years to discover the real Mount Sinai.
4	48272	19-Aug	1914.63	President Woodrow Wilson's neutrality speech in front of the Senate after WWI started in Europe.
4.25	51289	15-May	1906.37	May 1906: Signing of a Treaty between the British and Ottoman Empires determined the current-day Sinai boundary, resulting in the transfer of El Arish Rapha to Egypt.
5	60340	3-Aug	1881.59	Boers signed Convention of Pretoria: Transvaal semi-autonomous.
6	72408	19-Jul	1848.55	1st US women's rights convention (Seneca Falls NY).
6.5	78442	10-Jan	1832.029	Jews of Canada were accorded equal political rights with Christians, 1832.
7	84476	4-Jul	1815.51	The Barbary Treaties signed in Algiers 1815. In France, the British Field Marshal Prince Blucher, signed an agreement settling the military matters between the French army and the allies after Napoleon's defeat at Waterloo.
7.5	90510	27-Dec	1798.989	First Jewish censor was appointed by the Russian government, 1798, to censor all Hebrew books printed in Russia or imported from other countries. Month and day not determined.
				General Nepolian Bonaparte decides to invade Palestine from his position in Egypt. The Ottoman Empire declared war on France in November 1798. During the siege of Acre in Feb. 1799, Napoleon prepared a proclamation declaring a Jewish state in Palestine, though he did not

Cycle	Days from Cataclysm	Day/Month	Year	The 12,068-day cycles and the events occurring on those dates.
8	96544	20-Jun	1782.47	Congress Adopted Charles Thomson's Design for the Great Seal - The Eagle and Pyramid (1782). Emperor Joseph II of Austria issued an Edict of Toleration in 1782 which repealed most restrictions on Jews that had been imposed by the Church. Month and day not determined.
9	108612	5-Jun	1749.43	In Spain all male Roma (Gypsies) were rounded up in June 1749 and sent to penal establishments and mercury mines.
10	120680	21-May	1716.39	
11	132748	6-May	1683.35	In 1683, Pope Innocent XII extended his law abolishing Jewish loan-banks from Rome to Ferrara and other Jewish ghettos under his authority. Prohibited from shop-keeping and most trades and crafts, the Roman Jewish community shrinks, while the Jews of Northern Italy begin entering commerce and industry. In 1683, during the siege of Vienna in 1683 Islam seemed poised to overrun Christian Europe. Samson Wertheimer and Samuel Oppenheimer, Jewish imperial court agents, provided financial support to the Austrian army to defeat the invading Turkish army, thus strengthening Jewish ties to the community and preventing Islam from entering further into Europe.
12	144816	22-Apr	1650.31	September 3: England's Charles II is defeated at the Battle of Dunbar by Oliver Cromwell. As a result of Cromwell taking power, he let the Jews officially back into England on Dec. 14, 1655.
13	156884	7-Apr	1617.27	
14	168952	23-Mar	1584.23	Sir Walter Raleigh acquired Humphrey Gilbert's patent to explore North America. The agreement was signed on the 25th. This led to the colonization of North America, and in turn the creation of the United States.
15	181020	8-Mar	1551.19	The Grand Duke of Tuscany issued a charter to attract Sephardic Jewish merchants from the Balkans to Pisa. They traded using routes through Ancona and Pesaro.

Cycle	Days from Cataclysm	Day/Month	Year	The 12,068-day cycles and the events occurring on those dates.
16	193088	23-Feb	1518.15	3 February 1518 Pope Leo X ordered the Augustinian Order to discipline Martin Luther for his writings. Many scholars now consider his reformational break-through (reformatorisches Durchbruch), his theological vision crystallized towards the end of February 1518. The protestant reformation was the beginning of the end of the Catholic Church's strangle hold on northern Europe. February 1518: John Reuchlin, the Father of the Study of Hebrew among the Christians. He defended the Jewish Books from destruction. His first publication was published in February 1518.
17	205156	8-Feb	1485.11	Feb 1485: Henry VII's reign began. The Battle of Bosworth Field on 22 August 1485 ended the Wars of the Roses between Henry's family, the Lancastrians, and the Yorkists. The Lancastrians triumphed under the leadership of Henry Tudor. After winning the throne of England, he wed Elizabeth of York, the eldest daughter of the dead Yorkist king Edward IV. The Tudor dynasty was noted for Henry VIII Creating the Church of England and separating from The Catholic Church. During the reign of the Tudor's the Jews were allowed back into England. Feb 1485: Pope Innocent VIII began a crusade against the Turks but he organized the multiple crusading Christians groups to pass through Jewish towns and areas, to war against the Jews first. In 1485, Columbus' first proposal to sail to the orient by sailing west was to the King of Portugal. It has been acknowledged by many scholars that Columbus was also Jewish as well as most of his sailing crew.
18	217224	24-Jan	1452.06	Book production work begun on the 1st book published, Johann Guttenberg's Bible, Finished, publication date, September 30.
19	229292	9-Jan	1419.02	
20	241360	25-Dec	1385.98	
21	253428	10-Dec	1352.94	
22	265496	25-Nov	1319.9	

Cycle	Days from Cataclysm	Day/Month	Year	The 12,068-day cycles and the events occurring on those dates.
23	277564	11-Nov	1286.86	
24	289632	27-Oct	1253.82	
25	301700	12-Oct	1220.78	
26	313768	27-Sep	1187.74	Saladin took Jerusalem from the Christians.
27	325836	13-Sep	1154.7	
28	337904	29-Aug	1121.66	
29	349972	14-Aug	1088.62	
30	362040	30-Jul	1055.58	
31	374108	16-Jul	1022.54	
32	386176	1-Jul	989.5	
33	398244	16-Jun	956.46	
34	410312	1-Jun	923.42	
35	422380	18-May	890.38	
36	434448	3-May	857.34	
37	446516	18-Apr	824.3	
38	458584	3-Apr	791.26	
39	470652	20-Mar	758.22	
40	482720	5-Mar	725.18	
41	494788	19-Feb	692.14	
42	506856	4-Feb	659.1	
43	518924	21-Jan	626.06	On January 2, 626 Mohamed invaded the Jewish town of Badr, the Second time. The Jews escaped from town the night before.

Cycle	Days from Cataclysm	Day/Month	Year	The 12,068-day cycles and the events occurring on those dates.
44	530992	6-Jan	593.01	
45	543060	21-Dec	559.97	
46	555128	7-Dec	526.93	
47	567196	22-Nov	493.89	
48	579264	7-Nov	460.85	
49	591332	23-Oct	427.81	
50	603400	9-Oct	394.77	Rome split into two empires. Byzantium ruled in Palestine.
51	615468	24-Sep	361.73	
52	627536	9-Sep	328.69	
52.5	633570	4-Mar	312.17	Constantine conquered Italy and secured his realm as Emperor. He began to promote Christianity.
53	639604	25-Aug	295.65	
54	651672	11-Aug	262.61	
55	663740	27-Jul	229.57	
56	675808	12-Jul	196.53	
57	687876	27-Jun	163.49	
58	699944	13-Jun	130.45	Roman Emperor Hadrian changed the name of Jerusalem to Aelia Capitolina.
59	712012	29-May	97.41	
59.5	718046	22-Oct	80.89	Colosseum opened in Rome. It was built with the wealth the Romans stole from the Temple in Jerusalem, which was taken and destroyed in 70 C.E. Over a million Jews were shipped to Rome for gladiatorial combat and mass slaughter.
60	724080	14-May	64.37	

Cycle	Days from Cataclysm	Day/Month	Year	The 12,068-day cycles and the events occurring on those dates.
61	736148	29-Apr	31.33	
62	748216	17-Sep	-1.71	
63	760284	1-Oct	-34.75	
64	772352	16-Oct	-67.79	
65	784420	31-Oct	-100.83	
66	796488	15-Nov	-133.87	
67	808556	29-Nov	-166.91	Jewish Maccabean revolt against the Geeks. Two envoys of Judah Maccabee were the first Jews to travel to Rome.
68	820624	14-Dec	-199.95	
69	832692	29-Dec	-232.99	
70	844760	13-Jan	-266.04	
71	856828	27-Jan	-299.08	
72	868896	11-Feb	-332.12	Alexander the great captured Jerusalem.
73	880964	27-Feb	-365.16	
74	893032	12-Mar	-398.2	
75	905100	27-Mar	-431.24	
76	917168	11-Apr	-464.28	
77	929236	26-Apr	-497.32	
78	941304	10-May	-530.36	529 King Cyrus died. The first return of Jews to Jerusalem in 538 b.c.e.
79	953372	25-May	-563.4	
80	965440	9-Jun	-596.44	The first time Jerusalem fell to Babylon.

Cycle	Days from Cataclysm	Day/Month	Year	The 12,068-day cycles and the events occurring on those dates.
81	977508	24-Jun	-629.48	
82	989576	8-Jul	-662.52	
83	1001644	23-Jul	-695.56	
84	1013712	7-Aug	-728.6	
85	1025780	22-Aug	-761.64	
86	1037848	5-Sep	-794.68	
87	1049916	20-Sep	-827.72	
88	1061984	5-Oct	-860.76	
89	1074052	20-Oct	-893.8	
89.5	1080086	28-Apr	-910.32	Solomon's temple was plundered by Pharaoh Sheshonk
90	1086120	3-Nov	-926.84	
91	1098188	18-Nov	-959.88	
92	1110256	3-Dec	-992.92	
92.5	1116290	11-Jun	-1009.44	Estimated date when David became uncontested king after Saul's death.
93	1122324	18-Dec	-1025.96	
94	1134392	2-Jan	-1059	
95	1146460	17-Jan	-1092.04	
96	1158528	1-Feb	-1125.09	
97	1170596	15-Feb	-1158.13	
98	1182664	1-Mar	-1191.17	
99	1194732	16-Mar	-1224.21	

Cycle	Days from Cataclysm	Day/Month	Year	The 12,068-day cycles and the events occurring on those dates.
100	1206800	31-Mar	-1257.25	
101	1218868	14-Apr	-1290.29	
101.5	1224902	22-Oct	-1306.81	The Exodus. The date for the crossing at the Bay of Suez.
102	1230936	29-Apr	-1323.33	
103	1243004	14-May	-1356.37	
104	1255072	29-May	-1389.41	Aaron born.
105	1267140	12-Jun	-1422.45	
105.5	1273174	20-Dec	-1438.97	Tuthmoses III died. 11 of the 12 tribes of Israel were enslaved within 1 year.
106	1279208	27-Jun	-1455.49	
107	1291276	12-Jul	-1488.53	Joseph became Prime Minister of Egypt.
108	1303344	27-Jul	-1521.57	
109	1315412	10-Aug	-1554.61	
110	1327480	25-Aug	-1587.65	
111	1339548	9-Sep	-1620.69	Abraham was to sacrifice Isaac on Mount Sinai. Abraham's contract with God began on this date.
112	1351616	24-Sep	-1653.73	
113	1363684	8-Oct	-1686.77	
114	1375752	23-Oct	-1719.81	
115	1387820	7-Nov	-1752.85	
116	1399888	22-Nov	-1785.89	
116.5	1405922	31-May	-1802.41	Estimated birth day for Terah, Abraham's father.
117	1411956	6-Dec	-1818.93	

sheep to give birth to the lambs, and the lambs had to be weaned and strong enough to endure the long trip back to Midian. (All lambs in the northern hemisphere are born from March to April.) He then collected his family and traveled back to Mount Sinai where he met his brother, Aaron. They all then traveled into Egypt.

1021 B.C.E.: AN EARTHLY KING

In 1021 B.C.E., the Hebrews asked the prophet Samuel to choose an earthly king for them, to lead them in the constant battles against the Philistines. The End of Days started in 1996. The number of years between these two dates is 3,017 years (¼ of 12,068).

1021 B.C.E. THE DEATH OF DAVID

The Hebrew's request to have an earthly king occurred in 1021 B.C.E. The death of David happened in 971 B.C.E. The difference is 50 years (50 × 24.136 = 1206.80).

1021 B.C.E.: THE END OF A UNIFIED COUNTRY

The Hebrew's request to have an earthly king occurred in 1021 B.C.E. The death of Solomon and the end of the unified Hebrew kingdom occurred in 922 B.C.E. The number of years between these two events are 99+ years. My feeling is that if we knew the exact month and day, it would turn out to be the 36,204 days which make up 99.12 years.

1021 B.C.E.: THE END OF THE KINGDOM OF ISRAEL

From the 1021 B.C.E. to the fall of the Kingdom of Israel in 721 B.C.E. there are exactly 300 years (300 × 24.136 = 7,240.8 ÷ 12 = 603.4).

1200 B.C.E.: FROM THE JUDGES TO THE END

The end of the period of the Judges occurred in 1200 B.C.E. The Fall of Jerusalem and the end of the Judean kingdom happened in 587 B.C.E. The number of years between these two events is 613 years. If you remember from the previous Chapter, that was the elevation in meters of the valley where I believe the new Mount Zion will be built. 613 meters converts to 24,136 inches. Orthodox tradition states that there are 613 commandments listed in the entire Torah.

1004 B.C.E.: DAVID AND JERUSALEM

David chose Jerusalem to be his capital in 1004 B.C.E. The End of Days began in 1996. The number of years between the two events is 3,000 (3000 × 24.136 = 72,408 ÷ 12 = 6,034).

971 B.C.E.: DAVID'S DEATH TO GOD'S DAY OF JUDGMENT

I estimate that David died in 971 B.C.E. The date of the polar reversal, or God's Day of Judgment, is October 16, 2046. The difference is 3,017 years (¼ Of 12,068). I think David's name was used so many times in connection with the End of Days because it is a clue God gave us concerning when the cataclysm will occur.

597 B.C.E.: FROM THE FIRST EXILE TO THE CATACLYSM

The first time Nebuchadrezzar invaded Judea is estimated to have been in 597 B.C.E., when he installed Zedekiah as king of Judea. Shortly thereafter, the first exile of a major part of the population of Judea. The Judeans were forcibly removed to Babylon. Jeremiah and Baruch reckoned the Jewish exile from that date. I estimate the date was early June, 596 B.C.E. The number of days from that date to the polar reversal (October 16, 2046) is 965,440, which is 80 × 12,068. God linked these two events perhaps because He compares the first exile with the final return for salvation.

538 B.C.E.: FROM THE RETURN TO THE BEGINNING OF THE END OF JUDEA

The year of the first return of the Jewish exiles from Babylon to Judea is estimated to be 538 B.C.E. The first version of the book of Mark, in the New Testament, was the Ur Marcus, written in 62 C.E. The total years between these two dates is 600 (600 × 24.136 = 14,481.6 ÷ 12 = 1206.8).

JULY 4, 1776: THE BIRTH OF THE UNITED STATES WAS AN ACT OF GOD

Many historians, as well as the Founding Fathers, felt that the Almighty had a hand in the creation of the nation, because of all the "lucky" events that favored the American revolutionaries. But no one has been able to prove it. I believe I can. The two principal designers and authors of the Declaration of Independence were Thomas Jefferson and John Adams. Each man later became Presidents of the young nation. The most important document they both helped author was the Declaration of Independence, completed on July 4, 1776. Both men died within a few hours of each other, on July 4, 1826, exactly 50 years after their great work (50 × 24.136 = 1206.80).

JUNE 20, 1782: CONGRESS ADOPTED THE DESIGN FOR THE GREAT SEAL OF THE UNITED STATES

The importance of this event (listed in Table 11-1) is what was placed on the Great Seal and why it was put there. General Washington's financial advisor and assistant was Hyam Salomon, a Jewish merchant and financier. During the Revolutionary War, he loaned the young country over $600,000 of his own money as well as raising over £3,500,000 from the Sassoon banking family of Baghdad and the Rothschild Banking house of France. Both of them were Jewish owned firms. During the winter at Valley Forge American soldiers were freezing and running out of food. Hyam Salomon solicited Jews in America and Europe to give money to aid the stranded American troops. This act changed history because without this help, the Continental Army, and the fight for American Independence would have likely perished before they could be resupplied and rallied to defeat the British army.

Hyam Salomon and Robert Morris joined with Hamilton to provide the funds for the establishment of the Bank of North America, which funded Congress before the Constitution was established.

George Washington insisted that two elements be placed on the Great Seal of the United States. The first is the arrangement of the 13 stars above the Eagle's head is rendered in the six pointed Star of David to honor the Jews. The second feature is the 9 tail feathers on the American eagle If you turn the Eagle upside down you will see a configuration in the likeness of a Menorah. President Washington said we should never forget the Jewish people and what they have contributed towards the independence of the young country and the formation of the United States of America.

Following the Revolutionary War, Congress reneged on their promise to reimburse Hyam Salomon for bonds they had issued, and he died destitute and a pauper, getting out of paupers prison just in time to die.

Some scholars have commented that the creation of the United States was greatly helped by many acts of God. I have noticed from my research doing Table 11-1 that two prominent dates are directly related to the creation of the United States. The first being when Sir Walter Raleigh purchased the rights to explore and develop North America and the other being the adoption of the design of the Great Seal. There was one more that I did not included in this table because it was 8.25 cycles (March 21, 1774). In September 1774 was the first meeting of the Continental Congress. On October 20, 1774, they signed the Articles of Association. The colonial representatives must have started working of this agreement earlier that year. God must have cause events to occur that resulted in the formation of the Continental Congress and this agreement. It is my conclusion that God wanted the formation of the United States as a refuge for oppressed peoples who wanted personal freedom including the freedom of religion, which the Jews desperately needed. It is also my conclusion that as long as the United States does not forsake the Jews and Israel, God will not forsake the United States no matter what odds we might face.

1915 C.E.: THE LIFE OF GENERAL DAYAN

One of the great modern-day Jewish generals was the late general Moshe Dayan. He was the commanding general in the 1967 war, when Jerusalem was recaptured from the Jordanians. General Dayan was born on May 20, 1915 and died on October 16, 1981. That adds up to a total of 24,256 days (24,136 + 120). The 120-day remainder is a clue to the three instances when the number 120 years appears in the Torah. This is a clue connected with mount Sinai.[3]

AUGUST 25, 1939: FROM THE BEGINNING OF A WAR TO THE BEGINNING OF THE JEWISH NATION

On August 25, 1939, the British government signed an Agreement of Mutual Assistance between the United Kingdom and Poland:

> The Government of the United Kingdom of Great Britain and Northern Ireland and the Polish Government: Desiring to place on a permanent basis the collaboration between their respective countries resulting from the assurances of mutual assistance of a defensive character which they have already exchanged . . . [4]

This was a mutual defense treaty signed between these two countries to discourage Germany from invading Poland. On September 1, 1939, 7 days later Germany invaded Poland and the provisions of this treaty went into effect. Great Britain declared war on Germany and World War II began. If this treaty had never existed the Second World War would not have started then.

The next event linked to the start of World War II was the United Nations vote to create the Jewish State of Israel. The number of days from August 25, 1939 to November 29, 1947 when the U.N. vote took place is 3,017 (¼ of 12,068).

In Table 11-1 on the fourth cycle (48,272 days from the cataclysm), I showed that President Woodrow Wilson gave his neutrality speech in front of the U.S. Senate, coinciding exactly with a 12,068-day cycle. Because of his action World War I dragged on until November 1918. It seems logical that if President Wilson had decided to bring the United States into the war at the beginning the Kaiser would have put an early end to the war. Instead, the war dragged on for over four years.

Even before the end of World War I there was great interest in letting the Jews return to "Palestine" for the establishment of a Jewish national home. This led to the Balfour Declaration, signed November 2, 1917. It was a letter from Arthur James Balfour to Lord Rothschild stating that the British Government was in favor of a Jewish State in Palestine after the territory had been won from the Ottoman Empire, at the close of the war. On July 24, 1922, the newly created League of Nations passed the Palestine Mandate, instructing Britain to form a Jewish state in their Palestine protectorate. It was going to be all the land west of the Jordan River. Unfortunately, for the Jews, oil was discovered in Saudi Arabia, and the British had a change of heart. This British act of treachery did not go unnoticed by God, as you will see next.

The unreasonable treaty conditions put on Germany ending World War I, brought about the bankruptcy of the country in the 1920s and the eventual rise of Adolf Hitler, who brought us World War II. England lost big during and after World War II. Their cities were bombed. They lost over 60,000 civilians and 400,000 military personnel. Prime Minister Winston Churchill had to submit to President Franklin Roosevelt's demand, before the United States passed the Land-Lease act, that England would give up her colonies after the war. After the war England was almost bankrupt.

What these date relationships show is that God caused or permitted both World Wars in order to create the State of Israel. There were a lot of deaths just to fulfill the prophecy of the creation of the State of Israel, and the return of the Jews to their homeland, in order to fulfill God's promise to return His people to their land. As you are beginning to see, God writes his *signature* by embedding sacred numbers into events. It is very possible that there would not have been a World War II if the British Government had kept its promise to create a Jewish state in 1923. God would not have had to shape the conditions and personalities that brought us World War II and the Holocaust. After all, ask yourself," "What is the probability that a failed poor artist and corporal of a defeated army, could become leader of a major industrial democratic country?"

1947: THE CREATION OF THE MODERN STATE OF ISRAEL.

The vote in the United Nations to create a Jewish state, in the former British protectorate of Palestine, occurred on November 29, 1947. My first geological expedition to Mount Sinai took place in 1997. After I discovered these time cycles in our existence, I wanted to test them and prove to myself that God had a hand in the actions of man. Therefore, what I did was to let the Egyptians choose the date for the expedition and the time when we were to arrive at Mount Sinai. I gave them a two-month leeway. They chose November 29, 1997 to leave for the Mount and arrive there that evening before sundown. We arrived at the base of Mount Sinai exactly 50 years, to the day, after the U.N. voted to create the State of Israel ($50 \times 24.136 = 1206.80$).

THE PRESIDENTIAL ELECTION OF NOVEMBER 2000

After my two expeditions to Mount Sinai in 1997 and 1999, I next wanted to lead an archaeological expedition to research the site the proper way, and perhaps find the family cave and its contents. But after a series of major problems, including choosing the wrong person to head the archaeological team, I realized that God did not want me to try to dig it up until that territory changed hands again and was returned to Israel. In the beginning of 2000, no one dreamt that there would be another Middle East war. But I told my friends there would be another war just so the Sinai would change hands once again.

Before I explain why the election of 2000 is so important, I have to explain two other things first. The number seven represents Mount Sinai, which is why the number is mentioned 205 times in the Torah. The reason is there are seven distinct markings on one side of the Mount. So, whenever you see the number seven in the Bible and the Jewish Oral Tradition, it is usually referring to Mount Sinai. The other point to remember is that the number 6,034 represents the Torah because the Torah has 6,034 chapters and verses in it, and the two tablets which the Torah is engraved upon it is in the Ark of the Covenant. The Ark is in the family cave in Mount Sinai. Therefore, the number 6,034 represents both the Torah and the Ark.

The 2000 election between George W. Bush and Al Gore was the most contested election since Rutherford B. Hayes became the nineteenth president of the United States in 1876. The Florida State vote was the contested one, and the whole U.S. election rested on that one state. The news services reported on Friday, November 10, that the vote spread between Bush and Gore was only 300 votes. This was before they counted the absentee ballots. I immediately felt that 300 ($300 \times 24.136 = 7,240.80 \div 6,034$) was a message from God. The other fact that intrigued me was George Bush's name. Ask yourself: "Where did Moses first see God and in what form?" The answer is: "In the Cave and from a burning **Bush**!"

There was more to this story. At a Goodwill store in Redmond, Washington, in the week of November 10 to 17, 2000, a rare quilt was found with Rutherford B. Hayes's signature on it.[5] The probability is astronomical that such an historical artifact with Hayes's signature on it, would be discovered during the same fateful week. Hayes' election was also a closely contested election.

I know this is not much of a clue to God's intensions, but it was an indication that perhaps this George Bush was going to be President when the Ark would make another appearance. I already knew the Ark was prophesied to appear again; it was just a question of when? I also knew we were in the block of time called the End of Days, so it had to happen soon.

911: SEPTEMBER 11, 2001

On September 11, 2001 (commonly known as 911), 19 Arab Muslim terrorist hijackers flew two passenger planes into the Twin Towers of the World Trade Center in New York City, and one plane into the Pentagon in Washington D.C. The final reported death toll given on February 1, 2003 was that 2,823 people had died in the Twin Towers along with 184 who died in the Pentagon, for a total of 3,007 innocent people. This number was somewhat shocking to me because of what it implied. The number is made up of two important numbers, the 3,000 converts to 6,034[6] representing the Torah and the Ark—and the number 7 represents Mount Sinai. The message I got from the death toll was that the Ark is going to be rediscovered. It also implies that God kills us in numbers which amplify proof of His existence and that He is the ultimate controller of events.

There is a well-known Jewish expression which states: "God moves in mysterious ways." I do not wish anyone to think that God had nothing better to do on September 11, and decided to kill 3,007 people for no reason. The result of the 911 attacks was that the United States went to war against Afghanistan and later against Iraq. When the United States invaded Iraq it removed one more enemy of Israel.

God's second reason for 911 was to change American public opinion against the Arabs and in favor of Israel. It is too bad the State Department is so disconnected from God and the American People that it has failed to get the message.

By 2007 there was little doubt that there will be another war in the Middle East. In Chapter Ten, I presented two possible scenarios for how the war could start. It is just a matter of time before the right set of conditions present themselves.

THE ROAD MAP AND THE GAZA STRIP

The Bush administration's State Department came up with a "peace plan" for Israel and the Arabs in the west bank territories Israel has held since 1967. The west bank was originally part of the Jewish homeland all the way back to Joshua's conquest of Canaan starting in 1,264 B.C.E. The area was later known as Samaria and Judea. The Europeans, Russians and the United Nations supported the Peace plan, known as the Road Map. None of these parties has historically been friendly towards Jews or Israel. The United States pressured Israel to withdraw from the Gaza Strip and forcibly removed 10,000 Jews from their homes and towns in the Gaza. As I have shown in the previous section and the prophecies in Chapter 10, God manipulated events so the state of Israel would be created and the Jews could move back to their homeland. Since the forcible evacuation of the Gaza Strip was against the will of God, He punished the United States for the Bush administration's actions. The Evacuation was completed on August 29, 2005 when the last Jew left Gaza. 10,000 Jews were evicted from their homes. The total population of Israel is about 6,000,000 Jews $\div$ 10,000 = .00167 of the total population.

On August 29, the very same day, Hurricane Katrina hit the Louisiana and Mississippi coast and destroyed New Orleans (except for the historic French Quarter) forcing the evacuation of most of the people in the area, totaling about 500,000. The population of the United States is about 300,000,000 which when divided by 500,000 = .00167 of the total population. This is the same percentage of Jews forced out of Gaza in proportion to the total Israeli population. It appears God did not wait long to punish the United States for interfering with His plans. A book entitled *Eye to Eye* by William Konig, compiled over fifty catastrophes that cost the country tens of billions of dollars, for the same reasons. When the State Department and the Administration pushed a land for peace settlement on Israel, some catastrophe hit the United States the same day or the day after.

He Kills us in Numbers that Amplify His Message that He Exists

I thought for over a year about whether I should put the following in this book, because it is similar to what I discovered about the number of people who died on 911. But I feel that if it helps explain how God works, and proves His existence, then I should include it.

The Titanic

The height of the British Empire was before the start of World War I. The British were on top of the world, top dog, and all that. Unfortunately, their egos were also grossly inflated, which clouded their better judgment. The chief designer of the HMS Titanic was reported as saying "Even God cannot sink it." Advertisements and newspaper articles at the time echoed the same theme. On April 15, 1912 the Titanic hit an iceberg and sank in the north Atlantic. The first newspaper to report the sinking was the Boston Globe. Its headline read "1,500 Drowned," $(1,500 \times 24.136 = 36,204)$. I guess God can sink it if He wants to! It also implies that God will actively intervene if man's ego and stupidity warrant it.

The Holocaust

The common people tend to believe whatever their political and religious leadership tell them. Between the years 1933 to 1945 European Christians following their religious and political leadership, went from persecuting the Jews to outright extermination. It was not just the Germans who did this outrageous deed, but other Christian countries to the east. And I do not want to leave out the French who collaborated with the Germans. The British were too "polite" to do the dirty deed themselves, but they made sure the Jews were blocked from leaving the European continent in the 1930s and 1940s, and thereby condemned the Jews to their fate. Furthermore, beyond question, a major share of the blame for the Holocaust squarely rests on the heads of the Roman Catholic Church and the Pisos. No other explanation is necessary. Its leadership knows exactly what I mean. Hitler and his entire inner circle were raised as "good" Catholics and educated in Catholic schools. That is where they learned their hatred for Jews.

Six million Jews were murdered during the Holocaust from 1939 to April 1945. The calculation is simple: $6,000,000 \times 24.136 = 144,816,000 \div 12 = 12,068,000$. You are looking at the very heavy hand of God. I am not going to leave it there; God did not want these people murdered. Man and freewill were responsible for that. God merely made sure the death toll only went to six million so that the knowledge of His existence would be reaffirmed.

I mentioned in Chapter 10 that the worst decision the Hebrews ever made was deciding to have an earthly king instead of God (1021 B.C.E.). The following quote answers the question, "Why did God permit the Holocaust."

> [1 Samuel 10:18] "Thus saith the Lord, the God of Israel: I brought up Israel out of Egypt, and I delivered you out of the hand of the Egyptians, and out of the hand of all the kingdoms that oppressed you. [19] But ye have this day rejected your God, who Himself *saveth you out of all your calamities and your distresses*; and ye have said unto Him: Nay, but set a king over us. Now therefore present yourselves before the Lord by your tribes, and by your thousands." [Emphasis added.]

What this is saying is that before 1021 B.C.E., God actively intervened on the side of the Hebrews to prevent evil befalling them. After 1021 B.C.E., God no longer actively intervened against their enemies. He did tell the high priest whether they should do battle or not and He would also warn them, but He did not actively change events. There are exceptions throughout history, but generally He did not intervene. After the Ark was hidden away little intervention was exercised, except where it sited His purpose.

Active Programming

This Chapter shows that there is an unusually high number of important historical events which occurred on a 12,068-day cycle, or factor thereof, counting back from the estimated cataclysm date of October 16, 2046. Also shown is that the "holy number" shows up as the number of days or years between important events. This implies that events are linked together by God. Most of these events directly or indirectly have something to do with the Jewish people. What these dates and numbers show is that God is actively programming events to come out the way He wants them to. He is creating a set of circumstances in order to accomplish something in our future.

Time is like a fulcrum and a lever. The closer to the fulcrum, a force is applied to the lever, the less of an effect is created. The farther away from the fulcrum, a force is applied the greater the effect. The Hebrews made many mistakes, from the time of Moses to the fall of Jerusalem, but the farther away in time from the initial mistake, the greater the effect will be in the future. For example, if Joshua, Saul, and the other Hebrew kings had listened to God's order to remove or kill all of the other nations that were living in the Promised Land, the Jews of today would not have to deal with the recently created "Palestinian" people. "Palestinian" identity was created in the early 1970s for those Arabs who had moved to Israel from the surrounding Arab countries for jobs. This migration happened mostly since the British occupation after World War I.

The question still is: "How does God get people to do what He wants?" I believe I can speak from personal experience on this subject. Many times I would work on a problem and a thought would "float" into my head. For instance, it was on a Thursday that the thought came into my mind that the number 12,068 is somehow incorporated in the chapters and verses of the Hebrew Scriptures. By Saturday night, I had gone through four separate versions of the Bible and discovered the variations on the numbering system and how the chapters were grouped to total the holy numbers. The Question is: "Where did the thought come from?" How did I know the numbers would be revealed within the Hebrew Scriptures? That is the whole point. I think God places an idea into someone's mind, and maybe the person will then act upon it.

Free Will

Now comes the question of man's free will. Man does have free will to screw things up, and we do a very good job of it—thank you very much. The Hebrews demonstrated that, and look what happened to them. What none of us thought was that God would program from the end of time to the beginning of time, to make sure He obtains the results He wants. If man changed events because of free will, God would reprogram events from that point, to the end of this cycle, to accomplish the desired end result. The two World Wars and the Holocaust happened because that was the only possible way to achieve the primary result of creating the State of Israel. That may be a harsh thing to say, but once I discovered that the Woodrow Wilson neutrality speech fell on one of the 12,068-day cycles and that the number of days between the start of World War II and the creation of the state of Israel was 3,017 days, what other conclusion could I come to?

We may ask: "Why is it so important for God to create the State of Israel again?" The loss of life from the two world wars totaled over 70 million people. Is the cost of 70 million people, and all their suffering, worth the creation of little Israel? Ignorance is bliss, and before you picked up this book, your logical answer would have been "no." You have to look at this situation the way God does. Remember the prophecies from Chapter 10 about the new Mount Zion. For Zion to be built, the State of Israel has to exist first. So if you were God, it is either no State of Israel, no Jews, and most (maybe all) of the human race dead during the polar reversal—or lose 70 million people, create the State of Israel, that eventually Mount Zion will be built, and maybe billions of people saved. That is the decision.

My Conclusion

The main purpose of this book is to prove the existence of God and reveal what His relationship is to the Universe. I had to cover many subjects to accomplish this but I think I have done it, and I am sure I am the first in history to accomplish this feat. I hope I have convinced all of you that the 12,068 number, embedded within the Torah, is a clue to the polar reversal and the cataclysm. I also want to make it very clear that God, the Lord of Hosts, wants us to survive the polar reversal, and He will do anything He can to help man survive the event. But there is a big "but." It seems He uses the polar reversal period as a filter to cleanse society of false theologies, philosophies, customs, etc. In other words, *He saves us on His terms, not ours.* This is very important to understand.

In the beginning of this Chapter, I referred to the Greek philosophy of a man-centered Universe, where God took no part in the affairs of man. Well, it looks like the Greeks lost and God and the Hebrews won the argument. This chapter proves that God actively programs our existence, to the point of micromanaging the actions of those who are important in His overall plan. The lesson to be learned from this Chapter is that we may have free will but if you have the ultimate being, the Operating System of the Universe, reprogramming to fix whatever man screws up, you realize that free will outside His parameters is tenuous at best.

I also do not want you to think that God is an unemotional machine without compassion or feelings. That would not be correct. Since we are a reflection of the Universe we are part of, and since we certainly have emotions, then we can be sure that God, who is the Operating System of our Universe, also has these feelings. He certainly demonstrated feelings when the Hebrews rejected Him in favor of a human king.

This book has also proven that we are the product of a synchronous system (a computer), the Diehold. We live in a created reality whether we like it or not. The world we live in is *like* a hologram, but this hologram creates the matter world around us. That is the message of the Hebrew alphabet. Psychologically, this point is the hardest fact to accept, but after the evidence of the 12,068-year clock cycles, phantom-leaf effect of Kerlian Photography, the six blank periods found in space, you have no other logical conclusion. We are the product of information in a Diehold.

Chapter 12:
Why was the Universe
Created?

In the Beginning . . .

I began this book with six difficult and important questions. The first three questions have been answered in the previous chapters. The last three questions are the most important. They are: "Why was the Universe created?" "If God created the Universe, then who created God?" " Why was man created and what is his place in the Universe?" These have to be the most difficult questions in the entire Universe to answer—difficult, but not impossible. If a theory of existence can answer these three questions, it must be the secret of the Universe and how it really works. This chapter WILL answer these three questions.

Summary of what we think we know:

The previous chapters should have proven to you that there is a God in the Universe, who is also called "the Lord of Hosts." What he "hosts" and builds is the Universe, just as an operating system hosts and controls other programs in a computer. The "computer" which I have named "The Diehold,"[1] holds all the information of our Universe including the Operating System.[2] Our Diehold has to be constructed out of matter, otherwise information could not be contained within it, to be processed, and transmitted to create our existence. If it is constructed out of matter then the Theory of Multidimensional Reality states that it is transmitted into existence from another Diehold—just not ours but one above ours. I think you can see where I am going with this. If I have presented enough proof that the Universe is the product of information, then it logically leads you to the conclusion that a very highly advanced civilization must have built our Diehold and turned it on. I will refer to them as the "Builders." The action of turning on our Diehold created our Big Bang, the beginning of our Universe. However, that still does not tell us why they built our Diehold and Man's purpose, in the scheme of things. To answer that, we must delve deeper.

The Answer is in Genesis

In order to unravel this great mystery of what Man is in relation to the Universe we must start at the beginning, Genesis Chapter 2:8: "And the Lord God planted a garden eastward, in Eden; and there He put the man whom He had formed."

The Hebrew word for "formed" can also mean *pattern*. So what this verse reveals is that God created a *pattern* that represented man. In other words, He created a *program* that represented man, which the Diehold creates into existence. A program cannot "die" in a computer unless it is destroyed or turned off by the Operating System. Your soul is the program and it does not "die" after your physical body (receptacle) stops functioning.

Our next clue is found in Geneses 2:17: "But of the tree of the knowledge of good and evil, thou shalt not eat of it; for in the day that thou eateth thereof *thou shalt surely die*" [Emphasis added.] The verse implies that Adam's (Man's) previous state of existence was to live forever, but if he entered some device, allegorically represented as the "tree of knowledge of good and evil," Man would then, and only then, be mortal. The same theme is again repeated in Chapter 3:3: "But of the fruit of the tree which is in the midst of the garden, God hath said: 'Ye shall not eat of it, neither shall ye touch it, *lest ye die*' " [Emphasis added.] The message is loud and clear, these people were in a state of existence that they would live forever.

I have already explained in Chapter 4, on the Hebrew Alphabet, what Adam and Eve did and what they were. They made a conscious decision to de-evolve into fifth dimensional beings, but the Biblical story does not tell us why they did it, and that is the secret you are about to discover.

An Advanced Civilization—The Builders

To avoid confusion, I do not mean the advanced civilization that placed the advanced technology in Abraham's cave. I am referring to the "Builders" of our Diehold. What kind of civilization could build a computer to replicate their Universe? It would have had to be either a highly advanced fifth dimensional beings or six-dimensional beings who have been around for more time then I am willing to guess. Why would an advanced civilization want to build a replica of their own Diehold that they know they are in? There are two reasons The first reason is important, but the second reason is the ultimate, paramount reason.

The Simple Reason

The first reason I thought of was the obvious one. The principle of the experiment is: Do you get out what you put into it? There must be a mechanism within the Diehold, which enables the Operating System to report to the "Builders" of the Diehold. I thought of a simple experiment that would prove whether the Operating System permeated all levels of existence. Allow me to explain what I mean. Imagine you are standing between two parallel mirrors. When you look at one, you see yourself reflected an infinite number of times in the two mirrors. Another analogy is taking a computer program and copying it onto a compact disk (CD), loading it into another computer, and then copying the same program onto another CD and loading it into another computer. Let us say that you copy it over a million times. The millionth copy of the program will be the same as the first one. Do we care what "copy" we have? No. Do we care if we have the first copy or the 300th or the 300,000th copy? No. The experiment would be to see how long it would take for intelligent life, in your Diehold, to arrive at the correct programming for the Operating System of the Universe and to feed the result back to you in its entirety. If the Builders received back the same Operating System program they put in, that would prove the Operating System permeates all levels of existence. Most likely the spaces (distances) between the atoms, form the patterns for the program, and it permeates through all levels of existence. This could be related to the carrier wave, which could be carrying the Operating System.

What is profound about this experiment is that it proves the Universe started with a thought-form and that started it all. In other words, the Ultimate God started it all with a thought form with the correct pattern. This is not to diminish our God, because He is a reflection of the original thought form that created it all. By worshiping our God, we are indirectly worshiping the ultimate God who created the thought form that started everything!

Our Universe is thought to be about 18 billion years old, dating from the Big Bang, when it is supposed to have started. When I came up with the first reason for creating the Universe, I wondered: "How much time would it take for the Builders of our Diehold to get the answer from their Diehold? I figured it would take probably no more then 4.5 billion years of our time. I came up with this number by assuming that planets would form within half a billion years of the Big Bang, and it would take the rest of the time for a planet to create the right conditions to produce intelligent life. Then it was just a matter of the odds of a planet developing an advanced intelligent civilization on it. Of the many trillions of Solar Systems in the Universe, what percentage might produce life which would evolve to the fifth or sixth-dimensional level? I figured there would be quite a few. So, for this discussion, let us say the Builders of our Diehold would have gotten their first answer after 4.5 billion years of our time.

Now let us say their Diehold has a clock cycle of 100 megahertz, which is rather slow by today's standards. Remember that one clock cycle in our Diehold equals 12,068 years of our time. Let us now do a little calculating to see how long it would take the Builders to receive their answer to the experiment, in their time. Let us first convert our Earth time into clock cycles in the Diehold. We divide 4.5 billion years by 12,068 to get 372,887 clock cycles. Next, we convert to the Builders time standard by dividing 372,887 by 100,000,000, and we get .00373 seconds! Even if we assume it would take 18 billion years for a Universe to produce its first sixth-dimensional beings, it still converts down to only .0149 seconds! If the speed of the computer goes up, it will take less time.

I was rather taken aback after realizing that it would take so little time to get the answer. It would be like turning on your computer and receiving the answer before you could take your finger away from the ON button! It did not make any sense. Would any civilization go to the expense of building a Diehold for an answer they would get back almost immediately? I don't think so.

In summary, the first test or experiment is an important one but it seems that it is not the primary reason the "Builders" built our Diehold.

Man

Man is very special to the Diehold. In fact, I have to conclude it was designed for man to evolve in. When I say man, I mean fourth, fifth and six dimensional intelligent beings. Nevertheless, man is a special program within the Diehold, with a special function, which I will explain later.

The Ultimate Question: Why was the Universe Created?

I came up with the answer in the year 2000 and it left me emotionally flat. So I warn you, you may not like the answer but after you think about it, you will agree it is the ONLY ANSWER.

I have found that people and governments act on things for their own selfish reasons. There is nothing wrong with self-interest. After all, that is what competition is all about, which creates progress, not stagnation. Without progress and inventions, you wind up with a bronze-age civilization with only forty years left before the polar reversal, and they are ignorant of the events which are going to happen, and do not have the technology to save themselves. Oh—sorry! Our civilization fits into that category also. Oh well, it looks like we also did not evolve far enough before the impending polar reversal.

So there has to be another reason why our Diehold/Universe was built. The answer was written in the Genesis story of Adam and Eve. I previously showed that the civilization Adam and Eve belonged to was a sixth-dimensional society where people lived forever. The Torah and I mean that literally—these people exist forever. After all, if you are a computer program in the Diehold, there is no reason why you would die, unless the Operating System decided to destroy you, for whatever reason.

Now, since I have established that, I want you to think about LIVING FOREVER and remembering EVERYTHING you have ever done, seen, lived, read, loved, hated, etc. Astronomers have theorized our Universe will expand to a certain point, and then collapse down again. Let us say they are correct—and I agree, the Universe will expand to a certain point, then collapse down to a finite point, and then start all over again in a new cycle. We can actually estimate when this cycle would occur, if we knew the clock-speed of our Diehold. We assume in the example I gave that a 100-megahertz computer is employing a clock (CPU) speed that converts to 1,206,800,000,000 years for our Universe to go from the Big Bang to the ultimate collapse and start all over again.

Now let us get back to people who live forever. They may not merely have lived to see one of these cycles but a dozen or so. Now imagine that you have lived and reincarnated as different people, not hundreds or thousands of times, but millions-upon-millions of times. At some point you would have seen every imaginable life form. You would have seen every form of government, religion, science, technology, life style—good and bad. I believe you can see where I am heading with this. At some point in your existence the repetition and resulting boredom would become so unbearable that you would do anything to FORGET!— to forget everything you did in previous lives, who you knew, what you did, even the enjoyable events. It would be so painful to remember all of it, that you could be driven to madness. Many people say they wish they could remember what they were and did in a previous lifetime, but I believe that may be a mistake. What I believe now is that it is a blessing to forget. The reason why Adam and Eve decided to devolve to fifth-dimensional beings was so they could forget and live a life free of their past memories. They wanted to be human again. God was not tremendously upset with them. He honestly informed them what existence would be like after they took this lower form of life. I am beginning to believe that most sixth-dimensional beings devolve for the same reason. I would not term it entertainment for them, but more like "what you have to do to keep your sanity."

The Answer

To return to the Builders of our Diehold, I believe their civilization came to a critical point in their existence. They knew that they are also in a Diehold which creates their existence. They also knew that the Diehold they built (our Universe) is teeming with life and advanced civilizations within it, who are also building Dieholds to pose the same eternal question. That question is: "*Will someone inside their Diehold TURN IT OFF? . . . and if so how long would it take?*" Yes, I believe that when these computers are built there is a mechanism for a six-dimensional being to turn it off. It might take the form of an octahedron-shaped structure, which mirrors the actual Diehold computer, but is a created image with an equivalent software off switch.

The moral question for any advanced civilization is: "Do they have the moral right to shut it down?" I take the moral position that we do not have that right. The Universe is teeming with life. Once you realize that one Diehold eventually spawns untold billions of additional Dieholds within it, you realize that when a Diehold is turned on it is literally an explosion of life going on inside.

I am sure many of you will be shocked at my conclusion why our Universe/Diehold was created. Once you accept that we are in a created reality, we are created from information and the Torah clues say some form of Man can live forever, there is no other logical answer to the question.

Why did God become irritated with Adam? I believe that the Operating System/ God is programmed to encourage upward evolution, so as to test the question *"Will someone inside the Diehold, turn it off?"* That is why He gave us all the Laws of conduct in the Torah. He wants Man to evolve to an eventual understanding of how the Universe really works and what His relationship is to it.

How about Us?

We are only fourth-dimensional beings so we have a long way to go before we need to worry about this type of decision. Some of us may have devolved many times, and have chosen to become fourth-dimensional beings so we could spend a much longer time evolving upwards, thus forgetting longer. I believe God gave us the ten commandments, and many other rules of conduct, because He knows the best way to live and not make life a living hell. You cannot argue with an Entity that has seen and controls it all.

Conclusion

I have covered many subjects and philosophical issues in this book. I have delivered what I promised you in the introduction. You all now know there IS a God, and what his relationship is to the Universe. He may not be what you envisioned, or what you were taught. Keep in mind that if what you were previously taught was truly correct, then why are you within 39 years of the polar reversal and ignorant of the event? Why didn't your religious and academic institutions warn you of God's Day of Judgment? You had to read this book to find out what is going to happen to you and how the Universe really works.

Some of you who have decided that you probably will not be alive in 2046 and also have no children, or just do not give a damn for anyone else, I would like to give you a very good reason why you must care that some of us survive the cataclysm. Remember you are a soul that exists in the Diehold and your soul "lives" forever. Our bodies are nothing more than a receptacle for our souls. If, God forbid, no one makes it during the polar reversal and man becomes extinct, what new receptacle (a human baby) are you going to reincarnate into? Our souls may have to wait 10 to 50 million years until another mammal evolves to a humanoid body worthy for us to reincarnate into. I hope I have made my point. We have no choice but to care.

I want people to have hope. There is a way to survive this terrible polar reversal and resulting cataclysm. God has given us the clues to save ourselves, and we must—for the survival of the human race.

Appendix A:
Table of Stars, Open Clusters and Globular Clusters

Table of Stars in the Milky Way Galaxy

I include the following table because of its importance and so other researchers could follow in my path.

The following table lists the Open Clusters (O) and Globular Clusters (G) found in our Milky Way Galaxy. I did not list the stars closer than 9,400 light years even though they are part of my complete database. An Open Cluster is a loose collection of stars in a small area of space. They usually number from a few hundred to a few thousand stars. A Globular Cluster is a tightly packed collection of stars of all sizes. A globular cluster contains hundreds of thousands or millions of stars. There are over 100 globular clusters in our galaxy.

The following table is dominated by Open Clusters up to a distance of 13,400 light years (LY). After that there is a mixture of Open Clusters (O) and Globular Clusters (G). After 17,000 LY the database is dominated by Globular Clusters. This is probably because it was difficult for astronomers to reconcile single-light (*i.e.,* of the smaller open clusters) sources so far away.

Astronomers measure stellar distances in units of measure called parsecs. Each parsec is equal to the distance light travels in 3.262 years. You can see we are dealing with some very great distances and of course some subjective decisions astronomers have to make to determine these distances. What is *interesting* is that the six different astronomers who determined the distances for the stars at the beginning of the six blank periods had their calculations come out to reveal the 12,068 number. They, of course, had no idea of the importance of that number. There for the subjective part of their calculations were not influenced.

To make the stellar graph (Graph 3-1) display the blank periods correctly, I added two records, with a value of zero, just after the star cluster, and just before the next star cluster. The data comes from several sources but predominantly from *Sky Publishing's* 1982 directory of stars call Sky Atlas 2000, of Open Clusters and Globular Clusters. The star systems that were close to the time/distance periods I checked with other astronomical sources and picked the one that best fit my model. I know this is not what you would normally do in research, even though it is done all the time. It is like cherry-picking your results but remember I already knew there was a known error factor of 2.5% to 10%. The important thing to realize is that there is a degree of subjective interpretation of the raw results. Moreover, as I have said before, I think God influences man by putting ideas in our minds to lead us to the "desired" results. Therefore, I wanted to see if there were any astronomers who were "divinely" influenced, which made their results come out more correctly for my purposes. I only started looking for other distance results after I started seeing a pattern. The important constellations are footnoted as to their sources. At no time did I alter anyone's distance figures. All I did was convert their findings from parsecs into light years and graph them.

Constellation	Name	Type of Stars	Quantity	Dist./PC	Dist/LY
HOR	AM-1		1		0
MON	NGC 2324	O	70	2900	9458
OPH	PAL 6	G	50000	2900	9458
CEN	CR 272	O	40	2900	9458
PUP	NGC 2483	O	30	2900	9458
CAS	NGC 103	O	30	3000	9785
CAS	NGC 609	O	25	3100	10111
SGR	NGC 6656	G	100000	3100	10111
CEP	NGC 7510	O	60	3160	10306
NOR	NGC 6031	O	20	3200	10437
PER	BERK 68	O	60	3200	10437
PYX	NGC 2818	O	40	3200	10437
CAS	NGC 7790	O	40	3200	10437
CAS	BERK 65	O	20	3300	10763
CAR	NGC 3324	O	25	3300	10763
CAS	IC 166	O	120	3300	10763
PUP	NGC 2467	O	50	3400	11089
GEM	NGC 2266	O	50	3400	11089
MON	NGC 2236	O	50	3400	11089
CAR	NGC 3603	O	30	3500	11415
CYG	NGC 7067	O	20	3500	11415
CEP	NGC 7380 [1]	O	40	3600	11741
1 ST	**DARK AREA**		0		11742
1 ST	**END DARK AREA**		0		13045
CRU	RU 97 [2]	O	20	4000	13046
AUR	NGC 1893 [3]	O	60	4000	13046
AUR	BERK 19	O	40	4000	13046
OPH	NGC 6366 [4]	G	100000	4000	13046
SGR	NGC 6440	G	50000	4100	13372
PUP	RU 32	O	30	4100	13372
SGE	NGC 6838	G	50000	4100	13372
AUR	KING 8	O	30	4150	13535
CAR	PISMIS 17	O	25	4200	13698
CRU	HOGG 15	O	15	4200	13698
PAV	NGC 6752	G	100000	4300	14024
AQL	NGC 6760	G	40000	4300	14024
CIR	PISMIS 20	O	25	4400	14351
OPH	NGC 6254	G	100000	4400	14351
PUP	RU 55	O	12	4400	14351
ORI	NGC 2141	O	100	4400	14351
CMA	NGC 2204	O	80	4450	14514
TUC	NGC 104	G	50000	4600	15003
SGR	NGC 6544	G	100000	4600	15003
CMA	NGC 2243	O	100	4600	15003

Constellation	Name	Type of Stars	Quantity	Dist./PC	Dist/LY
MUS	NGC 4372	G	100000	4900	15981
GEM	NGC 2158	O	25	4900	15981
SGR	NGC 6603	O	50	4906	16001
MON	BIUR 10	O	20	5000	16308
VEL	NGC 3201	G	100000	5000	16308
CAR	WESTR 2	O	12	5000	16308
LYR	NGC 6791	O	300	5100	16634
CEN	NGC 5139	G	1000000	5200	16960
MON	DO 25	O	50	5200	16960
SGR	NGC 6809	G	100000	5200	16960
OPH	NGC 63047	G	40000	5400	17612
ARA	NGC 6352	G	50000	5400	17612
MUS	NGC 4833	G	75000	5500	17938
OPH	NGC 6218	G	100000	5500	17938
OPH	NGC 6171	G	100000	5900	19243
AUR	BASEL 4	O	15	5900	19243
SGR	NGC 6553	G	100000	5900	19243
OPH	NGC 6266	G	50000	6000	19569
SGR	NGC 6626	G	100000	6100	19895
SGR	NGC 6642	G	40000	6200	20221
SGR	NGC 6522	G	40000	6400	20874
PUP	RU 44	O	40	6600	21526
OPH	NGC 6401	G	40000	6800	22178
OPH	NGC 6355	G	40000	6800	22178
CRA	NGC 6541	G	100000	6900	22504
OPH	NGC 6333	G	100000	6900	22504
ARA	NGC 6362	G	100000	7100	23157
SCO	NGC 6453	G	20000	7100	23157
HER	NGC 6205	G	100000	7200	23483
OPH	NGC 6293 [5]	G	50000	7300	23809
2ND	**START DARK AREA**		0		23810
2ND	**END DARK AREA**		0		25439
SGR	NGC 6528 [6]	G	20000	7800	25440
SCT	NGC 6712	G	50000	7800	25440
HER	NGC 6341	G	100000	7800	25440
LUP	NGC 5927	G	100000	7800	25440
SGR	NGC 6712	G	50000	7800	25440
SGR	NGC 6638	G	40000	8000	26092
SCO	NGC 6144	G	70000	8100	26418
SGR	NGC 6569	G	40000	8200	26744
CAP	NGC 7099	G	100000	8200	26744
SCL	NGC 288	G	100000	8300	27070
SCO	NGC 6093	G	100000	8300	27070
SGR	NGC 6624	G	40000	8500	27723
SCO	NGC 6380	G	20000	8700	28375
OPH	NGC 6517	G	40000	8700	28375
SGR	NGC 6723	G	100000	8700	28375
SCO	NGC 6139	G	50000	8900	29027
CEN	NGC 5286	G	100000	8900	29027
OPH	NGC 6287	G	30000	9000	29354
MON	BIUR 8	O	70	9000	29354
SGE	PAL 10	G	30000	9000	29354
SCO	NGC 6496	G	50000	9000	29354

Constellation	Name	Type of Stars	Quantity	Dist./PC	Dist/LY
TUC	NEC 362	G	100000	9000	29354
SCO	LILLER 1	G	20000	9000	29354
CAR	NGC 2808	G	100000	9200	30006
SGR	NGC 6558	G	20000	9200	30006
PEG	NGC 7078	G	100000	9400	30658
LYR	NGC 6779	G	40000	9500	30984
SER	NGC 5904	G	10000	9600	31310
HYA	NGC 4590	G	100000	9600	31310
NOR	NGC 5946	G	50000	9600	31310
CVN	NGC 5272	G	100000	9900	32289
SGR	UKS1751-241	G	30000	10000	32615
OPH	NGC 6235	G	50000	10000	32615
OPH	NGC 6284	G	50000	10000	32615
SCO	GRINDLAY 1	G	30000	10000	32615
LUP	NGC 5986	G	100000	10200	33267
OPH	NGC 6402	G	100000	10200	33267
SCO	NGC 6441	G	50000	10300	33593
SGR	NGC 6637	G	50000	10300	33593
OPH	NGC 6273	G	100000	10600	34572
SGR	NGC 6681	G	50000	10800	35224
COL	NGC 1851	G	100000	10800	35224
AQL	PAL 11	G	20000	11000	35877
SER	NGC 6535 [7]	G	20000	11000	35877
3RD	**START DARK AREA**		0		35878
3RD	**END DARK AREA**		0		36854
AQR	NGC 7089 [8]	G	100000	11300	36855
OPH	NGC 6316	G	30000	12000	39138
ORI	BERK 21	O	40	12000	39138
LIB	NGC 5897	G	100000	12100	39464
PUP	NGC 2298	G	100000	12300	40116
APS	NGC 6101	G	100000	12300	40116
SER	IC 1276	G	50000	12900	42073
LEP	NGC 1904	G	100000	13300	43378
HOR	NGC 1261	G	75000	13400	43704
SCO	NGC 6388	G	50000	14500	47292
BOO	NGC 5466	G	100000	14500	47292
DEL	NGC 6934 [9]	G	40000	14700	47944
4TH	**START DARK AREA**		0		47945
4TH	**END DARK AREA**		0		48922
SGR	NGC 6652 [10]	G	20000	15000	48923
TEL	NGC 6584	G	50000	15000	48923
OPH	NGC 6342	G	20000	15000	48923
COM	NGC 5053	G	100000	15200	49575
SGR	NGC 6717	G	20000	16000	52184
OPH	NGC 6426	G	30000	16000	52184
OPH	NGC 6356	G	60000	17200	56098
COM	NGC 5024	G	50000	17200	56098
AQR	NGC 6981	G	40000	17300	56424
COM	NGC 4147	G	50000	17500	57076
SGR	NGC 6864 [11]	G	50000	18200	59359
5TH	**START DARK AREA**		0		59360
5TH	**END DARK AREA**		0		60663
APS	IC 4499 [12]	G	75000	18600	60664

Constellation	Name	Type of Stars	Quantity	Dist./PC	Dist/LY
CAP	PAL 12	G	20000	19000	61969
OPH	NGC 6325	G	30000	19400	63273
SER	PAL 5	G	70000	21400	69796
SGR	NGC 6715	G	100000	21500	70122
VIR	NGC 5634	G	150000	21600	70448
AQR	NGC 7492 [13]	G	50000	21900	71427
6TH	**START DARK AREA**		0		71428
6TH	**END DARK AREA**		0		77297
LUP	NGC 5824 [14]	G	50000	23700	77298
PEG	PAL 13	G	20000	24400	79581
SGR	PAL 8	G	40000	30800	100454
HER	NGC 6229	G	30000	31200	101759
HYA	NGC 5694	G	30000	32300	105346
DEL	NGC 7006	G	10000	34700	113174
AUR	PAL 2	G	25000	35000	114153
CEP	PAL 1	G	20000	46000	150029
OPH	NGC PAL 15	G	30000	70000	228305
HER	PAL 14	G	20000	75000	244613
LYN	NGC 2419	G	35000	93100	303646
UMA	PAL 4	G	25000	93300	304298
SEX	PAL 3	G	30000	96000	313104

LY is Light Years.

Footnotes

[1] Wil Tirion, *Sky Atlas 2000*, (Cambridge, MA, Sky Publishing Corporation, 1982), p. 284.

[2] *Ibid.* p. 280.

[3] *Ibid.* p. 274.

[4] *Ibid.* p. 294.

[5] *Ibid.* p. 294.

[6] Michael Rowan-Robinson, *The Cosmological Distance Ladder: Distance and Time in the Universe,* (New York: W. H. Freeman and Co, 1985), p. 321.

[7] Wil Tirion, *Sky Atlas 2000*, (Cambridge, MA, Sky Publishing Corporation, 1982), p. 294.

[8] *Ibid.* p. 294.

[9] *Ibid.* p. 294.

[10] *Ibid.* p. 294.

[11] *Ibid.* p. 294.

[12] *Ibid.* p. 293.

[13] *Ibid.* p. 294.

[14] *Ibid.* p. 293.

Appendix B: Gravitational Tests

Appendix B will prove that the "accepted theories" of gravity and time are incorrect. I will demonstrate that time and size can change dramatically, and this does not have to be as a result of proximity to a large dense mass. Appendix B is a summary of previously published research material, found in my 1996, book *Gravitational Mystery Spots of the United States.* The following includes on-site experiments and research I performed to test my *Theory of Multidimensional Reality.*

The Test Area

I investigated four sites during the week of June 1 to June 9, 1996. They include the Oregon Vortex® in Gold Hill, Oregon; Magnetic Hill, east of Rohnert Park, California; Santa Cruz Mystery Spot in Santa Cruz, California; and Confusion Hill in Piercy, California. For purposes of this book I will only briefly cover the results from the Oregon Vortex® in Gold Hill, Oregon, and the Santa Cruz Mystery Spot in Santa Cruz, California. My assistant at the Santa Cruz Mystery Spot was my friend Vic Ardelean. I was at each location for one to two days.

Traditional Explanation of Gravity

In 1916, Albert Einstein developed his General Theory of Relativity which incorporated his theories on gravitation. He assumed that gravity is a physical effect produced by space-time curvature in four-dimensions, with the fourth dimension being time. Where Newton had four index tensors, or vectors, to describe the curvature of space-time, Einstein used ten tensors for space-time geometry, developed by Riemannian.

Both approaches conclude that gravity is due to mass. Height or size of the object is directly proportional to the gravitational field the object occupies. With the General Theory of Relativity, time slows down and objects become smaller when they are in proximity to a strong gravitational field. The shrink-and-grow phenomenon, found at all three gravitational vortices, have been explained away by saying that there is some large meteor or metal mass buried in the ground. All three tourist attractions have reported greater than a 7% size change in their vortices. Such a size change would represent a mass of such great density or size, that it would be implausible for such a massive body to exist in the geological areas where these attractions are located. I will describe, using the results from my experiments, how the traditional theories of existence do not explain what gravity truly is, and what is happening in these unusual gravitational vortices.

Time Shifts

The General Theory of Relativity predicts that time slows down as an object approaches a large gravitational mass. Time would slow down on clocks placed near a black hole. The clock itself would also become much smaller. This is similar to the Special Theory of Relativity, which holds that time slows down and objects get smaller as they approach the speed of light.

Shrink-and-Grow Phenomenon

The one phenomenon that defines these special gravitational anomalies is the observation that objects including people, become smaller as they get closer to the center. This effect is reminiscent of the General Theory of Relativity. As objects get closer to a large mass, its gravitational field warps time and the size of objects in its vicinity. It is because of the *General Theory of Relativity* that scientists have assumed there is a meteor, or other massive metal object, buried under these tourist sites. The problem with this assumption is that the Oregon and Santa Cruz sites are located in areas of sediment, probably no older then 36,000 years, and all of the three locations present no evidence of iron or other heavy metals in the regions. Moreover, a meteor or other massive metal object, would have to be huge and very dense to create the size and time changes observed in all the sites—and this is just not the case, because the anomalies themselves are no greater then 165 feet in diameter. There is no evidence of meteor impact creators in the areas.

Experimental Procedures

My experimental approach was to test for a time-shift associated with the size changes reported. That would rule out any possibility of optical illusions or trickery in the "shrink-and-grow" phenomenon. To measure any time shifts, I used a 25 MHz freely-oscillating crystal, powered by a 9-volt battery. I used only Eveready® Energizers®, because they started with 9.4 volts and their discharge rate was consistent. The crystal frequency output averaged 24,997,980 over eight hours, plus or minus 6 hertz. Graph B-1 plots the frequency output, in normal time and space, over eight hours. Graph B-2 shows a more detailed graph covering the first hour and ten minutes. The chart only displays the last four numbers of the frequency. The frequency output was fairly stable for such a simple circuit. It was a simple circuit with no voltage-regulating or frequency-compensating circuitry. As long as the voltage was above six volts, the crystal operated within the frequency range. I tested and plotted the crystal oscillator in the Seattle area, which I considered "normal" space and time. The 35-MHz frequency counter displayed single hertz (cycles). This was important, because I wanted to measure the slightest change in time. The crystal was connected to the frequency counter by a 113-foot, 58-foot coax, or a 200-foot coax. I was also able to connect the two cables together if I needed extra length. The frequency counter was powered by a 12-volt 31-amp gel cell battery, or 110-power line voltage. The frequency counter was tested with both power sources, and there was virtually no difference in output readings or performance.

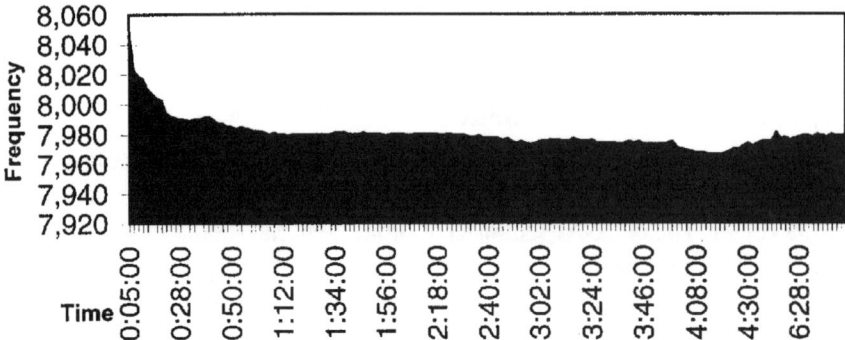

Graph B-1: Only the last 4 digits of the frequency output are displayed. The table shows the results of the 24,997,980 MHz crystal in normal time and space, over eight hours.

The procedure was the same for all of the locations. First, a new 9-volt battery was loaded into the crystal source. The frequency counter was turned on and both were left for over 30 minutes, so they would both reach ambient temperature, and the frequency would stabilize. Next, the crystal was brought into the vortex to see where the lowest or highest frequency was indicated, so the center could be located. I received some strange frequency readings from all three locations, when I first brought the crystal into the anomaly. What happened was the frequency counter recorded wide swings in output for about five minutes. I called this observation "the jell-o effect." At first I thought the crystal was stabilizing but it was not that.

Five years previous, I did some time experiments with the same crystal and frequency counter at the Oregon Vortex®. My observation then was that the

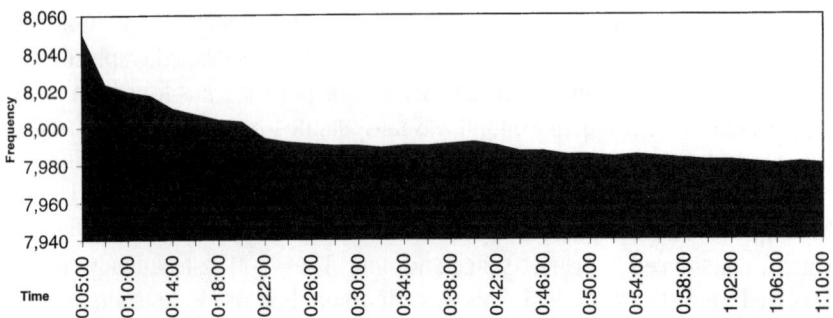

Graph B-2: The frequency output, in normal time and space, over 90 minutes. Only the last 4 digits of the frequency output are displayed.

frequency shifted up. This time, I observed frequencies shifting up and down. I also discovered that the number of people entering these anomalies also affected the frequency readings.

Proof of the Shrink-and-Grow Phenomenon

The problem photographing the "shrink-and-grow" phenomenon was making sure the camera and tripod were set up equidistant between both subjects, and perpendicular to the board they stood on. This was accomplished with two guide wires cut to the same length and connect to the two ends of the board and the camera location. The center of the anomaly had to be located, so the subjects would be in line with the center. There were problems at the Mystery Spot because the area next to the two-room cabin was not that level, and it was not a full 15′ from our board. Figure B-2 shows the six-foot measuring polls that were at the opposite ends of the 7.5-foot board. The picture has been spliced together to easily show the height distortion over just 7.5-feet. Figure B-1 shows a photograph of me years earlier at the Oregon Vortex®. The level platform was about 7-feet long and the center of the Vortex was to the left of the picture. You can see from the spliced photo-graph that I appear at least three inches shorter closer to the center than away from it.

Figure B-1: A spliced photo of the author at the Oregon Vortex® showing the shrink-and-grow effect.

The Oregon Vortex® Experiments on June 2, 1996

The time was 11:32 A.M. June 2, 1996 when a new battery was installed. I had the frequency counter turned on for about ten minutes before it was connected to the crystal, and both were outside of the vortex.

The crystal was placed in the approximate center of the vortex at 1:17 P.M. and the frequency counter recorded a frequency of 24,997,934 hertz ±4 hertz. Graph

B-3 displays only the last 4 digits (7,934 Hz), because the changes observed were always less then 1,000 hertz, except for the start-up period, when I observed the jell-o effect.

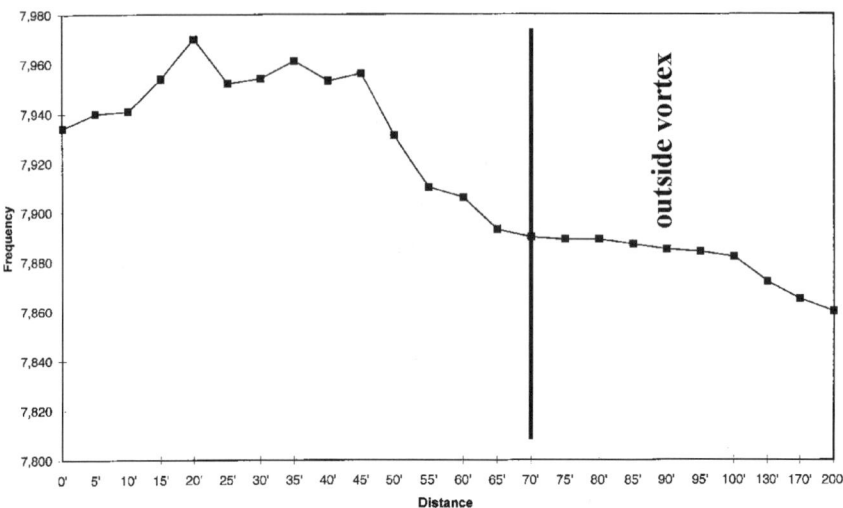

Graph B-3: Run number one, measuring frequency from the center (approx.) to 200' from the center. Start time: 1:17 P.M. Duration of the experiment: 47 minutes.

Frequency was higher near the center of the anomaly than outside of it. There was a 110-hertz difference between inside and out. At the 65-foot mark, it appeared that the frequency stabilized. This is the edge of the vortex but not the edge of all of its effect. There may be a corona, a certain distance around the anomaly. In addition, there are dynamic changes that go on inside these vortices.

The second run (Graph B-4), was measured from the frequency counter located at 170-feet from the center of the vortex. The frequency difference was 40 hertz. The reasons for the lower-frequency readings may be due to the overall gravity change within the vortex, which periodically occurs during the day.

A new battery was installed into the crystal oscillator at 2:42 P.M. After 22 minutes, the third run (Graph B-5) was begun. The frequency difference between inside and outside was 50 hertz. All three runs recorded frequencies higher in the center than outside of it.

Experiments of June 9, 1996

Dynamic Change

After testing Confusion Hill and the Mystery Spot the question was: "Were these gravitational anomalies dynamically changing over time?" At both of those

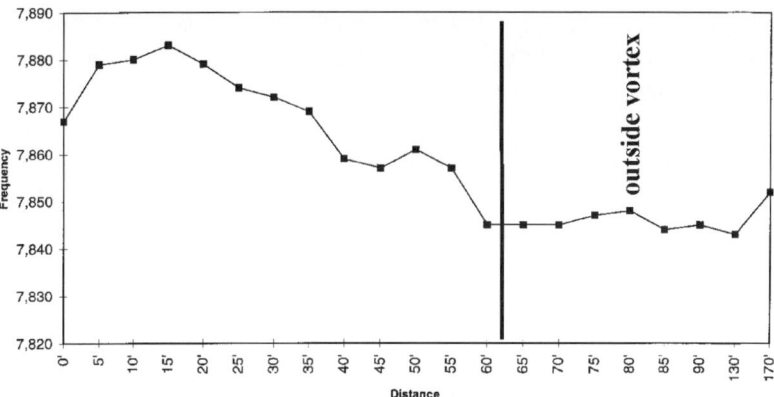

Graph B-4: Run number two: measuring frequency from the frequency counter to the center of the vortex. Start time: 2:04 PM. Duration of the experiment: 33 minutes.

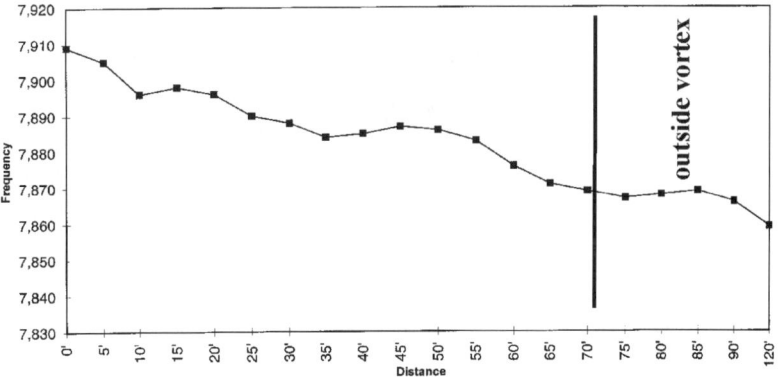

Graph B-5: Run three: Measuring frequencies from the 115-foot mark into the center of the vortex, using 313 feet of cable. Start time 3:04 PM. Duration of the experiment: 22 minutes.

locations I created a new test by leaving the crystal at one location for a number of hours, and recorded the frequency fluctuation over time.

On this visit I only wanted to check one thing: "What was the frequency change over time, if the crystal was left at one location?" The experiment started at 12:12 P.M., June 9, 1996. The crystal was placed at the 15-foot mark and frequency was recorded every two minutes. Only the 113-foot cable was used and the frequency counter was 125 feet away from the center. It was left there for 93 minutes. For test purposes, I also placed the crystal outside the vortex, 240 feet away from the center to see if the frequency counter properly recorded the frequencies. The crystal was placed outside the vortex for 10 minutes. The frequency stabilized at 24,998,073 ±2 hertz. After the crystal was put back at the

15-foot mark, it resumed the same frequency pattern it recorded before. It remained inside the vortex for 48 minutes. While the crystal was located at the 15-foot mark, there was a variation in frequency output. Over the two-minute interval, the frequency variation ranged beyond what I was comfortable with. That was one of the reasons I placed the crystal at the 240-foot mark, to see if it was the equipment or was it recording something real. At this point, I concluded that it was recording something real. For some time the frequency variation was

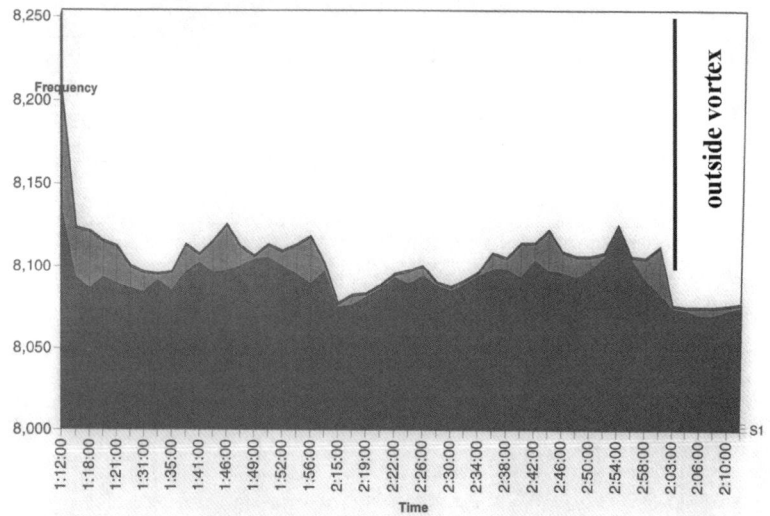

over 20 hertz. Graph B-6 shows the frequency variations recorded every two minutes over 50 minutes inside the vortex and 10 minutes outside.

Graph B-6: Measuring frequency from the 15-foot mark from the center of the vortex, using 113-feet of cable. At 2:02 elapsed time the crystal was moved 240-feet away from the center. Start time: 3:36 P.M. Duration of the experiment: 62 minutes.

During the middle of this experiment (12:40 PM), I decided to record the frequency output-per-second for one minute, to observe if a pattern would emerge. Graph B-7 shows a graph of the output frequencies for that minute. It looks as if a pattern was present.

Shrink-and-Grow Phenomenon

The shrink-and-grow phenomenon is the main attraction of these tourist sites. Usually two people will stand on opposite sides of a level plank or concrete slab. When they change positions, the observers, perpendicular to the subjects, see the change in their heights. The problem is that the perceived change can be

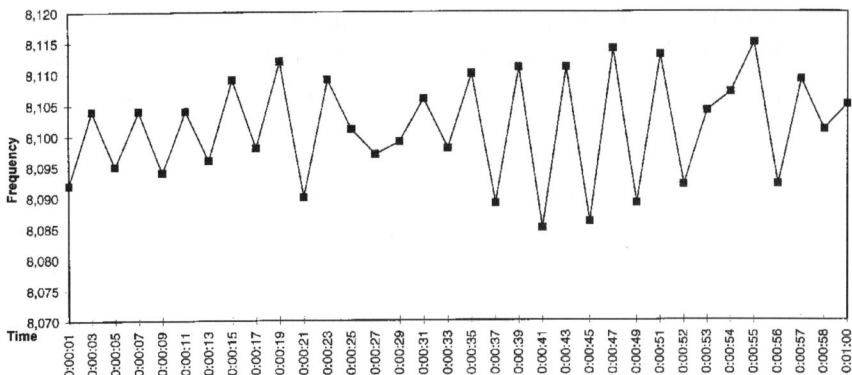

Graph B-7: Measuring frequency from the 15-foot mark for one minute. Start time: 12:40 P.M. Duration of the experiment: 1 minute.

the product of an optical illusion, because of such devices as a crooked building along with a sloping fence as a backdrop. The Oregon Vortex® has three platforms where people can see this phenomenon. I set up my camera and tripod at the cabin courtyard, which was level. For this experiment, I brought along two 7-foot measuring poles. I took numerous photographs and measured a 5.4% to 5.8% reduction in height for the pole closer to the center of the vortex.

I had been at the Oregon Vortex® previously, in June of 1991. I took a series of pictures of one of my seven-foot measuring poles over three hours. My camera was on a tripod in a stationary location 40 feet away. Years later, I decided to measure the height of the pole over time. I noticed that the height changed over time. That would explain why the frequency was changing over time. Something dynamic was occurring in these vortices.

The Santa Cruz Mystery Spot Experiments
The Start of the Experiment

We arrived at the attraction at 12:45 P.M. on June 7, 1996. A new 9-volt battery was placed into the crystal frequency source. It was 1:53 PM when the frequency counter, the 113-foot cable, and crystal were set up. The frequency counter was turned on for one hour before it was connected to the crystal via the cable. The frequency counter was placed behind the gift shop on a concrete path, and 20 feet outside the vortex. We spent about 53 minutes probing the vortex, looking for the center. I again noticed the same fluctuation in the frequency counter when the crystal was first introduced into the vortex. For the first three minutes the counter read a frequency swing of over 500-hertz. After a short while, it went from 24,998,042 to 24,998,146, over three minutes. After about 15 minutes of that we started probing the anomaly to locate the center. Over a distance of 113 feet from the counter the frequency went from 24,998,042 to 24,998,092 closer to the center (50-hertz difference). At this time I began to think that the "jell-O effect" was real. I began to develop a theory of why this was happening.

Time Fluctuation Results

After I determined that the center was on the other side of the crooked shack, we moved the equipment up the hill through some thick underbrush. We then went far enough up the hill, about 100 feet away. A new 9-volt battery was placed into the crystal source box. The frequency counter was placed 110 feet

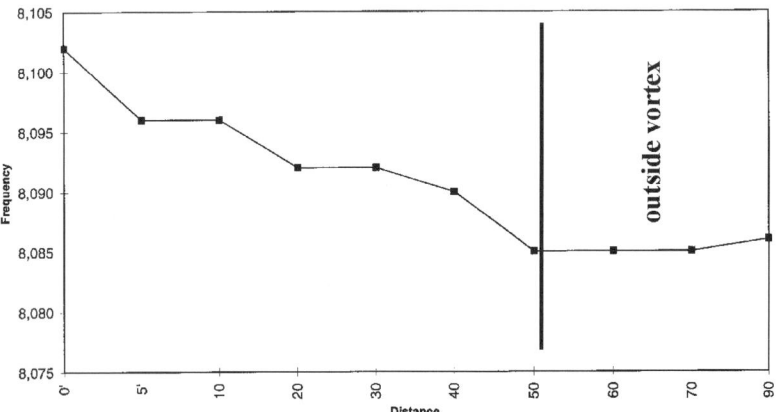

away from the center. Only the 113-foot coax cable was used. Graph B-8 displays the results.

Graph B-8: Run number 1: Measuring frequency from the 90-foot mark into the center of the vortex, using 113 feet of cable. Start time: 3:31 P.M. Duration of the experiment: 7 minutes.

The first test run indicated that the frequency inside the anomaly was higher than outside of it. The total adjusted frequency difference was 16 hertz. We located the approximate center near the base of a twisted eucalyptus tree. The center was about 10 feet west of the south end of the shack.

For the third run, we moved the equipment to the front of the attraction. This was about 20 feet from the edge of the vortex, where we were before. This was done because the working conditions on the backside were too difficult to deal with. We set it up again by 4:44 P.M. near the gift shop. I connected the two cables together, so I had 313 feet of coax. During run three (Graph B-9), the frequency went down towards the center, as it did during run 2. The direct distance to the center from the frequency counter was more like 150 feet, but we could not go straight, because there were fences and buildings in our way. The distance listed here is more like the amount of cable stretched out. We did have the 100-foot tape measure stretched out, but that went just so far. We had to mark off every ten feet with stakes to get the distances. The 170-foot mark was by the back fence near the center of the vortex, and the 180-foot mark was inside the shack. The total frequency difference was 18 hertz.

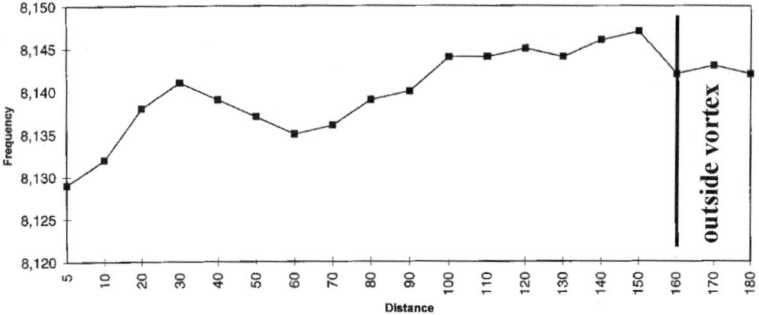

Graph B-9: Run number 3: Measuring frequency from the frequency counter to the center. Start time: 4:44 P.M. Duration of the experiment: 26 minutes. The outside of the vortex is at the 5-foot mark and the 170-foot mark is approximately where the vortex center is located.

Dynamic Change

The last frequency/time experiment was to leave the crystal 40 feet away from the center and record the frequency changes over a long period of time. The times were recorded randomly because we were doing other experiments at the same time. Graph B-10 shows the frequency changes over seven hours and thirteen minutes. At the 3:29 hour mark, I moved the frequency counter 50 feet further outside but the crystal remained. The employees let me plug the frequency

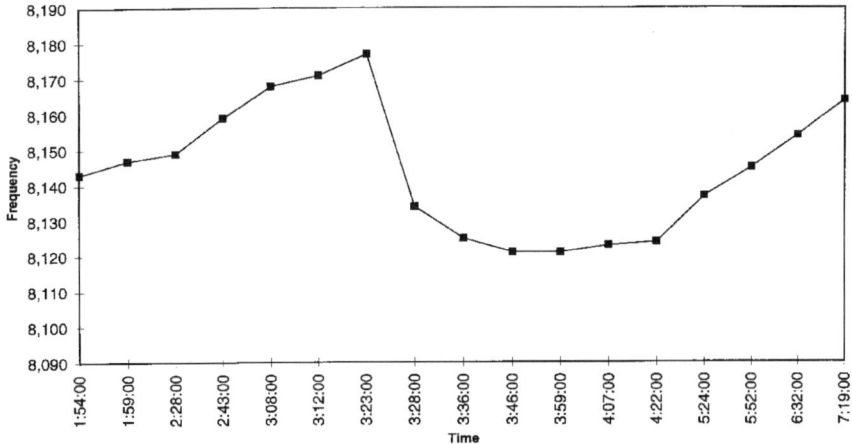

Graph B-10: Measuring frequency from the 40-foot mark from the center of the vortex using 313 foot of cable. The experiment started at: 5:14 PM. Duration of the experiment: 5 hours and 25 minutes.

counter into their power line. That may be why the frequency dropped down. It was not due to the frequency counter not being at its ambient temperature, but rather it may have been influenced by some time distortion caused by a coronal affect other researchers have mentioned. The frequency difference between high and low was 56 hertz. This experiment definitely showed there are frequency changes occurring over long periods of time.

The next morning at 8:03 AM, I placed a new 9-volt battery into the crystal frequency source and started reading the results for the next 69 minutes. Graph B-11 shows the frequency first rising over 35 minutes, then decreasing for the

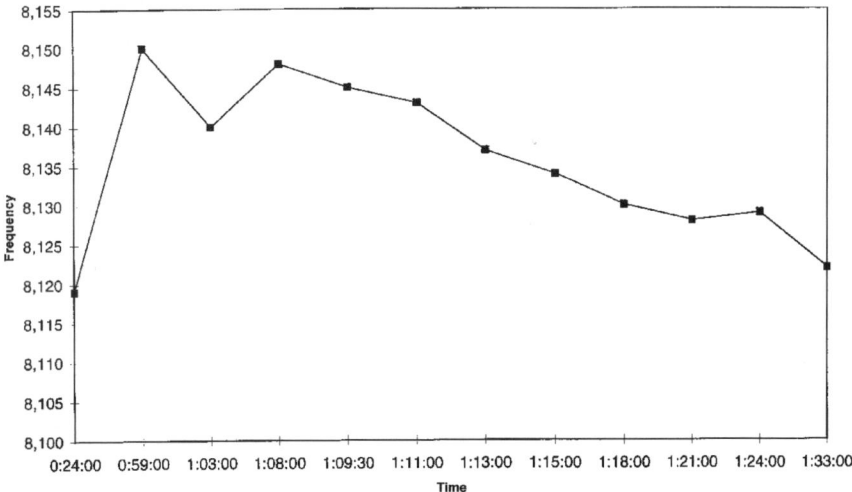

Graph B-11: Measuring frequency from the 40-foot mark from the center of the vortex using 313 feet of cable. The experiment started at: 8:03 AM. Duration of the experiment: 69 minutes.

rest of the time. This pattern shows that the frequencies are dynamically changing during the day, sometimes going down, other times reversing.

Size Changes in Two Poles.

The main attraction of the site is the "shrink-and-grow" phenomenon. The area of greatest height change occurred by the courtyard to the south of the shack. Figure B-1 shows a 5.24% shrink and grow affect over only 7.5 feet. To ensure that the tripod was placed perpendicular to the center of the board, I brought with me two wires of identical length that connected to the ends of the 7.5-foot board. Its apex was exactly center to the board. (I did check that the platform was level.) I estimated that the center of the vortex was about 15 feet away from the camera.

Conclusions

Time and Size Changes

The most obvious attractions at these gravitational anomalies are the size changes. The time changes are a new discovery, due to my experiments. The two phenomena are related, but not in the same ratio that the Lorenz transformations call for. With size changes of 5.8%, we should have seen time being affected much more than we did. If these vortices were caused by "natural" formations, then there should not have been any time or size fluctuations. Another interpretation could be that the actual time-size relationship using traditional gravitation equations do not work. In Chapter 3, time was explained as being the rate at which information is propagated from the Diehold into our existence. The speed of the information through the "CPU" is constant, therefore time and size stay constant. But the experiments found that time dynamically shifted up and down. The *General Theory of Relativity* says it should only slow down. Why is it shifting up? We know that the size percentage changes during the day, and from day-to-day. The only condition that can explain these observations is that something is buried below these vortices which has the ability to dynamically change the rate (time) at which information is modulated into our existence, and also independently control information density (size).

The Jell-O Effect

This was a real puzzling phenomenon inside these vortices. What was happening was that the frequency counter recorded wide swings in frequencies when the crystal was first introduced into the vortex. The phenomenon happened at all three locations. After some time, the frequency counter began to record more stable frequencies. I am sure about what was going on after my third trip to the Oregon Vortex®.

On the third trip, June 9, I was recording the frequency at one location for a long period of time. I had a clear view

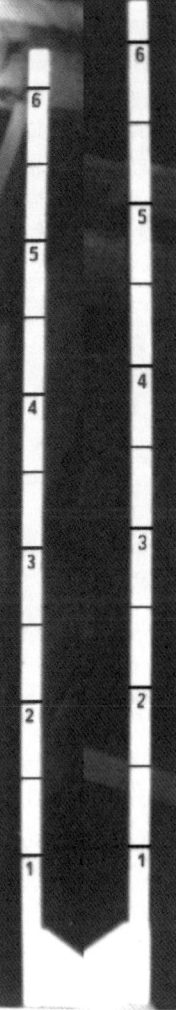

Figure B-2: Shows just the measuring poles that were only 7.5 feet apart attached to a level wooden plank.

of the whole vortex, and who was in it and who was not. Sometimes I noticed swings of 15 hertz over a short period of time. Sometimes the frequency was very stable. I correlated that the frequency was most stable when no one was inside the vortex. Once I watched as five young people walked into the vortex. Suddenly the frequency counter responded with a frequency fluctuation of 10+ hertz. Whatever was causing the gravitational and time anomalies was also dynamically responding to people walking into the vortex.

Conclusion

Are these gravitational anomalies caused by a large chunk of metal, such as a meteor, buried in the earth below? Absolutely not! The area of the vortex is too well defined in a small area, and the size of objects, as well as time, is changed over time (see videotape analogy). For a normal gravitational body to do this, it would mean that its mass would have to be changing dynamically, and that is impossible. If it was caused by such a huge mass of iron or other dense metal, then we would have observed a huge meteor crater like the one in Arizona, and we do not. Finally, the gravitational field seems to be interacting with people and frequency generating devices that enter it. Size and time seem to be changing independently. So, if "natural" forces do not cause this effect then what could cause it?

Remember my explanation of what gravity is—"concentrations of information directed to a point in time and space." What I believe causes these vortices is a computer-like device built by a highly advanced previous civilization that lived on this planet a long time ago. Perhaps it was the "Golden race of beings" mentioned in many mythologies. What is interesting is that it seems that these devices are still "working" and have the capacity to sense human presence by adjusting their electrostatic fields. Maybe if one is dug up we will know for sure. Could it be a working pyramid?

Complete Experimental results are found in the book:

Gravitational Mystery Spots of the United States

by Douglas Vogt

© 1996; ISBN 0-930808-04-5 (Paperback)
QB337.5.U5V64   1996; 96-39110 CIP

Published By:
Vector Associates
PO Box 40135; Bellevue, WA 98015
Cost $10.95 + $4.00 for domestic mailing.

Appendix C: Moses' Code System

Basic Background

The Torah is written like a woven garment. One part relates to another just as clues for one subject relate to another. Clues are scattered all through the Torah and the Prophets' portions. The Torah is the world's greatest codebook! But why have a code system unless there are subjects and names concealed within? Moses had a lot to conceal, so he created numerous code systems so later Hebrew leaders would discover the truth about their past and the true purpose of the Torah.

The subjects he concealed include:

1. The actual number of years between cataclysms (the 12,068-year intervals).
2. The nature of the people who hid the high technology in the family burial cave.
3. The location of Mount Sinai and the family cave.
4. The secret identity of Joseph in Egypt, and what he did there.
5. Why eleven of the twelve Hebrew tribes went into slavery.
6. The true ancestry of his family.
7. What the secret of redemption truly is.

As you can see, Moses had excellent reasons to develop code systems to conceal things without being too obvious. I have already revealed the secret number, 12,068, in Chapter 2. The other code systems involve the Hebrew letters.

It is important to know that there are at least six code systems used in writing the Torah and later books of the Hebrew Scriptures. While researching the Torah, I realized there was a tremendous amount of coded material and secrets yet to be uncovered. I have only found a small portion of the coded material. I kept my focus on the sections that helped prove my thesis so I did not try to uncover it all. One could spend a lifetime decoding the entire Hebrew Scriptures, but that was unnecessary to prove what I needed to in this book.

The Biblical Code Systems

Large Numbering and Small Numbering

Moses assigned number values to each Hebrew letter. Table D-1 shows the two numbering systems he used. The first is called *large numbering*, which is simply assigning a number value to each successive letter in the Hebrew alphabet. The second system was a little more difficult to discover. It is called *small numbering,* and it merely eliminates the zeros from large numbering. I state in Appendix D that Baruch created the design of the five final Hebrew letters. These final letter designs did not exist at the time of Moses, but I have found evidence that Moses did use the small numbers of the five final letters as codes. This may be where Baruch got the idea of redesigning those five letters. I sometimes do not use the final forms of the Hebrew letters כ נ מ פ צ (Ch, N, M, F/P, Ts), because they did not exist at the time of Moses. I do, however, use the Hebrew large-numbering values for those final letters, because Moses used them.

I believe Baruch had to create the five final letters because he knew, in the future, there would no longer be a High Priest to hand down this body of knowledge to his son. So Baruch made it easier for future generations to read the Torah and know where the word breaks were.

	Letter	Large Numbering	Small Numbering
1 alpha-silant	א	1	1
2 bet-b	ב	2	2
3 gimel-g	ג	3	3
4 daleth-d	ד	4	4
5 he-h	ה	5	5
6 vav-v, w	ו	6	6
7 zayin-z	ז	7	7
8 chet-ch	ח	8	8
9 tet-ta	ט	9	9
10 yod-y	י	10	1
11 caph-k	כ	20	2
12 lamed-l	ל	30	3
13 mem-m	מ	40	4
14 nun-n	נ	50	5
15 samek-s	ס	60	6
16 ayin-silant	ע	70	7
17 pe-p	פ	80	8
18 sadhe-ts	צ	90	9
19 koph-k	ק	100	1
20 resh-r	ר	200	2
21 shin-sh	ש	300	3
22 tav-y	ת	400	4
Final cH ח	ך	500	5
Final M מ	ם	600	6
Final N נ	ן	700	7

Final F/P פ	ף	800	8
Final TS צ	ץ	900	9

Table C-1: Table of the Hebrew numbering system developed by Moses.

Algebraic Principle

Moses and the later writers of the Hebrew Scriptures use a simple algebraic principle for one of the main code systems used. It is the *algebraic principle*, which states that if A = B and B = C, then A = C. Moses uses this technique all the time.

Composite Word Code

Another very commonly used code system I call *the composite word code* system. This is when the author creates a long word composed from two or more smaller words—but the reader must look at the smaller words in order to uncover the real meaning intended by the author. Moses hinted at the use of this system when he wrote the story of Joseph and his two sons. Joseph named the older son Manasseh (מנשה), "for God hath made me *forget* all my toil, and all my father's house."[1] The word נשה means "to forget." The younger son was called Ephraim (אפרימ), "for God hath made me *fruitful* in the land of my affliction."[2] The word פרי means "fruitful." Another Example is found in Genesis 41:43, when Joseph was made Prime Minister and the Pharaoh paraded him in a chariot before the people. The word the crowd shouted at him was *abrech*. The Hebrew word is אברכ, which is made up of two smaller words. The first is אב (ab), which means *lord, master, counselor, adviser*. The second word רכ, (rech) means *chariot* or *wagon*. In other words, the word echoes the description in the story line. It is important to note that Moses told us, with these three examples, that he was using a code system, whereby a name is composed of smaller words, which mean something important to the real meaning of the sentence. The meaning of the composite word might be very different from the surface story.

Letter Swapping

The next code system was the practice of swapping adjacent letters. Moses and the later priests would swap a letter for a sequential letter, such as an N נ for an M מ or an M מ, for an L ל. The following is a very important example of the *composite word code system* and *letter swapping*. The following example had been figured out by very few priests, and almost none over the past 1,000 years, except myself of course.

The most familiar scriptural passages which demonstrate these code systems are found in Genesis Chapter 37, verses 3, 23 and 33. All three verses have the verse *coat of many colors* or *long-sleeved coat*. The Hebrew words translated as "coat of many colors" are כתנת פסים as they appear in verse 3 and הפסים כתנה in verses 23 and 33. The word כתנה (ketanut) means shirt or tunic. But if you substitute the n (נ) for the adjacent m (מ or final ם), you get the word כתם (ketem), which means *fine gold*. The next word, פסים, had been the difficult word for translators. When you drop the Yud (י) and change the M (מ) to an L(ל), you get the word פסל (petel), which means a *carved or cast idol*. When you put the two words together, they translate to a *cast golden idol*. This is the real reason why Joseph's brothers hated him. His father had given him an important family heirloom. Rachel had stolen the golden idol (a golden calf) from her father's house when they left Haran. Since Laban was the son of Nahor, who was the brother of Abraham, we are dealing with the same family. The golden idol was most likely the original property of Abraham's father, Terah, and was handed down to Laban. This I believe is the real reason why his brothers hated Joseph. They may have felt that if Joseph had received such an important family treasure, then maybe he would also receive a much larger share of Jacob's inheritance. Therefore, out of jealousy and economic reasons, the brothers found it expedient to eliminate Joseph. I believe it is also an excellent possibility that the idol was a golden calf, and that is perhaps why the Hebrews kept going back to the worship of *Baal*, in Egypt known as Hathor, as represented by a golden calf. The golden calf idol was the symbol of the Canaanite fertility god of the Earth. Since Genesis 29:31 reveals that Rachel was barren, the idol most likely was that of Hathor/Baal. The Egyptian fertility god was the cow god, *Hathor.* To Moses the pagan god Hathor equaled Baal. It is possible that if Rachel had not stolen the idol the Hebrews would not have periodically slid back to pagan worship, and could have avoided future problems. Remember what I wrote in Chapter 11, that actions in the past have tremendous effects in the future. This is a prime example.

The Numbered-Verse Codes

I conclude this subject with the *numbered-verse code* system used here. The Chapter plus verses where "coat of many colors" appears total 40, 60 and 70 and should be synonymous with Joseph. That is why the number 40 is used 42 times in the Torah and 77 times in the rest of the Hebrew Scriptures. The number 60 is only used in the Torah 5 times but the number 70 is used 23 times in the Torah, and 38 times in the rest of the Hebrew Scriptures. The High Priests who wrote these books were giving clues that they also knew the great secret. The numbers 40, 60, and 70 can be seen as being synonymous with Joseph—the owner of the golden calf—what he did and who he was in Egypt.

Appendix D: The Prophet and Priest Baruch

Thanks to Baruch

We have much to thank Baruch for, because he redesigned the Hebrew alphabet closer to the original design. He rewrote the entire Torah with word breaks, using the newly designed alphabet, including the addition of the five final letters. He did that so it would be easier for everyone to read the Torah without memorizing where the word breaks were. He changed the Jewish religion from an Ark-centered, sacrifice-oriented religion to a Torah-study religion. He safely hid the Ark away, after the fall of Jerusalem, following God's instructions. God also told him that the Ark would remain hidden until the End of Days, when God would reveal its location again. This Appendix describes Baruch's ancestry and his many accomplishments.

Who was Baruch Related to?

Baruch was the grandson of Jeremiah aka Mahseiah aka Shaphan. Baruch's father was Neriah. I estimate that Jeremiah was at least 75-80 years old at the time of the fall of Jerusalem and, therefore, too old to have written Ezekiel, Joel, Obadiah, or the Psalms. Baruch most likely wrote the book Obadiah. I doubt Jeremiah lived much more then a few years after the fall of Jerusalem and the destruction of the Temple. The last few years of his life must have seen him complete most of the book Jeremiah, and his portions of II Kings and II Chronicles.

Neriah and his son Baruch

We do not know how much younger Neriah was than his oldest brother, Gemeriah. That is important to determine how old Baruch was when he began to show up in Jeremiah's writings on or about 605 B.C.E. I estimate he was about 16-to-18 years old when he accompanied his grandfather's brother, Gemeriah, and his uncle, Elasah, to Babylon, on orders from king Zedekiah.[1] What is interesting is that his name does not appear in the first account of the trip, but does show up in a much later account of the same trip.[2] In fact, the second account left out the name of Elasah, and inserted Seraiah who was Baruch himself. The second account states: "The word which Jeremiah the prophet commanded Seraiah the son of Neriah, the son of Mahseiah, when he went with Zedekiah the king of Judah to Babylon in the fourth years of his reign." It made it sound as though Baruch was then Jeremiah's personal representative, which was very unlikely because he was only in his late teens at the time. The responsibility probably fell on Elasah, Jeremiah's oldest son. The second account of the journey to Babylon tended to indicate that it was Baruch who finished the writing of the Book of Jeremiah, and that is why his name, Baruch, appears no less then 23 times.

Baruch, aka Seraiah, the prophets Ezekiel, Joel and Obadiah

I cannot emphasize strongly enough the importance of what Baruch did for the Jewish religion, and his importance in preserving the religion over the past 2,600 years. I can say with absolute certainty that he was the last prophet because he was the person who hid the Ark safely back into the family cave after the fall of Jerusalem. He was the scribe for Jeremiah. He sat in front of the Ark when Jeremiah was in prison. He wrote as the prophet Ezekiel, which describes, in code, how and where he hid the Ark. The proof that he wrote as the prophet Ezekiel is found in Chapter 1:1-2, where he placed himself in Babylon at the exact time as the trip Seraiah, Ahikam and Gemeriah took to Babylon.

I am sure that Baruch was the writer of the prophecies at the end of the Kingdom of Judah, and also some of the prophecies associated with the Babylonian period, because the only two priests God would speak to, and allow in front of the Ark, were Jeremiah and Baruch. The other priests, and even the king, had to ask one or both of them to consult with God. But at the end Jeremiah was in prison so the only one left was Baruch.

Baruch is most likely the writer of Joel because it contains prophecy about the End of Days, the cataclysm, as well as what happened to the Hebrew population after the fall of the first Temple.[3] There are many parallel verses between Ezekiel and Joel. Both books speak of God "talking" from Zion, a fountain of water flowing out of Zion watering the valley of Shittim,[4] and a great war of many nations against Israel in the future. More importantly, in Joel Chapter 1:13, Baruch wrote that there was no Ark to perform offerings to. That meant that the Book of Joel was written after Baruch had hidden the Ark.

There is also a high probability he wrote the Book of Obadiah, because again that book mentions events after the Temple was destroyed and it also describes the End of Days and Mount Zion. Additionally, Obadiah means "servant" and that name may be an allusion to the fact that Baruch was the servant to Jeremiah.

Baruch also wrote some chapters of Psalms. I conclude this because it contains many prophesies which state that God spoke to the writer. I also found a number of legends which describe what Baruch did and wrote:

> At first his [Baruch's] mourning over the misfortunes of Jerusalem and the people knew no bounds. But he was in a measure consoled at the end of a seven days' fast, when God made known to him that the day of reckoning would come for the heathen, too. Other Divine visions were vouchsafed him. The whole future of mankind was unrolled before his eyes, especially the history of Israel, and he learned that the coming of the Messiah would put an end to all sorrow and misery, and usher in the reign of peace and joy among men. As for him, he would be removed from the earth, he was told, but not through death, and only in order to be kept safe against the coming of the end of all time.

Thus consoled, Baruch addressed an admonition to the people left in Palestine, and wrote two letters of the same tenor to the exiles, one to the nine
tribes and a half, the other to the two tribes and a half. The letter to the nine
tribes and a half of the captivity was carried to them by an eagle.

Five years after the great catastrophe, he *composed a book* in Babylonia,
which contained penitential prayers and hymns of consolation, exhorting Israel and urging the people to return to God and His law. This book Baruch
read to king Jeconiah and the whole people on a day of prayer and penitence.
[Emphasis added.]

The book that best fits this description, containing "penitential prayers and
hymns of consolation," would be Psalms. There are some chapters in Psalms,
which I know he did not write. These were written in the First and Second Centuries of the Common Era—and I will describe them in another book.

The Book of Ezekiel

In my opinion, the Book of Ezekiel is the most important prophetic book in
the Hebrew Scriptures, aside from Deuteronomy. It was the last book, written by
the last real prophet who sat in front of the Holy Ark. It was the last time God
revealed what his long-term plans were for Israel as a country as well as for the
descendants of Jacob. The last chapters of Ezekiel are a mirror image of what is
in the Garden of Eden story in Genesis.

The genealogy he gives us in Chapter 1:3 is made up. He wrote that he was
the son of Buzi spelled בוזי, in Hebrew. There was no Buzi. He created the name
by reaching far back into Jewish history. He combined the names of two of his
ancestors, namely Bukki (בקי) and his son Uzzi (עֻזִּי). The *Bu* came from Buzi,
and the *zi* came from Uzzi.

The Book of Joel

The prophet Joel was another name Baruch wrote under to present
prophecies which God spoke to him after the fall of Jerusalem and when he was
in the cave with the Ark. The name Joel consists of two words: The first, יום
(yom), which means *day,* but Baruch dropped the m. The second word is אל
(El), which is another word for God. Together they mean *God's day,* or *the day
of God.* This meaning is reinforced in verse 1:15 when it says "For the day of the
Lord is at hand," and that is the subject of the book. The book starts out by
describing the season when "God's day of judgment" will occur, and we can
conclude it will be in the fall. Chapters two and three describes some of the
events which will occur. The last chapter describes a great battle, similar to the
one described in Ezekiel, Chapters 38-39. It sounds as though God is baiting the
nations surrounding Israel to attack, so He can pass judgment on them, all at
one time. The end of the chapter again describes what the sun and the heavens

will look like during God's Day of Judgment. It also mentions the new Mount Zion, and water coming forth from it, just as it is in the Book of Ezekiel.

Baruch's other writings

Baruch also wrote many chapters in The Psalms, as I mentioned earlier, as well as completing the last chapters of Jeremiah, II Kings and II Chronicles. In my opinion, Baruch, aka Seraiah, aka Ezekiel, Joel and Obadiah, was the most important prophet since Moses.

Endnotes

Chapter 1: Introduction

[1] The traditional date for the Exodus is 1313 B.C.E., but I calculated the exact date and matched it to the Exodus route and the correct calendar where the full and new moons match the Sabbath days and the story line. I was able to do this after I was in Egypt in 2002, and traced the Exodus route from Goshen to the battle of Amalek. This information and story will be for my next book.

[2] Malcolm Neaum, producer of a British television documentary (BBC March 1, 2003) on Isaac Newton, in 2003.

[3] A Canadian academic and researcher, Mr. Stephen Snobelen, found the notation on a scrap of paper. (University of King's College in Halifax, Nova Scotia)

[4] Exodus 32:16.

[5] He later moved his foundation to its current address: Meru Foundation, POB 503, Sharon, MA 02067 USA.

[6] Van Biema, David, *In Search of Moses*, Time Magazine, December 4, 1998. Sheler, Jeffrey, *The Fight for History*, U.S. News & World Report, December 24 2001.

Chapter 2: God's Unit of Measure

[1] The theory was presented in my first book, *Reality Revealed; The Theory of Multidimensional Reality,* (published by Vector Associates, Bellevue, WA, 1977).

[2] These discoveries were made in 1989.

[3] Ezekiel 40:5.

[4] Tompkins, Peter, *Secrets of the Great Pyramid,* Pg 209. (Harper Colophon Books, Harper & Row, Publishers, 1971).

[5] The exceptions are the first two sons of Adam, who were Cain and Able. We use the age of Adam when he had Seth, which is stated in Genesis.

[6] Genesis 6:3.

[7] If you are a scholar of Roman history or 1st and 2nd century Church history, you will understand exactly what I am saying.

[8] Arius Calpurnius Piso (pronounced Peso with a long e), Governor of Syria. In his own writings as Josephus, he uses the pseudonyms of Gessius Florus and Cestius Gallus.

[9] The generals at the siege of Jerusalem were Titus, Arius Calpurnius Piso, and Pliny the Elder.

[10] Reuchlin, Abelard, "The True Authorship of the New Testament", pg. 1, (Abelard Reuchlin Foundation, POB 5652, Kent WA 98064, $12.00, 1979).

[11] Ibid. pg. 3, Their leader was Yochanan ben Zakai.

[12] Ibid. pg .3.

[13] The King James Version of the Bible may be using some of these verse-numbering systems. The following books have a one number difference with the Hebrew Bible: I Samuel, I Kings, Isaiah, Joel, Malachi; and a 66-verse difference in Psalms, compared to the Jewish Old Testament count.

[14] Actually, there is another literary insertion by Piso in the Hebrew scriptures, but it appears the Rabbis still grouped it with the Jewish books. His insertion was Ch. 40 to 66 of Isaiah.

Chapter 3: Theory of Multidimensional Reality

[1] Plato, *The Sophist*, Vol. 7, pp. 371, Loeb Classical Library, Harvard University Press, 1967.

[2] Plato, *the Republic, Book VII*. Loeb Classical Library, Harvard University Press.

[3] Raymond A. Moody, Jr., M.D.; *Life after Life*; (Bantam Books, 1975).

[4] Random Access Memory.

[5] Both dates have very high sunspot numbers (over 355 and 342), areas of the sun spots, and other output.

[6] V. Bucha, "Changes of the Earth's magnetic moment and radiocarbon dating," *Nature*, Vol. 224 (Nov. 15 1969): p. 681.

[7] The study of short term frequencies found on the Sun.

[8] This total does not count the occurrences which appear in books that Arius C. Piso had inserted into the Hebrew books.

[9] M. Isaacson, et. al., "The Study of the Absorption and Diffusion of Heavy atoms on Light Elements Substrates by means of the Atomic Resolution STEM": Ultramicroscopy 1, p. 359-376, 1976.

[10] Cambridge University Quantum Theory web site: http://www.damtp.cam.ac.uk/user/gr/public/qg_ss.html.

[11] The latest evolutionary version of String Theory is called M-Theory that calls for 11 dimensions and combines a little from each of the five other String Theories. M-theory is more convoluted than the others and I will not cover it here.

[12] G. Setti, (ed.), *Structure and Evolution of the Galaxies*, (Holland, D. Reidel Publ., 1975).

[13] G. O. Abell, *Exploration of the Universe*, 3ed. ed. (Holt, Rinehart & Winston, N.Y., 1975).

[14] Thelma Moss and J. Hubacher, *Phantom Leaf Effect as revealed through Kirilian Photography* (UCLA unpublished, 1975).

[15] Delayed Radio Signals QST-America Radio Relay League, Vol. 55, #5, p.54-58, May 1971.

[16] Bill Kerrell, and Kathy Goggin, *The Guide to Pyramid Energy*, p.164 (Santa Monica, Pyramid Power V., 1975).

[17] *Ibid.*

[18] A friend of mine who wrote *The Guide to Pyramid Energy*.

Chapter 4: The Hebrew Alphabet

[1] The Exodus took place 1,306 B.C.E. and it was about 11 months after the Exodus that Moses presented the two tables to the congregation.

[2] Matityahu Glazerson, *Letters of Fire, Mystical insights into the Hebrew language,* (New York:. Feldheim Publishers, 1991), p. 13.

[3] *The Zohar,* Translated by Maurice Simon and Dr. Paul P. Levertoff , Vol. IV, (London, New York: The Soncino Press, 1933) p. 57.

[4] *Ibid.*

[5] "Earliest Alphabet A Canaanite Invention—Preserved in Sinai Mines"; *Biblical Archaeology Review*, (July/August 1984), pp. 46-54.

[6] John Noble Wilford, "Discovery of Egyptian inscriptions indicates an early date for origin of the alphabet"; *Forum on science in the news*; The New York Times Company, 1999.

[7] I do not capitalize the word "creator" here because I mean a highly advanced past civilization that left these tablets in the cave.

[8] My friend, Gary Sultan, and I discovered the equation in 1980.

9 Quote attributed to Dr. Smolin in 1977 and listed on the Open-Site Web page: http://open-site.org/Science /Physics/Modern/Theory_of_Everything/.

10 Louis Ginzberg, *Legends of the Jews, The Alphabet*, Vol. I, (New York, Jewish Publication Society, 1911).

11 H.W.F. Saggs, *People of the Past, Babylonians* from *Egypt to Afghanistan* (British Museum Press, 1995): 3. p142)

12 Moses and Aaron's family tree will be covered in another book on the Jewish religion.

13 http://www.10ticks.co.uk/s_codebreaker_letter.asp.

14 Michael Drosnin, *The Bible Code* (New York, Simon & Schuster, 1997). Dr. Jeffrey Satinover, M.D. *Cracking the Bible Code* (New York, William Morrow and Comp., 1997).

15 The sequence of letters that most of the time is used to spell Moses.

16 Karl H. Pribram, Marc Nuwer, and Robert Baron, " The Holographic Hypothesis of Memory Structure in Brain Function and Perception," *Contemporary Developments in Mathematical Psychology,* vol. 2, (1974): p. 416-457.

17 Michael McDonald, San Jose, California.

18 Exodus 23:21 and 23.

19 *Ibid* 23:20.

Chapter 5: The Cave

1 The special holy light that shown during creation, the light that came forth out of the Ark, and the light that will shine from Mount Zion when God dwells there.

2 Louis Ginzberg, *The Legends of the Jews,* Vol. 3, Moses in the wilderness, "The Contest of the Mountains;" Jewish Publication Society of America, 1909.

3 Genesis 23:8-10.

4 The first being the contract with God on mount Sinai.

5 Genesis 23:19.

6 Exodus 6:18.

7 II Samuel 21:18-22.

8 Genesis 13:18 & 35:27.

9 Genesis 23:17, 25:9, 49:30, & 50:13.

10 Genesis 23:19.

11 A more complete explanation and proof for this statement will be given in my next book.

12 This reference will have special meaning in the next chapter on Joseph.

13 In Exodus 33:6 the word is spelled חֹרֵב, which is a bit different. On this occasion, the whole Hebrew congregation was at mount Sinai. The tabernacle had been built already, and Moses had just come down from the mount and told the people to take off their ornaments.

14 Exodus 17:6.

15 Exodus 20:15-18.

16 I will try to explain why Moses said this in Chapter 10.

17 Louis Ginzberg, *The Legends of the Jews,* Vol. 3, Moses in the Wilderne, "The Thirteen Attributes of God."

18 Exodus 33:16.

19 Louis Ginzberg, *The Legends of the Jews,* Vol. 3, Moses in the Wilderness, "The Thirteen Attributes of God."

20 Louis Ginzberg, *Legends of the Jews,* Vol. 1, Adam, "Adam and Eve in Paradise."

[21] Samael was an evil angel.

[22] Louis Ginzberg, *Legends of the Jews*, Vol. I, "Noah," "The Holy Book."

[23] Louis Ginzberg, *Legends of the Jews*, Vol. III, "Moses in the Wilderness," "The long Route."

[24] Louis Ginzberg, *Legends of the Jews*, Vol. III, "Moses in the Wilderness," "The thirteen attributes of God."

[25] Louis Ginzberg, *Legends of the Jews*, Vol. I, "Adam," "Adam and Eve in Paradise."

[26] *Ibid.*

[27] To clarify this statement, what is meant is evidence of phonetic symbols representing writing from the roots of our civilization. Archaeologists have found symbols representing forms of writing that predate the last ice age 12,000 years ago but they have not been deciphered (American Antiquarian, 1:178, 1879).

[28] *Supplementary Study No. 1 of the Bulletin of the American School of Oriental Research*, 1945.

[29] Samuel N. Kramer, *Sumerian Mythology, A study of Spiritual and Literary Achievement in the Third Millennium B.C.*, The American Philosophical Society, 1944., p. 98.

[30] Genesis 9:29.

[31] Samuel N. Kramer, *Sumerian Mythology, A study of Spiritual and Literary Achievement in the Third Millennium B.C.*, The American Philosophical Society, 1944., p.55.

[32] *Ibid.* p. 55.

[33] Samuel N. Kramer, *History begins at Sumer,* (Thames & Hudson, London) p. 198.

[34] *Sargonic and Gutian Periods (2334 – 2113 BC); The Royal Inscriptions of Mesopotamia, Early Periods, Vol. 2.* Univ. of Toronto Press, 1993, p. 30.

[35] *Ibid.* p. 32.

[36] Sidney Smith, Chatto & Windus, *Early History of Assyria to 1000 B.C.E.*, 1928, p. 89.

[37] *Ibid.* p.89.

[38] Danniel D. Luckenbill, Ph.D., *Ancient Records of Assyria and Babylonia*, Vol. 2, Univ. of Chicago Press, 1927; p. 21, p. 36 inscription. 70; p. 46. Inscription 92; p. 52, inscription. 99; p. 102, inscription. 185.

[39] *Ibid.;* inscription 54, p. 26; inscription. 96, p. 48; inscription. 102, p. 54.

[40] *Ibid.* p.41.

[41] *Ibid.* p.185.

[42] *Ibid.* p.206.

[43] *Ibid.* p.220.

[44] *Ibid.* p.257.

[45] *Ibid.* p.374.

[46] Zecharia Sitchin, *The Stairway to Heaven*, Avon Books, 1983.

[47] He comes to the conclusion that extra-terrestrials over the last 12,000 years were the occupants of Tilmun and that the natives considered them the gods.

[48] Alexander Heidel, *The Gilgamesh Epic and Old Testament Parallels*, The University of Chicago Press, 1946. p. 16.

[49] *Ibid.* p. 57.

[50] *Ibid.* p63.

[51] *Ibid.* p57.

[52] *Ibid.* p. 48.

[53] *Ibid.* p. 64.

[54] Lewis Ginzberg, *Ledgends of the Jews*, Vol. 3., "Moses in the Wilderness," "Pharaoh Pursues the Hebrews."

[55] Alexander Heidel, *The Gilgamesh Epic and Old Testament Parallels*, The University of Chicago Press, 1946, p. 74.

[56] *Ibid.* p. 79.

Chapter 6: The Rod of God

[1] The cargo cult depicted in Gualtiero Jacopetti's pop documentary, *Mondo Cane*, filmed in New Guinea in 1959.

[2] Through the remainder of the book I capitalize the Rod because it is a very special object. Just as we capitalize the word Bible, Tabernacle, Torah or God, the word Rod deserves to be capitalized because of where it came from and what it is.

[3] Genesis 1:11.

[4] Exodus 7:10.

[5] This is not the Joshua, son of Nun, but an older Joshua. The proof for this will be presented in my next book.

[6] *Tanakh, The Holy Scriptures,* (The Jewish Publication Society, 1985), p89.

[7] Exodus 4:2.

[8] Angelo Rapport, *Legends of Ancient Israel; Vol. 2 At the Well of Median,* 1966, p.254.

[9] *Ibid.* p.278.

[10] Louis Ginzberg, *Legends of the Jews*; "Moses in the wilderness," "Miriam's Well," Jewish Publication Society, 1909.

[11] Numbers Chapter 16:1-19.

[12] The Pharaoh at the time of Moses was Amonhotep IV, who later changed his name to Akhenaton. The full description will be written in my next book.

[13] Louis Ginzberg, *Legends of the Jews*; "Moses in the wilderness," ":Moses Prays for Death," Jewish Publication Society, 1909, pp. 464.

Chapter 7: The Ark of the Covenant

[1] Their intellectual arrogance seems to prevent them from accepting a finding unless the discovery was made by one of their own peers. To become one of their peers, one must regurgitate whatever they teach, without question, or else one does not pass. When a student gets to the level of Ph.D. candidate, the student must accept any and all theories of the professors reviewing the student's Ph.D. thesis, or else the student will not graduate. The end result of this form of education is that universities institutionalize false ideas. That is why the old guard never accepts new ideas, because they are constantly defending their own theses.

[2] The local Bedouins have buried their dead on the top of the hill and removed a great number of stones from the altars.

[3] Exodus 20:22.

[4] Exodus 20:23.

[5] Exodus 26:31-2; and 36:35-36.

[6] A much more detailed description will be given in my next book.

[7] The English plural for Cherub is Cherubim but the translator sometime used Cherubim for singular and Cherubim for plural. I am going to use the accepted English usage.

[8] Exodus 261-6; and 36:8-13.

[9] Exodus 26:7-11; and 36:14-18.

[10] The translation of this animal is uncertain. Some translations say seal skins, but I do not know where the Hebrews could have acquired seal skins in the Sinai desert.

[11] Exodus 26:1-14; and 36:12-19.

[12] Exodus 26:31-33; and 36:35-36.

[13] Exodus 26:24, the King James version: "And they shall be coupled together beneath," which is in error. It should be: "And they shall be double breadth."

[14] Exodus 27:4-5; and 38:4-5.

[15] Exodus 27:1-2; and 38:1-2.

[16] Exodus 27:3-4. and 38:3-4.

[17] Exodus 27:6. and 38:3-4.

[18] Exodus 40:34-35.

[19] I Kings 14:25-26.

[20] Exodus 25:23-28 and 37:10-15.

[21] Louis Ginzberg, *Legends of the Jews, "Moses in the Wilderness", "The Ark with the Cherubim."* (New York, Jewish Publication Society, 1911).

[22] *Ibid.*

[23] *Ibid,* "Moses in the Wilderness", "The Interrupted Joy."

[24] *Ibid,* "Moses in the Wilderness", 'The Ark with the cherubim."

[25] *Ibid, "*Moses in the Wilderness", "The Revelations in the Tabernacle."

[26] Mr. Morandi and USC Engineering Department, James Joseph, *Flame Amplification and a Better Hi-Fi Loudspeaker;* (Popular Electronics; May 1968), pp 47-53.

[27] Logos Foundation, Center for Experimental Music Production, Postal adress: Kongostraat 35, Logos Tetrahedron concert hall: Bomastraat 26-28, 9000 Gent   Belgium.

[28] Jeremiah 51:59.

[29] Jeremiah 32:12.

[30] A complete description of all the priests' and the prophets' names they wrote under, will be in my next book.

[31] Jeremiah 36:26, Ezra 7:1.

[32] Jeremiah 23:16-32

[33] Louis Ginzberg, *Legends of the Jews,* "The Exile", "Baruch", (New York, Jewish Publication Society, 1911), p.232.

[34] First Samuel 10:5 & 10:8

[35] Louis Ginzberg, *Legends of the Jews.,* "Moses in the Wilderness", "Israel convinced of Aaron's Priesthood," (New York, Jewish Publication Society, 1911).

[36] *Ibid,* "The Exile", "Baruch."

[37] *Ibid,* "Moses in the Wilderness", "Israel convinced of Aaron's priesthood."

[38] *Ibid,* "The Judges", The First Judge."

[39] *Ibid,* "Moses in the Wilderness", "The Table and the Candlestick."

[40] *Ibid,* "The Exile", "The Temple Vessels."

[41] *Ibid.*

[42] The Dead Sea scrolls were also managed like this because Christian leadership was scared that the Jews of the first and second century may have written down what the Pisos had done to them and what Piso was forcing them to believe in and translate into Hebrew.

Chapter 8: The Polar Reversal and the Ice Age

[1] G. Sivjee, G. Romick, "Pleistocene-recent boundary and Wisconsin Glacial biostratigraphy in the northern Indian Ocean," *Science*, Vol. 159 (Mar. 1968): p. 1456.

[2] R. F. Flint, "Status of the Pleistocene Wisconsin stage in Central North America," *Science*, Vol. 139 (Feb. 1, 1963): p. 402.

[3] Roger Lewin, "Extinctions and the history of life," *Science*, Vol. 221 (Sept. 2, 1983): p. 935.

[4] AL Reddy, PG Fialkow and A Salo, "Ultraviolet radiation—induced chromosomal abnormalities in fetal fibroblasts from New Zealand in black mice," *Science*, Vol. 201 (Sept. 8, 1978): p. 920.

[5] Berger, A. L., "Support for the astronomical theory of climate change," *Nature*, Vol. 269 (Sept. 1, 1997): p. 44.

[6] D. Pollard, "An investigation of the astronomical theory of the ice ages using a simple climate-ice sheet model," *Nature*, Vol. 272 (March 16, 1978): p. 233.

[7] Dennison, Brian, Mansfield, V., "Glaciations and dense interstellar clouds," *Nature*, Vol. 261 (May 6. , 1976): p.32.

[8] D. Clark, W. McCrea & F. Stephenson, "Frequency of nearby supernovae and climatic and biological catastrophes," *Nature*, Vol. 265 (Jan. 27, 1977): p. 318.

[9] Broecker, Wallace, "Absolute dating and the astronomical theory of glaciation," *Science*, Vol. 151 (Jan. 21, 1966): p.299.

[10] T. M. Wigley, "Spectral analysis and the astronomical theory of climatic change," *Nature*, Vol. 264 (Dec. 16, 1976): p. 629.

[11] W. Ruddiman & A. McIntyre, "Oceanic Mechanisms for amplification of the 23,000-year ice-volume cycle," *Science*, Vol. 212 (May 8, 1981): p. 617.

[12] Donn, William; Ewing, Maurice "A theory of ice ages III," *Science*, Vol. 152 (June 24, 1966): p. 1706.

[13] G. Denton, C. Hendy, "Climate changes: Global as well as large and abrupt," *GSA Today*, (May. 1997): p. 3.

[14] W. Ruddimas, "North Atlantic Ice-Rafting: A major change at 75,000 years before the present," *Science*, Vol. 196 (June 10, 1997): p. 1208.

15 Beals, C., Stories Told by the Aztecs Before the Spaniards Came (N.Y., Abelard-Schuman, 1970). "Listen to me carefully, he whispered melodiously. 'Take your hearth fire and hide yourself in a cave in the nearby mountain.' He was the beneficent wind from the east, from the garden of paradise, but soon, he warned he would blow from the north and from the south as a furious hurricane and sweep over the entire world.

Whirlwinds and cyclones swept over the world, picking up sand, stones, rocks, waters and finally trees, houses and human beings. The snowy capes of the mountain peaks were whisked away, converting the whole world with an immense white sheet (snow).

The chosen man and woman, in their cave beside their red hearth fire, continued their conversation, unperturbed by the roar of the wind, not feeling the glacial cold that gripped the world."

[16] M. Stupavsky, Geological Society Of America; v.85, p.141, Jan. 1974.

[17] Opdyke, Glass, Hays and Foster, "Paleomagnetic Study of Antarctic Deep-Sea Cores," *Science*, Vol. 154 (Oct. 21, 1966): p. 349.

[18] A collapsing magnetic field creates voltage or potential 90 degrees out of phase from the magnetic field.

[19] Prinja, R. K., Knigge, C., Ringwald, F. A., and Wade, R. A. "Episodic Absorption in the Outflow of the Old Nova V603 Aquilae." Volume 318, Number 2, (Monthly Notices of the Royal Astronomical Society, October 2000): p. 368-374(7).

[20] R. Davis, D. Stannard & R. Conway, "Gamma-ray bursts from neutron star glitches," *Nature*, Vol. 251 (Oct. 4, 1974): p. 399. Even though the authors believe that the gamma-ray bursts come from neutron stars, the bursts are more likely to come from stars that have novaed, since neutron stars are only a theoretical idea, and none have ever been photographed.

[21] E. Witkin, "Radiation-Induced mutations and their repair," *Science*, Vol. 152 (June 3, 1966): p. 1345.

[22] H. Curtis & H. Smith, "Corn seeds affected by heavy cosmic ray particles," *Science*, Vol. 141 (July 12, 1963): p. 359.

[23] Day, R. A. and Johnson, Ronald, *General Chemistry* (New Jersey: Prentice-Hall, Inc., 1974): p. 34.

[24] The coldest temperature recorded was $-129°F$ on July 21, 1983 in Vostok, Antarctica.

[25] M. Stuiver, C. Heusser, I. C. Yang, "North American glacial history extended to 75,000 years ago," *Science*, Vol. 200 (April 7, 1978): p. 16.

[26] S. F. Singer, "Zodiacal Dust and deep-sea sediments," *Science*, Vol. 156 (May. 26, 1967): p. 1080.

[27] J. Lattimer, D. Schramm & L. Grossman, "Supernovae, grains and the formation of the Solar System," *Nature*, Vol. 269 (Sept. 8, 1977): p. 116.

[28] *Ibid.*

[29] *Ibid.*

[30] K. Knie, G. Korschinek, T. Faestermann, E. Dorfi, G. Rugel & A. Wallner "^{60}Fe anomaly in a deep-sea manganese crust and implications for a nearby supernova source," *Physical Review Letters*, Vol. 93 (Oct. 22, 2004).

[31] Hajdas, Bonani, Boden,, et. al., "Cold reversal on Kodiak Island, Alaska, correlated with the European Younger Dryas by using variations on atmospheric ^{14}C content," *Geology*, Vol. 26 (Nov. 1998): p. 1049.

[32] Minze Stuiver and Paul Quay, "Changes in atmospheric carbon-14 attributed to a variable Sun," *Science*, Vol. 207 (Jan. 4, 1980): p. 11.

[33] V. Bucha, E. Neustupny, "Changes of the Earth's magnetic field and radio-carbon dating," *Nature*, Vol. 215 (July. 15, 1967): p. 261.

[34] V. Bucha, "Changes of the Earth's magnetic moment and radiocarbon dating," *Nature*, Vol. 224 (Nov. 15 1969): p. 681.

[35] H. Kitagawa & J. van der Plicht, "Atmospheric radiocarbon calibration to 45,000 yr B.P.: Late glacial fluctuations and cosmogenic isotope production," *Science*, Vol. 279 (Feb. 20, 1998): p. 1187.

[36] 1 calorie = 9×4.184 joules; 1 joule = 1×10^7 ergs.

[37] Pre-sandstone.

[38] Mudstone, hardpan, etc.

[39] R. Thompson,, et. a*l.*, *Nature*, Vol. 263, (Oct. 7, 1976): p. 491.

[40] A. Cox, "Geomagnetic Reversals," *Science,* Vol. 163, (Jan. 17, 1969): p. 237.

[41] B. Warner, R. Mathewes, J. Clague, *Science*, Vol. 218, (Nov. 12, 1982): p. 675.

[42] J. England, R. Bradley, "Past glacial activity in the Canadian high Arctic," *Science*, Vol. 200 (April 21, 1978): p. 265.

[43] R. F. Flint, "Status of the Pleistocene Wisconsin stage in Central North America," *Science*, Vol. 139 (Feb. 1, 1963): p. 402.

[44] *Ibid.*

[45] E. Winkler, "Radiocarbon ages of postglacial lake clays near Michigan City, Indiana," *Science*, Vol. 137 (Aug. 17 1962): p. 528.

[46] M. Stuiver, C. Heusser, I. C. Yang, "North American glacial history extended to 75,000 years ago," *Science*, Vol. 200 (April 7, 1978): p. 16.

[47] J. Mercer, C. Laugenie, "Glacier in Chile ended a major re-advance about 36,000 years ago: some global comparisons," *Science*, Vol. 182 (Dec. 3, 1973): p. 1017.

[48] B. John, "A possible main Würm Glaciation in West Pembrokeshire," *Nature*, Vol. 207 (Aug. 7, 1965): p. 622.

[49] G. Coope, F. Shotton, and I. Strachan, *Phil. Trans. Royal Society*, Vol. 244 (1961): p. 379.

[50] J. Mercer, C. Laugenie, "Glacier in Chile ended a major re-advance about 36,000 years ago: some global comparisons," *Science*, Vol. 182 (Dec. 3, 1973): p. 1017.

[51] *Ibid.*

[52] B. John, "A possible main Wurm Glaciation in West Pembrokeshire," *Nature*, Vol. 207 (Aug. 7, 1965): p. 622.

[53] G. Denton, C. Hendy, "Climate changes: Global as well as large and abrupt," *GSA Today*, (May. 1997): p. 3.

[54] R. Newnham & D. Lowe, "Fine-resolution pollen record of late-glacial climate reversal from New Zealand," *Geology*, Vol. 28 (August. 2000): p. 759.

[55] J. Mercer, C. Laugenie, "Glacier in Chile ended a major re-advance about 36,000 years ago: some global comparisons," *Science*, Vol. 182 (Dec. 3, 1973): p. 1017.

[56] J. Peterson, G. Hope, "Lower limit and maximum age for the last major advance of the Carstensz Glaciers, West Irian," *Nature*, Vol. 240 (Nov. 3, 1972): p. 36.

[57] M. Williams & D. Adamson, "Late Pleistocene desiccation along the White Nile," *Nature*, Vol. 248 (April 12, 1974): p. 584.

[58] A. Hamilton, A. Perrott, "Date of deglacierisation of Mount Elgon," *Nature*, Vol. 273 (May 4, 1978): p. 49.

[59] D. Adamson, F. Gasse, F. Street & M. Williams, "Late Quaternary history of the Nile," *Nature*, Vol. 288 (Nov. 6, 1980): p. 50.

[60] *Ibid.*

[61] R. Fairbridge, "Global climate change during the 13,500-B.P. Gothenburg geomagnetic excursion," *Nature*, Vol. 265 (Feb. 3, 1977): p. 430.

[62] S. Falk, J. Lattimer, & S. Margolis, "Are supernovae sources of presolar grains?," *Nature*, Vol. 270 (Dec. 22, 1977): p. 700..

[63] S. F. Singer, "Zodiacal dust and deep-sea sediments," *Science*, Vol. 156 (May. 26, 1967): p. 1080.

[64] J. Reyse, Y. Yokoyama & S. Tanaka, "Aluminum-26 in deep-sea sediment," *Science*, Vol. 193 (Sep. 17, 1976): p. 1119.

[65] J. Wasson, B. Alder & H. Oeschger, "Aluninum-26 in Pacific sediment: Implications," *Science*, Vol. 155 (Jan. 273, 1967): p. 446.

[66] J. Higdon & R. Lingenfelter, "Sea sediments, cosmic rays, and pulsars," *Nature*, Vol. 246 (Dec. 14, 1973): p. 403.

[67] J. Lattimer, D. Schramm & L. Grossman, "Supernovae, grains and the formation of the Solar System," *Nature*, Vol. 269 (Sept. 8, 1977): p. 116.

[68] D. Tilles, "Implantation in Interplanetary dust of Rare-gas ions from Solar flares," *Science*, Vol. 153 (Aug. 26, 1966): p. 981.

[69] Ed Weyer, "Pole movement and sea levels," *Nature*, Vol. 273 (May. 4, 1978): p. 18.

[70] T. Ku, M. Kimmel, W. Easton & T. O'Neil, "Eustatic sea level 120,000 years ago on Oahu, Hawaii," *Science*, Vol. 1831 (March 8, 1974): p. 959.

[71] J. Marshall & B. Thom, "The sea level in the last interglacial," *Nature*, Vol. 263 (Sep. 9, 1976): p. 120.

[72] R. Lighty, I. Macintyre & R. Stuckenr, "Submerged early Holocene barrier reef southeast Florida shelf." *Nature*, Vol. 276 (Nov. 2, 1978): p. 59.

[73] W. Barnhardt, D. Belknap & J. Kelly, "Stratigraphic evolution of the inner continental shelf in response to late Quaternary relative sea-level change, northwest Gulf of Mane," *GSA Bulletin*, Vol. 109 (May, 1997): p. 612.

[74] K. O. Emery & L. Garrison, "Sea levels 7,000 to 20,000 years ago." *Science*, Vol. 157 (Aug. 11, 1967): p. 684.

[75] A. Merrill, K. Emery and Meyer Rubin, "ancient oyster shells on the atlantic continental shelf," *Science*, Vol. 147 (Jan. 23, 1965): p. 398.

[76] K. O. Emery & L. Garrison, "Sea levels 7,000 to 20,000 years ago." *Science*, Vol. 157 (Aug. 11, 1967): p. 684.

[77] W. Berger, "Deep-sea carbonate and the deglaciation preservation spike in pteropods and foraminifera." *Science*, Vol. 213 (July 17, 1981): p. 331.

[78] A. Bloom 7 M. Stuiver, "Submergence of the Connecticut coast," *Science*, Vol. 139 (Jan. 25, 1963): p. 332.

[79] H. Richards & W. Broecker, "Emerged Holocene South America shorelines," *Science*, Vol. 141 (Sep. 13, 1963): p. 1044.

[80] K. O. Emery & L. Garrison, "Sea levels 7,000 to 20,000 years ago." *Science*, Vol. 157 (Aug. 11, 1967): p. 684.

[81] H. Richards & W. Broecker, "Emerged Holocene South America shorelines," *Science*, Vol. 141 (Sep. 13, 1963): p. 1044.

[82] F. Shepard, "Sea level changes in the past 6,000 years: possible archaeological significance." *Science*, Vol. 143 (Feb. 7, 1964): p. 574.

[83] *Ibid.*

[84] P. Brown & J. Kennett, "Megaflood erosion and meltwater plumbing changes during last North American deglaciation recorded in Gulf of Mexico sediments," *Geology*, Vol. 26, no.7 (July, 1998): p. 599.

[85] Ed Weyer, "Pole movement and sea levels," *Nature*, Vol. 273 (May. 4, 1978): p. 18.

[86] T. Nardin, R. Osborne, D. Bottjer & R. Scheidemann Jr., "Holocene sea-level curves for Santa Monica shelf, California continental borderland." *Science*, Vol. 213 (July 17, 1981): p. 331.

[87] H. Josenhans, D. Fedje, R. Pienitz and J. Southon, "Early humans and rapidly changing holocene sea levels in the Queen Charlotte Islands—Hecate Strait, British Columbia, Canada," *Science*, Vol. 277 (July 4, 1997): p. 71.

[88] *Ibid.*

[89] T. Nardin, R. Osborne, D. Bottjer & R. Scheidemann Jr., "Holocene sea-level curves for Santa Monica shelf, California continental borderland." *Science*, Vol. 213 (July 17, 1981): p. 331.

[90] H. Josenhans, D. Fedje, R. Pienitz and J. Southon, "Early humans and rapidly changing holocene sea levels in the Queen Charlotte Islands—Hecate Strait, British Columbia, Canada," *Science*, Vol. 277 (July 4, 1997): p. 71.

[91] *Ibid.*

[92] Ed Weyer, "Pole movement and sea levels," *Nature*, Vol. 273 (May. 4, 1978): p. 18.

[93] B. Vuilleumier, "Pleistocene changes in the Fauna and Flora of South America," *Science*, Vol. 173 (Aug. 27, 1971): p. 771.

[94] M. Bothner & E. Spoker, "Upper Wisconsinan till recovery on the continental shelf southeast of New England," *Science*, Vol. 210 (Oct. 24, 1980): p. 423.

[95] H. Veeh, J. Chappell, "Astronomical theory of climate changes: support from New Guinea," *Science*, Vol. 167 (Feb. 5, 1970): p. 862.

[96] J. Hoyt & J. Hails, "Pleistocene shorelines sediments in coastal Georgia: deposition and modification," *Science*, Vol. 155 (March. 24, 1967): p. 1541.

[97] Frank C. Whitmore, Jr., K. O. Emery, H. B. S. Cooke, and Donald J. P. Swift, "Elephant teeth from the Atlantic continental shelf," *Science*, Vol. 156 (June 16, 1967): p. 1477.

[98] M. Bothner & E. Spoker, "Upper Wisconsinan till recovery on the continental shelf southeast of New England," *Science*, Vol. 210 (Oct. 24, 1980): p. 423..

[99] K. O. Emery, R. L. Wigley, Alexandra S. Bartlett, Meyer Rubin, and E. S. Barghoorn, "Freshwater peat on the continental shelf," *Science*, Vol. 158 (Dec. 8, 1967): p. 1301.

[100] *Ibid.*

[101] W. Ererichs, "Pleistocene-recent boundary and Wisconsin glacial biostratigraphy in the northern Indian Ocean," *Science*, Vol. 159 (March, 1968): p. 1456.

[102] *Ibid.*

[103] D. Adamson, F. Gasse, F. Street & M. Williams, "Late Quaternary history of the Nile," *Nature*, Vol. 288 (Nov. 6, 1980): p. 50.

[104] F. Street & A. Grove, "Environmental and climatic implications of late Quaternary lake-level fluctuations in Africa," *Nature*, Vol. 261 (June. 3, 1976): p. 385.

[105] *Ibid.*

[106] W. Watts & M. Stuiver, "Late Wisconsin climate of Northern Florida and the origin of species-rich deciduous forest," *Science*, Vol. 210 (Oct. 17, 1980): p. 325.

[107] W. Zeist & H. Wright, Jr., "Preliminary pollen studies at Lake Zeribar, Zagos mountains, southwestern Iran," *Science*, Vol. 140 (April 5, 1963): p. 65.

[108] D. Ericson, M. Ewing & Goesta Wollin, "Pliocene-Pleistocene boundary in deep-sea sediments," *Science*, Vol. 139 (Feb. 22, 1963): p. 727.

[109] Douglas Williams, R. Thunell and J. Kennett, "Periodic freshwater flooding and stagnation of the eastern Mediterranean Sea during the late Quaternary," *Science*, Vol. 201 (July 21, 1978): p. 252.

[110] R. C. Thunell, *Quat. Res.*, Vol. 6 (1976): p. 281.

[111] C. Emiliani, S. Gartner, B. Lidz, K. Eldridge,, et. *al.*, "Paleoclimatological analysis of late Quaternary cores from the northeastern Gulf of Mexico," *Science*, Vol. 189 (sep. 26, 1975): p. 1083.

[112] W. Holser, "Catastrophic chemical events in the history of the ocean," *Nature*, Vol. 267 (June. 2, 1977): p. 403.

[113] J. Thiede, "A glacial Mediterranean," *Nature*, Vol. 276 (Dec. 14, 1978): p. 680.

[114] *Ibid.*

[115] D. Twichell, H. Knebel & D. Folger, "Delaware River: Evidence for its former extension to Wilmington submarine Canyon," *Science*, Vol. 195 (Feb. 4, 1977): p. 483.

[116] J. Trunbull & M. McCamis, "Geological exploration in an east coast submarine canyon from a research submersible," *Science*, Vol. 158 (Oct. 20, 1967): p. 370.

[117] J. Robb, J. Hampson Jr., D. Twichell, "Geomorphology and sediment stability of a segment of the U.S. continental slope off New Jersey," *Science*, Vol. 211 (Feb. 27, 1981): p. 935.

[118] A. Hammond, "Paleoceanography: Sea floor clues to earlier environments," *Science*, Vol. 191 (Jan. 16, 1976): p. 169.

[119] Ball, R. H.,, et.al., *Variations in the Geomagnetic Field and in the rate of the Earth's rotation* (Santa Monica, The Rand Corp., Memorandum No. RM-5717-PR, Oct. 1968).

[120] R. Challinor, "Variations in the rate of rotation of the Earth," *Science*, Vol. 172 (June 4, 1971): p. 1022.

[121] J. Bahcall, "Neutrinos from the Sun," *Scientific American*, Vol. 221-1 (July, 1969): p. 29.

[122] Beverly Hartline, "in Search of Solar Neutrinos," *Science*, Vol. 204 (Apr.6, 1979): p.42.

[123] Roger Ulrich, "Solar Neutrinos and Variations in the Solar Luminosity," *Science*, Vol. 190 (Nov.14, 1975): p.619.

[124] Melvin Freedman, Charles Stevens, et, al., "Solar Neutrino: Proposal for a New Test," *Science*, Vol.193 (Sept.17, 1976): p.1117.

[125] John Bahcall, Raymond Davis Jr., "Solar Neutrinos: A Scientific Puzzle," *Science*, Vol. 191 (Jan.23, 1976): p.264.

[126] John Ross, Lawrence Aller, "The Chemical Composition of the Sun," *Science*, Vol. 191 (Mar.26, 1976): p.1223.

[127] Peter Toth, "Is the Sun a pulsar?," *Nature*, Vol. 270 (Nov.10, 1977): p.159.

[128] P.H. Scherrer, J.M. Wilcox,, et. *al.*, "Observations of solar oscillations with periods of 160 minutes," *Nature*, Vol. 277 (Feb.22, 01979): p.635.

[129] J.L. Snider, M.D Kearns, P.A. Tinker, "Evidence for long-period global oscillations of the sun," *Nature*, Vol.275 (Oct.26, 1978): p.730.

[130] A.B. Severny, V.A. Kotov, et, al., "Observations of solar pulsations," *Nature*, Vol.259 (Jan.15, 1976): p.87.

[131] GBLO, "Sun appears to be oscillating at many frequencies," *Physics Today*, (May, 1976): p. 17.

[132] R. H. Dicke "Is there a chronometer hidden deep in the Sun?" *Nature*, Vol. 276 (Dec. 14, 1978): p. 676..

[133] W. Kundt, "Are supernova explosions driven by magnetic springs?," *Nature*, Vol. 261 (June 24, 1976): p. 673.

[134] W. H. Lewin & P. Joss, "X-ray burst sources," *Nature*, Vol. 270 (Nov. 17, 1977): p. 211.

[135] D.W. Parkin, "Solar constant during a glaciation," *Nature*, Vol. 260 (Mar.4, 1976): p. 28.

[136] H. Johnson, "Infrared Stars," *Science*, Vol. 157 (Aug. 11, 1967): p. 635.

[137] R. Gomes, "On the edge of the solar system," *Science*, Vol. 286 (Nov. 19, 1999): p. 1487.

[138] H. E. Bond, A. Henden, Z.G. Levay,, et. *al.* "An energetic stellar outburst accompanied by circumstellar light echoes" *Nature* Vol. 422 (March 27, 2003), p.405.

[139] This nova has received a lot of attention from astronomers around the world. In just the first four journal articles, I found there were over 34 scientists listed, representing universities across the country, plus government agencies: Space Telescope Science Institute, US Naval Observatory, European Space Agency, and others.

[140] Tylenda, R., "On the light echo in V838 Mon.," *Astronomy & Astrophysics,* July 2004.

[141] *Ibid.*

[142] *Ibid.*

[143] H. E. Bond, A. Henden, Z.G. Levay,, et. *al.* "An energetic stellar outburst accompanied by circumstellar light echoes" *Nature* Vol. 422 (March 27, 2003), p.405.

[144] *Ibid.*

[145] Wisniewski, J.P., et. *el.*, Spectroscopic and Spectropolarimetric Observations of V838 Mon. *Astronomy & Astrophysics,* January 2003.

[146] Tylenda, R., "On the light echo in V838 Mon.," *Astronomy & Astrophysics,* July 2004.

[147] W. H. Lewin & P. Joss, "X-ray burst sources," *Nature*, Vol. 270 (Nov. 17, 1977): p. 211.

[148] *Ibid.*

[149] I had read on one of the science journals that they had detected over 150 per year but I cannot for the journal article as of yet.

[150] J. Wdowczyk, A. Wolfendale, "Cosmic rays and ancient catastrophes," *Nature*, Vol. 268 (Aug.. 11, 1977): p. 510..

[151] *Ibid.*

[152] D. Clark, J. Caswell & A. Green, "New Galactic Supernova remnants," *Nature*, Vol. 246 (Nov. 2, 1973): p. 28.

[153] F. Hoyle, N. Wickramasinghe, "Dust in Supernova explosions," *Nature*, Vol. 226 (April 4, 1970): p. 62.

[154] E. Leibowitz, "Determination of the time of the shell ejection in nova outbursts," *Nature*, Vol. 264 (Nov. 4, 1976): p. 41. Even though the author estimates only 500 kms, I have found articles estimating 5,000 kms. My estimate is based on what the Hindus saw and where the dust finally landed on the Earth.

[155] F. Hoyle, N. Wickramasinghe, "Dust in Supernova explosions," *Nature*, Vol. 226 (April 4, 1970): p. 62.

[156] J. O'Keefe, "Tektite Glass in Apollo 12 sample," *Science*, Vol. 168 (June 5, 1970): p. 1209.

[157] E. Roedder of the USGS & P. Weiblen, "Apollo 17 'Orange Soil' and meteorite impact on liquid lava," *Nature*, Vol. 244 (July 27, 1973): p. 210.

[158] N. Bhandari, S. Bhat, D. Rajagopalan, A. Tamhana & V. Venkatavaradan, "Super-heavy elements in extra-terrestrial samples," *Nature*, Vol. 230 (March 26, 1971): p. 219.

[159] *Ibid.*

[160] *Ibid.* p. 211.

[161] S. Tolansky, "Lunar Glass: Interferometric evidence for low-temperature shock," *Science*, Vol. 176 (May. 12, 1972): p. 671.

[162] S. Tolansky, "Interfermetric studies on Apollo 11 and 12 lunar glass objects," The Moon Symposium #47, *International Astronomical Union*, (1972): p. 249.

[163] G. Brown, J. Holland & A. Peckett, "Orange soil from the Moon," *Nature*, Vol. 242 (Apr.. 20, 1973): p. 515.

[164] M. Pugh, "Rotation of Lunar dumbbell-shaped globules during formation," *Nature*, Vol. 237 (May. 19, 1972): p. 158.

[165] Apollo 16 preliminary examination team, "The Apollo 16 Lunar Samples: Petrographic and chemical description," *Science*, Vol. 179 (Jan. 5, 1973): p. 23.

[166] C. Singer, "Supernovae and lunar melting" *Nature*, Vol. 272 (March 16, 1978): p. 239.

[167] Thomas Gold, "Apollo 11 observations of a remarkable glazing phenomenon on the lunar surface," *Science*, Vol. 168 (Sept. 26, 1969): p. 1345.

[168] *Ibid.*

[169] *Ibid.*

[170] G. Mueller & G. Hinsch, "Glassy particles in Lunar finds," *Nature*, Vol. 228 (Oct. 17, 1970): p. 254.

[171] J. Laul, R. Ganapathy & E. Anders, "Glazed Lunar rocks: origin by impact," *Science*, Vol. 172 (May 7, 1971): p. 556.

[172] E. Roedder of the USGS & P. Weiblen, "Apollo 17 'Orange Soil' and meteorite impact on liquid lava," *Nature*, Vol. 244 (July 27, 1973): p. 210.

[173] J. Green, "Origin of glass deposits in Lunar craters," *Science*, Vol. 168 (May 1, 1970): p. 608.

[174] E. Roedder of the USGS & P. Weiblen, "Apollo 17 'Orange Soil' and meteorite impact on liquid lava," *Nature*, Vol. 244 (July 27, 1973): p. 210.

[175] C. Cross, "Formation of glass spheres on the Moon," *Nature*, Vol. 233 (Sept. 17, 1971): p. 185.

[176] E. Dietz & P. Vergano, "response to Jack Green article," *Science*, Vol. 168 (May. 1, 1970): p. 609.

[177] C. Cross, "Formation of glass spheres on the Moon," *Nature*, Vol. 233 (Sept. 17, 1971): p. 185.

[178] *Ibid.*

[179] H. Faul, "Tektites are terrestrial," *Science*, Vol. 152 (June 3, 1966): p. 1341. This author used two dating methods which are unreliable for tektites. The two methods he cited are ^{40}Ar/^{40}K ratio testing and counting fission tracks in the tektite. The problem with both is, the nova creates the radioactive isotopes and it throws out massive cosmic rays that will leave fission tracks in the glass, so his dating will lead you to believe the tektites are older than they really are. The age of the sediment layer the tektites are found is more reliable a dating method.

[180] G. Bigazzi & F. Bonadonna, "Fission track dating of the obsidian of Lipari Island (Italy)," *Nature*, Vol. 242 (March 30, 1973): p. 322.

[181] B. Glass & B. Heezen, "Tektites and geomagnetic reversals," *Nature*, Vol. 214, (April 22, 1967): p. 372.

[182] *Ibid.* p. 373.

[183] G. Keller, S. D'Hondt & T. Vallier, USGS, *Science*, Vol. 221, (July 8, 1983): p. 150.

[184] B. Glass & B. Heezen, "Tektites and geomagnetic reversals," *Nature*, Vol. 214, (April 22, 1967): p. 372.

[185] R. Olsson, K. Miller, J. Browning, et. *al.*, "Ejecta layer at the Cretaceous-Tertiary boundary, Bass River, New Jersey (Ocean drilling program leg 174AX)," *Geology*, Vol. 25; No. 8 (Aug. 1997): p. 759.

[186] H. Nininger & G. Huss, "Tektites that were partially plastic after completion of surface sculpturing," *Science*, Vol. 157 (July 7, 1967): p. 61.

[187] *Ibid.*

[188] S. Durrani, "Are Microtektites the result of cometary impacts with the Earth?" *Nature*, Vol. 235 (Feb. 18, 1972): p. 383.

[189] F. Frey, C. Spooner, P. Baedecker, "Microtektites and Tektites: a chemical composition," *Science*, Vol. 170 (Nov. 20, 1970): p. 845.

[190] D. Pal, C. Tuniz, R. Moniot, T. Kruse & G. Herzog, "Beryllium-10 in Australasian tektites: Evidence for a sedimentary precursor," *Science*, Vol. 218 (Nov. 19, 1982): p. 787.

[191] J. O'Keefe, "Origin of Tektites," *Science*, Vol. 130 (July 18, 1959): p. 97.

[192] Y. Yokoyama, "Accretion rate of cosmic dust estimated from cosmogenic aluminum-26," *Nature*, Vol. 220 (Dec. 7, 1968): p. 1016.

[193] E. Chao, I. Adler, E. Dwornik, & J. Littler of the USGS, "Metallic Spherules in Tektites from Isabela, Philippine Islands," *Science*, Vol. 135 (Jan. 12, 1962): p. 97.

[194] J. O'Keefe, "Origin of Tektites," *Science*, Vol. 130 (July 18, 1959): p. 97.

[195] R. Ganapathy, J.W. Larimer, "A Meteoritic Component Rich in Volatile Elements: Its Characterization and Implications," *Science*, Vol. 207 (Jan.4, 1980): p.57.

[196] M. Maurette, P. Pellas & R. Walker, "Cosmic-ray-induced particle tracks in a meteorite," *Nature*, Vol. 204 (Nov. 28, 1964): p. 821.

[197] P. Fowler & A. Lang, "Giant holes in mica," *Nature*, Vol. 270 (Nov. 10, 1977): p. 163.

[198] N. Bhandari, S. Bhat, D. Lal,, et. *al.*, "Fossil tracks in the meteorite Angra dos Reis: a predominantly fission origin," *Nature*, Vol. 234 (Dec. 31, 1971): p. 540.

[199] P. Baedeckert, *Science,* Vol. 170, p.846, Nov. 20, 1970.

[200] L. Greenland, *Nature,* Vol. 196, p. 1195, Dec. 22, 1962.

[201] N. Rogers, "Granulite xenoliths from Lesotho kimberlites and the lower continental crust," *Nature*, Vol. 270 (Dec. 22, 1977): p. 677.

[202] N. Bhandari, *Nature,* Vol. 234, p. 541, Dec. 31, 1971.

[203] Z. Jaworowski, "Unusually radioactive fossil bones from Mongolia," *Nature,* Vol. 214, (April 8, 1967): p. 161.

[204] R. Fleischer & H. Hart, "Tracks from extinct radioactivity, ancient cosmic rays, and calibration ions," *Nature,* Vol. 242, (March 9, 1973): p. 104.

[205] J. Reyss & Y. Yokoyama, "aluminium-26 in a manganese nodule," *Nature*, Vol. 262 (July 16, 1976): p. 203.

[206] F. Guichard, J-L. Reyss & Y. Yokoyama, "Growth rate of manganese nodule measured with ^{10}Be and ^{26}Al," *Nature,* Vol. 272 (March 9, 1978): p. 155.

[207] E. Anders & T. Owen, "Mars and Earth: Origin and abundance of Volatiles," *Science*, Vol. 198 (Nov. 4, 1977): p. 453.

[208] Douglas Vogt and Gary Sultan, *Reality Revealed, the Theory of Multidimensional Reality,* (Vector Associates, Bellevue, WA, 1977).

[209] Hogland, R., *Monuments of Mars, A City on the Edge of Forever,* (North Atlantic Books, Berkeley, CA, 1985)

[210] S. Rasool, D. Hunten & W. Kaula, "What the exploration of Mars tells us about Earth," *Physics Today*, (July, 1977): p. 23.

[211] David Pieri, "Martian Valleys: Morphology, Distribution, age, and origin," *Science*, Vol. 210 (Nov. 21, 1980): p. 895.

[212] J. Curtts, K. Blasius, G. Briggs, M. Carr, R. Greeley and H. Masursky, "Mars Dynamics, atmospheric and surface properties: determination from Viking tracking data.," *Science*, Vol. 194 (Dec. 17, 1976): p. 1329.

[213] E. Anders & T. Owen, "Mars and Earth: Origin and abundance of Volatiles," *Science*, Vol. 198 (Nov. 4, 1977): p. 453.

[214] A. Nier, W. B. Hanson, A. Seiff, , et. *al.*, "Composition and structure of the Martian atmosphere: Preliminary," *Science*, Vol. 193 (Aug. 27, 1976): p. 786.

[215] Y. Yung & J. Pinto, "Primitive atmosphere and implications for the formation of channels on Mars," *Nature*, Vol. 273 (June. 29, 1978): p. 730.

[216] S. Rasool, D. Hunten & W. Kaula, "What the exploration of Mars tells us about Earth," *Physics Today*, (July, 1977): p. 23.

[217] *Ibid.*

[218] H. Alfven and A. Mendis, "Nature and origin of comets," *Nature*, Vol. 246 (Dec. 14, 1973): p. 410.

[219] D. Hughes, "Are comets dirty snowballs or dust swarms?," *Nature*, Vol. 270 (Dec. 15, 1977): p. 558.

[220] J. Feynman, "Solar, geomagnetic and auroral variations observed in historical times," Secular Solar and Geomagnetic Variations in the Last 10.000 Years (NATO Science Series C, *Kluwer Academic Publishers*, 1988): p. 141.

[221] P. Conti & R. McCray, "Strong stellar winds," *Science*, Vol. 208 (April 4, 1980): p. 9.

[222] Allen Hammond, "Solar Variability: Is the Sun an Inconstant Star?," *Science*, Vol. 191 (Mar.19, 1976): p.1159.

[223] T. M. Wigley, "The climate of the past 10,000 years and the role of the Sun," Secular Solar and Geomagnetic Variations in the Last 10,000 Years (NATO Science Series C, *Kluwer Academic Publishers* 1988): p. 209.

[224] Research news, "Solar variability: Is the Sun an inconstant star?" *Science*, Vol. 191 (March 19, 1976): p. 1159.

[225] T. M. Wigley, "The climate of the past 10,000 years and the role of the Sun," Secular Solar and Geomagnetic Variations in the Last 10.000 Years (NATO Science Series C, *Kluwer Academic Publishers* 1988): p. 209.

[226] *Ibid.*

[227] *Ibid.*

[228] J. Feynman & N. Crooker, "The solar wind at the turn of the century," *Nature*, Vol. 275 (Oct. 19, 1978): p. 626.

[229] *Ibid.*

[230] R. Kerr, "Did satellites spot a brightening Sun?," *Science*, Vol. 277 (Sept. 26, 1997): p. 1923.

[231] R. Willson, C. Duncan & J. Geist, "Direct measurement of solar luminosity variation," *Science*, Vol. 207 (Jan. 11, 1980): p. 177.

[232] R. Willson, H. Hudson & G. Chapman, "Observations of solar irradiance variability," *Science*, Vol. 211 (Feb. 13, 1981): p. 700.

[233] George Reid & K. Gage, NOAA, "The climatic impact of secular variations in solar irradiance," Secular Solar and Geomagnetic Variations in the Last 10,000 Years (NATO Science Series C, *Kluwer Academic Publishers* 1988): p. 225.

[234] G.W. Lockwood, "Planetary Brightness Changes: Evidence for Solar variability," *Science*, Vol. 190 (Nov.7, 1975): p. 560.

[235] G. Siscoe, "Solar-terrestrial influences on weather and climate," *Nature*, Vol. 276 (Nov. 23, 1978): p. 348.

[236] G. Reid, "Influence of solar variability on global sea surface temperatures," *Nature*, Vol. 329 (Sept. 10, 1987): p. 142.

[237] J. A. Eddy, *Climatic Changes*, Vol. 1 (1977): p. 173-190.

[238] G. Reid, "Influence of solar variability on global sea surface temperatures," *Nature*, Vol. 329 (Sept. 10, 1987): p. 142.

[239] J. King, "Solar Radiation changes and the weather," *Nature*, Vol. 245 (Oct. 26, 1973): p. 443.

[240] *Ibid.*

[241] George Reid & K. Gage, NOAA, "The climatic impact of secular variations in solar irradiance," Secular Solar and Geomagnetic Variations in the Last 10,000 Years (NATO Science Series C, *Kluwer Academic Publishers,* 1988): p. 225.

[242] *Ibid.*

[243] W. Broecker, "Climatic change: Are we on the brink of a pronounced global warming?" *Science*, Vol. 189 (Aug. 1, 1975): p. 460.

[244] W. S. Broecker. *et al., Impingement of man on the Ocean,* D. Hood, Ed. (Wiley, New York, 1971: p. 287.

[245] *Ibid.,* p. 462.

[246] I know Columbia University is a very liberal university and the agenda of most departments is left to far left. I mention this because it did not escape my attention that when the fall of the Soviet Union occurred in 1989, the ex-communists became environmentalists overnight, even though these same communists made an environmental sewer out of Russia. Their agenda was, and still is, to make people think that government is the only entity that can save "mommy" Earth from the big-bad industrialists, who are polluting the Earth. Man is the cause of global warming and only they will save us from ourselves—given enough governmental regulation, taxes and fines that ingratiates their coffers.

[247] W. Komhyr, R. Grass & R. Leonard, NOAA, "Total Ozone decrease at South Pole, Antarctica, 1964-1985,"*Geophysical Research Letters*, Vol. 13, No.12 (Nov. 1986): p. 1248.

[248] W. Komhyr, R. Grass, P. Reitelbach,, et. *al.,* "Total ozone, ozone vertical distributions, and stratospheric temperatures at South Pole, Antarctica, in 1986 and 1987," *NOAA research Paper1988.*

[249] G.M. Keating, "Relation between monthly variations of global ozone and solar activity," *Nature*, Vol. 274 (Aug.31, 1978): p.873.

[250] P. Fabian, J.A. Pyle *et, al,* "The August 1972 Solar proton event and the atmospheric ozone layer," *Nature*, Vol. 277 (Feb.8, 1979): p.458.

[251] Donald Heath, Arlin Krueger,, et. *al,* "Solar Proton Event: Influence on Stratospheric Ozone," *Science*, Vol.197 (Aug.26, 1977): p.886.

[252] F. Busse, "Core motions and the geodynamo," *Reviews of Geophysics and Space Physics*, Vol. 13, No. 3 (July, 1975): p. 206.

[253] E. Bullard & D. Gubbins, "Geomagnetic dynamos in a stable core," *Nature*, Vol. 232 (Aug. 20, 1971): p. 548.

[254] *Ibid.* p. 207.

[255] K. Verosub, "Alternative to the geomagnetic self-reversing dynamo," *Nature*, Vol. 253 (Feb. 27, 1975): p. 707.

[256] M. Krs, "Intensity of the Earth's magnetic field in the geological past," *Nature*, Vol. 215 (Aug. 12, 1967): p. 697.

[257] C. Russell, R. McPherron & P. Coleman Jr., "Fluctuating magnetic fields in the magnetosphere," *Space Science Reviews*, Vol. 12 (1972): p. 810.

[258] R. McPherron, C. Russell and P. Coleman, Jr., "Fluctuating magnetic fields in the magnetosphere," *Space Science Reviews*, Vol. 13 (1972): p. 411.

[259] C. Russell, R. McPherron & P. Coleman Jr., "Fluctuating magnetic fields in the magnetosphere," *Space Science Reviews*, Vol. 12 (1972): p. 810.

[260] *Ibid.* p. 833.

[261] J. A. Jacobs, "Reversals of the Earth's magnetic field," *Nature*, Vol. 230 (April 30, 1971): p. 574.

[262] J. Dunn, M. Fuller, H. Ito & V. Schmidt, "Paleomagnetic study of a reversal of the Earth's magnetic field," *Science*, Vol. 172 (May 21, 1971): p. 840.

[263] The 700,000 years is a result of an incorrect assumption as to the rate of sedimentation in past study areas. These dates were derived in the later part of the nineteenth and early twentieth century before carbon-14 and other dating methods.

[264] J.A. Jacobs editor, "Geomagnetism; 3- Behaviors of the Earth's magnetic field," (Academic Press, New York, 1989): p191.

[265] K. Verosub, "Paleomagnetic excursions as magnetrostratigraphic horizons: cautionary note," *Science*, Vol. 190 (June 12, 1975): p. 48.

[266] *Ibid.*

[267] Allan Cox, "Frequency of geomagnetic reversals and the symmetry of the nondipole field," *Reviews of Geophysics and Space Physics*, Vol. 13, No. 3 (July 1975): p. 38.

[268] A. Cox, "Geomagnetic Reversals," *Science*, Vol. 163, (Jan. 17, 1969), p. 238.

[269] *Ibid.* p. 243.

[270] *Ibid.* p. 242.

[271] Allan Cox, "Frequency of geomagnetic reversals and the symmetry of the nondipole field," *Reviews of Geophysics and Space Physics*, Vol. 13, No. 3 (July 1975): p. 35.

[272] J. Harwood & S. Malin, "Present trends in the Earth's magnetic field," *Nature*, Vol. 259, (Feb. 12, 1976): p. 469.

[273] M. Steiner & C. Helsley, "Reversal pattern and apparent polar wander for the Late Jurassic," *Geology*, Vol. 86 (Nov. 1975): p. 1537.

[274] Allan Cox, "Frequency of geomagnetic reversals and the symmetry of the nondipole field," *Reviews of Geophysics and Space Physics*, Vol. 13, No. 3 (July 1975): p. 38.

[275] P. J. Wyllie, *The Way the Earth Works*; p.148 (N.Y., John Wiley & Sons, 1976)

[276] J. Hollin, *Nature*, Vol. 202, (June 13, 1964): p. 1099.

[277] J. Ewing & M. Ewing, "Sediment distribution on the mid-ocean ridges with respect to spreading of the sea floor.," *Science*, Vol. 156 (June 23, 1967): p. 1590.

[278] R. Norris & P. Wilson, "Low-latitude sea-surface temperatures for the mid-Cretaceous and the evolution of planktic foraminifera," *Geology*, Vol. 26, No. 9 (Sep. 1998): p. 823.

[279] A. Hammond, "Paleoceanography: Sea floor clues to earlier environments," *Science*, Vol. 191 (Jan. 16, 1976): p. 168.

[280] E. Barron, S. Thompson & S. Schneider, "An ice-free Cretaceous? Results from climate model simulations," *Science*, Vol. 212 (May. 1, 1981): p. 501.

[281] A. Hammond, "Paleoclimate: ice age Earth was cool and dry," *Science*, Vol. 191 (Feb. 6, 1976): p. 455.

[282] The Rand Corp. did a study in the 1970s for the Air Force and their conclusion was that the Earth's rotation was slowing down because of the decaying Earth's magnetic field. Ball, R. H.,, et.*al.*, *Variations in the Geomagnetic Field and in the rate of the Earth's rotation* (Santa Monica, The Rand Corp., Memorandum No. RM-5717-PR, Oct. 1968).

[283] M. Rampino, S. Self & R. Fairbridge, "Can rapid climatic change cause volcanic eruptions?," *Science*, Vol. 206 (Nov. 16, 1979): p. 828.

[284] J. P. Kennett, R. C. Thunell, *Science*, Vol. 196, (June 10, 1977): p. 1231.

[285] J. Bray, "Pleistocene volcanism and glacial initiation," *Science*, Vol. 197 (July 15, 1977): p. 251.

[286] D. Ninkovich & W. Donn, "Explosive Cenozoic volcanism and climatic implications," *Science*, Vol. 194 (Nov. 5, 1976): p. 899.

[287] L. Prueher & D. Rea, "Rapid onset of glacial conditions in the subarctic North Pacific region at 2.67 Ma: Clues to causality," *Geology*, Vol. 26, No. 11 (Nov. 1998): p. 1027.

[288] J. P. Kennett, R. C. Thunell, *Science*, Vol. 187, (Feb. 14, 1975): p. 497-503.

[289] J. Kennett & N. Watkins, "Geomagnetic polarity changes, volcanic maxima and faunal extinction in the South Pacific," *Nature*, Vol. 227 (Aug. 29, 1970): p. 970.

[290] Allan Cox, R. Doell & G. B. Dalrymple, USGS "Geomagnetic polarity epochs: Sierra Nevada II," *Science*, Vol. 142 (Oct. 18, 1963): p. 382.

[291] J. R. Bray, "Volcanic triggering of glaciation," *Science*, Vol.260 (April 1, 1976): p. 251.

[292] D. Ericson, M. Ewing & Goesta Wollin, "Pliocene-Pleistocene boundary in deep-sea sediments," *Science*, Vol. 139 (Feb. 22, 1963): p. 727.

[293] N. Morner, "Low sea levels, droughts ad mammalian extinctions," *Science*, Vol. 271 (Feb. 23, 1978): p. 738.

[294] J. Hays & N. Opdyke, "Antarctic radioaria, magnetic reversals, and climatic change," *Science*, Vol. 158 (Nov. 24, 1967): p. 1001.

[295] N. Watkins & H. Goodell, "Geomagnetic polarity change and faunal extinctions in the southern ocean," *Science*, Vol. 156 (May 26, 1967): p. 1083.

[296] Definition from the Radiolaria organization, Norway.

[297] Roger Lewin, "Extinctions and the history of life," *Science*, Vol. 221 (Sept. 2, 1983): p. 935.

[298] Roger Lewin, "What killed the giant mammals?," *Science*, Vol. 221 (Sept. 9, 1983): p. 1036.

[299] A. Long & P. Martin, "Death of American Ground Sloths," *Science*, Vol. 186 (Nov. 15, 1974): p. 638.

[300] R. Fairbridge, "Global climate change during the 13,500-b.p. Gothenburg geomagnetic excursion," *Nature*, Vol. 265 (Feb. 3, 1977): p. 430.

[301] H. Laster, "Cosmic rays from nearby supernovae: Biological Affects," *Science*, Vol. 160 (June 7, 1968): p. 1138.

[302] D. Russell & W. Tucker, "Supernovae and the extinction of the dinosaurs," *Nature*, Vol. 229 (Feb. 19, 1971): p. 553.

[303] K. Terry & W. Tucker, "Biologic Affects of supernovae," *Science*, Vol. 159 (Jan. 26, 1968): p. 421.

[304] G. Hunt, "Possible climatic and biological impact of nearby supernovae," *Nature*, Vol. 271 (Feb. 2, 1978): p. 430.

[305] D. Russell & W. Tucker, "Supernovae and the extinction of the dinosaurs," *Nature*, Vol. 229 (Feb. 19, 1971): p. 553.

[306] I had discovered this by studying the sediments and coral formations from around the world. This research gave me a very accurate picture of what side of the Earth faced the Sun 12,068 years ago at the time of the nova. I also did a study of the mythologies from around the world, and their stories reinforced what I had theorized.

[307] Thomas Gold, *Science,* Vol. 168, (Sept. 26, 1969): p. 1345.

[308] G. Reid, I. Isaksen, T. Holzer & P. Crutzen, "Influence of ancient solor-proton events on the evolution of life," *Nature,* Vol. 259, (January 22, 1976): p. 177.

[309] W. Kundt, "Are supernova explosions driven by magnetic springs?," *Nature*, Vol. 261 (June 24, 1976): p. 673.

[310] Dale Russell, *Nature*, Vol. 229, (Feb. 19, 1971): p. 553.

[311] J. Higdon, *Nature*, Vol. 246, (Dec. 14, 1973): p. 403.

Chapter 9: Philosophies of the Universe

[1] The term "Jews" originally referred only to the tribe of Judah. Later, it was one of the two Kingdoms of Israel after the death of Solomon. At the time of Abraham, through Solomon, they were all referred to as either Hebrews or Israelites (descendants of the twelve sons of Jacob, also known as Israel).

[2] Ginzberg, L., *The Legends of the Jews*, pp. 8; The Jewish Publication Society, Philadelphia Society, 1913.

[3] *Ibid.* pp. 13.

[4] *Ibid.* pp. 14.

[5] *Ibid.* pp. 145.

[6] The Torah says it was 100 years. The 100 years is derived from the age when Noah had his son, Shem, at 500 years old and deducting Noah's age, 600 years old, when the flood occurred.

[7] *Ibid.* pp. 154.

[8] *Ibid.* pp. 162.

[9] Jowett, B., *The Dialogues of Plato*, 4th ed., Vol 3, p. 681 (Oxford, Clarendon Press, 1953).

[10] Plato, Timaeus, Vol. 9, pp. 51, (Loeb Classical Library, Harvard Univ. Press).

[11] *Ibid.* Pp. 57.

[12] *Ibid.* Pp. 75.

[13] *Ibid.* Pp. 117.

[14] *Ibid.* Pp. 123.

[15] *Ibid.* Pp. 137.

[16] Jowett, B., The Dialogues of Plato, 4th ed., Vol. 3, pp. 435 (Oxford, Clarendon Press, 1953).

[17] Velikovsky, I., *Worlds in Collision*, (Victor Gollancz Ltd., London, 1950).

[18] Jeremiah 43:8.

[19] Plato, in *Phaedo*, states, "Then I heard someone who had a book of Anaxagoras, as he said, out of which he read that *mind was the disposer and cause of all*, and I was quite delighted at the notion of this, which appeared admirable..." This statement is similar to the Hebrew philosophy and Plato's. In his *Apology*, Plato wrote: "Friend Meletus, you think that you are accusing Anaxagoras; and you have but a bad opinion of the judges, if you fancy them ignorant to such a degree as not to know that those doctrines are found in the books of Anaxagoras the Clazomenian, who is full of them. And these are the doctrines which *the youth are said to learn of Socrates*, ..." It seems to be saying that Socrates learned from the books of Anaxagoras.

[20] Plato, *Timaeus*, Loab Classical Library (Harvard Univ. Press. 1967).

[21] The conflict Greece had with the Hebrews is described in *The Parthenon Code; Mankind's History in Marble;* Robert B. Johnson, Jr.; Solving Light Books, Annapolis, MD, 2004.

[22] The Oxford companion to classical literature, 2nd ed., p. 579 (Oxford Univ. Press, 1989).

[23] Plato, *Meno*, Loeb Classical Library (Harvard Univ. Press. 1967), p. 303.

[24] D. A. Mackenzie, *Indian Myths and Legend*, (London, Gresham Publishing Co., 1913), p304.

[25] F. M. Muller, (ed.), *The Sacred Books of the East*, Vol. 34; "the Vendanta-Sutras," (Oxford,

Oxford University Press, 1890), p. 386.

[26] D. A. Mackenzie, *Indian Myths and Legend,* (London, Gresham Publishing Co., 1913),. p113.

[27] *Ibid.* p142.

[28] I. Velikovsky, *Worlds in Collision,* (London, Abacus ed., Sphere Books, Ltd., 1972), p. 43.

[29] D. A. Mackenzie, *Indian Myths and Legend,* (London, Gresham Publishing Co., 1913),. p. 105.

[30] F. M. Muller, (ed.), *The Sacred Books of the East,* Vol. 34; "the Vendanta-Sutras," (Oxford, Oxford University Press, 1890), p. 212.

[31] F. M. Muller, (ed.), *The Sacred Books of the East,* Vol. 43; "the Satapatha-Brahmana," (Oxford, Oxford University Press, 1890), p. 352.

[32] F. M. Muller, (ed.), *The Sacred Books of the East,* Vol. 12; "the Satapatha-Brahmana," (Oxford, Oxford University Press, 1897), p. 343.

[33] F. M. Muller, (ed.), *The Sacred Books of the East,* Vol. 15; "the Upanisads," (Oxford, Oxford University Press, 1900), p. 272.

[34] Ignatius. Donnelly, *The Destruction of Atlantis, Ragnarok; The Age of Fire and Gravel,* (Blauvelt, N.Y., Multimedia Publications , 1971), p. 132.

[35] Acknowledgment: R. Sahai, Jet Propulsion Lab) and B. Balick (Iniversity of Washington). These nebulas are also know as planetary nebula. I disagree with the traditional explanation for planetary nebula.

[36] Hichols, F. H., *Through Hidden Shensi* (N.Y. Charibners Sons, 1902).

[37] Bishop, J. F., *The Yangtze Valley and Beyond* (N.Y. G.P. Putnam and Son, 1900).

[38] Muller, F. M. (ed.), *The Sacred Books of the East*, Vol. 39, (Oxford, Univ. Press., 1891), pp. 287.

[39] *Ibid.* pp. 58.

[40] Mackenzie, D., *Myths of China and Japan*, (Gresham Publishing Co., London, 1923), p. 111.

[41] Donnelly, I., *Ragnarok: The age of Fire and Gravel*, p. 259 (N.Y., D. Appleton, 1885).

[42] Velikovsky, I., *Worlds in Collision*, (Victor Gollancz Co., London, 1950), p. 43.

[43] Mackenzie, D., *Myths of China and Japan*, (Gresham Publishing Co., London, 1923), p.268.

[44] Velikovsky, I., *Worlds in Collision*, (Victor Gollancz Co., London, 1950), p. 247.

[45] *Ibid.* pp. 247.

[46] Donnelly, I., *Ragnarok: The age of Fire and Gravel*, (N.Y., D. Appleton, 1885), p. 210.

[47] Mackenzie, D., *Myths of China and Japan*, (Gresham Publishing Co., London, 1923), p.59.

[48] Holmberg, U., *The Mythology of all Races*, Vol. 4; FinnoUgric, Siberian, (Marshall Jones Co., Boston, 1927), p. 369.

[49] Muller, F. M. (ed.), *The Sacred Books of the East*, Vol. 39; The Texts of Taoism, (Oxford Univ. Press, Oxford, 1891), p.14.

[50] Ibid. pp. 85.

[51] Ibid. pp. 243.

[52] Morgan, E., *Tao, the Great Lumina,* (Ch'eng-Wen Publishing Co., Taipei, 1966).

[53] Muller, F. M. (ed.), *The Sacred Books of the East*, Vol. 39; The Texts of Taoism, (Oxford Univ. Press, Oxford, 1891, p.315.

[54] Ibid. p. 315.

[55] Ibid. p. 119.

[56] Douglas B. Vogt and Gary Sultan (Introduction and part of Chapter 1), *Reality Revealed; The*

Theory of Multidimensional Reality, Ch 6 (Vector Associates, POB 40135, Bellevue WA, 98015, 1977).

Chapter 10: Prophecies of the End of Days

[1] Torah, Numbers, 12:6-8.

[2] These books were written by Arius Calpurnius Piso, and a few parts by his younger half-brother Pliny the Younger. Piso forced Rabbi Akiva at B'nai Barack to produce the Hebrew translations. This was all after the first Jewish revolt of 70-72 E.C. A full explanation is found in *The True Authorship of the New Testament,* By Abelard Reuchlin, Pub., 1979 (Abelard Reuchlin Foundation, P.O. Box 5652, Kent, WA 98064. cost $10.00).

[3] After Solomon's Temple was built the Tabernacle was no longer needed, so I believe that the priests secretly took it down to Sinai and placed it back in the Cave for safe keeping.

[4] Jeremiah 43:3 and 43:6.

[5] Over the past 10 years I have discovered that all the priests had three names: Their given name, their priestly name, and, if they were the High Priest, they would write under a prophet's name. This book will not cover the whole list of priests and prophets of the Temple period. That will be for my next book.

[6] Webster's New Universal Unabridged Dictionary, 6th edition, 1983 Simon & Schuster.

[7] This was the case in most of Europe.

[8] The seven are, going from west to east; El-Khandaq el-Gharbi Canal, Red branch of the Nile, El-Gharbiya el-Raiai, Bahr Shibin, The Nile, Bahr Saft Drain, and Bahr el-Baqar Drain.

[9] Yannis A. Sakellarakis, *Trade before the Flag? On the Principles of Phoenician Expansion in the Mediterranean*; Biblical Archaeology Today, 1990; Proceedings of the Second International Congress on Biblical Archaeology, Jerusalem, June-July 1990, (Israel Exploration Society, The Israel Academy of Sciences and Humanities), p. 340.

[10] In my research on the Torah, Jewish history, and Egyptian history, I discovered the Egyptian name that Joseph used in Egypt. I figured out much of it from the clues Moses gave in Genesis. I know who Joseph was, and what he did, and all of this will be in my next book, which will reveal the rest of this astonishing story.

[11] The area today called Provence France and was originally a very Jewish area. The Albigensian Crusade (1209 – 1229) was instigated by Pope Innocent III for the purpose of murdering the "heretics" which were of course were the Jews. The general of the northern French army was Simon de Montfort working for the King of Fance

[12] A complete explanation will be in my next book.

[13] Many nations have deported or confined hostile or unwanted populations living among them such as all the Jews living in Arab countries of the Middle East after the State of Israel was created. The mass exodus of the Muslim population from India and the Hindus from newly formed Pakistan and the deportation of ethnic Germans after WW II from Poland and Czechoslovakia. There have been many more throughout the world. The point being that the only country that seems to always get voted against in the United Nations is little Israel.

[14] The US State Department has been very anti-Semitic since well before the 1920s. They are overtly pro-Arab for the same reasons as the British, namely oil, money and theology. After President Truman recognized the new State of Israel his State Department placed an arms embargo on Israel, knowing that the British would continue supplying the Arabs with arms

and British officers to command the Jordian Army. I can only conclude that the State Department expected the Arabs to win and murder the remaining Jews.

[15] Genesis 41:55 and 41:57.

[16] Moses and Aaron were descendants of Ephraim, which will be proved in my next book.

[17] Numbers 16:1

[18] Louis Ginzberg; *The Legends of the Jews*; "Moses in the Wilderness," "Pharaoh Pursues the Hebrews;" 1910 Jewish Publication Society.

[19] Louis Ginzberg, *The Legends of the Jews*, "Moses in the Wilderness," "Eldad and Medad;" 1910 Jewish Publication Society.

[20] The composition of interplanetary dust particles (asteroidal and cometary) is of three major types: chondritic 60%, iron-sulfur-nickel 30%, and mafic silicates, which are iron-magnesium-rich silicates, (i.e. olivine and pyroxene), 10%. Gruen, E. "Interplanetary Dust and the Zodiacal Cloud", *Encyclopedia of the Solar System*, Academic Press, 1999, who uses this reference for Table III: Jessberger, E.K. et al., (1992), Earth Planet. Sci. Lett. p. 112, 91-99.

[21] The reason I put "Temple" in quotes is because I do not think it is really a temple but rather a functional device. I will explain this later in this chapter.

[22] Jeremy Shere, *A Very Holy Cow*; The Jerusalem Post, June 13, 1997.

[23] Dina Kraft, Associated Press Writer, *Red Heifer a Sensation in Israel,* Washington Post, May 28, 1997.

[24] The reason it was 28 days is because one lunar month is the period of mourning.

[25] Ezekiel 42:16 and 45:2. The sacred cubit is equal to 24.136", so multiply that measurement by the values to convert it into feet.

[26] Ezekiel 45:2

[27] These are Chapters 1 to 39 excluding 14 and 24.

[28] Isaiah 2:3.

[29] *Ibid.* 4:5

[30] Isaiah 35:8.

[31] Second Samuel 7:19.

[32] Louis Ginzberg, *Legends of the Jews*, "Moses in the Wilderness," "Curses turned to Blessings,"; 1910, Jewish Publication Society, p. 380.

[33] Ezekiel Chapters 40 to 42.

[34] *Ibid.* 48:15-17.

[35] *Ibid.* 43:13-17.

[36] *Ibid.* 44:3.

[37] Ezekiel 45:2.

[38] A land formation has to be over 500' tall to be classified as a hill.

[39] The name and the very concept of Satan is not Jewish. Jewish philosophy has it that there is only God and He does it all. Arius Calpurnius Piso created the name and concept of Satan, he played the part of Satan and inserted it into his various books, including the ones he forced the Jews to insert into their Hebrew Scriptures. A more detailed explanation will be reserved for another book.

[40] Isaiah 11:9.

[41] I believe he was 20 years old because, in the Torah, it gives the age of adulthood as 20 years. When Moses had to leave home he was officially on his own, and that is the sign of an adult, being on one's own.

[42] Deuteronomy 2:14.
[43] Numbers 10:11.
[44] *Ibid.* 10:33.
[45] *Ibid.* 11:1.
[46] *Ibid.* 11:3.
[47] *Ibid.* 11:31-35.
[48] *Ibid.* 12:1-13.
[49] *Ibid.* 16:32.
[50] *Ibid.* 16:35.
[51] *Ibid.* 17:9-14.
[52] Deuteronomy 32:22.
[53] God said they were to eat the birds that fell for 30 days but they were there one day before, for a total of 31 days.
[54] David John Oates, *Reverse Speech: Voices From The Unconscious;* Published by David John Oates; P.O. Box 678, Noarlunga Centre, SA 5168. Australia.
[55] Michael Drosnin, *The Bible code*, Simon & Schuster, New York, 1997, p175.

Chapter 11: Timelines and Patterns

[1] *The Parthenon Code; Mankind's History in Marble,* by Robert Bowie Johnson, Jr.; published by Solving Light Books, 727 Mount Alban Drive, Annapolis, MD 21401.
[2] Exodus 7:15-20.
[3] This clue will be revealed in a future book.
[4] Treaty: Agreement of Mutual Assistance between the United Kingdom and Poland. August 25, 1939; located at The Avalon Project at Yale Law School; http://www.yale.edu/lawweb/avalon/wwii/bluebook/blbk19.htm.
[5] The story of the discovery of the quilt is found in the Rutherford B. Hayes Presidential Museum. Their web site has a picture of the quilt and its story: http://www.rbhayes.org/quilt.html.
[6] 3,000 x 24.136 = 72,408 ÷ 12 = 6,034.

Chapter 12: Why was the Universe Created?

[1] I created the name *Diebold* in 1975 because there was no other name in the English language, or others that described a computer which holds all of existence within it.
[2] I have capitalized "Operating System" because the term represents God, and traditionally, all references to God are capitalized.

Appendix C: Moses' Code System

[1] Genesis 41:51.
[2] Genesis 41:52.

Appendix D: The Prophet and Priest Baruch

[1] Jeremiah 29:3.
[2] Jeremiah 51:59.
[3] Joel 4:4-6.
[4] Joel 4:16-18.

Bibliography

Some of the following are books were not referenced in the Endnotes section.

Dr. Karl Feyerabend, *Langenscheidt'S Pocket Hebrew Dictionary to the Old Testament*, (Berlin, Langenscheidt KG).

The Holy Scriptures According to the Masoretic Text, (Philadelphia: The Jewish Publication Society of America, 1917).

Rabbi Aryeh Kaplan, *The Living Torah, The Five Books of Moses and the Haftarot*, 2nd Edition (New York,: Maznaim Publishing Corp., 1981)

The Holy Bible containing the Old and New Testaments (Authorized King James Version) (Nashville: The Gideons International, 1972).

JPS Hebrew-English Tanaka, the Traditional Hebrew Text and the new JPS translation, 2nd Edition (Philadelphia: The Jewish Publication Society, 1999).

The Hebrew Bible with English Translation, The Five Books of Moses, 5 volumes (Tel-Aviv: Sinai Publishing, 1973).

Referenced in the Endnote section.

Douglas Vogt, and Gary Sultan, *Reality Revealed, The Theory of Multidimensional Reality* (Bellevue, WA, Vector Associates, 1978.)

Robert B. Johnson, Jr. *The Parthenon Code, Mankind's History in Marble* (Annapolis, MD: Solving Light Books, 2004).

Peter Tompkins, *Secrets of the Great Pyramid* (New York, Harper Colophon Books, 1971).

Flavius Josephus [pen name] (actual name was Arius Calpurnius Piso) *The Complete Works of Flavius Josephus* (Philadelphia: Porter and Coates, 1963).

Abelard Reuchlin, *The True Authorship of the New Testament* (Bellevue, WA: Abelard Reuchlin Foundation, 1979).

Index

The HebraicaIIC Fonts used to print this
work are available from Linguist's Software, Inc., PO Box 580, Edmonds, WA
98020-0580 USA tel (425) 775-1130.